Disinfection and Public Health 2013

Indianapolis, Indiana, USA
24 - 26 February 2013

ISBN: 978-1-5108-7146-5

Printed from e-media with permission by:

Curran Associates, Inc.
57 Morehouse Lane
Red Hook, NY 12571

CURRAN ASSOCIATES INC. proceedings.com

Some format issues inherent in the e-media version may also appear in this print version.

Printed by Curran Associates, Inc. (2018)

Water Environment Federation (WEF)
601 Wythe Street
Alexandria, Virginia 22314
USA

Phone: 1-800-666-0206
Fax: 1-703-684-1545

csc@wef.org

Additional copies of this publication are available from:

Curran Associates, Inc.
57 Morehouse Lane
Red Hook, NY 12571 USA
Phone: 845-758-0400
Fax: 845-758-2633
Email: curran@proceedings.com
Web: www.proceedings.com

TECHNICAL PROGRAM

Session 1: Evaluating UV Doses and Operations

Monday, February 25, 2013
10:45am - 12:15pm

Room: Cosmopolitan C
1.5 PDHs

Moderators: G. Elliott Whitby; UV Technologies
Paul Ropic; Aquionics

TECHNICAL PROGRAM

Session 2: Disinfection Basics

Monday, February 25, 2013
10:45am - 12:15pm

Room: Cosmopolitan D
1.5 PDHs

Moderator: Bree Trembly; Aquionics Inc.

2A	**10:45am**	**Disinfection Fundamentals** S. Jeyanayagam; CH2M Hill	N/A
2B	**11:15am**	**Introduction to Established Disinfection Technologies** K. Bourgeous; Carollo Engineers	38
2C	**11:45am**	**Innovations in Wastewater Disinfection Technology** K.Bell, A. da Silva; CDM Smith	39
	12:15pm	**Session Adjourns**	

TECHNICAL PROGRAM

Session 3: New Options for Disinfection in the Developing World

Monday, February 25, 2013 — **Room: Discovery**
10:45am - 12:15pm — **1.5 PDHs**

Moderator: Andrew Salveson; Carollo Engineers

3A **10:45am** **Point-of-Use Disinfection of Household Drinking Water with Copper/Zinc Ions and Oxide Nanoparticles: a Proof-of-Concept Lab Study** 48
A. Komandur, A. Malone, M. D. Sobsey; University of North Carolina

3B **11:15am** **Regulation of Water Treatment Options for Cholera Response in Haiti** N/A
D. Lantagne; Tufts University

3C **11:45am** **Optimizing Char-Based Filtration Materials for Fluoride Removal from Drinking Water** 68
L.R. Brunson, D.A. Sabatini; University of Oklahoma

12:15pm **Session Adjourns**

TECHNICAL PROGRAM

Session 4: UV Standards and Validation

Monday, February 25, 2013 **Room: Cosmopolitan C**
1:45pm – 5:00pm **2.5 PDHs**

Moderator: Ken Kershner; Aquionics Inc.

4A **1:45pm** **Current Observations on UV Validation Protocols, Processes, and Practices** 72
C. Shen, O. K. Scheible; HDR Hydroqual Inc.

4B **2:15pm** **Uniform Protocol for Wastewater UV Validation Applications** 79
G. E. Whitby[1], O. Lawal[2], P. Ropic[2], S. Shima[3], B. Ferran[4], [5]B. Dussert; [1]Calgon Carbon Corporation, [2]Aquionics Inc, [3]Severn Trent Services, [4]Ozonia North America LLC, [5]B. Dussert, Dussert Consulting, LLC

2:45pm **Networking Break**

4C **3:30pm** **The Evolution of UV Disinfection in Indianapolis: Balancing Between USEPA Guidance and Other Regulatory Frameworks** 92
B. R. Townsend[1], D. Ginn[1], D. Moran[2], X. Zhao[1]; [1]Black & Veatch, [2]Citizens Energy Group

4D **4:00pm** **Alternate Indicator Organisms for Reclaimed Water in North Carolina** 107
P. Drummey Stiegel[1], M. Mann[1], M. Dalton[2]; [1]Hazen and Sawyer, [2]City of Raleigh

4E **4:30pm** **Spot-check Bioassay of A Full-Scale Gravity-Flow Open-Channel Vertical UV System at The Russian River Treatment Plant in Sonoma County, California** 115
B. A. Ferran[1], Y. Borroum[2]; [1]Ozonia North America, [2]Stantec Consulting Services

5:00pm **Session Adjourns**

Alternate 1 **Validation of UV Reactors – Lessons Learned** 78
A. Salveson, K. Bourgeous, N. Fontaine, N. Goel, J. Bandy, M. Noibi, E. Mackey; Carollo Engineers

Please note that in addition to the above program, the conference steering committee is pleased to announce the first mobile session at the Disinfection and Public Health Conference. Attendees should please meet at the Registration Desk at 1:45pm to proceed through an informative interaction with key vendors onsite in the exhibit hall.

Networking Reception – Regency Ballroom
Immediately following the conclusion of this session, please join us for the Networking Reception held in Regency Ballroom from 5:00pm – 6:30pm

TECHNICAL PROGRAM

Session 5: Molecular Methods and Mechanistic Studies

Monday, February 25, 2013
1:45pm – 5:00pm

Room: Cosmopolitan D
2.5 PDHs

Moderator: Jenny Reina; CH2M HILL

5A	**1:45pm**	**Oxidation and Disinfection Processes as Barriers to the Dissemination of Bacterial Antibiotic Resistance Genes via Water and Wastewater** 131 P. Zhou, S. Li, M. C. Dodd; University of Washington
5B	**2:15pm**	**UV Or PAA For Wastewater Disinfection: A Comparison of the Impact on Virulence Genes** 136 B. Biswal[1], R. Gehr[1], R. Khairallah[1], K. Bili[1], A. Mazza[2], L. Masson[2], D. Frigon[1]; [1]McGill University, [2]Biotechnology Research Institute
	2:45pm	**Networking Break**
5C	**3:30pm**	**An Assay for Estimating UV Disinfection Dose by Direct Quantification of DNA Damage in Indigenous Microorganisms in Drinking Water** 155 J. Blyth[1], M. Templeton[1], B. Cairns[2]; [1]Imperial College London, [2]Trojan Technologies
5D	**4:00pm**	**From Systematic Process Optimization To Alternative Indicator Development, How MSDGC Has Utilized Innovative Approach to Optimize the Chlorination Disinfection Efficacy** 159 T. Lu[1], R. Welsh[2], D. Linn[2]; [1]Black and Veatch, [2]Metro Sewer District of Greater Cinncinati
5E	**4:30pm**	**The Influence of Extracellular Polymeric Substance Composition on Bacteria Disinfection By Monochloramine** 165 K. M. Coburn, Z. Xue, Y. Seo; University of Toledo
	5:00pm	**Session Adjourns**

Please note that in addition to the above program, the conference steering committee is pleased to announce the first mobile session at the Disinfection and Public Health Conference. Attendees should please meet at the Registration Desk at 1:45pm to proceed through an informative interaction with key vendors onsite in the exhibit hall.

Networking Reception – Regency Ballroom
Immediately following the conclusion of this session, please join us for the Networking Reception held in Regency Ballroom from 5:00pm – 6:30pm

TECHNICAL PROGRAM

Session 6: Community Health

Monday, February 25, 2013
1:45pm – 5:00pm

Room: Discovery
2.5 PDHs

Moderator: Justine Rayner; Tufts University

6A **1:45pm** **Transforming Community Health with Simple Yet Effective Drinking Water Disinfection in Central Andes of Perú** 172
T. R. Armitage[1], T. Jackson[2]; [1]Lieutenant Commander USPHS Commissioned Corps, [2]Missions, Immanuel Community Church

6B **2:15pm** **Challenges and Successes of the Gift of Water Project in Haiti** 178
L. Moehling; Gift of Water

2:45pm **Networking Break**

6C **3:30pm** **Accuracy and Usability of Free Chlorine Residual Testing Methods** 179
A. Murray, D. Lantagne; Tufts University

6D **4:00pm** **The Safe Global Water Institute: An Integrated, Collaborative Approach for Improving Drinking Water and Sanitation Globally** 184
A. M. Gall[1,2], V. Madadi[2,3], J. L. Shisler[1,2], B. J. Mariñas[1,2]; [1]University of Illinois at Urbana-Champaign, [2]Safe Global Water Institute, [3]University of Nairobi, Kenya

6E **4:30pm** **The Role Of Sanitation And Hygiene Practices In Contamination Of Household Drinking Water In Rural Households Of South Africa** 188
N. Potgieter[1], T. Barnard[2]; [1]University of Venda, [2]University of Johannesburg, South Africa

5:00pm **Session Adjourns**

Please note that in addition to the above program, the conference steering committee is pleased to announce the first mobile session at the Disinfection and Public Health Conference. Attendees should please meet at the Registration Desk at 1:45pm to proceed through an informative interaction with key vendors onsite in the exhibit hall.

Networking Reception – Regency Ballroom
Immediately following the conclusion of this session, please join us for the Networking Reception held in Regency Ballroom from 5:00pm – 6:30pm

TECHNICAL PROGRAM

Session 7: Effects of Upstream Processes on Disinfection

Tuesday, February 26, 2013
8:30am – 9:30am

Room: Cosmopolitan C
1 PDH

Moderator: Keith Bourgeous; Carollo Engineers

7A **8:30am** **Impact of Wastewater Disinfection on Reuse Applications** 193
M. LeChevallier, Z. Bukhari, L. Weinrich, P. Jjemba; American Water

7B **9:00am** **Removals of Enteric Viruses and Indicator Organisms In Membrane Bioreactor and Conventional Secondary Activated Sludge Wastewater Plants** 207
A. Williston[1], K. Riddell[2] T. Gellner[1]; [1]CT Consultants, [2]City of Delphos and Smith Environmental

9:30am **Session Adjourns**

Alternate **Lessons Learned: UV Wastewater Operations** N/A
O. Lawal; Aquionics Inc.

TECHNICAL PROGRAM

Session 8: Innovative Disinfection Methods

Tuesday, February 26, 2013
8:30am – 11:45am

Room: Cosmopolitan D
2.5 PDHs

Moderators: Sam Jeyanayagam; CH2M Hill
G. Elliott Whitby; UV Technologies

8A	8:30am	**Innocuous For Reuse: Scaling Sustainable Wastewater Treatment and Natural Disinfection Processes For Safe Water Recycling At Delhi, California USA and at Varanasi, Uttar Pradesh, India.** 214 F. B. Green[1], R. E. Dawyot[2], P. Husby[3], E. McNaughton[3], S. K. Mishra[4], V. B. Mishra[4]; [1]GO_2 Water, Inc., [2]Allied Engineers, Inc., [3]US EPA Region 9, [4]Sankat Mochan Foundation
8B	9:00am	**How UV-C LED Technology is Shaping Todays Water Treatment Solutions** N/A K. S. Kershner; Aquionics Inc.
	9:30am	**Networking Break**
8C	10:15am	**Low(est) Energy Disinfection using Pasteurization - Results of Demonstration Scale Testing in Ventura California** 217 A. Salveson[1], D. Pfeifer[2], S. Epstein[2], J. Willis[2], G. Ryan[3], N. Fontaine[1]; [1]Carollo Engineers, [2]City of Ventura, [3]Pasteurization Technology Group
8D	10:45am	**Mixed Oxidants: Alternative Disinfectant That Can Reduce Operational Costs and Improve Water Quality** 229 J. Deogracias; Parkson Corporation
8E	11:15am	**The Use of Ozone to Treat Wet Weather Flows in Springfield, MO** 300 G. Hunter[1], J.J. Neemann[1], N. Burns[1], J. L. Pruss[1], K. Green[2]; [1]Black & Veatch, [2]Springfield Clean Water Services
	11:45am	**Session Adjourns**

TECHNICAL PROGRAM

Session 9: Engineers Without Borders: Real World Projects and Lessons Learned

Tuesday, February 26, 2013 **Room: Discovery**
8:30am – 11:45am **2.5 PDHs**

Moderator: Kristen Jellison; Lehigh University

9A **8:30am** **Post-Implementation Assessment of Student Executed Clean Water Programs in Ecuador** 249
G. String; Tufts University

9B **9:00am** **Fighting Failure: Lessons Learned in Pueblo Nuevo, Honduras** 253
C. Dutt, J. Bryant, A. Baker, K. Hoogenboom; Lehigh University

9:30am **Networking Break and Poster Session**

9C **10:15am** **A Self-Contained, PV-Powered Toilet and Domestic Waste Water Disinfection System** 257
Y. Qu, A. Aryanfar, C. Cid, H. Zhang, M. Hoffmann; California Institute of Technology

9D **10:45am** **From the Amazon to the Andes: WASH Projects in Rural Peru** 263
J. Milliken, G. Stonebraker; George Mason University

11:15am **Panel Discussion**

11:45am **Session Adjourns**

Poster **A Self-Contained, PV-Powered Toilet and Domestic Waste Water Disinfection System** N/A
Y. Qu, A. Aryanfar, C. Cid, H. Zhang, M. Hoffmann; California Institute of Technology

TECHNICAL PROGRAM

Session 10: Wet Weather

Tuesday, February 26, 2013
10:15am – 11:45am

Room: Cosmopolitan C
1.5 PDHs

Moderators: Leonard Casson; University of Pittsburgh
Keith Bourgeous; Carollo Engineers

10A **10:15am** **First-Flush Disinfection: Challenges of Chlorinating Wet Weather Primary Effluent to Meet A Stringent *E. Coli* Requirement** 270
M. J. Scarborough[1], M. Arends[2], S. Jeyanayagam[1], L. Schimmoller[1], J. Assouline[1]; [1]CH2M Hill, [2]City of Omaha

10B **11:45am** **In-Line Disinfection of a CSO First Flush Basin in Detroit** 310
C. D. Courter[1], K. Al-Omari1, M. Rabbaig[2], N. M. Sajdak[1],; [1]Hazen and Sawyer, [2]Detroit Water and Sewerage Department

10C **11:15am** **Pilot Study Results for Peracetic Acid and Sodium Hypochlorite Disinfection of Combined Sewer Overflows** 321
J. M. Brady; Stanley Consultants

11:45am **Session Adjourns**

TECHNICAL PROGRAM

Session 11: Biosolids

Tuesday, February 26, 2013 **Room: Cosmopolitan D**
1:15pm – 5:00pm **3.0 PDHs**

Moderators: Robert Reimers; Tulane University
Kari Briolara; Louisanna State University

11A	**1:15pm**	**The Historical Impacts of Stabilization and Disinfection of Municipal Sludge** N/A J. Donovan; CDM Smith
11B	**1:45pm**	**Future Directions Biosolids Management of The Second Decade of the 21st Century** 341 R. S. Reimers[1], Y. Xu[1], S. D. Pillai[2], K. B. Fitzmorris-Brisolara[3], J. A. Oleszkiewicz[4], G. Seidl[5]; [1]Tulane University, [2]Texas A&M, [3]Louisanna State University, [4]University of Manitoba, [5]Headworks Inc.
11C	**2:15pm**	**Application of High Energy Electron Beam and Chemical Oxidants to Destroy Endocrine Disrupting Compounds and Inactivate (Pathogenic) Microorganisms in Wastewater Effluent and Biosolids** 364 C. Rambo, C. Praveen, P. Jesudhasan, R. S. Reimers[1], S. D. Pillai[2]; [1]Tulane University, [2]Texas A&M
	2:45pm	**Networking Break**
11D	**3:30pm**	**Novel BioElectro™ Process for Sterilization of Biosolids** 367 E. S. Takhtehfouladi[1], M. Elektrowicz[1], R. S. Reimers[2], J. A. Oleszkiewicz[3]; [1]Concordia University, [2]Tulane University, [3]University of Manitoba
11E	**4:00pm**	**Developments in Disinfection Using Advanced Alkaline Stabilization** 374 K. B. Fitzmorris-Brisolara[1], R. S. Reimers[2]; [1]Louisanna State University, [2]Tulane University
11F	**4:30pm**	**Development of Heat Dried Biosolids Product for Application in the Wetlands** 380 Y. Xu, R. S. Reimers, A. J. Englande; Tulane University
	5:00pm	**Session Adjourns**

TECHNICAL PROGRAM

Session 12: Peracetic Acid

Tuesday, February 26, 2013
1:15pm – 5:00pm

Room: Discovery
3.0 PDHs

Moderator: Alberto Garib; FMC

12A	**1:15pm**	**Regulatory Environment and Challenges in Wastewater Disinfection** N/A B. Bastian; USEPA
12B	**1:45pm**	**Toxicology Profile of PAA in Wastewater Discharge Applications** N/A J. Meakim; Peragreen Solutions
12C	**2:15pm**	**Non-Deterministic CFD Modeling of Peracetic Acid Disinfection in Contact Tanks** 401 D. Santoro[1], G. Raspa[2], F. Crapulli[3], M. Raisee[4], C. Haas[5]; [1]Trojan UV, [2]La Sapenza University, [3]Western University, [4]Teheran University, [5]Drexel University
	2:45pm	**Networking Break**
12D	**3:30pm**	**Design and Operation Considerations for PAA Disinfection Systems** 405 K. Bell, T. Archer; CDM Smith
12E	**4:00pm**	**PAA Pilot Demonstration to Determine PAA Efficacy as an Alternative Disinfectant to Chlorine and Sodium Bisulfite Technology for Wastewater** 409 J. T. Meakim[1], R. T. Freeborn[2], C. Murphy[3]; [1]Engineered Water Treatment, Engineers Plus, Manchester, MO, [2]PERAGreen Solutions, [3]Steubenville Water Treatment Plant
12F	**4:30pm**	**Lessons Learned from St. Augustine WWTP – Comparison of VigorOx (Peracetic Acid) and Sodium Hypochlorite** 450 M. Tran; FMC
	5:00pm	**Session Adjourns**

TECHNICAL PROGRAM

Session 13: Inactivation of Drinking Water Pathogens and Indicators

Tuesday, February 26, 2013 **Room: Cosmopolitan C**
1:15pm – 2:45pm **1.5 PDHs**

Moderator: Laura Boczek; US EPA

13A 1:15pm **Kinetics and Mechanism of Coxsackievirus B5 Inactivation by Free Chlorine for Safe Global Water** 457
T. Vonder Haar, B. Marinas, J. Shisler; University of Illinois at Urbana-Champaign

13B 1:45pm **Photochemical Activation of Free Chlorine to Reactive Oxygen Species for Enhanced Inactivation of Chlorine-Resistant Waterborne Pathogens** 477
J. E. Forsyth, Q. Mao, J. S. Meschke, M. C. Dodd; University of Washington

13C 2:15pm **Anticipating Challenges Associated with In-Building Disinfection for Control of Opportunistic Pathogens in Premise Plumbing** 485
W. Rhoads, C. Proctor, A. Pruden, M. A. Edwards; Virginia Tech

2:45pm **Session Adjourns**

Alternate 1 **Groundwater Coliform Occurrence and Disinfection Practices** 463
R. A. Ryder; Kennedy/Jenks Consultants

TECHNICAL PROGRAM

Session 14: DBPs and Trace Constituents

Tuesday, February 26, 2013
3:30pm – 5:00pm

Room: Cosmopolitan C
1.5 PDHs

Moderator: Naoko Munakata; Los Angeles County Sanitation Districts

14A 3:30pm **Effects of UV-Based Treatment on Water Chemistry in Chlorinated, Indoor Pools** 509
M. Zare Afifi, E. Blatchley; Purdue University

14B 4:00pm **Influence of Biofilm on DBP Formation in a Simulated Water Distribution System** 537
Z. Wang, K. M. Coburn, C. Jennings, Y. Seo; University of Toledo

14C 4:30pm **Dynamic Behavior of Volatile DBPs and Their Precursors in Chlorinated, Indoor Swimming Pools** 543
S. Weng[1], M. Z. Afifi[1], R. W. Staton[1], Z. Sun[1], R. Berry[2], X. Zhu[2], E. R. Blatchley III[1]; [1]Purdue University, [2]West Lafayette High School

5:00pm **Session Adjourns**

Additional Paper

Evaluation of Oxidants Products as Disinfectants Seeking Reuse of Refinery Wastewater 241
A. Veiga, Y. Vera, V. Santiago, L. Yokoyama, V. Santos

Water Environment Federation®
the water quality people™

Disinfection and Public Health 2013

February 24 – 26, 2013
Hyatt Regency Indianapolis
Indianapolis, Indiana

This conference is hosted by the Water Environment Federation in cooperation with the Indiana Water Environment Association, International Ultraviolet Association, National Water Research Institute, and the National Onsite Wastewater Recycling Association.

February 24, 2013

Dear Conference Participants,

Welcome to Indianapolis, Indiana, and the Water Environment Federation's Disinfection and Public Health 2013 Conference! This event has been made possible through the cooperation of WEF's Committees on Disinfection and Public Health, International Coordination, Students and Young Professionals, and in collaboration with several national, international, and local water quality organizations including: Indiana Water Environment Association, International Ultraviolet Association, National Water Research Institute, and the National Onsite Wastewater Recycling Association.

The conference will begin with two full-day workshops: the first highlights the newly released update to the NWRI/WRF Ultraviolet Disinfection Guidelines for Drinking Water and Water Reuse and the second on Ceramic Pot and Biosand Filters.

Our Opening General Session will feature Dr. Ernest R. Blatchley III, professor of civil engineering at Purdue University, and Monica LaBiche Brown, from Water for People, a global non-profit organization that supports long-lasting access to water and sanitation. They will explore the role of disinfection and other treatment processes in protecting public health around the world, and discuss projects in which water treatment has improved the health of communities.

Throughout the conference, we will hear from national and international experts on many aspects of disinfection, ranging from drinking water to wastewater to recycled water to biosolids; from introductory concepts to practical application to cutting edge research; from household systems to large-scale urban treatment facilities; and from science and engineering to regulatory and social issues affecting disinfection and public health. In addition to technical presentations, the conference includes workshops and an interactive mobile session on the exhibit floor. We hope that this breadth allows all of you, whether experts or new to the field, to gain useful insights into this important field.

Thank you for joining this discussion on improving disinfection and public health – we hope that you enjoy the conference and find it informative.

Best regards,

Naoko Munakata
Conference Co-Chair

Jay Swift
Conference Co-Chair

GO GREEN IN INDIANAPOLIS

Supporting WEF's ongoing mission to support sustainable programming, we would like to ask conference attendees to please consider their carbon footprint as you travel to, stay in, and depart from Indianapolis and WEF's Disinfection and Public Health Conference.

SustainIndy is a bold and innovative enterprise for making Indianapolis more sustainable. The initiative is an action-based organization focused on reducing the environmental impact of city operations, working with private partners and helping pave the way for green economic development. By using best practices to create lasting environmental, economic and community vitality- enhancing, *SustainIndy* has helped Indianapolis reduce the city's carbon footprint.

Consider alternative transportation:
The nature of public transportation lends itself to increased individual sustainability. In other words, public transportation is a responsible environmental choice, reducing energy use and carbon footprints.

As you plan your stay in Indianapolis, consider using *IndyGo* with 29 fixed routes and more than 4,000 bus stops throughout Indianapolis; *IndyGo* takes you where you need to be. So, hop a bus to catch your flight or just get out on the town.

Dine at local farm-to-table restaurants:
One trend that is beginning to define the Indianapolis dining scene is the farm-to-table movement. When it comes to dining out in Indianapolis, the farms that dot the Indiana landscape present a huge advantage and offer restaurants in Indianapolis a wonderful, fresh source of ingredients. While there are a ton of restaurants in the city that are taking part in the farm-to-table movement, below are a few local eateries that are taking the farm-to-table concept to new heights:

Cafe Patachou
4911 N. Pennsylvania St
317-925-2823

R Bistro
888 Massachusetts Ave.
317-423-0312

Goose the Market
2503 N. Delaware St.
317-924-4944

Recess
4907 N. College Ave.
317-925-7529

Bluebeard
653 Virginia Ave.
317-555-5555

The Villa Inn
1456 N. Delaware St.
317-916-8500

CONFERENCE COMMITTEE

CONFERENCE CO-CHAIRS

Naoko Munakata
County Sanitation Districts of Los Angeles

Jay Swift
Gray & Osborne

DISINFECTION AND PUBLIC HEALTH CONFERENCE COMMITTEE

Robert Bastian
US Environmental Protection Agency

Keith Bourgeous
Carollo Engineers

Clint Cantrel
AECOM

Leonard Casson
University of Pittsburgh

Curtis Courter
Hazen and Sawyer

Liping Han
URS

Gary Hunter
Black and Veatch

Kristen Jellison
Lehigh University

Samuel Jeyanayagam
CH2M Hill

Daniele Latagne
Tufts University

Garry MacDonald
Beca

John Maziuk
Solvay

Andrew Salveson
Carollo Engineers Inc.

Michael Whalley
Associated Engineering Alberta Ltd

G. Elliott Whitby
UV Technologies

REGISTRATION

All events are held at the Hyatt Regency Indianapolis.

Registration is located in the **Cosmopolitan Foyer** of the Hyatt Regency Indianapolis and will be open at the following times:

Hours:

Sunday, February 24	7:30 am – 5:00 pm
Monday, February 25	7:30 am – 5:00 pm
Tuesday, February 26	8:00 am – 5:00 pm

SPEAKER AND MODERATOR INFORMATION

All session speakers, moderators and assistant moderators are required to sign in at the Registration Desk.

Speaker/Moderator/Assistant Moderator Breakfast – Room: Studio 1

All speakers must attend their assigned speaker breakfast. If you are unable to attend your assigned breakfast, please notify your session moderator.

Speakers, moderators, and assistant moderators shall attend their assigned speaker breakfast. If you are presenting or moderating for multiple sessions, you are requested to attend each breakfast if on different days. A light breakfast will be provided as part of the orientation. The Speaker Breakfast schedule is as follows:

Monday, February 25 — 7:30am – 8:15am
*Sessions 1 through 6

Tuesday, February 26 — 7:30am – 8:15am
*Sessions 7 through 14

COMMITTEE MEETING

WEF Disinfection Committee Meeting
Tuesday, February 26
7:00am – 8:30am
Room: Studio 5

RECEPTION AND MEAL FUNCTIONS

Networking Reception, Monday, February 25 **5:00pm – 6:30pm**

Take a break after the first full day of sessions! Enjoy a libation and hors d'oeuvres, visit with colleagues, and spend some time with the exhibitors. The Reception will be held in the Exhibit Hall located in the **Regency Ballroom** of the Hyatt Regency Indianapolis.

Luncheons – Regency Ballroom

Lunch will be provided for all full conference, daily registrants and all exhibitors. Please advise WEF Staff as soon as possible if you have any special dietary requirements. The luncheons will be held:

Monday, February 25 12:15pm – 1:45pm
This luncheon will also feature the presentation of the **William W. Edgar Pioneer Award**

*Monday Luncheon Sponsored by **Trojan UV***

Tuesday, February 26 11:45am – 1:15pm

Networking Breaks - Regency Ballroom

The Networking Breaks will be held:

Sunday, February 24 – **Cosmopolitan Foyer**
Times vary for each Workshop.

Monday, February 25 – **Regency Ballroom**
10:00am – 10:45am and 2:45pm – 3:30pm

Tuesday, February 26 – **Regency Ballroom**
9:30am – 10:15am and 2:45pm – 3:30pm

CONTINUING EDUCATION

Participants who attend the conference and accurately document their attendance at technical sessions will be awarded Professional Development Hours (PDH) at a rate of 1.0 PDH for every hour of participation. Continuing Education Units (CEU) will be issued to workshop attendees only. CEUs will be issued at the rate of 1 CEU for 10 hours of participation. **Please note that conference participants are responsible for exploring continuing education requirements in their state to ensure that WEF credits are accepted by the appropriate authority.**

To receive the certificate, attendees must do the following:

- *Request a continuing education credit form from the Registration desk upon check in.*
- *Note each time you enter/exit a session room on the CEU/PDH form.*
- *Have CEU/PDH form signed by Federation staff beside each session attended.*
- *Return CEU/PDH form to the registration desk at the conclusion of the conference. You will be required to present your photo ID upon submittal.*

The certificates will be mailed 8-10 weeks after the conclusion of the conference.

PROCEEDINGS

Conference proceedings, consisting of manuscripts for each presentation, have been produced on USB drives. Full conference and daily registrants will receive the proceedings at the registration desk at the time of badge pickup or onsite registration.

Copies of proceedings may be ordered after the conference at the member rate of $75 USD; nonmember rate of $115 USD, plus shipping and handling. All orders will be processed immediately after conclusion of the conference. You may call 1-800-666-0206. Ask for Stock Number CPDF 302.

SPONSORSHIP INFORMATION

We would like to thank the following sponsoring companies for their contributions to the conference and program.

FMC
ENVIRONMENTAL SOLUTIONS

Conference Proceedings
www.environmental.fmc.com
Booth 304

HDR

Opening General Session
www.hdrinc.com

TROJANUV™

Monday Conference Luncheon
www.trojanuv.com
Booth 206

WORKSHOP A

Workshop A: 2012 NWRI UV Guidelines – New Updates to Challenging Issues

Sunday February 24, 2013
8:30am – 5:00pm

Room: Cosmopolitan C
0.6 CEUs

8:30am	**NWRI UV Guidelines** J. Mosher; NWRI	
8:50am	**Scope of 2012 NWRI Guidelines** G. Tchobanoglous; University of California, Davis	
9:30am	**Regulatory Perspective on 2012 Updates** B. Bernados; CDPH	
10:00am	**Networking Break**	
10:30am	**NWRI 2012: An End User's Critical Perspective** K. Bourgeous; Carollo Engineers	
11:15am	**Practical Experience with Validation and Spot Check Bioassays** A. Salveson; Carollo Engineers	
11:30am	**A Laboratory Perspective on Bioassay Validation** T. Hargy; Tetra Tech	
12:00pm	**Lunch**	
1:30pm	**Data Analysis under NWRI 2012** G. Tchobanoglous; University of California, Davis	
2:00pm	**Design and Analysis Problem - Breakout session**	
3:00pm	**Networking Break**	
3:30pm	**Real Life Experience: NWRI 2012 Validation Compared to 2003** C. Bokermann; Xylem Water Solutions	
4:00pm	**Future Improvements to 2012 NWRI Guidelines** Panel Discussion	
5:00pm	**Adjourn**	

WORKSHOP B

Workshop B: Ceramic and Biosand Filter Workshop

Sunday February 24, 2013
8:30am – 5:30pm

Room: Cosmopolitan D
0.6 CEUs

8:30am	**Optimizing the Biosand Filter for Household Water Treatment in Developing Countries** J. Napotnik; Lehigh University
9:00am	**Performance of Biologically-Active Sand Filters for Drinking Water Treatment: Exploration of Biophysical Factors and Consumer Social Behavior Trends in Eldoret, Kenya** C. Grady; Purdue University,
9:30am	**Networking Break**
10:00am	**Barriers in Accessing Safe Household Water and Basic Sanitation for Persons with Disabilities in Rural Cambodia** M. Macleod; Samaritan's Purse
10:45am	**Protozoa Cysts Removal By Slow Sand Filtration: Post Treatment Alternative of Wastewater** R. Medeiros; University of Sao Paulo
11:15am	**Lessons Learned from the Implementation of the Hydraid BioSand Water Filter in Rural El Salvador** S. Gregory; Samaritan's Purse
11:45am	**Conducting research on Biosand Filtration in the Laboratory Setting** K. Jellison; Lehigh University
12:15pm	**Lunch**
1:45pm	**Evaluation of the Effectiveness of Ceramic Water Filters for Removal of Cryptosporidium from Drinking Water** C. Schneeberger; Centers for Disease Control and Prevention
2:15pm	**Performance of Ceramic Water Filters Impregnated with Silver Nanoparticles** V. Craver; University of Rhode Island
2:45pm	**Investigation of the Impact of Manufacturing Variables on Ceramic Filter Efficacy** K. Jellison; Lehigh University
3:15pm	**Networking Break**
4:00pm	**Ceramic Pot Filters: Next Steps in Research** J. Rayner; Tufts University
4:30pm	**Development of a Cost Projection Model for a Ceramic Water Filter Project in Arusha, Tanzania** L. Katica; Safe Water Now
5:00pm	**Potters for Peace and the Ceramic Pot Filter** K. Waganer; Potters for Peace
5:30pm	**Adjourn**

OPENING GENERAL SESSION

Monday, February 25, 2013
8:30am – 10:00am

Room: Discovery
1.5 PDHs

Moderators: Jay Swift; Gray & Osborne, Inc.
Naoko Munakata; Los Angeles County Sanitation Districts

8:30am **Opening Remarks from the Conference Chairs**
J. Swift; Gray & Osborne, Inc.
N. Munakata; Los Angeles County Sanitation Districts

8:45am **Organization Welcomes**
S. Grimes, Executive Director; Indiana Water Environment Federation
K. Chandran, Board of Trustees; Water Environment Federation

9:00am **Water Supply in Developing Countries: Roles for Universities to Play**
Ernest R. Blatchley; Purdue University

9:30am **Water For People: Ending Water Poverty Forever**
Monica LaBiche Brown; Water For People

10:00am **Adjourn**

EXHIBITION GENERAL INFORMATION

Located in the Regency Ballroom A-F at the Hyatt Regency Indianapolis, the exhibition hall hosts morning and afternoon refreshment breaks, luncheons, and a networking reception. Also, look forward to the new featured mobile session, happening on Monday afternoon.

The exhibition offers networking opportunities with companies displaying the latest product and service solutions directly related to the conference's technical program.

Exhibit Hall Schedule

Monday, February 25 **10:00 am – 6:30 pm**

10:00 am –10:45 am .... Coffee Break
12:15 pm – 1:45 pm .... Luncheon
2:45 pm – 3:30 pm .... Refreshment Break
5:00 pm – 6:30 pm ...Networking Reception

Tuesday, February 26 **10:00 am – 3:45 pm**

9:30 am –10:15 am .... Coffee Break
11:45 am – 1:15 pm .... Luncheon
2:45 pm – 3:30 pm .... Refreshment Break

Children under 18 entering the Exhibition must be accompanied by a parent or guardian at all times. The parent or guardian assumes all risk and responsibility for the child's safety.

Photographing or video recording in the Exhibition is only allowed with exhibitor's permission.

EXHIBITION DIRECTORY

Aquionics, Inc. ....Booth 200
www.aquionics.com
Phone: 859.341.0710
1455 Jamike Avenue
Erlanger, KY 41018

Compact and efficient UV disinfection systems featuring high-output medium pressure or low-pressure high-output amalgam lamps in both closed vessel and open channel configurations. Systems can treat a wide range of flows and water qualities.

Calgon Carbon Corporation...Booth 308
www.calgoncarbon.com
Phone: 724.218.7001
2000 McClaren Woods Drive
Coraopolis, PA 15108-7768

Calgon Carbon Corporation (NYSE:CCC), a leader in ultraviolet treatment, offers Sentinel® drinking water systems for disinfection and advanced oxidation; and C3 Series™ wastewater disinfection systems.

EXHIBITION DIRECTORY

Aquionics, Inc. .... Booth 200
www.aquionics.com
Phone: 859.341.0710
1455 Jamike Avenue
Erlanger, KY 41018

Compact and efficient UV disinfection systems featuring high-output medium pressure or low-pressure high-output amalgam lamps in both closed vessel and open channel configurations. Systems can treat a wide range of flows and water qualities.

Calgon Carbon Corporation .... Booth 308
www.calgoncarbon.com
Phone: 724.218.7001
2000 McClaren Woods Drive
Coraopolis, PA 15108-7768

Calgon Carbon Corporation (NYSE:CCC), a leader in ultraviolet treatment, offers Sentinel® drinking water systems for disinfection and advanced oxidation; and C3 Series™ wastewater disinfection systems.

Electrolytic Technologies Corporation .... Booth 306
www.electrolytictech.com
Phone: 305.655.2755
19597 NE 10th Avenue
North Miami Beach, FL 33179-3578

Manufacturer of onsite systems for generation of chlorine gas and/or high strength sodium hypochlorite as safe and efficient alternatives to pressurized chlorine gas.

Engineered Treatment Systems (ETS), LLC .... Booth 205
www.ets-uv.com
Phone: 877.885.4628
238 Commercial Drive
P. O. Box 392
Beaver Dam, WI 53916

US manufactured, validated, medium pressure and amalgam closed vessel UV systems for wastewater, drinking water, and reuse applications. Also active in recreational water and industrial process.

FMC
ENVIRONMENTAL SOLUTIONS

FMC Environmental Solutions .... Booth 304
www.environmental.fmc.com
Phone: 866.860.4760
1735 Market Street
Philadelphia, PA 19103

FMC Environmental Solutions Division offers wastewater utilities VigorOx® WWT II, a more cost effective and environmentally benign alternative to traditional chlorine technology for wastewater disinfection.

EXHIBITION DIRECTORY

Grundfos Pumps Corporation ....Booth 203
www.grundfos.us
Phone: 913.227.3400
17100 W. 118th Terrace
Olathe, KS 66061-6599

As a recognized global leader in advanced pump solutions and a trendsetter in water technology, Grundfos continuously pioneers new and improved ways to answer water and energy challenges.

JCS Industries, Inc. ....Booth 202
www.jcsindustries.us.com
Phone: 281.353.2100
P.O. Box 90028
Houston, TX 77290-0028

Manufacturers of innovative vacuum feed chemical injection systems, featuring patented liquid feed technology.

Olin Chlor Alkali Products. ....Booth 302
www.olinbleach.com
Phone: 423.336.4870
490 Stuart Road NE
Cleveland, TN 37312

Olin Chlor Alkali Products is a leading manufacturer of chlorine and caustic soda and is North America's largest manufacturer of industrial bleach. Consider us for water and wastewater disinfection options.

OZONIA North America ....Booth 201
www.ozonia.com
Phone: 201.676.2240
600 Willow Tree Road
Leonia, NJ 07605

Ozonia is one of the world's leading suppliers of UV and ozone disinfection systems. With a long history and extensive experience, the company is active in both municipal and industrial markets.

Parkson Corporation ....Booth 204
www.parkson.com
Phone: 1.888.PARKSON
1401 W Cypress Creek Road
Suite 100
Fort Lauderdale, FL 33309

EXHIBITION DIRECTORY

seepex Inc. .... Booth 207

www.seepex.com

Phone: 937.864.7150

511 Speedway Drive

Enon, OH 45323

seepex Inc., is an ISO 9001-2008 certified company located in Enon, USA that designs, manufactures, and globally markets progressive cavity pumps, macerators, and control systems.

Siemens Water Technologies .... Booth 300

www.siemens.com/water

Phone: 724.772.0044

181 Thorn Hill Road

Warrendale, PA 15086

Siemens combines the industry's broadest portfolio of technologies with technical expertise and proven installation experience to help consulting engineers develop best treatment options for municipalities.

Solvay Chemicals .... Booth 208

www.solvaychemicals.us

Phone: 713.525.6815

P. O. Box 27328

Houston, TX 77227

Solvay Chemicals is a world leading producer of essential chemicals inclusing soda ash, hydrogen peroxide, percarbonates, peracetic acid and special chemicals such as fluorinated products, ultra-fine fillers, high purity barium and strontium.

TROJANUV

Trojan Technologies .... Booth 206

www.trojanuv.com

Phone: 519.457.3400

3020 Gore Road

London, ON N5V 4T7, Canada

The leader in UV treatment solutions: protecting communities from damaging effects of microbiological/chemical contaminants in water. For 35 years, Trojan has set the standard in UV.

INDEX OF EXHIBITORS BY PRODUCT CATEGORY

PRESENTER AND MODERATOR DIRECTORY

Thomas Armitage, US Public Health Service, **Session 6**

Jeff Bandy, Carollo Engineeers, **Session 1**

Bob Bastian, USEPA, **Session 12**

Katherine Bell, CDM Smith Inc., **Session 2 & 12**

Brian Bernados, **Workshop A**

James Blyth, Imperial College London, **Session 5**

Laura Boczek, US EPA, **Session Moderator 13**

Christian Bokermann, Xylem, **Workshop A**

Keith Bourgeous, Carollo Engineers Inc, **Session 2, Session 7 Moderator, Session 10 Moderator, Workshop A**

Kari Briolara, Louisanna State University, **Session 11, Session 11 Moderator**

Laura Brunson, University of Oklahoma, **Session 3**

Leonard Casson, University of Pittsburgh, **Session 10 Moderator**

Kimberly Coburn, University of Toledo, **Session 5**

Curtis Courter, Hazen and Sawyer, **Session 10**

John Deogracias, Parkson Corporation, **Session 8**

Michael Dodd, University of Washington, **Session 13**

John Donovan, CDM Smith, **Session 11**

Patricia Drummey Stiegel, Hazen and Sawyer, **Session 4**

Cassandra Dutt, Lehigh University, **Session 9**

Maria Elektorowicz, Concordia University, **Session 11**

Bruno Ferran, Ozonia North America, **Session 4**

Robert Freeborn, PERAGreen Solutions, **Session 12**

Aimee Gall, University of Illinois at Urbana-Champaign, **Session 6**

Alberto Garibi, FMC, **Session 12 Moderator**

Ronald Gehr, McGill University, **Session 5**

Bailey Green, GO2 Water, Inc, **Session 8**

Samuel Gregory, McGill University, **Workshop B**

Tom Hargy, Tetra Tech, **Workshop A**

Vincent Hill, CDC, **Workshop B**

Gary Hunter, Black & Veatch, **Session 1, 8, 10, Workshop A**

Kristen Jellison, Lehigh University, **Session 9 Moderator, Workshop B**

Sam Jeyanayagam, CH2M HILL, **Session 2, Session 8 Moderator**

Ken Kershner, Aquionics Inc, **Session 8, Session 4 Moderator**

Daniele Lantagne, Tufts University, **Session 3**

Oliver Lawal, Aquionics Inc, **Session 7**

PRESENTER AND MODERATOR DIRECTORY

Mark LeChevallier, American Water, **Session 7**

Ting Lu, MSDGC, **Session 5**

John Meakim, Peragreen Solutions, **Session 12**

Raphael Medeiros, University of São Paulo - Brazil, **Workshop B**

James Milliken, George Mason University, **Session 9**

Laura Moehling, Gift of Water, **Session 6**

Jeff Mosher, National Water Research Institute, **Workshop A**

Naoko Munakata, Los Angeles County Sanitation District, **Session 14 Moderator**

Chuck Murphy, Steubenville Water Treatment Plant, **Session 12**

Anna Murray, Tufts University, **Session 6**

Jan Napotnik, Lehigh University, **Workshop B**

Vinka Oyanedel-Craver, University of Rhode Island, **Workshop B**

Brian Petri, Trojan Technologies, **Session 1**

Natasha Potgieter, University of Venda, **Session 6**

Yan Qu, California Institute of Technolgy, **Session 9**

Charlotte Rambo, Tulane University, **Session 11**

Justine Rayner, Tufts University, **Session 6 Moderator, Workshop B**

Robert Reimers, Tulane University, **Session 11, Session 11 Moderator**

Jenny Reina, CH2M HILL, **Session 5 Moderator**

William Rhoads, Virginia Tech, **Session 13**

Paul Ropic, Aquionics, **Session 1 Moderator**

Robert Ryder, Kennedy/Jenks Consultants, **Session 13**

Andrew Salveson, Carollo Engineers Inc, **Session 4, 8, Session 3 Moderator, Workshop A**

Domenico Santoro, Trojan UV, **Session 12**

Matthew Scarborough, CH2M HILL, **Session 10**

Chengyue Shen, HDR, **Session 4**

Mark Sobsey, University of North Carolina, **Session 3**

Gabe Stonebraker, George Mason University, **Session 9**

Gabrielle String, Tufts University, **Session 9**

George Tchobanoglous, UC Davis, **Workshop A**

Bryan Townsend, Black & Veatch, **Session 1 & 4**

Minh Tran, FMC, **Session 12**

Bree Trembly, Aquionics Inc., **Session 2 Moderator**

PRESENTER AND MODERATOR DIRECTORY

Theresa Vonder Haar, Water Environment Federation, **Session 13**

Kaira Wagoner, Pottor for Peace, **Workshop B**

Zhikang Wang, University of Toledo, **Session 14**

Shihchi Weng, Purdue University, **Session 14**

Michael Whalley, Associated Engineering, **Session 4 Moderator**

Elliott Whitby, Calgon Carbon Corporation, **Session 4, Session 1 Moderator, Session 8 Moderator**

Ashlie Williston, CT Consultants, **Session 7**

Yue Xu, Tulane University, **Session 11**

Mehrnaz Zare Afifi, Purdue University, **Session 14**

Peiran Zhou, University of Washington, **Session 5**

SESSIONS AT-A-GLANCE / EDUCATION CREDITS

Session #	Title	Credits	Room
SUNDAY			
Workshop A	2012 NWRI UV Guidelines – New Updates to Challenging Issues	0.6 CEUs	Cosmopolitan C
Workshop B	Ceramic And Biosand Filter Workshop	0.6 CEUs	Cosmopolitan D
MONDAY			
	Opening General Session	1.5 PDHs	Discovery
1	Evaluating UV Doses and Operations	1.5 PDHs	Cosmopolitan C
2	Disinfection Basics	1.5 PDHs	Cosmopolitan D
3	New Options for Disinfection in the Developing World	1.5 PDHs	Discovery
	Featured Mobile Session	2.5 PDHs	Exhibit Floor
4	UV Standards and Validation	2.5 PDHs	Cosmopolitan C
5	Molecular Methods and Mechanistic Studies	2.5 PDHs	Cosmopolitan D
6	Community Health	2.5 PDHs	Discovery
TUESDAY			
7	Effects of Upstream Processes on Disinfection	1.0 PDHs	Cosmopolitan C
8	Innovative Disinfection Methods	2.5 PDHs	Cosmopolitan D
9	Engineers Without Borders: Real World Projects and Lessons Learned	2.5 PDHs	Discovery
10	Wet Weather	1.5 PDHs	Cosmopolitan C
11	Biosolids	3.0 PDHs	Cosmopolitan D
12	Peracetic Acid	3.0 PDHs	Discovery
13	Innovation of Drinking Water Pathogens and Indicators	1.5 PDHs	Cosmopolitan C
14	DBPs and Trace Constituents	1.5 PDHs	Cosmopolitan C

NOTES

Investigating Polychromatic UV Complexity: Models and Experiments to Determine Polychromatic Correction Factors

Brian Petri, Trojan Technologies
Steve McDermid, Trojan Technologies
Wenjun Sun, Trojan Technologies
Chengyue Shen, HDR Hydroqual
Karl Scheible, HDR Hydroqual
Yuri Lawryshyn, University of Toronto

Trojan Technologies, 3020 Gore Rd.
London, ON, Canada N5V 4T7
(519) 457-3400

ABSTRACT

A bioassay was conducted on a set of polychromatic UV reactors to measure the impacts of polychromatic variables (sleeve spectra, water spectra, microbial action spectra), reactor design, and operating variables (flow, UVT, power, disinfection kinetics). The testing was modeled using a state of the art computational disinfection model, using inputs from the actual testing. The model agreed well with the measured results ($y=0.90$ x, $r^2=0.985$). The bioassay was very repeatable with low COVs (the majority of fully replicated tests had COVs < 10%). Direct comparisons of polychromatic variables showed differences in measured MS2 ranging from -4% to 32%, depending upon the compared conditions. It is noteworthy that doped sleeves did not always result in large dose reductions compared to synthetic quartz sleeves, that Superhume did not always result in larger correction factors than LSA, and long-path-length reactors did not always result in smaller correction factors than short-path-length reactors.

KEYWORDS

polychromatic UV, correction factors, model validation

INTRODUCTION

UV systems are validated for disinfection performance by measuring the inactivation of challenge microbes over a range of operating variables (flow, UV transmittance, power). The UV systems can then be installed with confidence, knowing the performance at the design and operating conditions.

The validation and installation of polychromatic UV systems is complicated by additional polychromatic variables. Polychromatic lamps emit UV across a range of germicidal wavelengths, different waters absorb those wavelengths differently, and different microorganisms are sensitive to those wavelengths differently. Thus the UV dose measured during the validation of a UV system in a given water with a given microbe, may differ from the UV dose delivered during the operation of the same reactor in a different water for a different microbe.

In order to properly account for dose delivery differences between validation and operation for polychromatic systems, mechanistic UV disinfection models must be used. Those models must be validated: they must accurately predict dose delivery, across the various variables they claim to account for, both operational variables (flow, UVT, power, microbe resistance) and spectral variables (sleeve spectra, water spectra, microbe action spectra). We bioassay validated two full scale polychromatic UV reactors, and compared the results with estimates from a proprietary state of the art computational disinfection model.

METHODOLOGY

We employed a disinfection model that predicted hydraulic flow through a reactor using computational fluid dynamics, calculated polychromatic light intensity distributions within the reactor, and applied disinfection kinetics to determine dose distributions and a final integrated dose for populations of microbes travelling through the reactor.

We conducted a set of full scale UV reactor trials at various flows and UVTs, using 2 different commercial UV reactors, 3 microbes (MS2 bacteriophage, *Bacillus pumilis* spores, *Aspergillus brasilliensis* spores), 3 different waters (dechlorinated municipal drinking water modified with either Superhume, lignin sulfonic acid, tea), and 2 different types of quartz sleeves (synthetic and doped quartz). Triplicate inlet and triplicate outlet samples for each organism were measured for microbiological enumeration by GAP Enviromicrobial (London, Canada). Absorbance scans (200-400 nm) for inlet and outlet water samples for each test were made on fresh samples using a Cary 100 spectrophotometer, verified with absorbance standards (Starna). Flow was measured using a calibrated magnetic flowmeter, and power consumption was verified using a calibrated power meter (Candura). Collimated beam dose-response curves were determined for each day of testing for each microbe tested that day. Sleeve spectra were measured throughout the bioassay period. All of the test-specific spectra, operating conditions, and kinetic parameters were used in the disinfection model. Various microbial action spectra were used in the models, in order to gauge the sensitivity of these inputs to the final results. These action spectra are the only inputs that cannot be reasonably measured during each test effort. All aspects of the testing were witnessed by University of Toronto, and the project was overseen by HDR Hydroqual.

RESULTS AND DISCUSSION

Approximately 30% of the bioassay test conditions were replicated in full repeat tests with MS2. The coefficients of variation (COVs) for 17 replicate bioassay results ranged from 0.3 to 15.6%, with 16/17 COVs < 10.4% and 12/17 COVs < 5.7%. The model accurately predicted dose delivery across all operating variables and polychromatic variables, for all tested reactors. For one reactor configuration, a full set of operating variables was tested, and the correlation of measured and calculated results for MS2 was $y = 0.90\ x$ ($r^2 = 0.985$). Given the high linearity, the model can be easily "calibrated" to align the measured and predicted results.

Paired tests with different polychromatic variables (different sleeve, different water) were done in a short-path-length reactor and a long-path-length reactor, at the same flow and power (within each reactor, but different flows between reactors), but at both high (92-93%) and low (72-73%) UVT. Comparing paired results between the modifiers Superhume and tea (a validation additive compared to a "natural" humic), the MS2 dose dropped between 6 and 15% for the short-path-length reactor, and between 8 and 32% for the long-path-length reactor, depending upon the sleeve and the UVT. Between the modifiers LSA and tea, the MS2 dose dropped between 1 and 32% for the short-path-length reactor, and between 2 and 23% for the long-path-length reactor, depending upon the sleeve and the UVT.

Comparing paired tests with different sleeves, the MS2 dose dropped between 5 and 24% for the short-path-length reactor, and between -4 and 12% for the long-path-length reactor, depending upon the water and the UVT.

It is noteworthy that LS2 (in comparison to Superhume) did not always result in the smallest correction factors in comparison to a "natural" water (tea), as held by conventional wisdom. It is also noteworthy that the difference in dose between synthetic quartz and doped quartz, could vary highly depending upon reactor design, water spectra and UVT, and that the reduction in dose for doped sleeves was sometimes negligible depending upon these conditions. Forcing UV systems to be designed or operated with doped sleeves does not necessarily impart the level of "safety" that is assumed. Finally, conventional wisdom holds that long-path-length reactors will always have lower correction factors than short-path-length reactors (all other variables equal), but this was not always the case from our results.

Clearly, polychromatic UV disinfection is very complex when anticipating the changes in performance between validation and operating conditions. Mechanistic disinfection models, based on theoretical principles and used with the best quality inputs, and validated against actual test results, are the best options for generating meaningful correction factors.

On-Line UV Sensors in Wastewater Applications: Why They're Needed and How They Benefit Operations, Compliance, and Public Health

Jeff Bandy, Carollo Engineers
Harold Wright, Carollo Engineers

ABSTRACT

Although the regulations governing the application of ultraviolet (UV) disinfection in drinking water treatment plants emphasize the importance of accurate on-line UV intensity sensors, current wastewater treatment regulations, particularly those that apply to recycled water systems, have not done the same. Many of the UV disinfection systems currently installed in wastewater treatment plants (WWTPs) do not use calibrated UV intensity sensors to directly monitor UV lamp intensity and instead simulate UV sensor signals via other control parameters (such as UV transmittance (UVT) and ballast power setting). Systemic permit violations by underperforming UV disinfection equipment can often be linked to the lack of robust control via online UV sensors. This paper presents a case study of the benefits of online UV sensors at two WWTPs in the Southeast.

KEYWORDS

UV disinfection, water reuse, UV intensity sensors

BACKGROUND

North American drinking water treatment plants commonly install ultraviolet (UV) disinfection reactors to achieve *Cryptosporidum parvum* and *Giardia lamblia* credits per the 2006 United States Environmental Protection Agency (USEPA) UV Disinfection Guidance Manual (UVDGM) as part of the Long Term 2 Enhanced Surface Water Treatment Rule (LT2ESWTR). The UVDGM provides guidance for validation, installation, and operation of UV disinfection equipment and places particular emphasis on the need for accurate system monitoring via calibrated flowmeters, UV transmittance (UVT) monitors, and on-line UV sensors, which measure the intensity of one or several UV lamps. The benefits of using calibrated UV sensors far outweigh the capital costs, particularly the ability to modulate lamp output via ballast power adjustments in order to meet a target UV dose while minimizing operational and maintenance costs. Using on-line UV sensors not only allows for this type of "dose pacing" but also empowers operations staff to schedule quartz sleeve cleaning events and other maintenance tasks only when required.

Although the benefits of using UV sensors for performance monitoring is well-established in the drinking water disinfection industry, the wastewater industry has historically shown some reluctance to incorporating them into open-channel UV systems. For instance, the Third Edition of the National Water Research Institute (NWRI) Guidelines for Water Resuse only vaguely specifies monitoring UV intensity signals. However, recent UV reactor validations have shown that modern UV sensor technology can reliably make the transition to low-UVT waters, thereby providing the aforementioned O&M benefits and related market advantages to UV equipment vendors. Furthermore, experience has shown that utilities that run wastewater UV systems without reliable UV dose monitoring can have systemic difficulties in meeting their discharge permits.

Examples of recent UV equipment validations for wastewater and combined sewer overflow applications include several Xylem/Wedeco (Herford, Germany) systems, particularly the new low-pressure high-output (LPHO) open-channel Duron UV system (which features high-wattage LPHO UV lamps and a unique lamp arrangement), and Calgon Carbon Corporation's (Pittsburg, PA, USA) C3500-D LPHO open-channel UV system (which includes novel flow-disrupting "delta wings" that tighten and increase the UV dose distribution).

This paper presents the results of an audit of two existing, validated UV disinfection systems at a WWTP in the Southeast US, performed by Carollo Engineers over the course of 2012. The UV system consisted of parallel banks of horizontally-aligned medium pressure (MP) UV lamps and frequently produced off-specification events and permit compliance problems. A combination of short-circuiting, outdated programming, and lamp aging and quartz sleeve fouling issues contributed to the issues, many of which would have been avoided with reliable online UV sensors.

UV SITE A UV AUDIT

UV Site A, located at a WWTP in the Southeast US, has a design flow of 7.5 mgd and an annual average daily flow (AADF) of 6.0 mgd. The wastewater is treated using conventional secondary activated sludge followed by DynaSand filtration. The filter effluent is disinfected by UV light before being discharged to either a surface water or to the public reuse system: the former effluent is treated by vertically-aligned low-pressure (LP) UV lamps, and the latter effluent is treated by horizontally-aligned MP UV lamps

The UV Site B has a permitted annual average daily flow of 3.0 mgd. The wastewater is treated using a sequential batch reactor (SBR) activated-sludge process followed by cloth-media disk filters. The effluent is disinfected by vertically-aligned low-pressure (LP) UV lamps before being discharged to a spray-irrigation public-access reuse system.

Table 1 summarized fecal coliform permit exceedances observed from October 2011 to August 2012 at UV Site A with the reuse effluent being treated by the MPUV system. The plant exceeded the 25 CFU/100 fecal coliform maximum criterion nine times and the 75th percentile monthly average criterion five times over this nine-month period. Over this time period, the river discharge being treated by the LP UV system met these criteria.

To identify the root cause behind these permit violations, the client contracted with Carollo Engineers Inc., (Carollo) to conduct a UV system audit at UV Site A and also requested that Carollo inspect the UV system at the UV Site B, where fecal coliform concentrations exceeded both the maximum and the monthly average values during July 2012. This technical memorandum summarizes the findings of these evaluations.

PROJECT HISTORY

The MPUV system sat UV Site A was installed in 2001. The system was originally designed to provide a UV dose of 140 mJ/cm^2 at a flow of 2.7 mgd and UVT of 55 percent. With the adoption of the Ultraviolet Disinfection Guidelines for Drinking Water and Water Reuse (NWRI, 2000) by the Florida reuse disinfection rules, the UV system was re-rated to provide a UV dose of 100 mJ/cm^2 at a flow of 5.7 mgd and UVT of 55 percent.

The NWRI Guidelines specify a UV dose of 100 mJ/cm^2 and a default UVT of 55 percent as design criteria for UV disinfection systems using granular media filtered wastewaters for water reuse applications. The Guidelines state that the UV system can be designed based on the 10th percentile of triplicate daily samples collected over at least a 6 month period.

Table 1 UV Site A Permit Exceedances Observed with the Reuse Effluent		
Date	Fecal Coliform	Violation
Oct 12, 2011	70 CFU/100mL	Exceeds 25 CFU/100 mL max
Dec 3, 2011	130 CFU/100mL	Exceeds 25 CFU/100 mL max
Dec 10, 2011	1600 CFU/100mL	Exceeds 25 CFU/100 mL max
Dec 12, 2011	58 CFU/100mL	Exceeds 25 CFU/100 mL max
Dec 31, 2011	64.3 %	Exceeds 75th percentile monthly average
Jan 25, 2012	27 CFU/100mL	Exceeds 25 CFU/100 mL max
Jan 26, 2012	100 CFU/100mL	Exceeds 25 CFU/100 mL max
Jan 31, 2012	74.2 %	Exceeds 75th percentile monthly average
Mar 31, 2012	74.2 %	Exceeds 75th percentile monthly average
Jun 8, 2012	72 CFU/100mL	Exceeds 25 CFU/100 mL max
Jun 30, 2012	71.4 %	Exceeds 75th percentile monthly average
Jul 7, 2012	36 CFU/100mL	Exceeds 25 CFU/100 mL max
Jul 31, 2012	69.6 %	Exceeds 75th percentile monthly average
Aug 28, 2012	60 CFU/100mL	Exceeds 25 CFU/100 mL max

The MP UV disinfection system consists of two reactors in series (with a total of 72 lamps) located in a concrete channel. Each lamp is contained within a quartz sleeve, which is cleaned using automatic wipers. Figure 1 shows top and side views of the UV system. Effluent from the DynaSand filters enters the UV reactor channel via a concrete influent box. The UV reactor channel is approximately 5 feet wide. The length of channel upstream of the first reactor is approximately 9 feet. The length of channel between the two reactors is approximately 11 feet. The length of channel between the second reactor and the downstream channel effluent weir is also approximately 11 feet. The depth of water within the channel is maintained by the height of the effluent weir and the head loss through the UV reactors, and is at least 2.5 feet. Each channel segment is equipped with a 6-inch floor drain to allow draining of the channel for maintenance of the channel and the reactor.

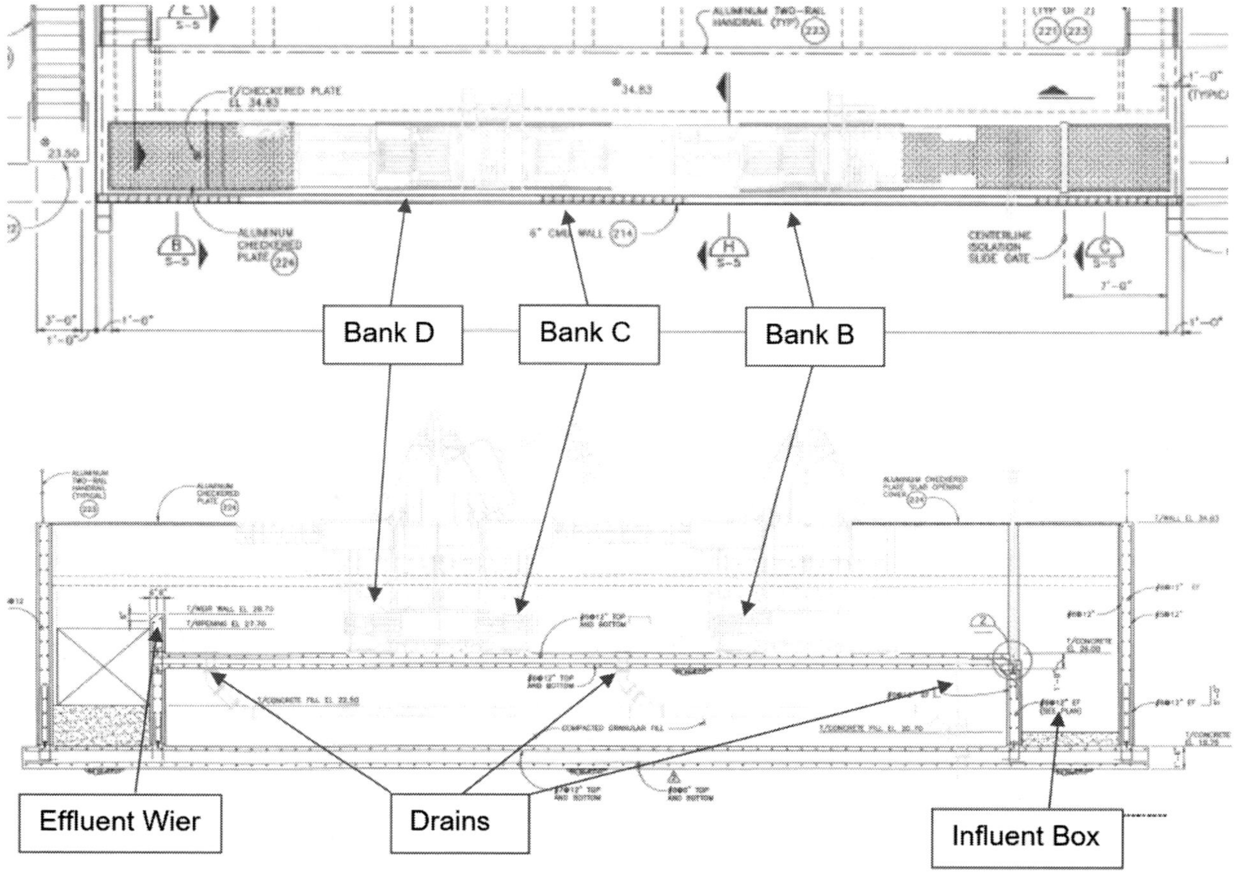

Figure 1 Top and Side Views of the MPUV system Used by the UV Site A

Figures 2, 3, and 4 show UV dose, UV system flow rate, and water UVT with the MPUV system recorded by SCADA from April 2011 to August 2012. Water UVT was measured by an online UVT monitor manufactured by HF Scientific. UV dose ranged from 120 to 4800 mJ/cm^2 with an average value of 361 mJ/cm^2. UV system flow ranged from 0.1 to 4.2 mgd with an average value of 1.74 mgd. The water UVT ranged from 55 to 78 percent with an average value of 68 percent and a 10th percentile value of 63 percent. UV dose, UV system flow rate, and water UVT were all within the re-rated UV system design parameters.

Figure 5 shows the UV system effluent fecal coliform concentrations recorded over the same time period. The fecal coliform concentrations ranged from below detection limits (< 2 CFU/100 mL) to 1,600 CFU/100 mL with a 75th percentile of < 2 CFU/100 mL. The fecal coliform concentration exceeded the 25 CFU/100 mL maximum criterions with 3 percent of the measurements. There was no significant correlation between fecal coliform concentrations and UV Dose, flow rate, and UVT.

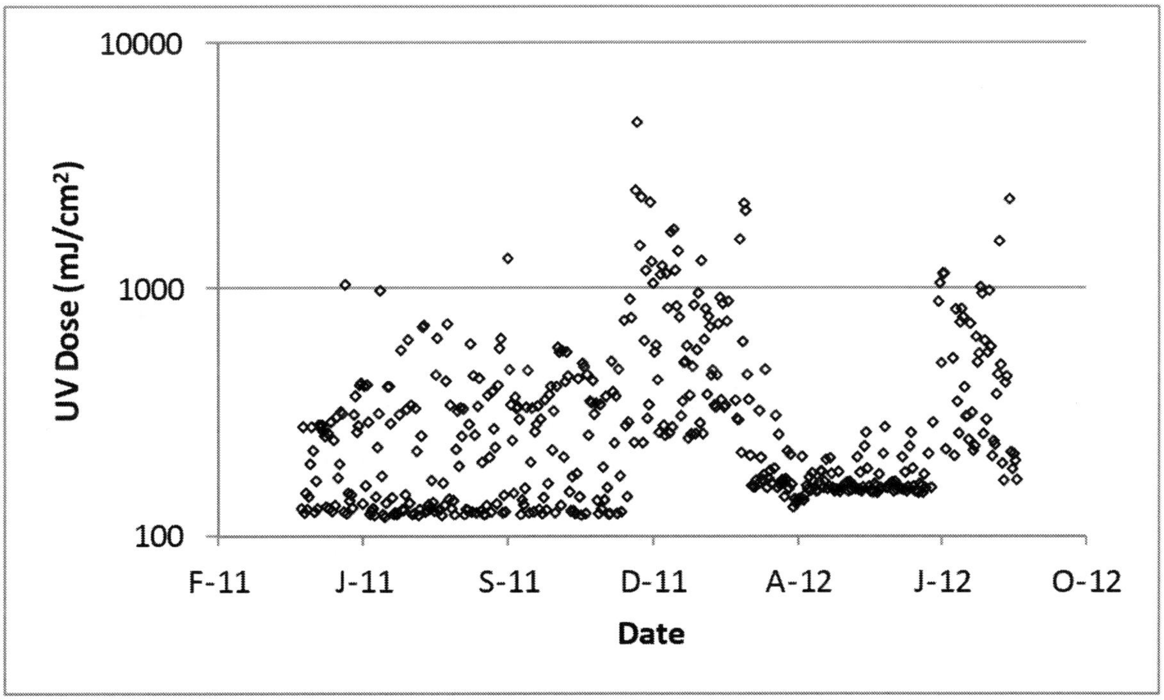

Figure 2 UV Dose Delivery by the MPUV system Used by UV Site A

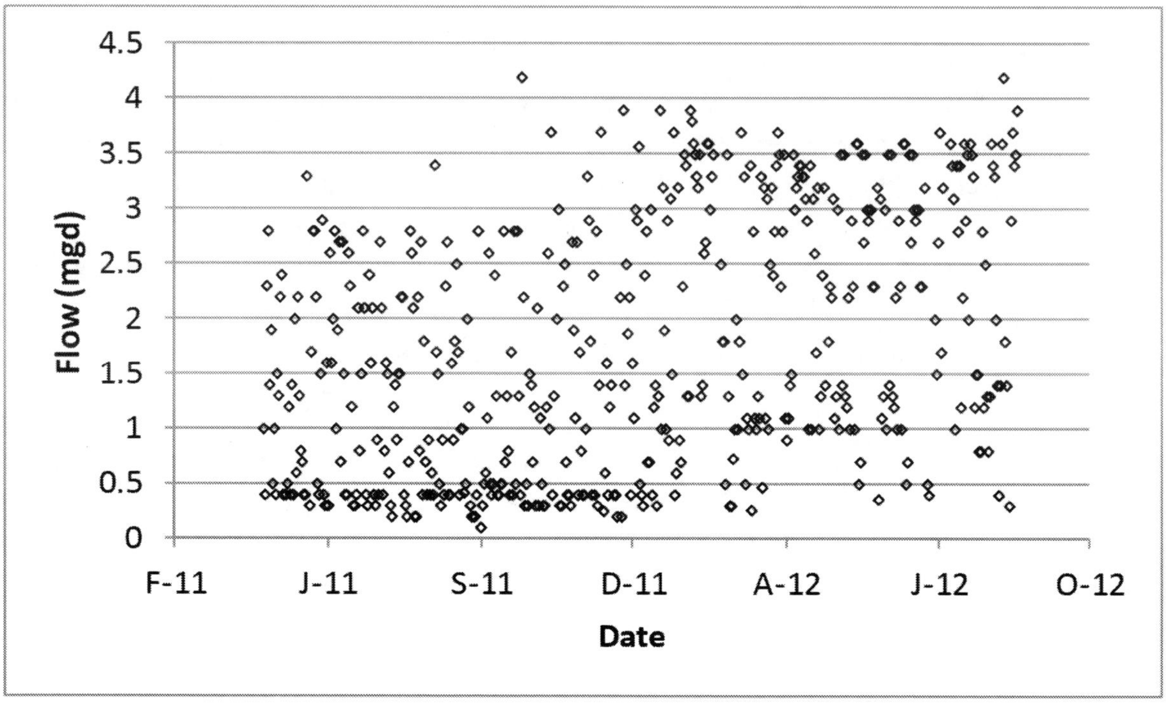

Figure 3 UV System Flow through the MPUV system Used by UV Site A

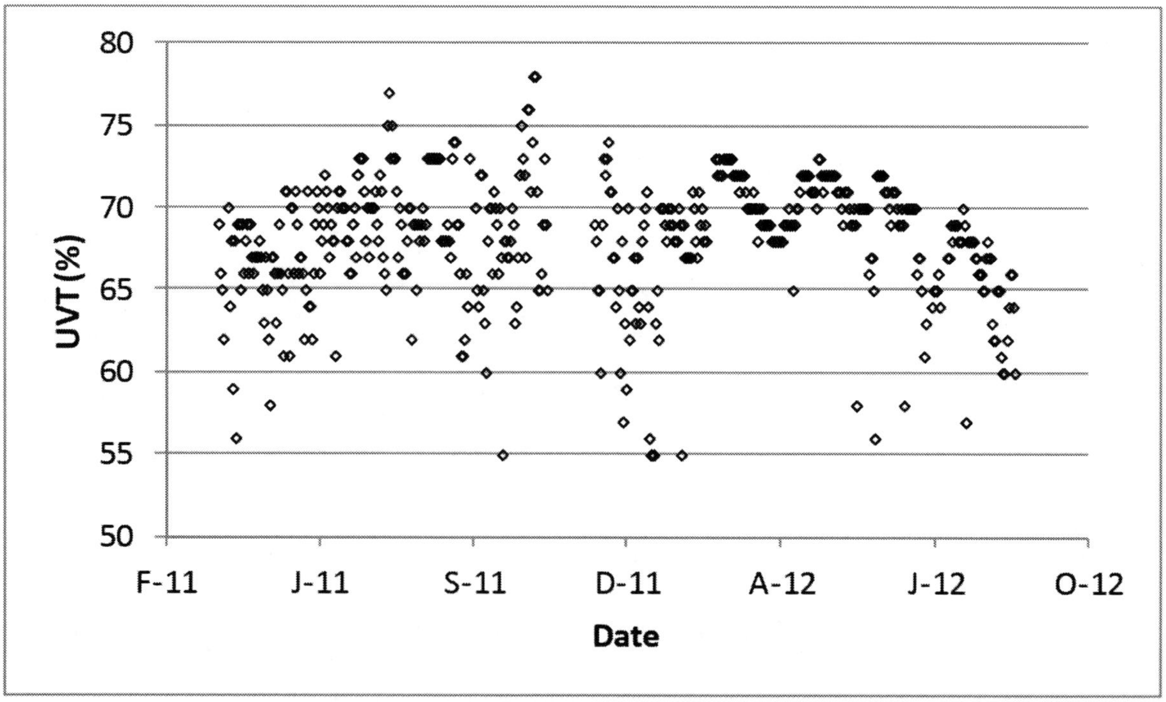

Figure 4 Water UVT with the MPUV system Used by UV Site A

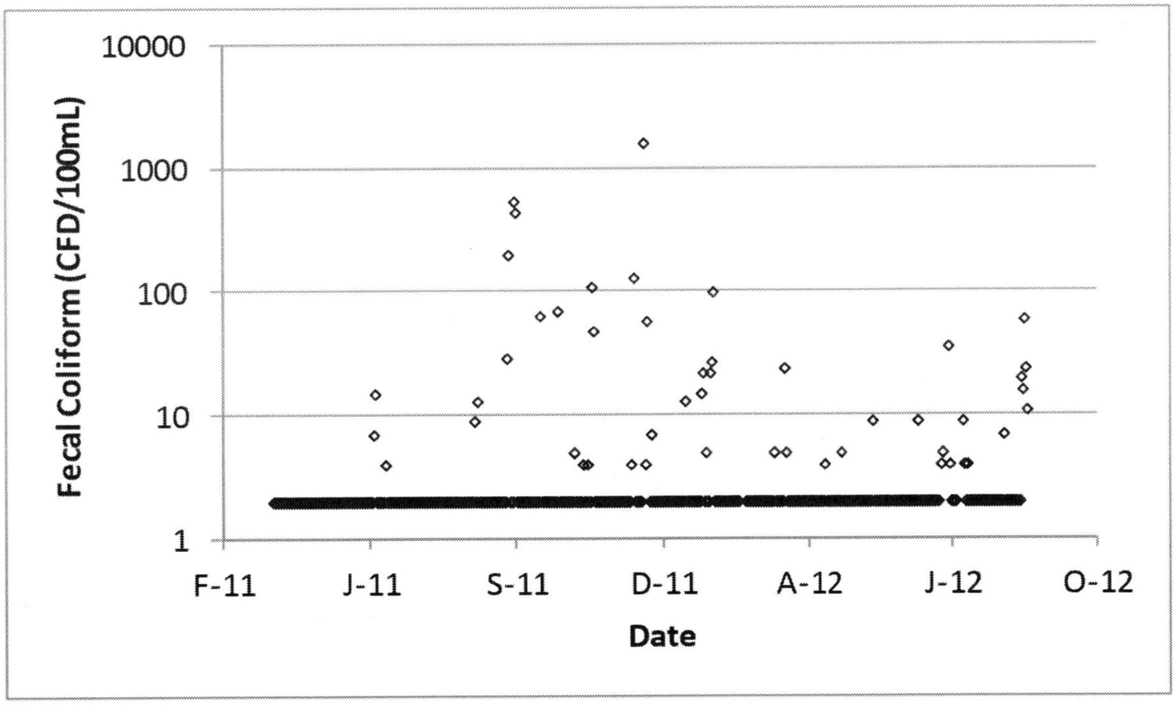

Figure 5 Effluent Fecal Coliforms with the MPUV system Used by UV Site A

RESULTS AND ANALYSIS

Carollo conducted the UV system audit at UV Site A on October 2 to 4, 2012. Results of the audit are presented in the following sections.

MAINTENANCE RECORDS

The fecal coliform exceedances began at UV Site A during September 2011, when the display on the MPUV system HMI failed. The plant staff replaced the HMI and transferred the The vendor software from the failed unit to the new unit.

Plant staff arranged for the vendor to conduct a site evaluation during February 2012. The vendor corrected a software issue related to the scaling of flow by the HMI, and verified the signals to the PLC, the proper operation of the PLC software, and the expected power consumption by the lamp ballasts. The vendor reported that the sleeves were seven years old and showed signs of external surface etching that impacted cleaning by the automatic wipers and seal failures that would lead to water leakage into the sleeves and premature lamp failure. The vendor also reported worn wiper seals that would also impact sleeve cleaning. The vendor recommended replacement of all sleeves. The vendor also observed significant algae buildup on the modules and recommended weekly cleaning of the modules and regular cleaning of the channels.

Following the site evaluation, plant staff replaced all lamps with the newer 12,000 hour, replaced lamp sleeves, and drastically upgraded the preventative maintenance schedule. However, coliform exceedances continued to occur, even after July 2012 when the UV system was operating with three banks on at 100 percent power.

VISUAL INSPECTION OF THE MPUV SYSTEM

During the site visit, the modules with all three banks were lifted for visual examination. The external surfaces of the quartz sleeves did not show any visual evidence of fouling over the length of the quartz sleeve that is cleaned by the automatic wipers. While a few sleeves showed evidence of internal fouling, the interior surfaces of the sleeves were also relatively clean.

Many lamps showed "end darkening," a phenomenon that is caused by the sputtering of material off of the lamp's electrodes, which then deposits on the internal surface of the lamp envelope. The lamp end darkening reduces the UV output from the lamps. The rate of end darkening is impacted by how the lamp is operating, such as the number of lamp on/off cycles.

DIRECT MEASUREMENT OF SLEEVE FOULING AND LAMP AGING

During the site visit, sleeves and lamps were removed for direct examination using a custom optics bench shown in Figure 6. The optics bench consists of a low-pressure mercury lamp source contained within an aluminum box, a research radiometer (International Light IL 1700) to measure the light from the lamp source, and supports to allow precise placement of a quartz sleeve and lamp in the light path from source to radiometer. This device allows direct measurement of the transmittance of UV light at 254 nm wavelength through clean and fouled sleeves and new and aged UV lamps. The UV transmittance of the sleeve and lamp is calculated using:

$$\mathrm{T} = \sqrt{\frac{I_1}{I_2}}$$

where T is the UV transmittance, I_1 is the UV intensity measured with an aged and fouled sleeve or lamp, and I_2 is the UV intensity measured with a new and clean sleeve or lamp. The square root function accounts for the fact that the UV light passes through both sides of the sleeve or lamp. A UV transmittance of 80 percent means that 20 percent of the UV light emitted by the lamps at 254 nm is absorbed by the sleeve or lamp envelope quartz and any internal and external fouling of that quartz.

Figure 6 Custom Optic Bench for Direct Measurement of Sleeve Transmittance And Fouling

Table 2 shows the results of UV transmittance measurements made on sleeves from module 1 of bank D. UV transmittance measurements were taken at five points along the length of the sleeve that correspond to where the lamp arc would be located inside the sleeve. Measurements were also made through horizontal and vertical sections of each sleeve. The UV transmittance with sleeves 1 and 2 were 88 and 96 percent, respectively. Sleeve 2 was the sleeve with internal brown-colored fouling shown in Figure 8. In contrast, sleeves 3 and 4 had a UV transmittance of 100 percent, indicating no fouling.

Table 2 UV Transmittance of Sleeves from Module 1 of Bank D Relative to a New Sleeve

Lamps	UVT (%)
1	88
2	96
3	101
4	100

Figure 7 shows the new and aged lamps evaluated using the optics bench. Note that to facilitate UV transmittance measurements with the lamps, the lamp wires and spacers were disconnected from the lamp.

Figures show the UV transmittance of the envelopes of the three lamps measured along the length of the lamp at 11 locations and through horizontal and vertical sections. As expected based on the end darkening of the lamps, the UV transmittance has peak values near the lamp center and drops significantly close to the electrodes. The average UV transmittance values of the three lamps were 65, 55, and 66 percent. According to the UV system HMI, the lamp hours with these lamps were 9,500, 8,828, and 8,828 hours, respectively.

UV systems are designed to deliver the required UV dose with aged lamps and fouled sleeves. The vendor specifies a fouling factor for the MPUV system of 0.95. The data in Table 2 indicates that the UV system at UV Site A is on average meeting this value. The vendor has also specified a lamp-aging factor for the a new UV lamp technology of 0.76 at 12,000 hours. This factor was later updated to 0.85 at 12,000 hours. The UV transmittance data measured with the lamps indicates that the lamps are not meeting these values.

Lamp 2, Module 5, Bank B

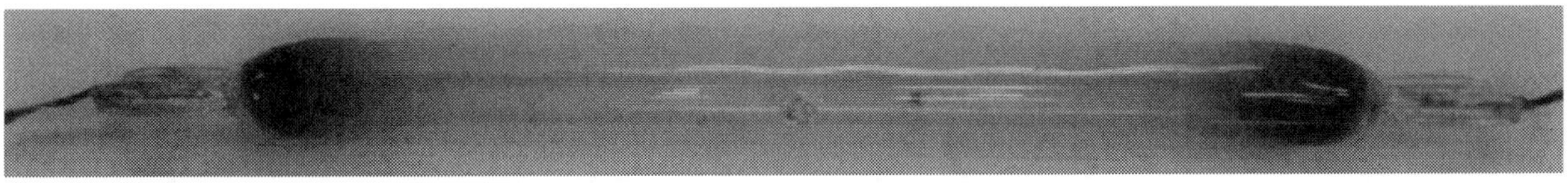

Lamp 4, Module 1, Bank D

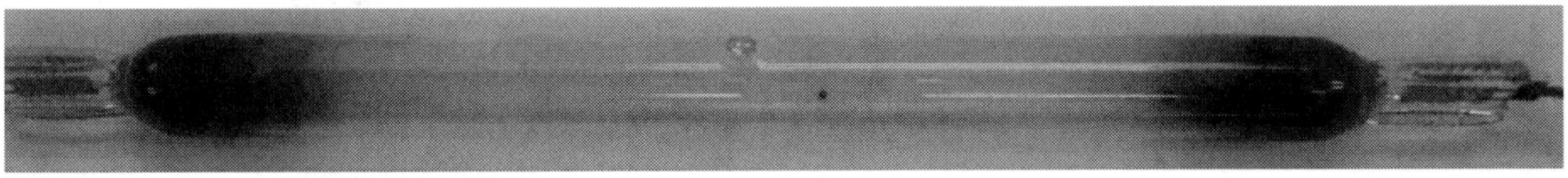

Lamp 3, Module 1, Bank D

New Lamp

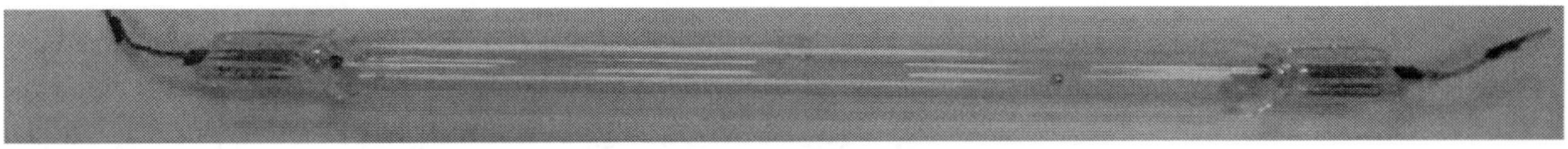

Figure 7 New and Aged Lamps Evaluated Using the Optics bench.

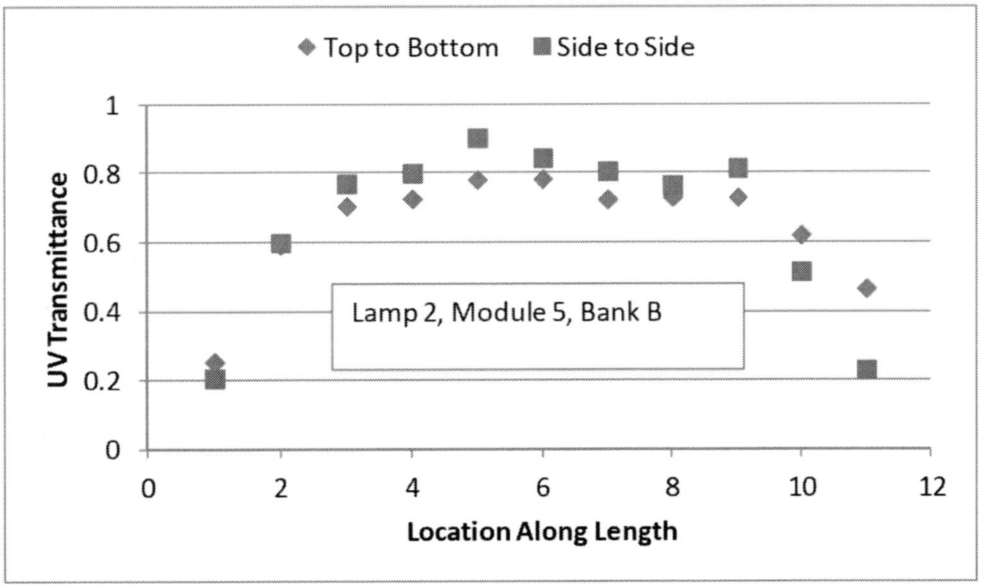

Figure 8 UV Transmittance of the Lamp Envelope at 254 nm Relative to a New Lamp with lamp 2 of Module 5, Bank B

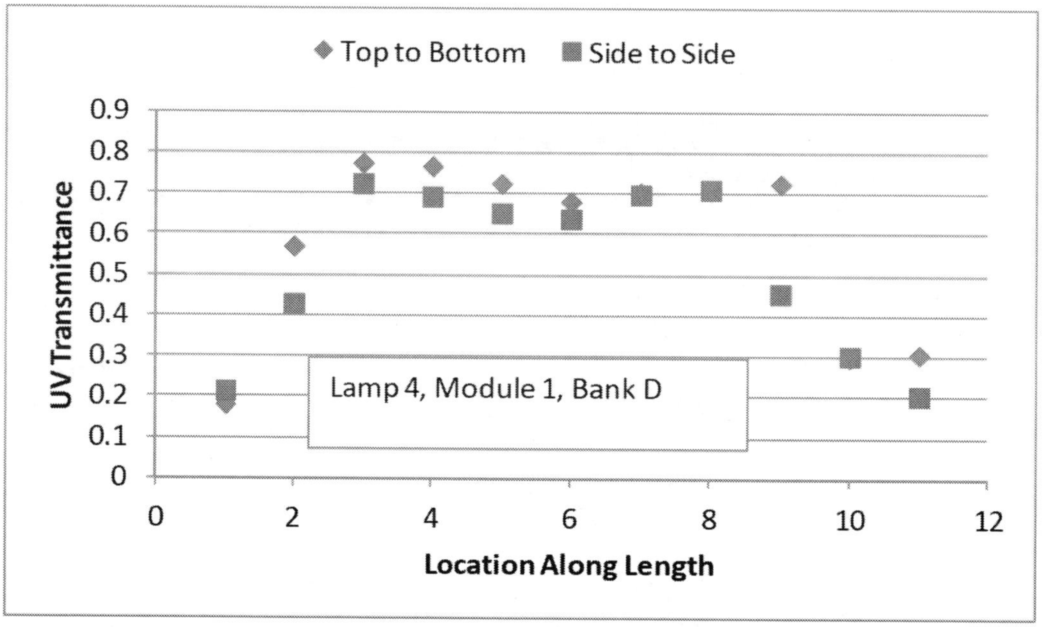

Figure 9 **UV Transmittance of the Lamp Envelope at 254 nm Relative to a New Lamp with lamp 4 of Module 1, Bank D**

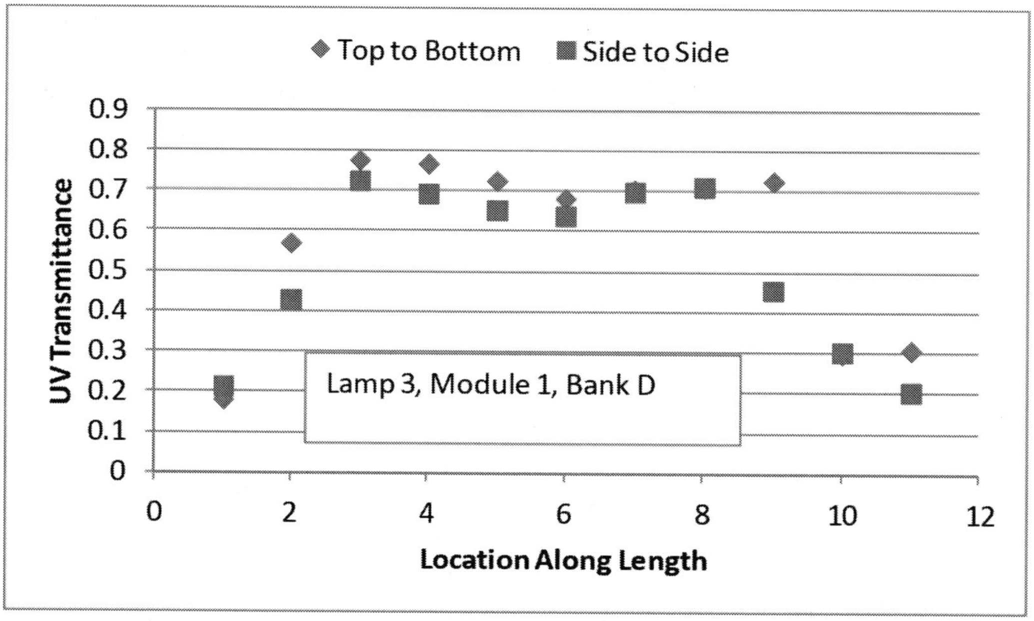

Figure 10 **UV Transmittance of the Lamp Envelope at 254 nm Relative to a New Lamp with lamp 3 of Module 1, Bank D**

UV DOSE MONITORING ALGORITHM

The MPUV system is equipped with a PLC that calculates UV dose delivered by the UV reactors. The UV system can operate in automatic mode turning on and off banks of lamps and adjusting the lamp ballast power setting so that the UV system meets a UV dose target without significant over dosing.

Understanding the UV dose monitoring and control algorithm used by the PLC is key towards understand overall UV system performance. The UV dose-monitoring algorithm used by UV system control PLC is typically not described in detail in the UV system documentation provided by the manufacturer. Hence, data were collected during the site visit to reverse engineer the UV dose-monitoring algorithm. The UV system was operated with various entered values of UVT, lamp power setting, and lamp hours. The flow rate was varied by adjusting the scaling factor used by the UV system's PLC to convert the flowmeter 4 to 20 mA signal to a flow measurement in mgd.

A key finding of this evaluation is the displayed UV intensity is a simulated value as opposed to measured by the UV sensor. The displayed UV intensity is calculated as a function of ballast power setting, lamp hours, and UVT. There is no relation between the displayed UV intensity and the real degree of lamp aging and fouling that occurs in the UV system. If lamp aging and fouling in the reactor is greater than the relative intensity predicted as a function of lamp hours, the UV dose predicted by the PLC will be greater than the UV dose actually delivered by the reactor.

Another key finding is that UV dose is calculated as the product of the UV intensity and residence time. This calculation assumes that the flow through the reactor can be modeled as turbulent plug flow with no short-circuiting. UV disinfection research over the last decade shows that UV dose delivery is impacted by the hydraulic flow patterns through the reactor. The hydraulics through the reactor impacts the residence time of microbes within the reactor and whether the microbes travel closer to the lamps in high UV intensity zones or further from the lamps in low intensity zones. Because of these effects, each microbe passing through the reactor may receive a different UV dose and UV dose delivery by UV reactors is best described by a UV dose distribution.

Because UV reactors do not have ideal turbulent plug flow, UV dose calculated as the product of average UV intensity and exposure time over states performance. The differences are greater at lower UVT and for reactors with poor hydraulic design, and can be as great as a factor of 2. To address this issue, regulations and guidance, such as the NWRI/AwwaRF UV Guidelines used by many States for reuse UV applications, specify application of UV reactors with UV dose delivery that has been measured using validation testing and documented in a validation report. With validation, log inactivation of a test microbe, such as MS2 bacteriophage, is measured as a function of flow, UVT, and lamp power settings. The UV dose associated with a measured log inactivation is calculated using the known UV dose-response curve of the test microbe. The data is analyzed to define the performance curves for the reactor, which are used for sizing the UV system for a UV application and should be used for dose monitoring.

The audit also found that UV intensity data were collected using a constant entered UVT of 50 percent and ballast power setting of 100 percent. The analysis shows that the displayed UV intensity is calculated using a linear relationship between the relative UV intensity (I/I_0) and the lamp age in hours. The analysis shows the PLC is programmed with a lamp-aging factor of 85 percent at 12,000 hours. The lamp-aging factor used by the PLC at 8,828 hours is 89 percent, notably greater than the values indicated by the UV transmittance measured through the lamp envelope.

UV systems provide better performance when they use a UV dose algorithm based on validation data. The algorithm will allow the plant to better define UV dose delivery, reducing the risk of coliform levels exceeding permit levels while providing better UV dose pacing and power cost savings.

FECAL COLIFORM DOSE-RESPONSE TESTING

Samples were collected during the site visit upstream and downstream of the UV reactors for fecal coliform analysis. Samples collected upstream of the UV system were used to measure the UV dose-response curve of the fecal coliform. Samples collected downstream of the reactor were used to determine the fecal coliform inactivation by the reactor and evaluate the impact of sampling location. The two sample locations evaluated were the compliance sampling location located immediately downstream of the UV reactors and a location immediately upstream of the effluent weir but downstream of the six inch drain location.

Because the concentration of fecal coliform at UV Site A was often below the detection limit of the standard method (i.e., < 2 CFU/100 mL), the standard method was modified to filter a larger volume of water, namely 500 mL. By using a larger volume of water, the detection limit was lowered to 0.2 CFU/100 mL.

Figure 11 shows the results of the UV dose response curve of fecal coliform measured in the lab using a collimated beam apparatus. The UV dose response curve is typical of wastewaters, being characterized by an initial rapid rate of inactivation at UV doses less than 20 mJ/cm^2 followed by a tailing zone where the inactivation rate decreases significantly at doses between 25 and 80 mJ/cm^2. The initial rapid inactivation is related to the UV inactivation of individual disperse coliform while the tailing is related to the inactivation of coliform associated with wastewater TSS.

The UV dose-response data was fit using a double exponential function, and the data was also analyzed to define a 75th percentile kill. The analysis shows that on average, the effluent meets the non-detect fecal coliform limit of < 2 CFU/100 mL on average with a UV dose of 46 mJ/cm^2 and 75 percent of the time with a UV dose of 58 mJ/cm^2. The required UV dose of 100 mJ/cm^2 should result in a fecal coliform concentration of 0.15 CFU/100 mL as an average and 0.23 as a 75th percentile, values that are an order of magnitude below the detection limit of the fecal coliform assay used for compliance monitoring.

Fecal coliform inactivation by the UV reactor was evaluated with four test conditions. Table 5 provides UV system operating data with each test condition. Table 6 provides data on the fecal coliform concentrations measured with each test condition.

The first two test conditions (tests 1 and 2) were designed to evaluate fecal coliform inactivation with low UV doses, obtained by operating the UV system at a high flow rate of 3.0 mgd with one bank on at a 30 percent power setting.

With test 1, Bank C was operating with all lamps on. The UV dose displayed by the HMI was 17.4 mJ/cm^2. The measured fecal coliform concentration was on average 12 CFU/100 mL at the compliance sampling location. Based on the UV dose-response curve (Figure 23), a fecal coliform concentration of 12 CFU/100 mL is obtained on average using a UV dose of 9.7 mJ/cm^2, which is 44 percent lower than the UV dose indicated by the HMI. The difference between these two UV dose values is likely due to two causes. First, the HMI is using a lamp aging factor of 89 percent whereas the lamp envelope transmittance data indicated a lower UV output more on the order of 55 to 65 percent. Second, the UV dose algorithm does not account for any UV dose short circuiting through the reactor.

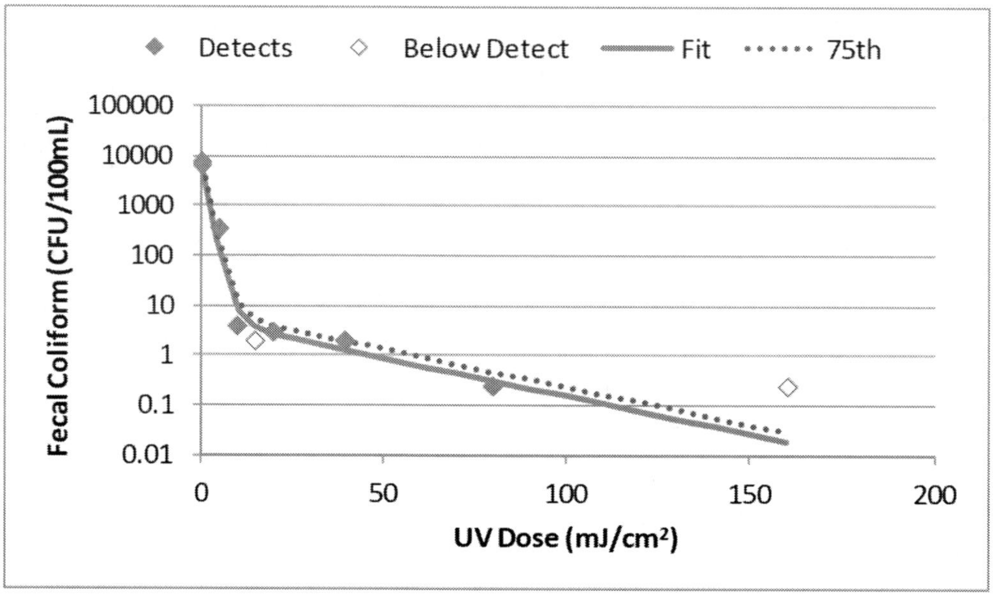

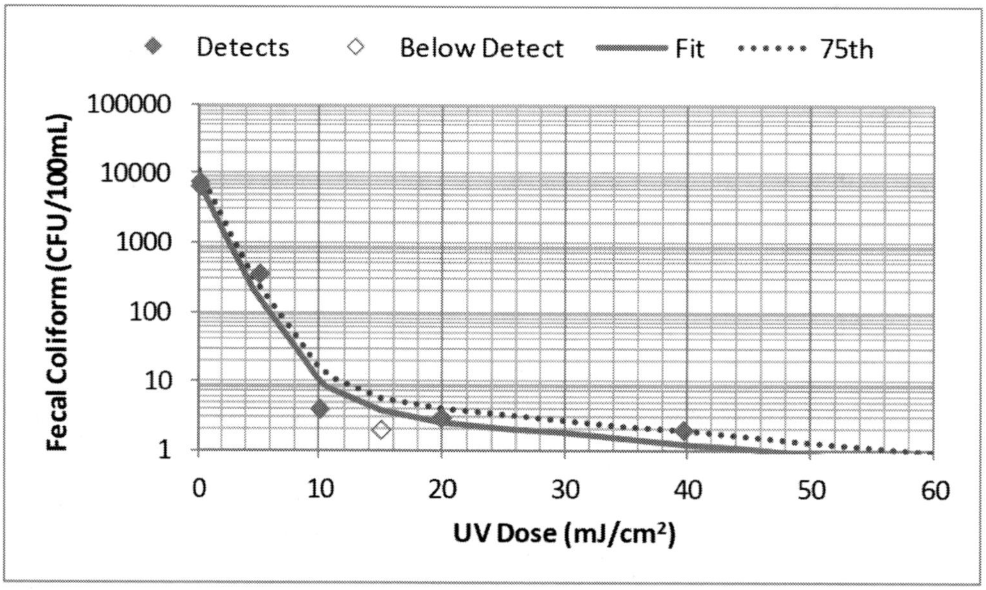

Figure 11 Fecal Coliform UV Dose Response

With test 2, Bank B was operating with one failed lamp. The UV dose displayed by the HMI was 16.7 mJ/cm^2. The measured fecal coliform concentration was on average 202 CFU/100 mL at the compliance sampling location. With one lamp out of service, the concentration at the compliance sampling location with test 2 is expected to be greater than test 1. The NWRI Guidelines for reuse UV disinfection allows UV systems to operate with a few lamps out, provided that there are at least two operating banks of lamps in series and the failed lamps are not adjacent to each other. With two operating banks of lamps in series, coliform surviving inactivation through one bank of lamps are inactivated by the second.

Test conditions 3 and 4 were designed to evaluate fecal coliform inactivation at high UV dose values. Both conditions were evaluated at a low flow rate of 0.3 mgd with one bank of lamps operating at 30 percent power, resulting in a displayed UV dose of ~173 mJ/cm^2. The coliform concentrations with both test conditions at the compliance sampling point was 8.4 and 7.1 CF/100mL, respectively. Based on the UV dose-response curve (Figure 23), a fecal coliform concentration of 8.4 and 7.1 CFU/100 mL is obtained on average using a UV dose of 10.7 and 11.3 mJ/cm^2, respectively. These UV dose values are more than an order of magnitude less than the displayed UV dose.

With all four test conditions, the coliform concentration downstream of the drain line was notably greater than at the compliance sampling point. With test conditions 3 and 4, the coliform concentration exceeded the upper limit of detection of the method. With test conditions, 1 and 2, the coliform concentrations downstream of the drain line were 977 and 947 CFU/100 mL. The high coliform concentrations downstream of the drain line are likely caused by undisinfected wastewater bypassing the UV reactors via the drain line. With test conditions 1 and 2, the estimated bypass flow is about 0.3 mgd, or 10 percent of the total flow. The analysis indicates that the drain lines between Banks B and C and downstream of Bank D needs to be plugged for the effluent to meet reuse criteria.

It is possible that the fecal coliform bypassing the UV system are causing back contamination at the compliance sampling point. This would explain the fecal coliform exceedances where the concentrations were greater than 10 CFU/100mL. Such high counts indicate the presence of disperse coliform. Since disperse coliform are easily inactivated by UV light, it is unlikely that they passed through the UV reactors.

Table 3	Test Conditions Used to Evaluate Fecal Coliform Inactivation (Equal UVT, ballast power, lamp hours)					
ID	Bank	Flow (mgd)	Int. (mW/cm^2)	Lamps Out per Bank	PLC UV Dose (mJ/cm^2)	Avg. *Coli.* (CFU/100mL)
1	C	3.0	43.1	0	17.4	12
2	B	3.0	41.4	1	16.7	202
3	C	0.3	43.1	0	174	8.4
4	B	0.3	41.4	1	173	7.1

SUMMARY AND CONCLUSIONS

The UV system audit summarized above showed that accurately monitoring online UV intensity sensor readings is a key component of making sure that not only is the equipment performing as expected but also that UV system components are performing within their warrantees. Furthermore, UV intensity sensors allow utilities to actively schedule their system maintenance and equipment replacement events, based on performance data. Based on the finding of this UV system audit, a selected number of recommendations included the following:

1. The UV lamps showed significant lamp aging characterized by end darkening and resonance distortion of the lamp envelope. The average UV transmittance at 254 nm of the quartz envelopes of three lamps ranged from 55 to 66 percent after about 9,000 hours of operation, notable less than the 89 percent assumed by the PLC software. As such, the lamps did not appear to be meeting lamp performance criterion specified by the vendor (i.e., 85 percent output after 12,000 hours). It was recommended that the three lamps evaluated during the audit be sent to a lamp lab for a third party assessment of UV output. The data obtained can be provided to the vendor as proof that the lamps are not meeting warranties and should be replaced.

2. It was recommended that the PLC be programmed with a lamp aging formula that agrees with the actual performance of the lamps. If the lamps show premature aging, the PLC software should be modified to reflect that level of aging. A recommended lamp aging equation based on the audit data would predict a lamp output of 55 percent after 9,000 hours.

3. The UV dose algorithm programmed within the PLC was based on theoretical UV dose delivery that does not account for short-circuiting of flow through the

reactor. The PLC software should be upgraded to a monitoring algorithm based on biodosimetry. The vendor recommended this upgrade with their UV system evaluation conducted February 2012.

4. Use large volume assays whenever conducting UV system performance assessment. Large volume assays lower the fecal coliform detection limit providing a better quantification of UV system performance. If the UV system is delivering UV doses equal to or greater than 100 mJ/cm^2, the large volume assay should result in a fecal coliform concentration of 0.15 CFU/100 mL as an average and 0.23 as a 75th percentile.

Operations of Wastewater UV Systems: Keys for Success

Gary Hunter [(1)*], Elliott Whitby [(2)], Oliver Lawal [(3)], and Ken Kershner [(3)]

[(1)] Black & Veatch Corporation, Kansas City, Missouri, 64114, USA
[(2)] UV Technologies, Calgon Carbon Corporation, 7100 Woodbine Ave, Suite 310, Markham, Ontario, Canada L3R 5J2,
[(3)] Aquionics, Inc., 1455 Jamike Ave. Ste. 100, Erlanger, KY 41018 USA
[*]To whom correspondence should be addressed. Email: huntegl@bv.com

ABSTRACT

Over the past twenty years, ultraviolet (UV) technology has become a commonly applied method of disinfecting wastewater effluent. UV disinfection involves passing wastewater effluent through a confined chamber with rows of underwater lamps emitting UV energy. Viruses and bacteria become inactivated upon exposure to dosages of UV light, thereby disinfecting the wastewater.

The UV system is defined by the physical (electrical and mechanical) requirements of the equipment. These requirements must be matched with the water quality of the wastewater plant to provide effective disinfection. For the purpose of trouble-shooting UV equipment, the UV system can be divided into three components: process, electrical, and mechanical. The process system deals with the quality of the water to be treated by the UV system and the disinfection goals that must to be attained. The electrical system consists of the lamps, wiring, and control system. The mechanical system includes the quartz sleeves, frames, cleaning mechanism, and reactor configuration.

When the UV system functions as intended, it is easy for operations staff to maintain compliance with the treatment facility's NPDES permit requirements. A survey was conducted of existing wastewater UV facilities to determine the key operational practices that treatment plant operators deemed critical. In addition, a review of recent literature was conducted to assess the most recent published information on operating UV systems.

KEY WORDS

UV disinfection, operations, troubleshooting

INTRODUCTION

Ultraviolet (UV) disinfection has several advantages when compared to other disinfection alternatives. Unlike chlorination and ozonation, UV disinfection is a physical process which does not involve the addition of chemicals. As a result, UV disinfection does not produce any known toxic residuals or form known byproducts which pose a risk to humans or aquatic organisms. In contrast, many municipalities are being required to remove residual chlorine before discharging treated wastewater due to concerns over chlorine's toxicity to aquatic life, requiring the addition of sulfur dioxide or other reducing agents (resulting in additional costs). Another benefit of UV disinfection is that it eliminates the need to transport, store and handle potentially dangerous chemicals. While this is primarily a safety issue, there are also costs associated with such transport and storage. The 1988 Uniform Fire Code (and subsequent revisions) includes requirements for the control of accidental discharges of gaseous chlorine as well as sulfur dioxide that are frequently used in the dechlorination process. The scrubbers required to meet these codes increase capital costs for chlorine disinfection. Since UV disinfection facilities do not require scrubbers and occupy only minimal building space and basin volume, UV disinfection is often found to be a viable, cost-effective alternative to chlorination and ozonation.

In 2006, WEF published a manual on the operation and maintenance of disinfection facilities. As part of this manual, information on UV disinfection systems was presented. This paper highlights only the maintenance and trouble shooting activities. Readers are encouraged to obtain the WEF manual for further information on UV systems.

As an update to the WEF manual, a survey was conducted to determine the operational activities conducted by staff at wastewater treatment plants. The goal of this survey was to identify key practices being implemented that allowed for UV systems to stay in compliance with discharge requirements. In addition a review was made of available information from various manufactures.

WEF WASTEWATER DISINFECTION MANUAL

In 2006 WEF published the Wastewater Disinfection Training Manual which presents a number of recommendations for typical operational tasks for UV systems. Table 1 presents these tasks that treatment plant staff should perform to ensure proper operations of the UV system.

Table 1
Recommended Operational Tasks for UV Systems
Adapted from EPA 2003

Frequency	Recommended Tasks
Daily	• Perform overall visual inspection of UV reactors • Ensure system control is on automatic mode (if applicable) • Check control panel display for status of system components and alarm status and history • Ensure all on-line analyzers, flow meters, and data recording equipment are operating within parameters • Review effluent data
Weekly	• Initiate manual operation of wipers (if provided)
Monthly	• Check lamp run time values. Consider changing lamps if operating hours exceed guaranteed life or UV intensity is low.
Semi-annually	• Check ballast cooling system for leaks or unusual noise from cooling fans. • Check operation of automatic and manual valves.

In addition to these items, effluent water quality needs to be examined on a regular basis. The manual identifies the following items as key considerations in operating a wastewater UV system:

- Intensity
- Flow Rate
- Flow rate per lamp
- Dose
- Wastewater effluent quality

Intensity
UV intensity is one of the key parameters used in the operation and control of a UV system. The UV intensity sensor can be used to determine the frequency of cleaning and to modulate the delivered dose in response to water quality changes and lamp aging. Intensity is also a function of the reactor design including the spacing of the lamps.

Flow Rate
Flow rate has a large influence on the time that bacteria are exposed to UV light. As flow rate increases, the exposure time decreases. In order to maintain the required UV dose, the intensity of the UV lamps will need to be increased or the flow through the reactor decreased. If either of these strategies cannot be implemented, then the delivered UV dose will be decreased. Reactors are therefore sized to deliver the target design UV dose at the peak design flow rate through the reactor.

Flow Rate per Lamp
Design engineers will often look at the maximum flow rate per lamp to deliver the target design UV dose for any particular wastewater quality. Division of the peak design flow rate by the flow rate per lamp for that water quality gives an indication of the total number of lamps needed to achieve the required level of disinfection. These calculations are generally built into the equipment control device (programmable logic controller [PLC]) that automatically makes appropriate lamp adjustments depending on changes in flow rate and water quality.

Dose
Dose is the key parameter that relates to the level of inactivation of the target microorganisms, and dose is a function of both intensity and exposure time. As the retention time decreases because of increased flow, the dose is decreased. For proper design, a UV system should be designed to deliver the target design UV dose at peak design flow rate. A lower flow rate than the peak design flow will increase the UV dose and allow the opportunity to reduce lamp output to save energy.

The NWRI guidelines (2002) define different doses as shown below.

- Delivered dose – The dose that is assigned to the UV test reactor based on reactor bioassay validation testing. The delivered UV dose is equivalent to that measured with the collimated beam apparatus for the same degree of inactivation of the challenge microorganism used in the bioassay validation (see Performance Validation section for more details).

- Design dose – the delivered UV dose required for a specific log inactivation of the target microorganism (specified indicator, challenge, or pathogen) the design dose is used for sizing of the UV disinfection system.
- Operational dose – The UV dose that is established based on the results of the equipment validation testing. The operational dose can be used to make the most efficient use of the UV disinfection system (reduce power demand, reduce number of reactors or reactor trains on line) while maintaining the design UV dose.

Dose requirements can be variable depending on the treated water quality and discharge requirements; therefore, there is not a typical dose requirement. Based on testing of various types of effluent, the following estimates may be used:

- Primary treated effluent - 600-800 J/m^2
- Secondary treated effluent – 200-300 J/m^2
- Reuse applications - 500-1000 J/m^2 or more (depending on upstream process .See NWRI 2000)

Wastewater Effluent Quality

Wastewater quality can have a major impact on the operation and performance of the UV system through its impact on intensity within the reactor. For instance, if the transmittance is higher than the design value then the intensity is higher, the dose is higher, and therefore less power may be needed to achieve the same level of performance. If the transmittance is lower, more power or other modifications (such as a reduction in flow) may need to be made to achieve the same level of performance. Flow measurement is critical to system performance because UV systems must operate within the validated flow range.

Water quality parameters that impact UV disinfection are listed and discussed below:

- Total Suspended solids
- UV Transmittance
- Metals
- Hardness
- Organics
- Age of UV lamps
- Lamp spacing
- Water level control
- Headloss

Total Suspended Solids. Total suspended solids (TSS) in the wastewater effluent can affect performance by scattering and absorbing UV light, or by shielding interior organisms within the solids from the UV light. In addition to TSS concentration, particle size distribution (PSD) should be determined. PSD provides an indication of the extent of particle associated coliforms in the wastewater. Typically, UV light inactivates free-swimming microorganisms readily. Higher UV doses are required for particle associated microorganisms because of the shielding effect. Scattering of UV light may not deteriorate reactor performance, but scattering can lead to erroneous UV transmittance measurements. The main concern with particles is the shielding of interior organisms and the impact on "tailing". Tailing is when disinfection becomes less effective (fewer organisms are rendered inactive) as higher doses of UV light are applied.

Transmittance. The greatest factor affecting the performance and design of a UV system is UV transmittance. UV transmittance is the ability of the UV light to penetrate water. UV transmittance is unique to every wastewater treatment plant because it is a function of the physical and chemical makeup of that water and the wastewater treatment plant. In general, values between 40 and 65 percent are common for biologically treated wastewater, with higher values corresponding to higher wastewater quality. Transmittance does not necessarily determine the target design dose that is usually related to the number and quality of UV-resistant particles in the wastewater, but the UV transmittance does influence the size of equipment necessary to achieve the target design dose. If two wastewaters have the same target design dose, but one has a lower UV transmittance, this latter wastewater will require more equipment to deliver the same dose. A UV scan (UV absorbance of the effluent over the germicidal range) should be conducted, particularly if medium-pressure (MP) lamps are used. The scan will indicate information over the UV spectra that may impact disinfection.

Metals – Dissolved Iron. Dissolved iron can readily absorb UV energy, preventing it from reaching the target organism. The higher the iron concentration, the lower the transmittance will be. In an aerobic environment, the iron could be oxidized to its ferric form, and the insoluble ferric hydroxide could aggregate along with the biological flocs as they form. Iron is a UV-absorbing element, and the aggregates of iron with particles makes disinfection of flocs even more difficult. A rule of thumb is that when iron is more than 3% of the TSS, then disinfection problems can be anticipated. Iron is sometimes added for phosphorus removal as a reduced form (ferrous iron), as for example in pickling liquor from the steel industry. The rate of iron oxidation is dependent upon the pH and the availability of oxygen. Ferrous iron is more soluble than ferric iron, and iron could remain in solution and pass through the UV reactor. In this case, iron would have an impact on the UV transmittance as shown in Figure 1.

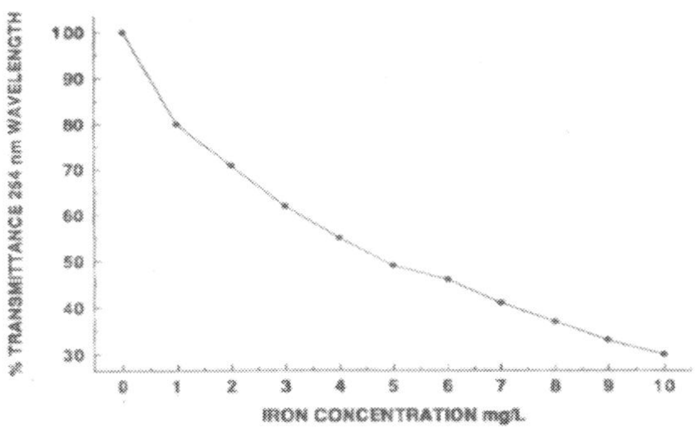

Courtesy of Trojan Technologies

Figure 1 Impacts of soluble Iron on Transmittance

In addition to impeding disinfection, iron deposits onto the quartz sleeves will increase the cleaning frequencies.

Metals – Manganese. Manganese can react in a similar fashion to iron in UV systems; however, it does not plate out on the quartz sleeves as fast as iron.

Hardness. High levels of magnesium and calcium carbonates in the effluent can lead to coating of the quartz tubes, reducing the UV dose applied to the water

Non-Biodegradable Organics. Dissolved organics and inorganics absorb energy from the system, reducing the amount of energy available for disinfection. Humic substances, some industrial organics such as textile and paper dyes, and some industrial inorganic compounds such as thiosulfate from the photographic industries can contribute to lower UV transmittance

Age of UV Lamps

The output intensity of the lamp decreases as the number of operating hours of the lamps increase. This means that the output of lamps needs to be checked on a periodic basis. The lamps should be replaced after output decreases beyond a preset amount, usually 70 percent as specified by the equipment supplier as being the end of lamp life and as used as the end of lamp life during validation and design of the equipment.

With older UV technologies, lamps do not fail before they reach a point where the output of the lamp will not deliver the desired dose, because UV systems are typically designed based on end of lamp life output. New internal lamp coatings have significantly improved lamp life. It has been shown that lamp output can be 80% or higher, compared to a new lamp, at the end of lamp life. Furthermore, these lamps can typically achieve the 80% output value right up to the point where they extinguish permanently.

With a lamp that is coupled with variable-output ballast, the lamp generally is operated at a lower output condition early, and then the output increased to maintain the same dose over the life of the lamp. UV lamps must be replaced soon after the output reaches the supplier specified end of lamp life value.

Lamp Spacing

Lamp spacing (or more appropriately water layer) and the resulting impact on intensity within a reactor is a function of water quality and reactor design. Generally the better the water quality, the wider the spacing (the greater the water layer) can be tolerated and achieve compliance.

Water Level Control Device

UV systems use some form of automatic level control (ALC) device to maintain the liquid level being treated by the system above the top of the upper UV lamps in the system. The most popular water level control devices include downward-opening penstocks, slide gates and weirs. Generally, after initial adjustments, facilities do not report any operations or maintenance issues related to the water level control device. Operation of the downward opening penstock, weighted gates, and slide gates is a function of programming code developed by the manufacturer and incorporated into the programmable logic controller (PLC). Minor adjustments, such as changing the weight on a weighted gate can be made by staff; however, no facilities reported having to make this adjustment.

Weirs are generally used where required downstream hydraulic conditions are limited. A survey of US UV installations that use weirs revealed that no changes to the weirs were necessary after construction was complete. In addition, plant staff indicated that weirs do need to be cleaned on a periodic basis.

Headloss

Headloss relates to the force needed to push water through the UV system. A number of factors may impact the operating headloss of the system. For instance, debris or algae may increase the headloss and not allow the water to move through the UV reactor. Headloss also increases with an increase in flow rate. The available head is an important consideration when designing a system, and must be known for proper design. If the flow rate through the system creates excessive headloss, there could be a wastewater backup in the plant, or water will flow over and/or around the lamp with a resulting loss of disinfection.

RESULTS OF SURVEY

Recently Black & Veatch conducted a survey of wastewater UV system. A questionnaire was developed that highlighted a number of design and operational activities relative to wastewater UV systems. Utilities were then selected based on the various lamp technologies and configurations, upstream unit processes, and flow being greater than 100 mgd. Phone conferences were held with 0perations staff to ensure that their practices were obtained via a telephone survey. Key findings of the survey included the following:

- Operations staff must have knowledge of wastewater characteristics of industrial users that are connected to the wastewater collection system.

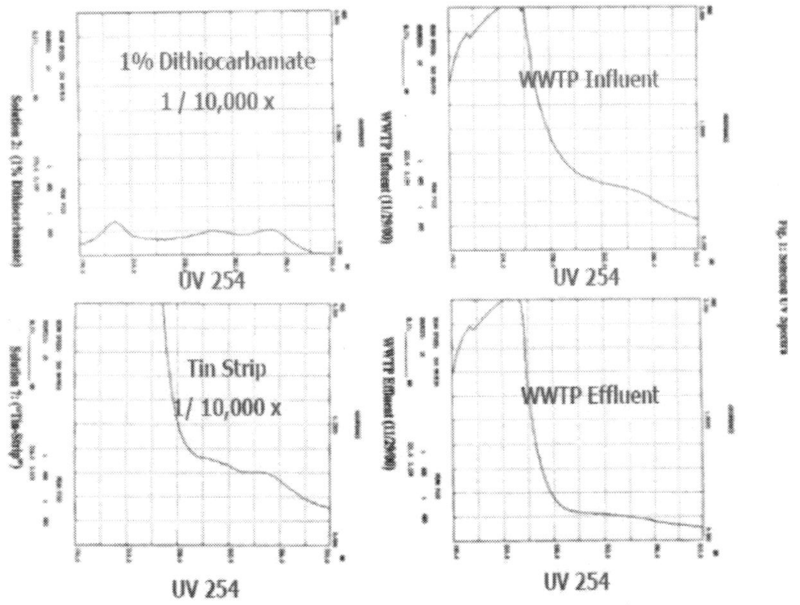

Figure 2 Industrial impacts on UV Transmittance

- To achieve effective results from UV disinfection, upstream processes also must be effectively operated. Figure 3 shows a facility where flow decreased to zero during the low point with the diurnal resulting in pumping process changes.

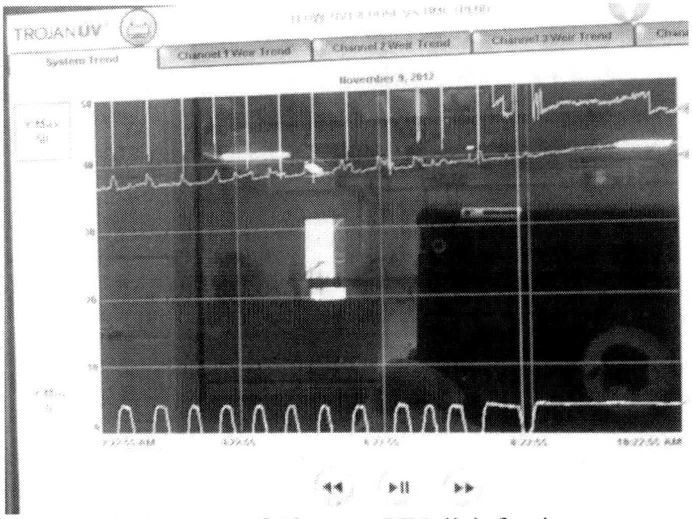

Figure 3. Impact of Flow on UV disinfection system

- Sampling and analytical testing need to follow established operating practices.
- Training of operations staff by manufacturers and others often was inconsistent; however as operations staff became familiar with the equipment it became easier to make the key adjustments at critical operation times.

The results of the survey were the key to understanding what was important to operations staff and not the design engineer or manufacturer

UV PROCESS CONTROL SYSTEMS

Overview of Types of Control Systems

The key to process control is the ability to deliver the correct dose for the wastewater quality to ensure compliance with the discharge requirements. Flow, intensity, and transmittance are used to control the dose that is emitted from the UV system. UV system controls must be operational only over the range of water quality and operating conditions for which the technology has been validated.

Flow-Paced Control

With flow paced control, the number of lamps energized increases or decreases with the flow being discharged from the facility. This type of process control does not account for decreased output of the lamp from aging or changes in water quality. Flow-paced is often used alone on LP systems; however, with LPHO and MP systems flow-paced control is integrated with other forms of control. This type of system usually assumes that the lamps are at their lowest output due to aging and quartz sleeve fouling so there may be excessive power usage compared to dose paced control.

Dose (or Intensity)-Paced Control

Controls based on a calculated dose derived from flow rate, online monitored UV transmittance and lamp power (including lamp aging and intensity sensor input) can be used in UV systems that have the potential of modulating both flow rate and lamp power level.. These control strategies are most commonly found with technologies using LPHO and MP lamps.

SENSOR TECHNOLOGY

UV Transmittance

UV transmittance is one of the most critical operational parameters related to a UV system. A decrease in transmittance may result in non-compliance with permit or guidance requirements. UV transmittance can change with the diurnal cycle of water quality entering the plant, with different days or seasons depending on the level of residential and industrial discharge, and over a longer time frame given growth and relocation of industrial users. UV transmittance data needs to be collected on a regular basis, but at least once per week. It is useful to take samples at that time during the diurnal cycle when the UV transmittance is lowest. For initial design purposes, a longer history of UV transmittance at the site is useful. In some cases on-line UV transmittance data is used to adjust the delivered dose.

UV Transmittance measurement devices (either bench scale or on-line) need to be calibrated on a regular basis as recommended by the equipment manufacturer.

Intensity Sensors

Intensity sensors are generally used to determine cleaning frequency or delivered dose. These sensors respond to changes in UV light intensity as a result of lamp aging, lamp sleeve fouling and changes in UV transmittance. Appropriate strategies can resolve the contributions of these factors and the intensity-related information can be used in establishing low intensity alarm set points or calculation of UV dose. It should be emphasized, however, that typically intensity readings will respond linearly with lamp output and exponentially with UV transmittance, but non-linearly with dose delivered.

Intensity sensors need to be calibrated on a regular basis as recommended by the equipment manufacturer. When multiple sensors are used in a system, a comparison between sensors will help identify anomalous behavior of any one sensor, and a comparison of operational sensors with a reference sensor can be used to identify out-of-calibration sensors. After manufacturer's estimate of useful life the sensors may need to be replaced.

Temperature

Temperature sensors are critical for the protection of UV systems that use lamps in reactors that can overheat if the flow is interrupted. If the temperature rises above a preset target, a signal from the sensor will be used to turn off the UV system and protect the components within the reactor. Temperature measurement devices need to be calibrated on a regular basis as recommended by the equipment manufacturer.

Flow Meters

Flow meter data is used to determine the dose delivered from the UV system. In simple systems, the operator can determine the delivered dose from the system validation data (see below) that plots dose against flow rate for various UV transmittances. This may simply be a manual performance monitoring activity that is not associated with any control response. In other systems, as the flow rate decreases below the peak design flow rate, the worst case validated dose at low UV transmittance is used to calculate a new dose and the amount of lamps in service are adjusted accordingly by the controller. In still other systems, the flow rate is only one of several inputs for calculation of the delivered dose by the control technology that then makes adjustments in lamp power and lamps in service accordingly. A variety of flow metering approaches can be used if the if manufacturer's installation recommendations are followed.

Flow meters need to be calibrated on a regular basis as recommended by the equipment manufacturer.

MANUFACTURERS INFORMATION

Since each manufacturer provides different information relative to its system it was difficult to develop a series of general recommendations. Even though it was a challenge, a number of key operating practices were observed during a review of manufacturer's operating manuals. A summary of these observations include the following:

- The control system set point should be established by the design engineer.
- Control systems need to be reviewed to ensure that the appropriate validation equation is being used to operate the system.
- On-line sensors such as intensity, transmittance, flow, water level all need to be monitored on a regular basis. Validation of sensors is a key operating practice.
- The operating practice for ballasts, ballast cooling systems, cleaning systems needs to be adjusted to meet the specific needs of the treatment plant.

MAINTENANCE ACTIVITIES

A number of maintenance activities need to be completed to ensure the UV system will continue to function adequately. Operations staffs need to review these activities to ensure that the UV system will continue to function. EPA (2003) recommends the maintenance activities for UV systems in Table 2 be completed.

During the startup of the UV system, operation staff need to review activities presented in Table 4 with their vendor representatives. Vendors have specific requirements for their respective system. This may range from determining lamp replacement frequency to cleaning to maintenance of the reactor. Treatment plant staff needs to understand that activities like cleaning and calibration of on-line instrumentation is critical to the successful operation of the UV system.

While not directly related to operation and maintenance, staffs need to understand the importance of following correct procedures for sampling and analytical procedures. Sampling of UV systems needs to be completed properly to ensure that samples are not contaminated. Clean sampling techniques may have to be implemented if the UV system has been designed to achieve high level disinfection requirements. In addition, analytical procedures need to be followed (including quality control and assurance practices). Microbial testing is the key to monitoring the performance of any disinfectant. If appropriate sampling procedures (40 CFR 136) are not followed either in the field or in the laboratory, results can be misleading. Samples should be collected in certified, clean sterile containers using grab techniques and should be kept at a temperature of 4° C in the dark during transportation to the laboratory. Analysis should start within 6 hours of the time of collection. In some cases, rosolic acid may need to be added to the sample to ensure accurate counts. After collection, filled sample bottles must be kept in the dark at all times. This can be accomplished by wrapping the bottles in aluminum foil.

WWTP operating staff should receive training in the proper operation of the UV equipment. The system manufacturer should provide the training, as each manufacturer has issues that are specific to its system as shown in Table 3.

The plant staff should review the operation and maintenance manuals for the equipment before the training sessions begin. Manuals should also be reviewed following training to determine what materials may be missing or need to be clarified before the manuals are finalized.

TROUBLESHOOTING

A number of activities can be implemented to allow a UV system to achieve the desired compliance requirements. These include process control modifications

Process Control Modifications

The effectiveness of a UV disinfection system depends on influent wastewater characteristics and on the upstream treatment processes. Industrial users may discharge substances such as UV-absorbing organics or inorganics that inhibit the performance of the UV system, and upstream operations may contribute additional solids or organic material that interferes with disinfection. Table 4 presents a checklist of factors that could influence the performance or perceived performance of an UV system and should be understood before modifications to the process are undertaken.

Table 2
Recommended Typical Maintenance Activities for UV Systems
Adapted from EPA 2003

Frequency	Task/ General Guideline	Action
Weekly	Check on-line UV Transmittance analyzer calibration (if applicable)	Calibrate UV Transmittance analyzer when manufacturer's guaranteed measurement uncertainty is exceeded
Monthly	Check sleeves, and wipers for leaks	Replace sleeves, seals, o-rings, or wiper seals if damaged or leaking
Monthly	UV Intensity Calibration Check	Check sensor calibration at the lamp power output utilized during routine operating conditions
Monthly	Check Cleaning Efficiency	• Record UV intensity sensor reading • Examine sleeves for streaks or remaining deposits • Record UV intensity sensor reading after cleaning and compare with pre-cleaned value. • Re clean or replace sleeves if intensity is not restored to pre-cleaned value.
Semi-annually	Check cleaning fluid reservoir (if applicable)	• Replenish solution if the reservoir is low. • Drain and replace solution if the solution is discolored.
Annually	Test Trip GFI	Maintain ground fault interrupt (GFI) breakers in accordance with manufactures recommendations.
Lamp/Manufacturer specific	Replace lamps	Replace lamps when any of the following conditions occur: • Initiation of low UV intensity after verifying that condition is caused by low lamp output. • Initiation of lamp failure alarm • System not achieving permit requirements after exceeding guaranteed lamp run time.
Manufacturer's recommended frequency	Check flow meter calibration	If effluent weir or flume is available, manually check depth and compare to primary device measurement. Primary device should be calibrated at recommended frequency or when measurement uncertainty is exceeded.
When lamps are replaced	Properly dispose of lamps	Send spent lamps to a mercury recycling facility or back to the manufacturer.
When quartz sleeves are replaced	Properly dispose of quartz sleeves	Replace sleeves as recommended by manufacturer or when damage, cracks, or excessive fouling occurs that would impede UV intensity.
Manufacturer's recommended Frequency	Clean/Calibrate Transmittance Monitor (if installed)	Clean/Calibrate according to manufacturer recommended frequency.

Table 3. Training Highlights
1. Design flows and characteristics.
2. Alarm system.
3. Mechanical checks.
4. Electrical checks.
5. Control logic for PLC.
6. Replacement of Basic Components

Table 4. Process Checks
1. Effluent transmittance
2. Effluent total suspended solids concentration
3. UV spectra
4. Effluent color
5. Industrial dischargers.
6. Algae.
7. Iron/manganese/hardness.
8. Microbial testing procedures.

Electrical System Modifications

Table 5 lists items to be checked by plant operations if the problems appear to be related to the electrical system.

Table 5. Electrical Checks
1. Lamps are energized
2. Lamps are connected
3. Useful lamp life
4. Ballast output

Mechanical System Modifications

Mechanical systems are critical to the operation of UV systems and some of the checks are listed in Table 6. Quartz sleeves have to be cleaned on a regular basis. The primary maintenance task required in the operation of UV systems is cleaning the surfaces of the quartz sleeves due to the tendency of constituents in wastewater to foul the surfaces of the quartz sleeves. Consequently, mechanical and chemical cleaning systems are used to periodically clean the sleeves. Cleaning requirements are based on the reduction in measured intensity caused by the impedance of the transfer of UV light into the bulk wastewater.

Table 6. Mechanical Checks
1. Cleaning system to ensure acceptable performance.
2. Proper delivery of chemical.
3. On-line transmittance measurements.
4. On-line intensity measurements.
5. Flow meter calibration.
6. Ballasts closed loop-cooling system.

SUMMARY

The proper operation and maintenance is critical if the wastewater treatment plant is going to meet their wastewater disinfection compliance. Operation staff, if taught properly can successfully operate and maintain UV systems. The procedures described above are simple and can provide staff with the key items to successfully operate a wastewater UV systems.

REFERENCES

1) USEPA, Ultraviolet Design Guidance Manual, 2003.

Introduction to Established Disinfection Technologies

Keith Bourgeous [1]

[1] **Carollo Engineers** kbourgeous@carollo.com

Disinfection is typically the last step in the wastewater treatment process. The performance of the disinfection process greatly depends on the performance of the upstream treatment processes (i.e., primary treatment, biological treatment, secondary sedimentation, and tertiary filtration). The purpose of disinfection in the wastewater treatment is the destruction or inactivation of disease causing pathogens (e.g., protozoan, bacteria, and viruses) to protect public health and the environment. As part of the Water Environment Research Foundation Project 04-HHE-4, a survey of the disinfection processes of 4450 publically owned treatment works (POTWs) was conducted. As a result of this study, it was determined that 75.3 percent of the POTWs in the United States utilized chlorine disinfection as the main disinfection process. The authors of this study also found that 20.6 and 0.2 percent of the POTWs located in the United States utilized ultraviolet light (UV) disinfection and ozone disinfection, respectively. It was determined that 3.9 percent of the POTWs did not disinfect the treatment plant effluent at all. This paper will provide a brief description of the three established groups of disinfection technologies used to disinfect wastewater (i.e., chlorine, UV, and ozone). A brief description of each disinfection technology as well as the mechanisms of pathogen inactivation of chlorine, UV, and ozone disinfection will be described in the presentation. The advantages and disadvantages of using each type of disinfection technology will be discussed. Pictures of the disinfection equipment utilized in each system as well as a comparison of the effectiveness of each technology will be presented.

Innovations in Wastewater Disinfection Technology

Katherine Y. Bell, CDM Smith[1], Ph.D., P.E., BCEE
Allegra da Silva, CDM Smith, Ph.D., P.E.
[1]CDM Smith
830 Crescent Centre Drive, Suite 400
Franklin, TN 37067
(615) 807-7800

ABSTRACT

With uncertainty regarding the requirements for future use of chlorine gas because of safety concerns as well as potential future challenges associated with chlorinated disinfection by-products (DBPs), wastewater facilities are proactively looking for options to replace both gas and liquid chlorine with other disinfection alternatives. The recent increased interest in alternatives to chlorine for disinfection has driven advancements in other mature alternative technologies (ultraviolet (UV), and ozone disinfection) and sparked investigation into other disinfection methods that have not historically been used in municipal applications (pasteurization, peracetic acid (PAA), and ferrate disinfection). This paper will describe how technological advancements in mature and emerging technologies are driving down capital and energy costs and addressing DBP formation, while also providing ancillary benefits such as oxidation of contaminants of emerging concern (CECs).

KEYWORDS

Peracetic acid, ferrate, UV LEDs, wastewater disinfection, disinfection by-products, pasteurization, energy

INTRODUCTION

With uncertainty regarding the requirements for future use of chlorine gas and potential future challenges associated with chlorinated DBPs, wastewater facilities are proactively looking for options to replace both gas and liquid chlorine with other disinfection alternatives. The increased interest in disinfection alternatives has pushed technology and efficiency advances in both mature technologies and innovative technologies that have been used in other industries. This paper will provide an overview of some of the recent technological advances that are making mature disinfection technologies more cost effective as well as some of the most promising innovative disinfection technologies.

ADVANCES IN MATURE DISINFECTION TECHNOLOGIES

Mature technologies, including UV and ozone disinfection, have been a focus of both fundamental and applied research, resulting in significant improvements in the efficiency of these systems. In addition, approaches that have long been used in food and beverage production have been increasingly transferred to wastewater disinfection.

Advances in UV Lamp Technologies

Improvements in UV lamp technology have greatly reduced the capital and operating cost of UV disinfection. UV lamps produce light as a result of electrical current flow through the mercury vapor between the electrodes. Lamps used in wastewater applications can generally be categorized into three groups: low-pressure low output (LPLO), low-pressure high output (LPHO), and medium pressure (MP). Each lamp type outputs radiation at different wavelengths, uses different operating parameters, and presents different advantages and disadvantages. Table 1 provides a summary of some of the important characteristics of conventional lamp technologies. In summary, the main disadvantage of the low pressure lamps is the need for a large number of lamps and correspondingly large footprint; however, the development of the LPHO lamps mitigates this issue to some degree. The energy demand of LPHO systems is similar to that of LPLO systems, ranging from 1.8 to 3.8 kW input power per MGD for disinfecting secondary effluent to 200 fecal coliform per 100 mL, which is significantly lower than the 6.8 to 14 kW input power per million gallons per day (MGD) required by medium-pressure UV systems. As a result, LPHO have become the favored lamp technology for wastewater disinfection applications.

Table 1. Characteristics of conventional UV lamp technologies (modified from EPA, 2006).

Lamp Characteristic	**LPLO**	**LPHO**	**MP**
Spectrum	Monochromatic	Monochromatic	Polychromatic
Input Power (watts/lamp)	70 - 90	200 - 400	1,300 – 5000
Germicidal UV (% of input)	35 - 40	30 - 42	15 – 20
Temperature (C)	40 - 60	100 - 200	600 – 900
Lamp Life (hours)	8,000 - 15,000	8,000 - 12,000	5,000 - 8,000
Number of Lamps (relative to MP)	10 - 15	4 - 8	1
Footprint	Large	Medium	Small

Although MP systems are much more energy intensive than LPHO systems, the lower installation cost, because fewer lamps are required, often drive use of this technology for applications such as facilities with high wastewater flows, stormwater overflows, or at space-limited sites.

<u>Microwave UV Lamps</u>

In evaluation of a UV disinfection system, regardless of the type of lamp system used, the cost of replacement lamps is a substantial fraction of the lifecycle cost of a UV disinfection system. As a result, one UV equipment supplier has developed an electrode-less mercury-argon UV lamp that is excited by microwave energy generated by a magnetron to irradiate the gas filled lamps. This radiation excites the gas and produces UV light. Because this technology does not use electrodes, the lamp life can be extended far beyond the expected lamp life of conventional electrode mercury arc lamps (about 12,000 – 15,000), with a 5 year lamp life claimed by the manufacturer.

There are a number of installations of this technology in the US, with the largest installation in Kingsport, Tennessee. The facility is sized to treat a peak flow of 35 MGD from a secondary trickling-filter activated sludge process. The UV facility, completed in early 2011, was one of the first in the US. Severn Trent Systems, the equipment manufacturer, continues to support the project and works with facility staff to obtain feedback and data for future equipment improvements.

High Wattage, Low Pressure Amalgam Lamps

In response to the challenge of providing high UV output at a high efficiency for large flow facilities, two major UV manufacturers have developed systems that utilize high power input (> 600W) lamps. Previously, LPHO lamps were limited to about 400W input due to the decreased electrical efficiency at high input. New high output LPHO lamp technologies have recently been incorporated into two systems that are currently available, with more expected to enter the marketplace. These include the TrojanUVSigna, which is based on a 1000W Solo Lamp that can be dimmed from 100 to 30% power in order to optimize power consumption; and the Wedeco Xylem DURON system, based on a 600W LPHO amalgam lamp. Both manufacturers are providing a guaranteed lamp life in excess of the conventional LPHO lamps. These lamp technology improvements have the potential to extend UV disinfection to wastewater treatment plants (WWTPs) that were previously considered "too large" because treatment of flows in the hundreds of MGD required very large numbers of LPHO lamps, or these facilities were left using medium pressure systems that have lower efficiency resulting in higher energy costs.

UV Disinfection and UV LEDs

In contrast to traditional UV disinfection lamps which employ an electrical arc across a mercury vapor chamber to produce light, light emitting diodes (LEDs) produce light by generating a voltage across a solid state semiconductor device to emit energy in the form of photons. Near-UV emitters at wavelengths around 375–395 nm are already cheap and often encountered, such as black light lamps used for inspection of anti-counterfeiting UV watermarks in documents and paper currency. With nitride semiconductors containing aluminum, most often AlGaN and AlGaInN, wavelengths in the germicidal range (250–270 nm) are achievable.

Shorter-wavelength diodes, while substantially more expensive, are commercially available for wavelengths down to 247 nm. The first UVC (covering wavelengths from 200 and 290 nm) LED water treatment device has already been released. Sensor Electronic Technology (SETi) has developed a UVC chip for germicidal applications. The technology used in this chip was developed under the Defense Advanced Research Project Agency (DARPA) Compact Mid-Ultraviolet Technology program. The chip delivers an output of 9.8 milliwatts (mW) at 278 nm and when driven at a forward current of 100 mA, it produces 30 mW and is applicable for effective disinfection of drinking water.

While the electrical efficiency of this system is low (8%), the research-level GaN-on-sapphire device represents a significant improvement in technology performance, and demonstrates that solid-state devices are capable of challenging the incumbent mercury lamp technology. While more research is necessary before large-scale applications are achievable, Aquionics has released a new UVC LED technology device, the UV-Pearl (**Figure 1**). The device is capable of producing a UV dose of 60mJ/cm^2 at a flow of 0.5 gallons per minute with only 50 mW of optical power. Disinfection has been verified by two independent laboratories using MS2 virus. Benefits of UV LED technology include instant full intensity on start-up, mercury-free lamps and unlimited cycling without lamp degradation. The system is not affected by water temperature and can continue to operate with no flow, unlike conventional mercury arc lamps which require flow to manage the lamp temperature.

Figure 1. The UV-Pearl, recently introduced to the water treatment market by Aquionics.

Ozone

The detection of CECs including pharmaceutically active and endocrine disrupting compounds in wastewater has resulted in an increased interest in the application of ozone disinfection. Ozone is a mature disinfection technology with secondary benefits of removal of CECs as well as color removal. While ozonation has substantial benefits, as of 2010, it was used at fewer than a dozen wastewater treatment plants in the United States, of which two were specifically reuse applications: El Paso, Texas, and Gwinett County, Georgia. Because of the strong market presence of ozone disinfection in drinking water treatment, the recent improvements in energy efficiency in ozone disinfection have arisen from the drinking water industry. The key developments include improvements in oxygen production for feed gas, improvements in generator technologies, and injection technologies that allow high mass transfer efficiencies. Ozone generators make up the biggest portion of the process power draw; however, improvements in dielectric technology and materials of construction, power supply components and control method, and ozone generator production control logic have made the ozone concentrations of 10 - 16 percent by weight economical for production with specific energy requirements, including ancillaries, as low as 4 – 5.5 kW-hr/lb ozone produced. This value is less than half of the energy for systems from two decades ago and when the secondary benefits of ozone is considered, begins to compare with hypochlorite which has a specific energy requirement of approximately 2.5 kW-hr/lb chlorine produced.

In addition to managing power consumption, improved process control programming allows the cost of power to be balanced with the cost of liquid oxygen (LOX) when LOX is used as the feed gas supply. As an ozone generation facility is operated at higher ozone concentrations, the specific energy requirement goes down but the LOX cost goes up. For most ozone projects in the US, the optimal operating point is around the 10 - 12 percent ozone concentration range. However, since LOX costs range from $60 - 150 per ton an operational optimization calculation should be performed for each site to evaluate both the LOX and power costs to determine the most cost effective operational point.

INNOVATIVE DISINFECTION ALTERNATIVES

While there have been significant advances in UV and ozone disinfection, there are sometimes challenges of high capital costs and consistent disinfection performance, particularly for facilities with high peaking factors or with low UVT or high TSS (frequently exceed 20 mg/L). In addition to minimizing capital costs, utilities are looking for options that have a low energy footprint. Two novel chemical disinfectants provide these potential advantages. In addition to chemical disinfection options, another disinfection that has been used in the food and beverage industries has made a successful transition to the wastewater market. Recently, California Title 22 approval was granted for use of pasteurization for recycled water disinfection and for some applications, this technology can be an energy efficient method for providing chemical free disinfection.

Peracetic Acid

Peracetic acid (PAA) is a wastewater disinfection alternative that is gaining interest due to its ability to provide bacterial inactivation performance at costs competitive to other mature technologies. PAA is a chemical oxidant that has been applied to the food, beverage, medical and pharmaceutical industries as a disinfectant for many years. PAA has been demonstrated to be an effective wastewater disinfectant over a wide range of effluent qualities, requiring low doses of chemical to achieve bacterial inactivation. PAA has very fast kinetics, requiring very short contact times for complete disinfection. Additionally, capital costs for retrofit of existing chlorine facilities is very low. In addition to potential cost savings, PAA is able to provide treatment for challenging effluent while meeting stringent limits for halogenated disinfection by-products (DBP) such as trihalomethanes (THMs).

Nonetheless, PAA faces a hurdle of limited use when compared to other more mature disinfection technologies such as chlorination, ozone and UV disinfection. The United States Environmental Protection Agency (EPA) has recently approved PAA for use as a disinfectant to treat wastewater; however, there is currently only one full-scale municipal WWTP application in the US permitted to use this technology. The full-scale PAA application was implemented as an alternative to UV or ozone to address compliance issues with halogenated DBP formation.

There are a number of ongoing pilot studies that are aiding in identifying engineering design criteria for PAA. The roles of factors such as suspended solids, temperature, pH and other site specific characteristics are being established to help identify dosing requirements for various wastewater types. Typically, for secondary effluent, initial design dosing is 1 - 2 mg/L of PAA. There are ongoing case study projects being conducted to demonstrate that a peak flow contact time of 5 minutes is adequate for achieving bacterial inactivation compliance. To respond to growing interest in PAA, it is necessary to further develop engineering design criteria that will aid decision-makers in understanding cost implications as well as obtaining local regulatory support for implementation.

Ferrate

Ferrate, or iron (VI), is an oxidant that has recently gained interest as an alternative disinfectant for municipal wastewater treatment. Ferrate was considered in the 1970s as a replacement for chlorine, but the high costs of previous manufacturing methods prevented its widespread municipal or industrial use. Little has been published about the disinfection mechanism of ferrate, although studies have shown that ferrate can oxidize *E. coli* DNA polymerase I and degrade deoxyribonucleosides. **Figure 3** provides a comparison of oxidation–reduction potentials of ferrate with other common chemical oxidants, many of which are used as disinfectants in water and wastewater disinfection. Higher oxidation-reduction potentials imply that a disinfectant is a more potent oxidant, and therefore disinfectant.

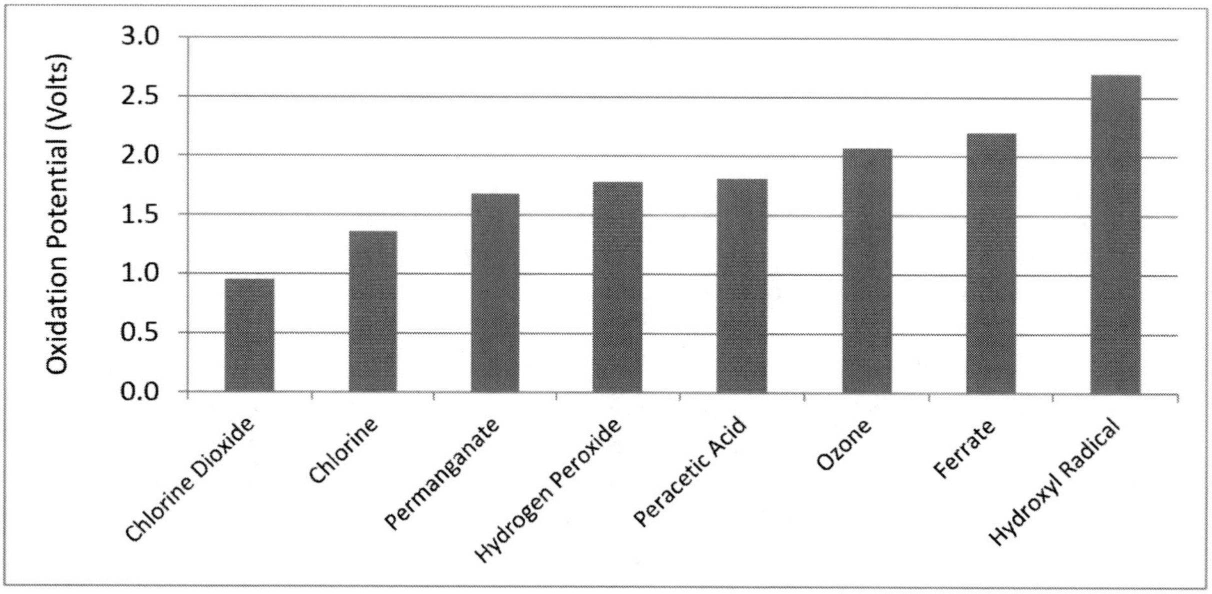

Figure 3. Oxidation potential of various chemical oxidants.

Sodium ferrate

Liquid sodium ferrate is a relatively new formulation of ferrate that can be used in wastewater disinfection. A new on-site synthesis process that produces a liquid sodium ferrate product, versus purchase of ferrate salt, makes ferrate a potentially cost-competitive alternative to chlorine and other disinfection technologies. On-site generation of liquid sodium ferrate is necessary because ferrate, in its dissolved form is very reactive and has a short half-life. Ferrate Treatment Technologies, LLC, the patent holder of this synthesis technology, has conducted a number of pilot studies in the US with variable results. In wastewater disinfection applications, doses of 3 – 15 mg/L have been successfully tested, however the effective dose is a function of effluent quality.

The on-site generation of liquid sodium ferrate requires bulk caustic, bulk ferric chloride, and bulk liquid sodium hypochlorite solutions. The components of a ferrate disinfection system are similar to that of a liquid hypochlorination system with the exception of the addition of the on-site liquid sodium ferrate generation system. Additional solids are produced in ferrate disinfection, so solids handling may be an additional component of a ferrate disinfection system. Site-specific testing must be conducted to determine the required disinfection dose.

While there have been numerous laboratory and pilot-scale investigations, the first full-scale installation of ferrate was slated for the 100 MGD East Bank treatment plant in New Orleans, Louisiana, but the project has yet to be implemented. The technology was selected for this application due to its advantages over other technologies, including the fact that it can provide oxidation and disinfection in the same application, similar to ozone or chlorine. This would allow the disinfection process to also provide degradation of endocrine disrupting compounds, which were a concern, since the effluent will be reused at the East Bank WWTP for wetlands restoration (AWWA, 2010).The operating cost for treatment is $3 – 4/pound of ferrate when the costs of power and raw materials are considered.

Potassium ferrate

A more stable form of ferrate is solid potassium ferrate. Solid potassium ferrate once cost about $100 per gram but a new manufacturing process pioneered by Battelle and St. Louis-based Ferratec, LLC has made the same compound available for a fraction of that price, and in a more pure form. The product is sold under the trade name Ferratec™. This new product is a step toward making this chemical acceptable for use outside its traditional role as a lab chemical. Currently, the product is still about $2 per gram, but the advance in the manufacturing process represents a technical leap that will hopefully lead to a scalable manufacturing process.

Pasteurization

Pasteurization has recently been introduced to the wastewater disinfection market. Pasteurization is a process of applying heat to a substance to inactivate pathogenic or spoilage microorganisms. The process was discovered by Louis Pasteur in 1864 and has since become standard practice in the food industry. Pasteurization has also become accepted practice in sewage sludge processing, with the goal of inactivating pathogens to achieve Class A Biosolids standards. Thermal inactivation of microorganisms depends on characteristics of the organism, stress conditions for the organism, growth stage, characteristics of the medium (e.g., heat penetration, pH, presence of protective substances like fats and solids, etc.), and temperature and exposure time combinations (Salveson et al., 2011).

Pasteurization has been demonstrated at the city of Santa Rosa's Laguna Wastewater Reclamation Plant, where validation testing was conducted as part of the California Department of Public Health (CDPH) program to review new technologies for wastewater treatment for water reuse and provide conditional approval (often referred to as "Title 22" approval) (Salveson et al., 2007). Based upon this and other work, the CDPH approved pasteurization to meet the stringent "tertiary recycled water criteria," with specific requirements for minimum pasteurization times and temperature.

The economic value of pasteurization is favorable when waste heat can be captured and transferred for disinfection. Heat exchangers can be used to recapture heat from hot disinfected water to preheat cooler undisinfected water, also cooling the disinfected effluent to just a few degrees above the influent undisinfected water. Other sources of waste heat include exhaust heat from a turbine fueled by natural gas, digester gas, or hot water. Favorable economics for pasteurization has been demonstrated in Ventura, Calif., where a 400 gpm (25 L/s) demonstration system (**Figure 2**) has been constructed and is in continuous operation. Because of the high cost of power at this utility, pasteurization is projected to save several million dollars in lifecycle costs compared to UV disinfection.

Figure 2. Pasteurization demonstration system in Ventura, CA (*photograph courtesy of Andy Salveson*).

More recently, the northern California town of Graton has signed a contract to purchase and install the patented pasteurization disinfection system to handle all the city's sewage treatment needs. By switching from a chlorine-based wastewater disinfection system, Graton anticipates that they will realize significant operational savings and meet or exceed state water quality standards. The disinfection and cogeneration system is part of an $8.5 million project to update and upgrade the 36-year-old WWTP, which serves 1,700 residents. The new disinfection system, which is less than one tenth of the total project cost, will process up to 576,000 gallons per day and is planned for completion by January 2014 (http://www.waterworld.com/articles/2012/12/ca-city-selects-ptgs-chemical-free-wastewater-disinfection-system.html).

SUMMARY

There is a tremendous amount of literature and experience with wastewater disinfection alternatives, with many new technologies entering the marketplace. It is difficult for wastewater professionals to sift through all of the available information, especially for relatively newer technologies and there are few resources that provide a comprehensive discussion of which technologies are advancing most quickly. The information presented in this paper is a summary of only a few of the most promising technical options for wastewater disinfection, and highlights mature technologies that are making significant advances and technologies that have significant potential to overcome compliance, cost or energy challenges issues associated with current technologies.

The information presented is intended to be a cursory review of disinfection advances. When an individual facility is evaluating disinfection options, it is important to consider site-specific factors such as compliance goals, operations and maintenance, the willingness of a particularly facility to take on risks associated with implementation of a new technology, as well as costs.

REFERENCES

American Water Works Association (AWWA). 2010. "New Orleans plans to restore wetlands using effluent have 'wow' factors," *Streamlines*, 2(24).

U.S. Environmental Protection Agency (EPA). 2006. *Process Design Manual: Land Treatment of Municipal Wastewater Effluents*. EPA/625/R-06/016. Environmental Protection Agency. Washington, D.C.

Salveson, A.; G. Ryan and N. Goel. 2011. "Pasteurization – Not Just for Milk Anymore." *Water Environment and Technology*. 23(3).

Salveson, A., Atapattu, K., Linden, K.G., Robinson, K., Bowman, R., Thurston, J., Cooper, R. (2007). Innovative Treatment Technologies for Reclaimed Water: Ozone-Hydrogen Peroxide Pilot Test Report at DSRSD. 22nd Annual WateReuse Symposium, Tampa, FL.

U.S. Environmental Protection Agency (EPA). 2012. *Guidelines for Water Reuse.* EPA/600/R-12/618. Environmental Protection Agency. Washington, D.C.

Point-of-Use Disinfection of Household Drinking Water with Copper/Zinc Ions and Oxide Nanoparticles: a Proof-of-Concept Lab Study

Abhinav S.R. Komandur, BSPH, University of North Carolina-Chapel Hill
Alyson M. Malone, BS., MSPH, University of North Carolina-Chapel Hill
Mark D. Sobsey, BS, MS, PhD, University of North Carolina-Chapel Hill
Dept. of Env. Sci. and Eng., Gillings School of Global Public Health
Chapel Hill, NC 27599-7431
919-966-7303

ABSTRACT

Water in households of the developing world is often microbially unsafe and the cause of waterborne disease. Therefore, simple, effective and convenient treatment methods are needed to disinfect such household water. Disinfecting water with antimicrobial metals as ions or nanoparticles has been inadequately studied for this application. This research quantified the disinfecting efficacy of low (0.2 mg/L) and high (2 mg/l) doses of combinations of ionic copper and zinc and copper and zinc oxide nanoparticles in model buffered water that are considered safe for consumption. Kinetics of disinfection were quantified for three test microorganisms: *Bacillus cereus* spores, *Escherichia coli*, and MS-2 coliphage in pH 7.3, 0.04M HEPES-buffered water with and without 10 mg/L added natural aquatic organic matter (NOM); concentrations of culturable microorganisms were assayed at 0, 20, 60, 180, and 360 minutes of contact time. Disinfection conditions that could meet the World Health Organization (WHO) "protective" performance guideline levels of 2 $\log_{10}$ bacteria and parasite reduction (for *E. coli* and spores of *B. cereus, respectively*) and 3 $\log_{10}$ virus reduction (for MS-2 coliphage) within 12 hours were investigated for kinetic modeling. Several test conditions of metal form (ion or nanoparticle oxide) and concentrations were efficacious against virus, bacteria and protozoan parasite surrogates within the 12 hour contact time specified. Both copper and zinc ions and nanoparticles oxides were effective in achieving microbial reduction targets in buffered test water lacking added NOM. None of the test conditions were effective in meeting microbial reduction targets in test water lacking added NOM. Additional studies are needed to better document the effect of NOM and other water quality parameters (such as turbidity, pH and chemical behavior of dissolved solids) on the ability of these metal ions and nanoparticles to disinfect drinking water.

KEYWORDS

Disinfection, kinetics, copper, zinc, ions, nanoparticles, viruses, bacteria, spores

INTRODUCTION

Background

The disease burden and mortality of vulnerable individuals, particularly the poor, women, children, and rural residents, who suffer from illness as a result of drinking contaminated water, is of a great magnitude. According to the World Health Organisation (W.H.O.) in 2012, nearly 800 million did not have access to water from improved sources. The term “improved” sources also doesn’t indicate microbial safety as many improved sources are still microbially contaminated in developing countries.

Annually, about 1.5 million children die of diarrhoea, of the nearly 2.5 billion childhood diarrhoea cases that occur. Waterborne illness is estimated to accounts for 5-10% of all deaths and disease worldwide; and diarrhoea is considered the second leading contributor to the global burden of disease (W.H.O., 2010).

Much of this diarrheal disease and death results from inadequate provision of safe drinking water in the household. The work presented here attempts to address the lack of microbially safe water access at the the household level. Point-of-use (POU) household water treatment systems are now recognized as a valid and appropriate means of disinfecting drinking water (Sobsey, 2002; Fewtrell, 2005).

However, currently available methods for POU water treatment have limitations.. Chlorine is an effective disinfectant against bacteria and viruses but it also has undesirable qualities such as chemical instability, unpleasant odour and taste, and potential health effects either directly or when combined with aqueous organic matter. Physical disinfection methods, such as boiling, portable UV irradiation from lamps, or filters (ceramic, biosand and membrane) also have disadvantages, such as the fuel costs for boiling, the need for electricity to power UV lamps and the relatively high initial cost of household water filters. Methods that combine physical removal and chemical disinfection are often effective and can overcome consumer dissatisfaction due to taste or olfactory effects, but some are priced out of reach for those most in need (Sobsey, 2002; Sobsey et al, 2008).

Certain potentially antimicrobial metals have the potential to address many of the needs for POU household water disinfection. The addition of chemically stable metals to water, in consumable amounts, would meet such criteria as convenience, ease of use, stability, mobility, and safe protection of water during storage. Metal doses low enough to be consumable can potentially be effective and yet cause no objectionable taste, colour, and smell of treated waters.

Uncharged metal nanoparticles in particular have been considered for use as general biocides or even as POU water treatment alternatives in the past. Uncharged silver has been repeatedly investigated as a control agent for infectious microorganisms (Qin, 2005), particularly in the context of hospital use (Atiyeh et al, 2007). Titanium dioxide has also been investigated as a drinking-water treatment both in its pure form (Ilyas and Oazi, 2010) and when doped with silver (Liga, 2011). If effective, uncharged copper and zinc, when used in combination, would meet the criteria for low cost and capital investment burdens, mobility, and simplicity of use. Metallic salts have also been considered for use as POU water treatment alternatives. Metals like copper and silver ions are known to have bactericidal properties in biological systems and have been used commercially for disinfection of drinking water, swimming pools, sewage effluents, and water stored in industrial cooling towers (Clarke, 1983; Wood 1984). The killing effect of copper alone and in combination with other metals on bacteria such as Escherichia coli, Salmonella typhi, Vibrio cholera, coliforms, and Legionella has been previously reported (Sudha, 2009; Kim, 2003; Domek, 1984; Blanc, 2005; Lin. 1996).

Uncharged copper, used alone, has been investigated for point of use disinfection of bacteria in water (Sharan et al., 2010; 2011; Sudha et al, 2009). The ionic forms of copper alone, as well as copper and zinc in combination have also been preliminarily investigated by our group (Armstrong 2011, Komandur 2011). Furthermore, ionic copper in combination with other metals as a biocide has been the subject of numerous studies (e.g. Lin et al 1996, Cervantes 1994, Thurman et al, 1989). Previously established uses of copper as a biocide, desirable electrochemical properties, the use of zinc as an oral anti-diarrhoeal supplement, and the known propensity of zinc to selectively inactivate certain types of virus all suggest a possible use for the two elements in combination. Copper ions and zinc ions used together, particularly in combination with other methods of water treatment, have been previously studied (e.g. Abad 1994, Pyle 1992, Yahya 1990). However, the efficacy and disinfection kinetics of uncharged copper and zinc, used together, on a variety of organisms and in various chemical environments, remain relatively uncharacterised.

The research described here aimed to further characterise the potential for uncharged copper and zinc, as well as copper and zinc ions, to be used as drinking-water disinfectants. Copper and zinc nanoparticles as oxides, and charged copper and zinc ions were investigated for their effects on the viability of three different organisms, in two different aqueous environments at different metal nanoparticle and metal ion doses. Batch disinfection experiments were performed, whereby known quantities of microorganisms in test waters were simultaneously subjected to various disinfection conditions, and the measurable quantity of surviving microbes was quantified over time. The research here, though an experimental set-up under laboratory conditions, was performed in such a manner as to approximate typical household use conditions as much as possible.

Objectives

This research aimed to test the hypothesis that a metal additive comprised of uncharged copper and zinc oxide nanoparticles or the ionic forms of these metals could disinfect microbes in water. Disinfection performance was evaluated for different doses of the metal additives at concentration suitable for safe daily consumption. Microbiocidal activity was evaluated for a representative range of microorganisms, including an enteric bacterium, a bacterial spore and an enteric indicator virus. Efficacy of disinfection was determined over contact time periods 18 to 24 hours, consistent with daily household use.

METHODS AND MATERIALS

Experimental Approach

Metal Nanoparticles and Salt Solutions

A focus of this research was to evaluate the use of uncharged copper and zinc particles. Any biocidal properties have been theorised to reside in the surface chemistry (reactivity) of these metals, and their propensity to release metal ions in aqueous solution. In order to maximise the amount of potential surface chemistry avaibale for reaction with test microbes, the nanoparticle forms of these metals were chosen. However, nanoparticles in general tend to be very highly reactive and potentially unstable in terms of their physical state in water (Shatkin, 2008). A certain amount of stability is desirable to assure the particles remain in suspension and do not extensively aggregate. Pure uncharged nanoparticle copper and zinc, though available, have very poor shelf life and storage capacity. Upon exposure to ambient air, they very readily form mono- or occasionally poly-oxygenated species, and react with water particles, environmental contaminants, and other heterogeneous airborne compounds to create complexes of unknown size, chemical activity, and prevalence.

For these experiments, the nanoparticle forms of copper and zinc were stabilised prior to use by employing oxide forms of these metals. Nanoparticles of cupric oxide (CuO) and zinc oxide (ZnO) ranging in size between 25 and 45 nanometres were used. One particle of metal oxide was considered, for the purposes of this experimentation, to be one particle of "deliverable metal". The particles were stored in separate colloidal solutions, using water without other nanoparticles, or without compounds that would obscure, bind with, or induce precipitation in the desired nanoparticles, under conditions of darkness at 4 degrees C. For copper and zinc metal ions, concentrated stock solutions of $CuCl_2$ and $ZnCl_2$ were prepared in reagent grade water at concentrations that were 250X and 2,500X their final concentrations in test waters.

Metals Doses in Test Waters

Maximum copper and zinc concentrations of 2 milligrams per litre of test water were used. These maximum concentrations in test water correspond to The W.H.O. and the United States Food and Drug Administration recommended levels of copper and zinc for drinking water (W.H.O., 2004, W.H.O., 2007, F.D.A., 1994). The recommended maximum concentration of 2 milligrams of zinc per litre of water, is based on aesthetic factors such as smell and taste, and was designated as the maximum concentration in this research (W.H.O. 2007). Therefore, the recommended maximum aqueous concentrations for both copper and zinc for this research were equal, and were taken to be 2 milligrams per litre. These are denoted henceforth as the "100%" dose.

In addition to the 100% dose, a tenfold lower concentration of metal additive, corresponding to 0.2 milligrams per litre, was also used Control samples having no addition of metal additive, were also evaluated and were treated identically to samples containing the 10%, or the 100% metals dose in order to facilitate scientifically valid and statistically rigorous comparisons.

Test Microorganisms and Added Concentration to Test Water

The three test microorganisms used in these experiments were *Bacillus cereus* spores, *Escherichia coli B,* and MS-2 bacteriophage. *Bacillus cereus* spores were used as a surrogate for protozoan parasites and other drinking water pathogens that are considered generally resistant to disinfection, such as *Cryptosporidium* oocysts. *Escherichia coli* vegetative cells were used as experimental surrogates for Gram-negative intestinal bacterial pathogens and MS-2 coliphage was used as the surrogate for enteric viruses, such as enteroviruses, Hepatitis A and E viruses, noroviruses or astroviruses.

E. coli B (American Type Culture Collection) from frozen stock (0.5 mL) was added to 30 mL of tryptic soy broth (TSB) in a 125-mL shaker flask and incubated with shaking at 37 °C for at least 6 hours until log-phase was reached. Log-phase *E. coli* cultures (1 mL) were centrifuged for 25 minutes at 3000rpm. The liquid layer was immediately decanted, and the *E. coli* cell pellet was then re-suspended in 1 mL of pH 7.0 to 7.2 Dulbecco's phosphate buffered saline, (DPBS) by vortex mixing for use in experiments. *B. cereus* was used in its spore form was stock spore suspensions from the American Type Culture Collection (ATCC), and were standardised at a concentration of 10^6 spores per 0.1 millilitre. MS-2 coliphages were used in the form of a thawed stock suspension. The stock suspension was prepared initially by harvesting top agar from a double layer plaque assay on host *E. coli* F*amp* that had confluent lysis into phosphate buffered saline (PBS), extracting with chloroform and then centrifuging at 5,000 xg for 15 min. to remove agar and cell debris. (Armstrong, 2011); MS-2 stocks were kept frozen at -81 degrees centigrade and thawed immediately prior to use. Test organisms were added to test waters prior to disinfection experiments atinitial concentrations of about $1\text{x}10^6$/mL to enable a potentially measurable 4 $\log_{10}$ (99.99%) reduction.

Test Waters

Two types of test waters were therefore used in this research. one was chemically-defined, autoclaved, deionised water, buffered to a pH near 7.5, with the "Good buffer" 4-(hydroxyethyl)-1-piperazineethanesulfonic acid (HEPES), is a zwitterionic buffer with a pKa of 7.55 (Good et al, 1966). HEPES was used in test water at a concentration of 10 grams per litre (0.042 M) in order to prevent complexation reactions with the metal additives, such as those observed with phosphate and carbonate buffers. With HEPES, photo-toxicity has been observed in buffered solutions (Masson et al, 2008). In order to prevent such photo-toxic effects, all solutions buffered with HEPES were kept in conditions of darkness, save for brief periods of time corresponding to sampling. Control vessels, also buffered with HEPES, were furthermore monitored for any consistent decrease in microorganism concentration over time, an effect which could have manifested in the case of photo-toxicity of HEPES.

The second test water was identical to the first, but with the addition of of dehydrated and then resuspended solids collected from waters from a highly colored surface water, Catfish Lake, Croatan National Park, NC The dehydrate was suspended in 10 millilitres of Catfish Lake water per litre of test solution (1% W/V). Test waters were autoclaved for 20 minutes and then allowed to cool to room temperature. Test waters were stored overnight in darkness in a tightly sealed container until the time of the experiment. The organic test water prepared as described above had a measured Total Organic Carbon (TOC) concentration of 11.8 mg/L. Test organic waters were kept in dark conditions during the course of the experiment to avoid any phototoxic effects that the HEPES buffer may impart on the test organisms.

Time Points and Sampling Methods

In order to approximate typical daytime use of stored water in low-resource settings, an overall time period of six hours (360 minutes) was selected for experiments. Sampling occurred at 0, 20, 60, 180, and 360 minutes, or 0, 0.33, 1, 3, and 6 hours. Intervals between time points were chosen to facilitate semi-log (i.e. log-linear) regression analysis.

At each sampling time, the sampled volume was immediately diluted 10-fold in Dulbecco's Phosphate-Buffered Saline (DPBS) containing ethylene diamine tetraacetic acid (EDTA) a concentration of $5x10^{-4}$ M in the dilution tube. Disodium EDTA was added as a chemical "neutraliser" in order to ensure that any bioactive form of the metal additive present in the sample was chemically bound, and thereby rendered unavailable to react with the test microbes. Preliminary experimentation using copper and zinc metal ions (Armstrong, 2011) indicated that disodium EDTA was a suitable chemical neutraliser at the concentration used, and one that did not exhibit statistically significant biocidal properties under these experimental conditions. Upon this initial sampling, subsequent dilutions were made in ten-fold series.

Bacterial and Viral Assay Techniques

B. cereus spores and *E. coli* were assayed on an agar medium of custom design via the spread-plate method. The medium, dubbed "Awesome Agar" is comprised of standard Difco Trypticase Soy Agar (TSA) with the addition of lactose, and Neutral Red (Toluylene Red), a chemical commonly used in histology. *B. cereus*, unlike *E. coli,* does not metabolically ferment lactose, a property that was used as a means of differentiating colonies of *E. coli* and of *B. cereus*, upon plating. Because the by-products of metabolic fermentation of lactose are acidic, *E. coli* bacterial colonies producing such by-products are therefore stained pink upon assay in medium containing neutral red. *B. cereus* colonies not producing such lactose fermentation by-products remain their original colour (usually white). Initial studies verified that *E. coli* and *B. cereus* spore detection on this medium was comparable to their detection on specific media for each one alone (data not shown).

Disinfectant Concentration Measurement

In order to measure the concentration of disinfectant at 0 minutes, and at 360 minutes, Inductively-Coupled Plasma Mass Spectrometry (ICP-MS) was used. In order to analyse the effect, if any, of the presence of microorganisms on the concentration of disinfectant, samples taken from organism-positive test waters were compared with those simultaneously taken from identically-prepared organism-negative test waters. These samples were in turn standardised against a sample taken from the control (metal-free, organism-positive) observation and against a sample of test water prior to analysis. In this fashion, the effect over time of disinfectant concentration was obtained, and the effect of the presence of organisms on the concentration of both copper and zinc aqueous species was inferred.

Summary of Independent and Dependent Variables

This research had four independent variables overall: (1) type of metal additive (whether the copper and zinc were delivered as uncharged nanoparticle oxides or as charged metallic salts), (2) metal additive dose, at nil (as a negative control), 0.2 mg/L, and 2 mg/Leach of "delivered" metal, (3) the type of microorganism as four possible values, namely, none (as a control), *Bacillus cereus* spores, *E. coli* strain B strain, and coliphage MS-2 and (4) the type test water, either sterile, buffered, heavy-metal-demand-free (HMDF, de-ionised), pH ~7.5, or the same test water supplemented with dissolved organic matter (DOM) from natural lake water at 11.8 mg/L TOC.

For the metal tested, masses of the negative counter-ion of the metal salt were not considered "delivered" metal, as these counter-ions would separate from the metal shortly after addition. For the metal oxide nanoparticles the masses of the oxygen component of the metal oxide nanoparticle were included, as the oxygen component would not separate from the particle to a large degree after addition. The total amount of metal added therefore varied between nil (as a control), 0.4 mg/L (0.2 milligrams each of "delivered" copper and zinc), and 4 mg/L (2 milligrams each of "delivered" copper and zinc). Throughout the course of this study, the highest dose was dubbed the "100%" dose, the middle dose the "10%" dose, and the lowest dose as the "control" dose.

The two dependent variables of this study were: (1) the measurable concentrations of culturable microbes, taken at various time points over the course of each disinfection trial and (2) the measurable concentrations of disinfectant, taken initially (at time $t = 0$ minutes) and at the conclusion of the disinfection trial (at time $t = 360$ minutes). The experimental time points used were 0, 20, 60, 180, and 360 minutes (0, 0.33, 1, 3, and 6 hours), and the concentration of culturable microbes at each time point were quantified as "colony-forming units" (CFUs) per unit volume or "plaque-forming units" (PFUs) per unit volume for bacteria, and viruses respectively.

Evaluation of Disinfection Efficacy: The World Health Organisation Guidelines

For the purpose of this experimentation, an adequate performance of disinfection was considered to meet or exceed the guideline set by the World Health Organization (WHO, 2011). For enteric bacteria, modelled here by *E. coli* strain B , and for the protozoan parasite surrogate modelled here by *B. cereus* spores, the recommended performance target is at least a 99% (2 log_{10}) reduction in culturable organisms as a "protective" level of performance. For enteric viruses, modelled here by MS-2 coliphage, the "protective" performance levels is at least a 99.9% (3 log_{10}) reduction in culturable organisms.

Experimental Procedures

All batch disinfection experiments were conducted in sterile 50 mL polypropylene (PP) tubes. Storage of test waters, storage of sample tubes, and further sample dilutions were all performed in tubes of lime glass. Test waters were prepared and subsequently autoclaved..

Nanoparticle-based metal additives as stocks of colloidal solutions of CuO and ZnO at 2 G/L were agitated, and sonicated for five minutes prior to use and then added as stock nanoparticle suspensions 50 mL test water volumes in polypropylene reactor vessels.

Metallic salt stock solutions were prepared by adding 0.053 grams of $CuCl_2$ and 0.106 grams of $ZnCl_2$ to 100 mL of autoclaved, deionized water in separate sterile 125-mL bottles respectively and mixed thoroughly; a Cu(II) solution concentration of $3.935x10^{-3}$ Molar, and a Zn(II) solution concentration of $7.75x10^{-3}$ M..

Two sets of test waters were prepared, one supplemented with microorganisms and one remaining microorganism-free. Five mL of *B. cereus* stock containing a total of $5x10^7$ spores were added per 500 millilitres of organism-positive test water, along with 1.5 millilitres of log-phase, washed, and re-suspended *E. coli* (an amount experimentally determined to contain $5X10^7$ CFUs). Prior to addition, stock suspensions of frozen MS-2 coliphage with approximate infectivity titre of $1x10^{12}$ PFU per millilitre were diluted ten-thousand-fold, and passed through serial 2 and 0.08 micrometre pore size polycarbonate membrane filters pre-washed with 0.1% Tween-80 solution and then pre-rinsed with heavy-metal-demand-free (de-ionised) waters, respectively, in order to eliminate aggregates of virus particles from the test water. Approximate titre targets for all microorganisms were $5x10^7$ culturable microorganisms/L. Test waters were dispensed amongst reactor vessels as shown below, tested to ensure appropriate pH, and initially sampled for 0-minute data. They were subsequently dosed with metal additives according to their respective assigned concentrations.

Organism-Positive Test Waters	**Organism Negative Test Waters**
Vessel 1 (0.2 mg/L metals concentrations) 5 μL of CuO Nanoparticle Stock Suspension 5 μL of ZnO Nanoparticle Stock Suspension	Vessel 4 (0.2 mg/L metals concentrations) 5 μL of CuO Nanoparticle Stock Suspension 5 μL of ZnO Nanoparticle Stock Suspension
Vessel 2 (2 mg/L metals concentrations) 50 μL of CuO Nanoparticle Stock Suspension 50 μL of ZnO Nanoparticle Stock Suspension	Vessel 5 (2 mg/L metals concentrations) 50 μL of CuO Nanoparticle Stock Suspension 50 μL of ZnO Nanoparticle Stock Suspension
Vessel 3 (metals free-control) No metals	[redacted]

Table 1: Reactor Vessels and Nanoparticle Additive/Microorganism Dosing

Organism-Positive Test Waters	**Organism Negative Test Waters**
Vessel 1 (0.2 mg/L metals concentrations) 10 μL of CuO Nanoparticle Stock Suspension 10 μL of ZnO Nanoparticle Stock Suspension	Vessel 4 (0.2 mg/L metals concentrations) 10 μL of CuO Nanoparticle Stock Suspension 10 μL of ZnO Nanoparticle Stock Suspension
Vessel 2 (2 mg/L metals concentrations) 100 μL of CuO Nanoparticle Stock Suspension 100 μL of ZnO Nanoparticle Stock Suspension	Vessel 5 (2 mg/L metals concentrations) 100 μL of CuO Nanoparticle Stock Suspension 100 μL of ZnO Nanoparticle Stock Suspension
Vessel 3 (metals free-control) No metals	[redacted]

Table 1: Reactor Vessels and Ionic Metal Additive/Microorganism Dosing

The time was recorded upon the dosing, and used as a reference to calculate sampling times. At each sampling time, including the initial 0-minute sampling, a 500 microliter sample was taken from reactor vessels 1-3 and added to the corresponding dilution tube with 4.5 mL of diluent to make a ten-fold dilution. Serial dilutions were made until a final dilution of 10^{-4} was achieved.

At 0 and at 360 minutes, samples were taken for ICP-MS analysis. Separate sterile 5 mL pipettes were used to take 3 millilitre samples from vessels 1-6 (with 6 being the control vessel, denoted by the blacked-out box in the table above). These samples were dispensed into sterile 15 mL polypopylene centrifuge tubes. Microorganism-positive samples were centrifuged at 3000 rpm for 15 minutes and decanted, in order to avoid analytical errors associated with aggregated particles, such as larger microorganisms, being present in the sample. Sample dilutions used during the course of the experiment were subsequently assayed for bacterial colony-forming units (CFUs) and viral plaque-forming units (PFUs). From each sample dilution, at each time point, two separate samples of 100 microliters were spread plated onto two separate Awesome Agar plates. After drying, plates were incubated at 37 °C for 24 hours prior to counting. Infectivity assay for MS-2 PFU counts were performed by the single agar layer plaque assay following the U.S. EPA Method 1602 (EPA, 2001).

Doisinfection Kinetics

Chick-Watson Kinetics

The Chick-Watson disinfection kinetic model was considered in order to characterize the experimental results for inactivation of the indicator organisms tested. This model considers the survival ratio of the microorganisms, the disinfectant concentration and contact time with the disinfectant. At disinfectant conditions where the disinfection kinetics was found to be log-linear, regression analysis of the data were used to fit the Chick-Watson kinetic model.

The first-order reaction between the microorganism and the disinfectant of interest is decribed by the Chick-Watson model as follows:

$$\frac{dN}{dt} = -kCN$$

N is the number of surviving organisms
t is contact time, the period of time the disinfectant is in contact with the microorganism
k is the rate constant, depending on the type of microorganism and disinfectant
C is the concentration of the disinfectant

When this rate law is combined with the survival ratio $\frac{N}{N_0}$ and contact time *t*, in a batch system it yields the following relationship:

$$\ln\left(\frac{N_t}{N_0}\right) = -kCt$$

N is the number of surviving organisms at time *t* and time zero

Watson accounted for differing disinfectant concentrations by adding the coefficient of dilution (n) (Watson 1908) to yield the Chick-Watson model:

$$\ln\left(\frac{N}{N_0}\right) = -kC^n t$$

The rate constant (*k*) corresponds to a specific microorganism and set of experimental conditions. When the coefficient of dilution (n) is equal to one, the reaction is said to be first-order and follows exponential kinetic decay so that the plot of $\ln\left(\frac{N_t}{N_0}\right)$ versus time (*t*) is linear. When the kinetics of disinfection are exponential or log-linear, the mathematical product of C (in mg/L) and t (in minutes), known as the Ct-value, remains constant or fixed regardless of the disinfectant concentration used. That is, a lower concentration of disinfectant would require a longer contact time to achieve the same extent of microorganism reduction. t Ct-values are specific per organism and per disinfectant and can also be influenced by temperature, pH, water quality, and sunlight exposure. In this study Ct-values were calculated for different experimental conditions and compared to determine if they were the same, as would be expected if the kinetics were first-order (exponential).

Deviations from First-Order Kinetics

To account for inactivation that deviated from first-order, log-linear kinetics, additional nonlinear models were considered to better describe more complex multi-hit, multi-site inactivation kinetics or the inactivation kinetics of mixed populations of specific microorganisms. The Hom model accounts for temporal variability as well as microbial concentration changes over time. The One-Hit Two-Population model accounts for the presence of two populations of microorganisms, each exhibiting an independent rate constant of inactivation. Using these models, the shouldering or tailing of the log-linear disinfection regression line as deviations from exponential Chick-Watson kinetics can be described and characterized. The following equation defines the Hom Model:

$$\log\left(\frac{N_t}{N_0}\right) = -kt^m$$

For the One-Hit Two-Population model, each subpopulation displays its own rate of inactivation and each rate is calculated independently to reflect two rate constants, k_1 and k_2. The fraction of the population that is more susceptible to the disinfectant is annotated by the variable f and the remaining fraction that is less susceptible is annotated by (1-*f*). The following equation defines the One-Hit Two-Population model:

$$\frac{N_t}{N_0} = fe^{-k_1 * t} + (1-f)e^{-k_2 * t}$$

The models calculated for each data set of disinfection kinetics for the experiments of this study were used for extrapolation, in order to determine the length of time required, under any particular set of experimental conditions, to reach the W.H.O. recommended performance target values of 2 $\log_{10}$ (99%) reduction of bacteria and protozoan parasites and 3 $\log_{10}$ (99.9%) reduction of viruses for drinking-water treatment POU technologies (W.H.O., 2011)

However, by definition, the One-Hit Two-Population model cannot be used for extrapolation to reach performance targets, owing to a residual fraction of microbes unaffected by the disinfectant (i.e. a population that had a rate constant, k_2, of 0). Therefore, all extrapolations were performed using the Hom model. The following equations to calculate the time required to reach 2 $\log_{10}$ (99%) and 3 $\log_{10}$ (99.9%) reductions were $t = \left(\frac{2}{k}\right)^{\frac{1}{m}}$ and $t = \left(\frac{3}{k}\right)^{\frac{1}{m}}$. The first was applied to the bacteria *B. cereus* and *E. coli* and the second to the virus, coliphage MS-2.

Cross-Model Analysis

Statistical analysis of disinfection kinetics data

For the purpose of analysing the effect of metal additive dose difference, or the presence of natural organic matter (NOM) in buffered test water, cross-model analysis was used instead of the matched-pairs t-test. Though valid, the matched-pairs t test suffers from overbroad margins of error in the case of data sets with low sample size. All data sets analysed in this research consisted of five data points, leading to large margins of error expressed and low generalizability when matched-pairs t-tests were used to analyse the significance of the low versus the high metal additive dose, or the significance of buffered water versus NOM-amended water. Accordingly, cross-model parametric analysis was performed between data sets differing by only a single variable (e.g., metal dose, or presence of NOM in water). Two different parametric techniques were employed, for linear and non-linear cases.

Baseline parametric analysis was used as the preferred method for cross-model comparisons. Associated error with baseline parametric analysis becomes large as the number of variables analysed increase; for this reason, its use is typically constricted to the case of univariate analysis. In this research, this analysis consisted of arbitrarily assigning one data set to the role of "baseline", and the data set with which it must be compared, the role of "test". In the case of the testing of dose significance, the 0.2 mg/L dose was assigned to be the "baseline" data set, and the 2 mg/L dose was assigned to be the "test" set. In the case of the testing of NOM significance, the waters without NOM were assigned to be the "baseline" data sets, and the waters with NOM were assigned to be the "test" set. Assignments were made in order to frame the significant effects (if any) of dose and NOM in terms of their increase.

The "baseline" data set was fit with the classic log-linear Chick-Watson model and designated as a function of t, as shown below:

$$f_{baseline}(t) = -k * t$$

and the rate constant k noted. Subsequently, the "test" data set was fit with the baseline parameters, and an additional parameter β, the target of significance testing, as shown below:

$$f_{test}(t) = f_{baseline}(t) + \beta * t$$

When fit, the test function yielded a 95% confidence interval for the estimate of β. A statistically insignificant result was yielded if the 95% confidence interval included 0, and vice versa. If the value for β is negative, that indicates that (with respect to the y-axis) the "test" result exhibits a significant decrease from the "baseline". For the purpose of this research, this implies an increase in disinfection efficacy. If the value for β is positive, that indicates a significant increase from the "baseline" with respect to the y-axis. In this research, this implies a decrease in disinfection efficacy.

When baseline parametric analysis cannot be accurately performed, as is often the case in models resulting from data sets with one, or both, of low *n*/multiple parameters of fit, analysis of parametric joint confidence spread can be used. This well-established technique is discussed in detail elsewhere (e.g. Box et al, 1978). Essentially, the spread of the 95% confidence intervals of model parameters are compared between models and their regions of intersection (if any) identified. Intersection of the confidence intervals of every single parameter is required for two models to have a statistically insignificant difference. Therefore, following the principle of the "one-check fail test", even a single deviant parameter implies a statistically significant difference exists between the models.

RESULTS

Concentrations of Metal Ions and Metal Oxide Nanoparticles in Test Waters

The measured average concentrations of copper and zinc ions were somewhat lower than the intended target doses in most test conditions. There were only two cases where the measured metal ion dose was higher than the intended dose, namely dissolved copper concentration in test water without natural organic matter at the 2 mg/L target dose in vessels without test microorganisms (measured dose was higher by 5%), and dissolved zinc concentration in synthetic test water without natural organic matter at the 0.2 mg/L target dose in vessels containing microorganisms (measured dose was higher than target by 50%).

The extent to which measured metal ion dose was lower than the target dose ranged over the different types of test water. In general, measured dose was within 30% of the target dose in vessels that contained microorganisms, except for the aforementioned condition and also for dissolved copper ion concentration in test water without NOM at the 2 mg/L target dose, for which the measured dose was lower by 60%. Therefore, the measured total dissolved copper and zinc ion concentrations were found to remain more or less constant over time for all trials and doses in both types of test water.

Although the measured concentrations of metal ions and metal particles in test waters were generally somewhat lower than the theoretical target doses added from stock solutions or nanoparticle suspensions, there were no significant changes in free copper concentrations between samples taken initially, at time $t = 0$ minutes, and finally, at time $t = 360$ minutes, for waters with and without dissolved organic matter added (data not shown). Some changes in concentrations of free zinc were observed over time (data not shown). However, such changes were essentially random as they had no consistent direction one way or the other, nor were they consistent with respect to dose. Overall, the data on metals concentrations in test waters over time indicated that the mass-balance of the added metal disinfectant remained about the same and did not decline appreciably from the beginning of the experiment to the end of the experiment, although the measured metal concentrations were generally somewhat lower than the target metal doses.

Disinfection Kinetics and Microbial Reduction Data Presentation

Using the methods of extrapolation mentioned in the Methods and Materials section, those experimental conditions that met, or were on track to meet, the W.H.O. "protective" performance levels within 12 hours of contact time were identified and fit with the standard Chick-Watson model of log-linear disinfection kinetics. Of these, those that showed substantial deviation from the Chick-Watson model were fit with the Hom model of non-log-linear disinfection kinetics. The best kinetic model for each data set was used to extrapolate the amount of time required to reach the W.H.O. "protective" performance levels. Those effective experimental combinations of disinfection conditions were then further considered on the basis of meeting W.H.O. "protective" performance levels for at least two of the three classes of test microbes. Such analysis was done because meeting this criterion is considered by the W.H.O. to be essential for classification as an "interim" POU drinking-water treatment, if further supported by field epidemiological evidence of significant protection from diarrheal disease. The key curve-fitting findings are summarized below in Table 2.

Metal Additive Tested	**Organism**	**Dose (mg/L)**	**Chick-Watson Fit**	**Hom Fit**	**Fit Type**
CuO/ZnO	***E. coli***	**2**	**97%**	**99%**	**Shouldering**
CuO/ZnO	***B. cereus***	**2**	**97%**	**98%**	**Shouldering**
Cu^{++}/Zn^{++}	***E. coli***	**2**	**97%**	-	**Linear**
Cu^{++}/Zn^{++}	**MS-2**	**2**	**89%**	-	**Linear**
ZnO	***E. coli***	**2**	**97%**	**99.73%**	**Tailing**
ZnO	***B. cereus***	**2**	**98%**	-	**Linear**
Cu^{++}/Zn^{++}	***E. coli***	**0.2**	**99.55%**	-	**Linear**
Cu^{++}/Zn^{++}	**MS-2**	**0.2**	**99.51%**	**99.99%**	**Shouldering**

Table 2: Selected Curve-Fitting Results for Ionic Metal and Metal Oxide Nanoparticle Inactivation Kinetics of *B. cereus* spores, *E. coli*, and MS-2 Coliphage in Tests Waters Over Time

Performance of Selected Metal Oxide Nanoparticle Disinfectants Against B. cereus and E. coli Over Time

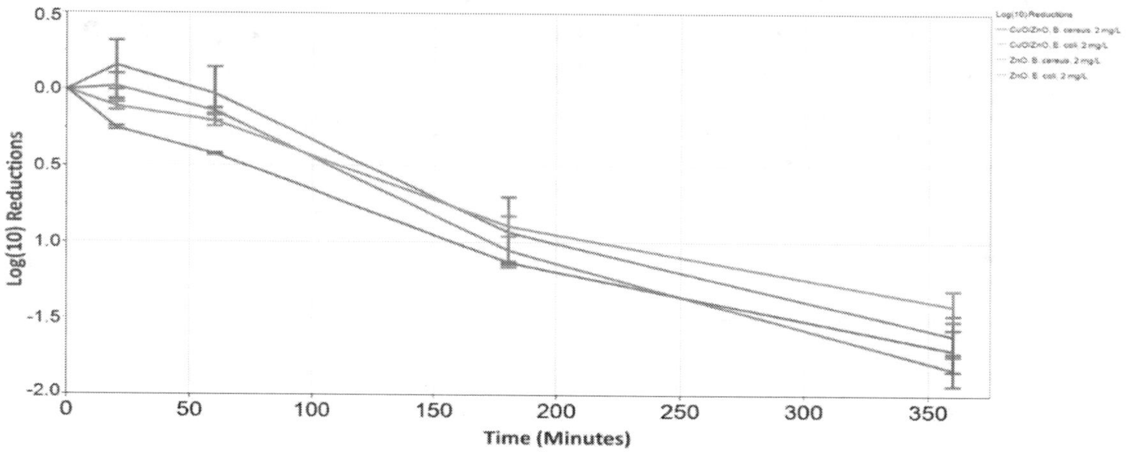

Performance of the 2 mg/L Metal Ion Disinfectant Against E. coli and MS-2 Coliphage Over Time

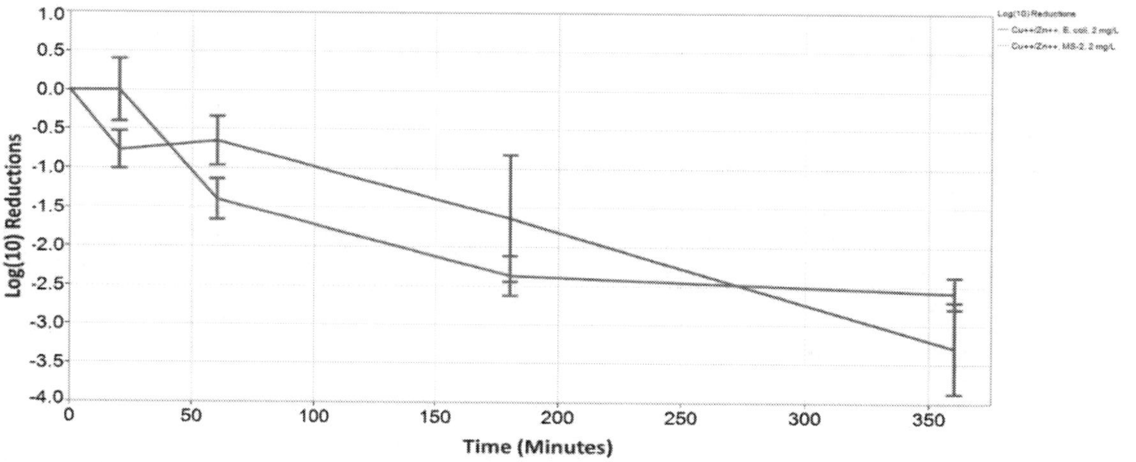

Performance of the 0.2 mg/L Metal Ion Disinfectant Against E. coli and MS-2 Coliphage Over Time

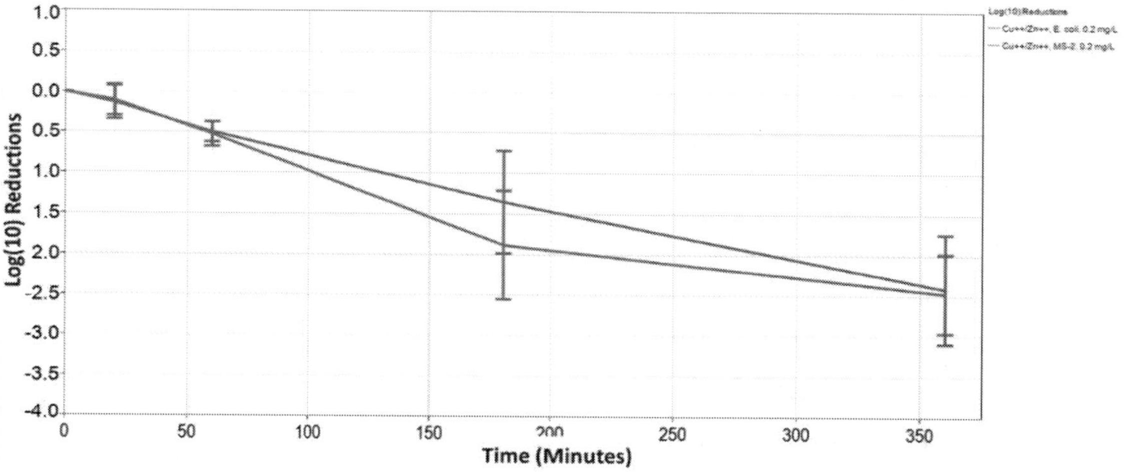

Figure 1: Log$_{10}$ Reductions of Culturable Microbes over Time for Selected Experimental Conditions

As shown in Figure 1, top panel, 2 mg/L (100%) doses of zinc oxide nanoparticles alone or zinc oxide and copper oxide particles together were capable of reducing concentrations of *E. coli* and *B. cereus* spores by about 1.5 $\log_{10}$ in 6 hours. These measured $\log_{10}$ reductions do not meet the 2 $\log_{10}$ reduction performance target recommended by W.H.O. for the "protective" level of POU treatment performance. However, based on the observed microbial inactivation kinetics, a 2 $\log_{10}$ reduction may be achievable with some additional contact time, as shown by the data in Table 3 below. Based on the results in the middle panel of Figure 1, 2 mg/L doses of the combination of zinc ions and copper ions reduced *E. coli* concentrations by about 2.5 $\log_{10}$ and MS-2 concentrations by about 3.3 $\log_{10}$ by 6 hours. These observed $\log_{10}$ reductions of *E. coli* and MS-2 meet the recommended 2 $\log_{10}$ bacteria reduction target and 3 $\log_{10}$ virus reduction target recommended by W.H.O for the "protective" level of POU treatment performance. According to the results in the bottom panel of Figure 1, even 10% or 0.2 mg/L doses of the combination of zinc ions and copper ions reduced *E. coli* and MS-2 concentrations by about 2.5 $\log_{10}$ by 6 hours. Hence, the observed $\log_{10}$ reduction of *E. coli* met the recommended 2 $\log_{10}$ bacteria reduction "protective" performance target recommended by W.H.O for POU treatment. The observed 2.5 $\log_{10}$ reduction of MS-2 did not meet the 3 $\log_{10}$ virus reduction "protective" performance target recommended by W.H.O., but based on the observed virus inactivation kinetics, this 3 $\log_{10}$ virus reduction performance target may be achievable with some additional contact time, as shown in Table 3 below.

Disinfectant	Organism	Dose	Target Reduction	Time (mins)	Time (hrs)
CuO/ZnO	***E. coli***	**2 mg/L**	**2 $\log_{10}$**	**372**	**7**
CuO/ZnO	***B. cereus***	**2 mg/L**	**2 $\log_{10}$**	**414**	**7**
Cu^{++}/Zn^{++}	***E. coli***	**2 mg/L**	**2 $\log_{10}$**	**217**	**4**
Cu^{++}/Zn^{++}	**MS-2**	**2 mg/L**	**3 $\log_{10}$**	**398**	**7**
Cu^{++}/Zn^{++}	***E. coli***	**0.2 mg/L**	**2 $\log_{10}$**	**291**	**5**
Cu^{++}/Zn^{++}	**MS-2**	**0.2 mg/L**	**3 $\log_{10}$**	**508**	**9**

Table 3: Selected Inactivation Kinetics Performance of Ionic Metal and Metal Oxide Nanoparticle Disinfectants against *B. cereus* spores, *E. coli*, and MS-2 Coliphage over Time, Compared with the W.H.O. "Protective" Performance Benchmarks

Based on the results in Table 3 above, three of the original eight unique metallic disinfectant conditions proved effective as " POU drinking-water treatments by meeting the "protective" performance levels of the W.H.O. for two of the three classes of microorganisms tested here within a 12-hour time period. However, it is important to note that the effective POU disinfectant performance conditions occurred only in waters without natural organic matter (NOM). Apparently, NOM in water reduced the performance efficacy of the disinfectants to the point where none could meet the "protective" microbial reduction performance targets recommended by W.H.O. Of the suitable interim POU disinfection conditions, one was comprised of copper and zinc oxide nanoparticles. These nanoparticles were efficacious (at the W.H.O. "protective" level) only against surrogates for bacteria and protozoan parasites, and only at the 100% (2 mg/l) dose, within the contact time period specified

The remainder of the disinfection conditions that met the recommended "protective" performance targets were comprised of ionic copper and zinc ions as metallic salts. These metallic salts were efficacious (at the W.H.O. recommended "protective" performance level) only against bacteria and enteric virus surrogates, but they were efficacious at both the 10% (0.2 mg/l) and 100% (2 mg/L) doses, within the contact time period specified.

DISCUSSION

Chemical Reactivity of Nanoparticle Copper and Zinc Oxides

For reasons described above, a particularly stable form of uncharged copper and zinc nanoparticles were used, namely the oxide form. Other forms or compositions of nanoparticles of these metals that were stable when stored but more reactive under the conditions of actual use, were not tested in these experiments. Such physically stable but more reactive particles may yield greater microbial reductions than observed here and further studies of such nanoparticles are recommended.

Physical State of Test Microbes and Disinfection Kinetics

The effects of disinfection on test microbes and their inactivation kinetics are potentially influenced by the physical state of the microbes. "Single" or dispersed microbes (i.e. isolated bacterial cells/spores or isolated virions) present different targets for chemical disinfectants than do "grouped" or aggregated microbes. The physical states of the test microbes were not specifically determined in this study, nor were the potential physical associations of the microbes with dissolved or colloidal organic matter in the NOM-amended test water. The goodness of fit to the experimental data of the Hom model and the One-Hit Two population model could be explained in terms of microbial aggregation. For example, "shouldering" or lag in disinfection kinetics in the Hom model could be a reflection of initial presence of aggregates that subsequently disperse into the test solution. Any "tailing" or tapering off disinfection kinetics in the Hom model could be a reflection of the formation of aggregates after a certain amount of time or their initial presence at low concentrations.

Implications for Point-of-Use Household Drinking Water Treatment

The findings presented here indicate that copper and zinc oxide nanoparticles and ions in solution, in concentrations within the limits for consumable copper and zinc as designated by the World Health Organisation, have the potential to be used as point-of-use household water treatments, because they meet W.H.O.'s "protective" performance benchmarks for microbial reductions under some experimental conditions. The metal additives studied here met the W.H.O. recommended performance targets for two out of the three microbe categories in a time period appropriate for same-day or overnight usage,under some test conditions in waters without dissolved organic matter. Therefore, waters low in organic matter could be reasonably treated at the point-of-use by copper and zinc oxide nanoparticles or ions at allowable concentrations of the metal in drinking water. However, further study is needed to determine if copper and zinc oxide nanoparticles or ions at doses within concentration limits for copper and zinc in drinking water are effective disinfectants in real-world waters as POU household drinking water treatments.

ACKNOWLEDGEMENTS

We thank Emily Bailey, Michael Clear, Mary Cooper, Mehrin Islam, Nikita Jathan, Hannah Leker, Hunter Motte, Jani Radhakrishnan, Prianca Reddi, Jess Richardson, Kathleen Rothwell, and Tucker Witsil, as well as Duke University's Centre for the Environmental Implications of Nanotechnology for their technical assistance in this study. We are grateful to Doug Wait, senior analyst and lab manager, for his logistical and technical support. This project was funded by the P3 Program of the US EPA, P3 Award # SU836019, to the University of North Carolina.

REFERENCES

Abad FX, Pintó RM, Diez JM, Bosch A., Disinfection of human enteric viruses in water by copper and silver in combination with low levels of chlorine., Appl Environ Microbiol Jul;60(7):2377-83, 1994

Armstrong, Andrew. "Characterization of Ionic Copper for Disinfection of Stored Drinking Water." Masters Technical Report, The University of North Carolina at Chapel Hill (2011) Print.

Atiyeh Bishara S et al, Effect Of Silver On Burn Wound Infection Control And Healing: Review Of The Literature, BURNS Mar;33(2):139-48, 2006

Blanc, D.S., et al. "Water disinfection with ozone, copper and silver ions, and temperature increase to control Legionella: seven years of experience in a university teaching hospital." Journal of Hospital Infection 60 (2005): 69-72. Print.

Box, G.E.P., W G. Hunter, and J S. Hunter. Statistics for Experimenters. Wiley, NY: N.p., 1978. Print.

Cervantes, Carlos, and Felix Gutierrez-Corona. "Copper resistance mechanisms in bacteria and fungi." FEMS Microbiology Reviews 14 (1994): 121-138. Print.

Clarke, N A. "Disinfection, Sterilization, and Preservation." Disinfection of drinking water, swimming pool water, and treated sewage effluents. Ed. S S. Block. Philadelphia: Lea & Febiger, 1983. 524-541. Print.

Domek, Matthew J., M W. LeChavallier, S C. Cameron, and G A. McFeter. "Evidence for the

Role of Copper in the Injury Process of Coliform Bacteria in Drinking Water." Applied and Environmental Microbiology 48.2 (1984): 289-293. Print.

EPA. Method 1602: Male-specific (F+) and Somatic Coliphage in Water by Single Agar Layer (SAL) Procedure. April 2001. EPA 821-R-01-029. Office of Water. Washington, D.C. 20460

Fewtrell L, Kaufmann R, Kay D, Enanoria W, Haller L , Colford J , Water, sanitation, and hygiene interventions to reduce diarrhoea in less developed countries: a systematic review and meta-analysis, Lancet Infect. Dis. 5:42-52, 2005

Ilyas Hassan, Oazi Ishtiaq, Titanium Dioxide Nanoparticles in Water Treatment: Phenol degradation using advanced oxidation process employing titania nanoparticles [Paperback], VDM Verlag, 2010

Kim, Jaeeun, et al. "Control of bacterial growth in water using synthesized inorganic disinfectant." Chemosphere 55 (2004): 775-780. Print.

Komandur, Abhinav. "Copper and Zinc Salts as E. coli Antagonists in Water." Senior Honors Thesis, The University of North Carolina at Chapel Hill (2011): Print.

Liga Michael Vincent, Virus inactivation by silver doped titanium dioxide nanoparticles for drinking water treatment. , Proquest, 2009

Lin, Yu-sen E., et al. "Individual and combined effects of copper and silver ions on inactivation of legionella pneumophila." Water Research 30.8 (1996): 1905-1913. Print.

Pyle BH, Broadaway SC, McFeters GA, Efficacy of copper and silver ions with iodine in the inactivation of Pseudomonas cepacia, J Appl Bacteriol. Jan;72(1):71-9. , 1992

Qin Y, Silver-containing alginate fibres and dressings, International Wound Journal2 (2): 172–6, 2005

Sharan, Riti, Chhibber S, Attri S, Reed RH. "Inactivation and sub-lethal injury of Escherichia coli in a copper water storage vessel: effect of inorganic and organic constituents." Antonie van Leeuwenhoek 98 (2010a): 103-115. Print.

Sharan, Riti, Sanjay Chhibber, and Robert H. Reed. "Inactivation and sub-lethal injury of salmonella typhi, salmonella typhimurium and vibrio cholerae in copper water storage vessels." BMC Infectious Disease 11 (2011): 204. Print.

Shatkin Jo Anne, Nanotechnology: Health and Environmental Risks (Perspectives in Nanotechnology), CRC Press, 2008

Sobsey MD, Managing Water In the Home: Accelerated Health Gains from Improved Water Supply, World Health Organisation, 2002.

Sobsey, M. D., C. E. Stauber, L. M. Casanova, J. M. Brown and M. A. Elliott. (2008) Point of use household drinking water filtration: A practical, effective solution for providing sustained access to safe drinking water in the developing world. Environmental Science & Technology.

42(12):4261-4267.

Sobsey, Mark D. "Inactivation of Health Related Microorganisms in Water by Disinfection Process." Water Science and Technology 21.3 (1989): 179-195. Print.

Sudha, V.B. Preethi, et al. "Killing of enteric bacteria in drinking water by a copper device for use in the home: laboratory evidence." Royal Society of Tropical Medicine and Hygiene 103 (2009): 819-822. Print.

Thurman, R, and C Gerba. "The molecular mechanisms for copper and silver ion disinfection of bacteria and viruses." Environmental Science and Technology 18.4 (1989): 295-315. Print.

Watson, H. "A note on the variation of the rate of disinfection with change in the concentration of the disinfectant." Journal of Hygiene 8 (1908): 536-542. Print.

WHO "Effects on Humans." Zinc in Drinking Water: Background document for the development of WHO guidelines for drinking water quality. 2003. 3. Print.

WHO "Guideline Value." Copper in Drinking Water: Background document for development of WHO Guidelines for Drinking-water Quality. 2004. 17-18. Print.

WHO (2010) Global Annual Assessment of Sanitation and Drinking Water, WHO, Geneva.

WHO , Global Water Supply and Sanitation Assessment, , 2000

WHO , Production of Zinc Tablets and Zinc Oral Solutions, , 2007

WHO , World Health Report: 2012, , 2012

WHO , Zinc in Drinking Water, Guidelines for drinking-water quality, 2nd ed. Vol. 2. , 1996

WHO Guidelines for drinking water quality. 3rd ed. Geneva: World Health Organization, 2008. Print.

Wood, J M. "Metal Ions in Biological System." Evolutionary aspects of metal ion transport through cell membrane. Ed. H Sigel. 18th ed. New York: N.p., 1984. 223-237. Print.

Yahya MT, Landeen LK, Messina MC, Kutz SM, Schulze R, Gerba CP., Disinfection of bacteria in water systems by using electrolytically generated copper:silver and reduced levels of free chlorine, Can J Microbiol. Feb;36(2):109-16, 1990

Optimizing Char-Based Filtration Materials for Fluoride Removal from Drinking Water

Laura R. Brunson, Ph.D. Candidate, University of Oklahoma
D.A. Sabatini, PE, Professor, University of Oklahoma
University of Oklahoma
CEES; 202 W. Boyd St.
Room 334
Norman, OK 73019
405-255-9622,
Fax – 405-325-4217
lbrunson@ou.edu

ABSTRACT

There is a global drinking water crisis. Even as the Millennium Development Goal for drinking water has been achieved, there are over 780 million people without access to improved drinking water sources. Microbial contamination in water is the issue that causes the largest amount of drinking water-related illnesses, but geogenic water issues, such as fluoride, arsenic and other chemicals in water are the second key issue reducing access to safe drinking water. One of the geogenic issues that affects the most people is fluoride, found naturally occurring in the ground water of many parts of the world, which causes a variety of detrimental health affects. Filtration materials such as bone char and activated alumina have been studied for many years to assess their effectiveness at fluoride removal from drinking water. However, in many situations these are not the most helpful fluoride removal options due to costs or cultural issues. Therefore, it is helpful for researchers to continue assessing improvements for these technologies as well as alternative technologies and their implementation strategies.

KEYWORDS

Drinking water, fluoride, bone char, wood char, aluminum amendments

INTRODUCTION

Access to safe drinking water is a struggle faced by many around the world. The Millennium Development Goal for drinking water has been achieved, but WHO and UNICEF (2012) suggests there are still approximately 784 million people without access to improved drinking water sources. Further, research by (Onda et al. 2012) suggests that beyond those without an improved water source, another 1.2 billion people may be consuming water from sources that, despite being improved sources, still contain potential health risks. This suggestion is supported by the WHO and UNICEF (2011) paper, *Drinking Water: Equity, Safety and Sustainability,* which reports that the number of people with access to sustainable drinking water is actually much lower than those considered to have access to an improved source. Having access to safe drinking water is an integral factor in overall community development and a lack thereof contributes to many global problems including: education, gender equality and economic development. Many people are focusing on solutions for this issue, but making it more difficult is the fact that approximately 2.6 billion people are living on less than $2.00 (US) per day. Therefore, solutions to increase access to safe drinking water must be inexpensive and sustainable (The World Bank, 2008).

Microbial contamination in water is the issue that causes the largest amount of drinking water-related illnesses and is also the issue being most heavily addressed. Geogenic water issues, such as fluoride and arsenic, affect less people but are still significant problems affecting approximately 200 and 100 million people, respectively (Amini et al., 2008a; Amini et al., 2008b). In some ways these geogenic issues can be difficult to tackle as these chemicals in water cannot be seen, tasted or smelled and the resulting negative health affects have a delayed appearance as opposed to the acute effects seen when consuming water contaminated by most harmful microbes. Fluoride, found naturally occurring in drinking water, exists above the World Health Organization recommended limit of 1.5 mg/L in many areas including several large areas in the Rift Valley of Africa (Ethiopia, Kenya and Tanzania), China and India (World Health Organization, 2011). The detrimental health effects of fluoride include dental and skeletal fluorosis. Skeletal fluorosis can cause bones to become deformed or stiff to the point where mobility is limited and/or painful and dental fluorosis causes a darkening or mottling of the teeth, which, while not particularly painful, can cause social and economic issues due to aesthetic reasons. Research has been conducted on fluoride removal technologies, but a good method that meets most of the treatment criteria, such as being readily available, inexpensive, easy to maintain, and effective at removing fluoride at a suitable drinking water pH, does not exist for many regional areas. Thus, research on fluoride removal technologies that will fit these criteria is important.

Bone char has been proven to be an effective fluoride removal technology; while effective, in some communities bone char is not considered appropriate for water filtration due to cultural or religious reasons and bone char has not demonstrated the same effectiveness as some aluminum-based materials. Therefore, it is helpful to investigate the principles that make bone char effective for fluoride removal, namely surface area and surface chemistry, and apply these to different materials. This may be helpful in finding fluoride filtration materials that are locally available and inexpensive. Limited work has been done with various char materials, but preliminary data suggests that some of these do not work well for fluoride removal. One way to alter this may be to add aluminum to high surface area char materials and test for improved effectiveness. This work will discuss this type of research including the production and fluoride removal of several novel materials including: two types of aluminum impregnated eucalyptus wood char and aluminum coated fish bone char, as well as results of some possible pre-treatment methods that are hypothesized to alter the point of zero charge of materials. Results of continuous flow column studies of several more promising materials will be presented.

PRELIMINARY RESULTS

For one set of experiments fish bone meal was charred at 500 °C, rinsed and then tested for fluoride removal capacity in a batch isotherm. A portion of that bone char was treated after charring with aluminum sulfate based loosely on the procedure described in Levya-Ramos et al. (1999). For this amendment an aluminum sulfate solution was combined with bone char in a solution adjusted to pH 3.5 and shaken for five days. After that the char was dried and rinsed and then used in a batch test. Isotherm starting concentrations for the batch test ranged from 0 to 100 mg/L fluoride. Results shown in Figure 1 indicate that the aluminum sulfate amended bone char obtained a much higher fluoride removal rate (approximately 8 mg/g as compared with 5 mg/g adsorption). This an exciting result as the preparation process is simple and may be applicable to other char materials found locally available in areas of elevated fluoride. Additional results of aluminum treated materials will be shared during this presentation.

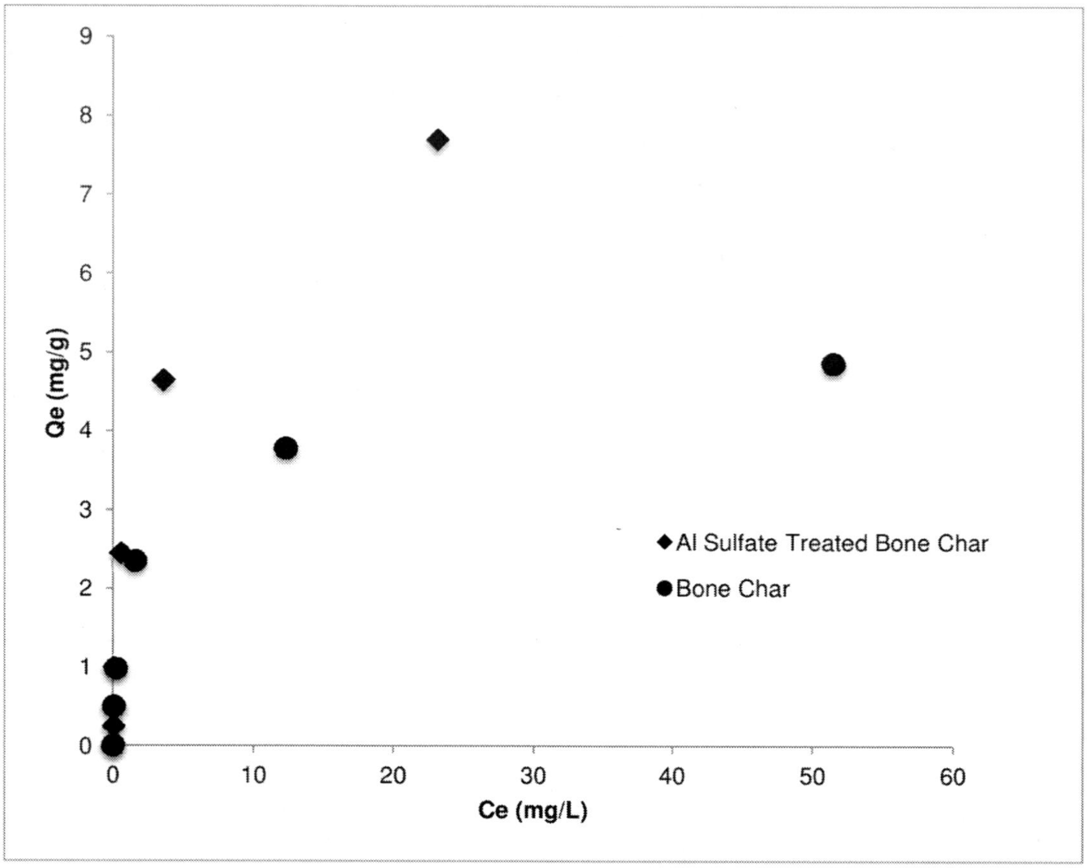

Figure 1: Isotherm results who fluoride removal of bone char and aluminum sulfate treated bone char. The X-axis shows C_e, the equilibrium fluoride concentration and the Y-axis shows Q_e, the milligrams of fluoride removed per gram of media.

While appropriate technologies are imperative to sustainable water treatment implementations, these solutions also require knowledge about communities as well as useful implementation strategies. Unfortunately, many water implementation projects fail within one to three years. Therefore, a brief discussion of a water organization that has avoided some of these pitfalls and research conducted on possible market-based solutions for fluoride removal technologies will also be discussed as part of this work.

REFERENCES

Amini, M., Mueller, K., Abbaspour, K., Rosenberg, T., Afyuni, M., Møller, K., Sarr, M., et al. (2008). Statistical Modeling of Global Geogenic Fluoride Contamination in Groundwaters. *Environmental Science & Technology*, *42*(10), 3662–3668.

Amini, Manouchehr, Mueller, K., Abbaspour, K. C., Rosenberg, T., Afyuni, M., Møller, K. N., Sarr, M., et al. (2008). Statistical Modeling of Global Geogenic Fluoride Contamination in Groundwaters. *Environmental Science & Technology*, *42*, 3662–3668. doi:10.1021/es071958y

Levya-Ramos, L. R., Ovalle-Turrubiartes, J., & Sanchez-Castillo, M. A. (1999). Adsorption of Fluoride from Aqueous Solution on Aluminum-Impregnated Carbon. *Carbon*, *37*(4), 609–617.

Onda, K., LoBuglio, J., & Bartram, J. (2012). Global Access to Safe Water: Accounting for Water Quality and the Resulting Impact on MDG Progress. *International Journal of Environmental Research and Public Health*, *9*(3), 880–894.

The World Bank. (2008). *2008 World Development Indicators:Poverty Data a supplement to World Development Indicators 2008*. Washington DC.

WHO and UNICEF. (2011). Drinking Water: Equity, Safety and Sustainability.

WHO and UNICEF. (2012). *Progress on Drinking Water and Sanitation 2012 Update*. New York, NY.

World Health Organization. (2011). *Guidelines for Drinking-Water Quality - 4th Edition*. World Health Organization.

Current Observations on UV Validation Protocols, Processes, and Practices

Chengyue Shen, O. Karl Scheible
HDR|HydroQual, 1200 MacArthur Blvd., Mahwah, NJ, USA, 07430

ABSTRACT

The Ultraviolet Disinfection Guidance Manual (UVDGM, USEPA, November 2006) and the Ultraviolet Disinfection Guidelines for Drinking Water and Water Reuse (National Water Research Institute [NWRI]/American Water Works Association Research Foundation [AwwaRF], May 2003) are the most commonly used UV validation protocols in North America. Since the release of these protocols, UV-related technologies and the industry's understanding of the UV disinfection process have both advanced significantly in the UV community. New issues have been raised and discussed to improve the validation process and the subsequent data analysis and reporting to ensure public health protection. The purpose of this article is to present and discuss new trends, new issues and regulatory requirements that affect UV validation. In particular, how these potential changes affect the uniformity of protocols for testing, their analysis, and how to bridge these interpretations to studies conducted in the past.

KEYWORDS

Disinfection, Ultraviolet, UV, Validation, Water, Wastewater

INTRODUCTION

The USEPA Long Term 2 Enhanced Surface Water Treatment Rule (LT2ESWTR, January 2006) recognized germicidal Ultraviolet (UV) light as an effective disinfectant to improve control of *Cryptosporidium* and other microbial pathogens. However, unlike chemical disinfectants (e.g., chlorine and ozone) whose dose can be quantitatively determined by a CT value, UV doses delivered by UV disinfection systems have to be determined and credited through full-scale validation testing. In current practice, the Ultraviolet Disinfection Guidelines for Drinking Water and Water Reuse (National Water Research Institute [NWRI]/American Water Works Association Research Foundation [AwwaRF], May 2003) is typically followed for validations dealing with wastewater UV disinfection for water reuse, while the Ultraviolet Disinfection Guidance Manual (UVDGM, USEPA, November 2006) for the LT2ESWTR primarily directs validation in drinking water applications.

Since the release of these UV validation protocols, UV-related technologies and the industry's understanding of the UV disinfection process have both advanced significantly in the UV community. New issues have been raised and discussed to improve the validation process and the subsequent data analysis and reporting to ensure public health protection. For example,

- According to recent research results, validations following the UVDGM (2006) protocol may overestimate the disinfection credit for UV systems using medium-pressure lamps with polychromatic UV output on *Cryptosporidium* disinfection.
- Recent practice has moved to use representative microbial surrogates for high dose UV validation to meet adenovirus disinfection requirements.
- The revised NWRI protocol (3rd edition, August 2012) has been released, and the changes in this revision are expected to be adopted in validation testing, especially for wastewater or water reuse UV disinfection.

- Field operations for UV validations have suggested that conducting UV validation in a recirculation mode is an acceptable approach, which has been endorsed in the revised NWRI protocol (2012). This is beneficial to UV testing facilities and their clients by reducing the amount of water required for testing.

LOW-WAVELENGTH MEDIUM-PRESSURE UV DISINFECTION

Recent research indicates that the difference in action spectra is more substantial between popular biological surrogates (MS2, T1UV, etc.) and target pathogens in drinking water (*Cryptosporidium*) at lower wavelengths (200 to 240 nm) than had been used to estimate the action spectra correction factor in the UVDGM (2006). In addition, output spectra of modern MP UV lamps are also significantly different from what was used in the UVDGM. As such, using MS2 (or T1UV) as the validation surrogate may overestimate the disinfection credit for *Cryptosporidium*, with potential public health consequences.

The AWWA/IUVA Working Group has announced its recommendations regarding this issue: revalidation in a manner that eliminates or fully accounts for these discrepancies, application of a blanket factor, or generating a site- and system-specific custom Action Spectra Correction Factor (ASCF). Among these, revalidation is a costly option and is unlikely to be favored by UV vendors. Application of a blanket factor can be overly conservative in most cases. The more attractive option within the UV community is that a custom ASCF based on CFD-I simulations could provide appropriate and representative correction factors for UV systems being validated with different water transmittance properties. However, the amount of effort and the level of knowledge required for conducting CFD-I modeling are generally substantial, and can be a burden for UV vendors or utilities. In addition, to obtain a site-specific ASCF, the CFD-I simulation has to be conducted for each commissioned UV system by incorporating the local water properties from the treatment plant.

A simplified Excel spreadsheet-based approach was developed. Even though this approach still requires a one-time CFD-I modeling based on the test conditions and water properties used in validation, it provides a screening tool to assess whether or not a site-specific custom ASCF is necessary. This tool conducts an analysis based on the difference in absorption spectrum between waters at the validation facility and the commissioning plant. The output spectra of UV lamps and transmittance spectra of the quartz sleeves for the UV systems are incorporated into the analysis. The output of this approach indicates if the ASCF determined based on the validation conditions can provide enough safety. This is a beneficial option for UV vendors, who can avoid the costly CFD-I modeling as well as avoid being overly de-rated by a blanket correction factor. It was noted that over 50% of the treatment plants evaluated in this study can benefit from this simplified approach.

SURROGATES FOR HIGH DOSE VIRUS DISINFECTION

Recent work has been directed to identifying and evaluating organism that could serve as surrogates for high-dose inactivation of adenovirus. Studies have been conducted to optimize the protocols for growth of *Bacillus pumilus* and its spore propagation behavior. The UV resistivity of *Bacillus pumilus* spores was manipulated by adjusting the $MnSO_4$ concentration in the growth medium, and a batch of *Bacillus pumilus* spores with UV resistance of about 100 mJ/cm^2/LI was prepared.

Field bioassay testing using the prepared *Bacillus pumilus* spores was conducted at the UV Validation and Research Center of New York (the UV Center, Johnstown, NY) on two full-scale UV systems: one with low-pressure (LP) amalgam lamps and one with medium-pressure (MP) mercury lamps. Multiple operating conditions were tested for each UV system, and some of the conditions were conducted in duplicate to evaluate the repeatability of assays using *Bacillus pumilus* spores. Field sample analysis showed a maximum 0.09-log standard deviation in log concentration among the triplicate influent or effluent samples. This implies the enumeration of *Bacillus pumilus* spores was in good control. Test conditions conducted in duplicate also demonstrated good agreement, indicating that a stable and repeatable RED or LI using *Bacillus pumilus* spores can be expected in field-scale UV validations. Cross-comparison among REDs measured using *Bacillus pumilus* spores on the LP UV system against measured REDs using other challenge surrogates (e.g., MS2 and *Aspergillus brasiliensis* conidia) at similar test conditions were observed with good agreement as well.

It had been reported in a previous study that the action spectrum of *Bacillus pumilus* spores is similar to that of adenovirus, whose sensitivity at shorter UV wavelength (e.g., below 230 nm) is much higher than that of MS2 phage. This was demonstrated by comparing the measured *Bacillus pumilus* spores REDs with predicted MS2 REDs at corresponding operating conditions. The *Bacillus pumilus* spores REDs were compared against the predicted MS2 REDs in linear regressions, which showed that the MP UV lamps were more efficient in reduction of *Bacillus pumilus* spores than LP lamps because of the polychromatic UV radiation; consistent with previous observations about the action spectrum of *Bacillus pumilus* spores.

Another high dose challenge surrogate, *Aspergillus brasiliensis* conidia, was used in previous validations for UV systems with LPHO lamps. Multi-surrogate bioassays were conducted using MS2, T1UV, T7, as well as *Aspergillus brasiliensis* conidia to cover a wide UV sensitivity range from as low as 1.5 mJ/cm^2/LI for *Cryptosporidium*/*Giardia* validation credit to 250 mJ/cm^2/LI for virus disinfection. The validity of using *Aspergillus brasiliensis* conidia in UV validation was confirmed by intercomparison with the dose measurements by MS2 and *Bacillus pumilus* spores.

IMPORTANT CHANGES IN NWRI 2012 REVISION

The NWRI guidelines provide an established protocol for validation of UV systems that are applied to treated reuse waters. Most recently, a 3rd edition was released by NWRI that modifies certain aspects of the 2nd edition issued in 2003, and is cited by the California Department of Public Health (CDPH) as the protocol to be followed for new validations and for Check Point assays that are implemented on commissioned systems to verify that the system meets or exceeds its validated RED performance for virus inactivation. Several points are noted about this recent issuance of the reuse UV guidelines:

1. The approach has been shifted to development of an RED expression, covering the operating range of the UV system with respect to flow, UVT, input power and lamp

sections. This is an improvement from the more prescriptive protocol that required three performance replicates at a given UVT, which were assigned by the level of upstream treatment. Thus an RED formulation would be developed at 55%, 65% and/or 80% UVT. The more expansive approach is in tune with UVDGM protocols.

2. The new guidance firmly establishes the practice of conducting checkpoint assays on a commissioned system as the means to verify that the performance of the full-scale process will equal or exceed that predicted by the generic UV equipment validation. It eliminates the requirement to characterize velocity profiles as part of the validation process, which would then have to be matched or improved in the commissioned system. The checkpoint assay process is discussed in detail in the 3rd edition, with inclusion of examples of the process and how it is interpreted. Basically, it calls for generating MS2 coliphage RED at a minimum of 8 test conditions representative of the expected operating envelope for the commissioned system. These results are analyzed and adjusted as needed with respect to attenuation factors and other modifying conditions, and then compared to the RED that is predicted by the system's validated operating equation. Success is defined as meeting or exceeding the predicted performance with at least 85% of the test conditions. If not, corrective action can be taken, or the system can be de-rated on the basis of the check point results relative to the expected results.
3. The new guidance eliminates the requirement that the validation should be conducted with the system lamp output attenuated to reflect end-of-lamp-life UV output. Instead, a factor is applied mathematically. This is also the practice with the check point assay protocols.
4. The requirement to conduct collimated beam dose response tests on the MS2 coliphage remains for both validation and checkpoint assays. However, the guidance now presents this as a QA task, whereby the dose-response relationship of the MS2 culture used for the assays must fall within boundaries found in the guidance. These boundaries are unchanged from previous editions, and do not attempt to reflect the large data base developed for coliphage surrogates in the past 10 years. Additionally, it now requires that the RED developed by translating the log-inactivation observed in the field to dose based on the organism's dose-response behavior, must use a "standard" dose-response formula that is based on means of the earlier data set used to establish the confidence bounds. This is a serious change, and one that may need adjustment in the future. It contradicts all practices to date that use the dose-response equation developed with the surrogate actually used for testing on a given test series or day. Using the standard equation assumes that the dose-response of MS2 is consistent and constant, which is not the case. Translating to RED may favor or penalize a system depending on where its dose-response data fall with respect to the standard curve. Note that previous validations conducted under earlier guidance are grandfathered; however, checkpoints that are conducted with these same systems are not.

There are other factors that affect validations and checkpoint assays. Those mentioned here are notable for their impacts on system verification. More work needs to be done within the industry to assure that these impacts are understood and addressed in the best interests of public health.

TESTING IN RECIRCULATION MODE

The previous version of the NWRI/AwwaRF Ultraviolet Disinfection Guidelines for Drinking Water and Water Reuse (2nd edition, 2003) requires that UV validation testing should be conducted with the challenge water in a single-pass flow-through mode as opposed to having the flow in a recirculation mode. A second guidance in the 2003 NWRI protocol identifies the potential impact that particles may have on disinfection performance, and prescribes testing in waters of the same quality. It is implied that "quality" is defined in terms of the presence of particles, and it is also implied that testing must be done in waters of the same or lesser quality. However, although particles do impact disinfection via particle-associated microorganisms being more resistant to disinfection, testing is not done with these indigenous microorganisms but rather with injected challenge microorganisms. Reactor testing with injected challenge microorganisms accurately reflects reactor performance (UV dose delivery), and these performance results can be applied to any water. In other words, the impacts of particles are manifested in a UV dose-response curve and are accounted for in setting the design dose; particles have no impact on the ability of a UV reactor to deliver a given dose. The only water quality parameter that significantly impacts UV reactor dose-delivery performance is UVT, not the presence of particles.

The updated NWRI protocol (3rd edition, 2012) is now allowing bioassay tests in a recirculation mode. However, the requirement of using waters of the same or lesser quality in validation remains unchanged. Field testing was conducted at the UV Center) on a LPHO UV system, by implementing bioassay tests to compare dose delivery performance verification with a) Flow-through mode versus Recirculation mode, and b) Potable water versus Media-filtered reclaimed water. Five test conditions were selected for each test scenario. Field test conditions, such water temperature and turbidity which may impact the UV system's performance were well controlled, and the test conditions were reproduced as close as possible among different test scenarios. The bioassay test results were then analyzed statistically for their likelihood or difference, and the results support the fact that validation bioassay tests conducted with potable water and in recirculation mode have no statistical differences from the ones using wastewater and flow through.

Although not stated directly, it was implied in the UVDGM protocol that flow-through mode is required for validation bioassay tests. Field practices at the UV Center were conducted in recirculation mode in events when high quality water is required (> 98% UVT) but the availability of such water requires significant effort due to seasonable variation in source water quality. These tests typically constitute a small portion of the overall test matrix, but the benefit of using this approach is sometimes critical. Results from tests done in recirculation mode were compared to the rest of data base on the same validated UV system, and no evidence was found to weaken the validity of using recirculation mode in bioassay tests for drinking water UV disinfection systems.

REFERENCES

USEPA *Ultraviolet Disinfection Guidance Manual for the Final Long Term 2 Enhanced Surface Water Treatment Rule,* United States Environmental Protection Agency, Office of Water, EPA-815-R-06-007, November, 2006.

National Water Research Institute (NWRI)/AWWA Research Foundation (AwwaRF), *Ultraviolet Disinfection Guidelines for Drinking Water and Water Reuse*, Second Edition. Fountain Valley, CA, May 2003.

National Water Research Institute (NWRI)/Water Research Foundation (formerly AwwaRF), *Ultraviolet Disinfection Guidelines for Drinking Water and Water Reuse*, Third Edition. Fountain Valley, CA, August 2012.

IUVA, *Low Wavelength Medium Pressure UV Disinfection Update*, August 2012

Validation of UV Reactors – Lessons Learned

Andrew Salveson, Keith Bourgeous, Nicola Fontaine, Nitin Goel, Jeff Bandy, Morayo Noibi, Erin Mackey Carollo Engineers

The last 15 years has seen the evolution of UV systems, from 80-watt low pressure (LP) systems to high wattage medium pressure (MP) systems to the latest generation of ultra high intensity LP systems. Alongside this evolution of UV systems was the comprehension that a more rigorous method was needed for analyzing and sizing UV systems. The National Water Research Institute has played a leadership role in the development of UV Guidelines to meet this need, including the release of the latest version of their Guidelines (*Ultraviolet Disinfection Guidelines for Drinking Water and Water Reuse, Third Edition*).

Carollo Engineers has been instrumental in the development of these Guidelines and the use of these Guidelines. Over the last 10 years alone, Carollo has led 61 pilot and full-scale wastewater and water reuse UV tests following the NWRI Guidelines. We have learned a great deal related to test methodologies, safety factors, shortcuts (to avoid!), and long term impact on future UV system efficiencies. Our presentation highlights many of these important issues.

Uniform Protocol for Wastewater UV Validation Applications

G. E. Whitby, Calgon Carbon Corporation
O. Lawal, Aquionics Inc
P. Ropic, Aquionics Inc
S. Shima, Severn Trent Services
B. Ferran, Ozonia North America LLC
B. Dussert, Dussert Consulting, LLC

ABSTRACT

The treatment objective of an ultraviolet disinfection system used in a wastewater application is to protect the aquatic environment and users of water that receive discharges from wastewater treatment plants. To ensure this objective is adequately met it is important to validate, or verify equipment performance for a specific application. The widely accepted method for completing this validation is by determining the UV dose delivery performance using biodosimetry. Until now a standard bioassay did not exist for UV equipment that is used to disinfect treated wastewater from a biological treatment plant that is achieving an average effluent quality of less than 30 mg/L BOD/TSS and disinfection requirements of 126 cfu/100 mL *E. coli* over a 30 day geometric mean or 200 cfu/100 mL fecal coliforms over a 30 day geometric mean. A uniform bioassay for wastewater UV equipment that can be widely adopted by industry and regulatory bodies was developed and approved by the International Ultraviolet Association. This bioassay for UV equipment is described.

KEY WORDS

Ultraviolet, UV, wastewater, bioassay, protocol, disinfection

INTRODUCTION AND BACKGROUND

The treatment objective of an ultraviolet disinfection system used in a wastewater application is to protect the aquatic environment and users of water that receive discharges from wastewater treatment plants. To ensure this objective is adequately met it is important to validate, or verify equipment performance for a specific application. The widely accepted method for completing this validation is by determining the UV dose delivery performance using biodosimetry. Whilst several protocols exist for completing biodosimetry tests, or bioassays, for different applications, only two methods are in wide scale use in the industry worldwide.

The first protocol is the Ultraviolet Disinfection Guidelines for Drinking Water and Reuse, 2nd Edition, published by the National Water Research Institute (NWRI) in collaboration with the Water Research Foundation (WRF previously AwwaRF). The two most important chapters are chapter two; Water Reuse and chapter three; Protocols. Hereafter referred to as NWRI/AwwaRF (National Water Research Institute and Awwa Research Foundation 2003). The second protocol is the Ultraviolet Disinfection Guidance Manual, published by the US EPA. Hereafter referred to as UVDGM (US EPA 2006).

Both guidelines follow similar formats (Table 1) and are in wide scale use by UV manufacturers, engineering consultants and regulators. However, neither specifically makes reference to the particular challenges associated with completing bioassays in wastewater applications as defined below. The term wastewater applications means for the purpose of this document a biological treatment plant that is achieving an average effluent quality of less than 30 mg/L BOD/TSS and disinfection requirements of 126 cfu/100 mL of *E. coli* over a 30 day geometric mean or 200 cfu/100 mL of fecal coliforms over a 30 day geometric mean or 33 enterococci cfu/100 mL over a 30 day geometric mean. These disinfection limits are not meant for wastewater reuse. Many stakeholders within the UV industry have called for such a uniform protocol for wastewater UV applications that can be widely adopted by industry and regulatory bodies.

Table 1. Format and Equivalency of Two Key UV Validation Protocols

US EPA UVDGM	NWRI/AwwaRF Guidelines
Planning and Preparation	Introduction
	Test Facilities Requirements and Set-up
Testing	Microbiological Testing
	Testing and Sampling Requirements
Validation Data Analysis	Data Analysis and Reporting
Compare Reactor Hydraulics Using CFD	
Reporting	

In an effort to provide a positive contribution to the industry in this matter, the International Ultraviolet Association (IUVA) Manufacturers Council formed a task force in 2007. The objectives of this group were to evaluate the existing protocols to identify aspects that could be of use for a uniform wastewater protocol. To facilitate discussions with both regulators and engineering consultants on the issue of a uniform wastewater protocol and to outline a position on a potential solution for a uniform wastewater protocol. After undertaking reviews and discussions with interested parties, this document represents the final portion of the task force objectives. A proposed protocol format is detailed in Table 2, following a similar format to that used in the UVDGM.

Table 2. Proposed Format for Uniform Wastewater UV Validation Protocol

Section Number	
1.0	Planning and Preparation
2.0	Microbiological Testing
3.0	Validation Data Analysis
4.0	Additional Analysis using Advanced Tools and Existing Data
5.0	Reporting

The treatment objectives for the wastewater and the UV dose or fluence that is required must be determined before this bioassay method can be used to size the UV equipment. This may be done through long term measurements of the flow rate, UVT, TSS etc., as suggested in the NWRI/AwwaRF protocol. The UV dose requirements may also be determined by doing studies with a collimated beam to determine the required UV dose or fluence with the lowest quality of wastewater.

PLANNING AND PREPARATION

In all cases it is important to understand clearly what are the goals of the testing, how they will be completed and within what limitations. This section describes the key elements of the planning and preparation stage of validation testing. The details of the test plan in addition to the final validation report must be reviewed by a third party so that they conform to the bioassay protocol or regulatory requirements. Validation testing of UV equipment for wastewater applications can be conducted at a wastewater treatment plant (WWTP) or a dedicated testing facility (manufacturer owned, or third party).

Test UV System Characteristics

A typical wastewater bioassay test stand is shown in Figure 1. In general the following criteria should be followed in relation to the test equipment configuration:

- The test unit must be equivalent in configuration and operation to the commercial unit, both in terms of components, i.e. lamp, ballast, quartz sleeve, sensor, control systems, and automatic cleaning device and other fixed or moving devices, that is, baffle, support bars, etc.
- The test unit must be hydraulically scale able or a commercially available full-scale module as per the NWRI/AwwaRF recommendations. However, it is recognized that additional analysis using field measurements and/or advanced tools as described in the section (Additional Analysis using Advanced Tools and Existing Data) could be used to justify operational variations. Closed vessel UV systems may not be hydraulically scaled.
- A single reactor can be used (equivalent to one bank) for validation.

Challenge Organism

It is critical that the challenge microorganism have the following properties:

- Is non pathogenic to humans.
- Should be easy to grow in high concentrations and simple to enumerate.
- Demonstrates repeatable results and stability over long time periods.
- Has a known action spectrum that correlates to the target pathogen or indicator organism.
- Is pre-defined, to ensure valid comparisons between bioassays.
- Is analyzed only within the linear region of the UV dose-response curve.
- Must have UV inactivation kinetics that are similar to the indicator organisms or pathogens or two organisms must be used that span the UV inactivation kinetics of the indicator organisms or pathogens.
- Must not be an indigenous organism since the UV inactivation kinetics vary from site to site.

Figure 1. Typical Layout of Wastewater Bioassay Test Stand

Based on the above criteria the following challenge microorganisms are permissible and it is preferable to use two that span the UV inactivation kinetics of the indicator organisms or pathogens of concern:

- T1
- Qβ phage
- MS2 phage

The microorganisms must be prepared and used in accordance with the NWRI/AwwaRF and UVDGM procedures.

Water Source

Key Characteristics:

- Finished water supply (potable water) and/or filtered (cloth or granular) wastewater treatment plant effluent (secondary or tertiary), or a blend of both.
- Turbidity < 2 NTU or < 5 mg/L total suspended solids in all cases. Since suspended solids are different in every wastewater treatment plant a filtered effluent should be used or potable water.
- Must not contain any disinfectant residuals that could affect the microorganisms used for the testing of the UV dose.
- Any quenching agents or their byproducts should not impact UV transmittance in the wavelength range of 200 to 300 nm, and the quenching agent should not affect the challenge organism.
- The water supply must be de-chlorinated (e.g. with thiosulfate or biosulfite) before being used for the bioassay with residual chlorine levels non-detectable with a 0.05 mg/L detection limit.
- pH after UVT and de-chlorination adjustment should be within ± 0.5 pH units of the initial pH; otherwise buffering is required.
- Impact of additives on polychromatic absorbance shall be measured and documented. It is recommended not to use additives that may have an adverse effect on the polychromatic absorbance.

Absorbing Chemical

Where a UV absorbing chemical is used to simulate the range of UVT values defined in the test plan, it is critical that it have the following properties:

- An absorbance spectrum similar to the background filtered effluent/water used for validation. (Note a full UVC spectral scan is required if polychromatic lamps are used.)
- Known and uniform impact on all relevant parameters.
- Not be toxic to the bio-surrogate ('challenge microorganism'; see note above).
- Known spectral absorbance in the UVC wavelength range.
- Solubility should not be affected by lamp heat dissipation.

Based on the above criteria the following UV absorbing compounds are permissible:

- Coffee
- Lignin Sulfonic Acid (LSA)
- Humic Acids such as Superhume™

MSDS sheets should be included in the report for these compounds.

Mixing and Sampling

It is critical that any UV absorbing chemical or challenge microorganism injected into the flow stream be uniformly dispersed at both the influent and effluent sampling points, therefore the following guidance is provided:

- The effluent sampling point must be far enough downstream of the reactor exit, so that all fluid streamlines exiting the reactor have had the opportunity to fully mix/disperse with each other, so that the effluent samples are representative of the bulk of the post-reactor effluent.
- In-line mixing, with one mixer before the influent sample point and one mixer before effluent sample point is required for closed vessel systems. The same is preferred for open channel systems; however, if a mixer is only used before the bank of UV lights then sampling must take place after the level control device.
- The effluent sample point, particularly for open channel systems, should be in such a location so as to eliminate free surfaces and wall edge effects.

Verification of mixing should be completed in accordance with the method described in the UVDGM and be fully documented in the final validation report.

Lamp Variability and UV Sensor Port Window Testing

To account for lamp variability the UV system supplier must include the NWRI/AwwaRF test results for end of lamp life testing. It is recommended that the designer use the average lamp output. Only lamps that have been running in quartz sleeves under water and tested in the same condition will be acceptable. Air testing is not acceptable. Lamps that have not been tested or operated this way should not be grandfathered. Closed vessel systems should follow the UVDGM procedures. UV sensor port window testing should follow the UVDGM.

Measurement Equipment

It is critical that all key process parameters are monitored and recorded. This includes: flow, UVT, electrical power consumption, power input to the lamps if possible but power to the lamps and ballasts can be substituted, UV intensity, water temperature, pressure (for enclosed vessels) and head-loss. The methods described in the UVDGM are comprehensive and should be adopted.

Inlet/Outlet Piping

The configuration of the inlet and outlet conditions must be documented in the validation report as per UVDGM for closed vessels and as per NWRI/AwwaRF for open channel systems.

If the site specific installation is shown to be 'different' than the validation testing, then the velocity profile or, preferably the UV dose and or UV dose distribution should be shown to be equivalent or better than that observed for the bioassay validation. This should be completed by one or more of the following methods, as appropriate:

- Velocity profile as described in the NWRI/AwwaRF guidelines
- CFD as per the section of this document called "Additional optional analysis using advanced tools and existing data".
- Check-point bioassay (Salveson 2011).

Test Lamps

For wastewater applications all UV lamps should be documented to have been seasoned/burnt-in for a period of at least 100 hours. Inlet power and UVT parameters should be adjusted to account for lamp aging and quartz sleeve fouling. Equipment verification protocols for these two variables are discussed in the section of this document called "Related equipment verification".

Test Conditions and QA/QC Samples

The validation test conditions should reflect as many variables as possible with respect to the wastewater and UV equipment. Some of these are described in the NWRI/AwwaRF Guidelines and UVDGM. Therefore, the test matrix should be designed to a specified range of water qualities (defined by UVT), flows and powers regardless of the ultimate operating philosophy: UV dose pacing, intensity pacing or confirming existing validation equations (check-point bioassay).

The challenge microorganism should be injected into the flow upstream of the UV reactor under steady state conditions. In general good sampling practices should be followed to collect at least three (3) influent and three (3) effluent samples for each test condition. An example would be three samples over 15 minutes or three volume changes of the UV system. The following process measurements should be taken for each sampling event:

- Flow rate
- UV intensity from all sensors
- Calculated UV dose
- Percent lamp current or power
- UV transmittance
- Electrical power

Where any of the above parameters change in value by more than the error of measurement (See Section 5.5 of UVDGM for error of measurement), over the course of each test condition, the test should be repeated.

The following standard quality control samples should be taken.

- Reactor Controls: With lamps off; Log Inactivation equivalent to a reduction equivalent dose (RED) < 3% of lowest RED (US EPA 2006) tested.
- Reactor Blanks: Without surrogate addition take one sample per day. The measured densities should be negligible.
- Trip Controls: One sample bottle of challenge microorganism stock solution should travel with the stock solution used for validation testing from the microbiological laboratory to the location of reactor testing and back to the laboratory. The change in the log concentration of the challenge microorganism in

the trip control should be within the measurement error. The change in concentration over the test run should be negligible. This is typically on the order of 3 to 5 percent. e.g., at 1×10^{11}, Log = 11. +/- 5% equivalent to ~10.5 to 11.5.

- Method Blanks: Normal laboratory blanks generated by plating with distilled reagent water.
- Stability Blanks: Purpose is to show no degradation of the surrogate with time in the UVT and water matrix.

Third Party Oversight

An independent third party must provide oversight to ensure that validation testing and data analyses are conducted in a technically sound manner and without bias. A person independent of the UV reactor manufacturer should oversee validation testing. Individuals qualified for such oversight include engineers experienced in testing and evaluating UV reactors and scientists experienced in the microbial aspects of biodosimetry. Appropriate individuals should have no real or apparent conflicts of interest regarding the ultimate use of the UV reactor being tested. A qualified third party should be present for and direct all testing, analyze data and author a detailed report. The final report should include the names and qualifications of all persons involved in the testing and their role. The independent third party should review the validation report before its release. When appropriate, the third party should rely on additional outside experts to review various aspects of UV validation testing, such as lamp physics, optics, hydraulics, microbiology, and electronics.

MICROBIOLOGICAL TESTING

The UVDGM contains a more comprehensive microbiological testing protocol than the NWRI/AwwaRF guidelines and reflects the latest understanding of UV disinfection technology; therefore it is proposed that this guideline be adopted in its entirety for the uniform validation testing of wastewater applications. However, it is recognized that specific unique challenges apply to microbiological testing with wastewater and therefore the following issues should be considered:

Preparing the Challenge Microorganism

- Stability - Stability should be checked, and consistent recovery from seeded effluent should be confirmed, particularly in a treated wastewater effluent matrix. The challenge microorganism concentrations should be stable over the holding time between sampling and completion of the assays. If they are not stable, the data collected will be unusable because distinguishing the sources of inactivation—exposure to UV light and die-off in holding—will be impossible. Stability verification can help ensure that the bioassay and challenge microorganism samples will be viable and the data useable.
- Refer to the previous section on challenge organisms for a list of recommended microorganisms.

Verifying UV Reactor Properties

- The water temperature must be measured during the bioassay. Since the water temperature cannot be varied during the testing the UV manufacturer must submit UV intensity testing by a third party of the same lamp, ballast and quartz sleeve combination at water temperatures from 5 to 30 °C.
- For medium-pressure (MP) systems, a temperature sensor and safety cut-off switch to prevent over heating should be provided by the manufacturer.

UV Intensity Sensors Performance

- Higher variability should be permitted if additional operational safety factors (i.e. set points) are included.

Measuring UV Dose Delivery

- UVT should be within ± 2 % of the target UVT.
- Water temperature variability shall be within 0.5 °C.
- Sampling shall not proceed until a minimum of five (5) total void volume exchanges have passed. This flush volume is calculated between the microorganism's injection point and the effluent sampling point.
- At least three (3) influent and three (3) effluent samples for each test condition should be collected. Influent and effluent samples are not collected at the same time, but collected in an alternating sequence at times that approximate the time of travel across the system. There should be at least one volume exchange between samples.
- Influent samples must be taken from the batch tank, the tap from the feed pipe, or from the channel; effluent samples must be taken from the reactor outflow or the channel after the effluent weir, or a sample tap, which is representative of the entire outflow.
- Plating should be at a minimum of two dilutions, with at least two plates per dilution.
- The following parameters should be measured and recorded: flow rate through the reactor, UV intensity, on-line UVT, calculated UV dose values both before and after the samples are collected, UVT as measured by a UV spectrophotometer with each influent sample, electrical power consumed by the lamps and or ballasts, ambient air temperature, water temperature for each test.
- Sampling for UVT measurements should be separate; measurements should be taken within 24 hours of collection.
- The concentrations of the challenge microorganisms before and after exposure to UV light should generally be measured within 24 hours of sample collection, unless stability studies indicate that the samples can reliably be considered stable over longer periods of time. Samples that are not assayed immediately should be stored in the dark at 4°C. Exposure of samples to visible light should be avoided.

Collimated-Beam Testing

- The protocols for collimated beam testing should follow those in Bolton and Linden (2003) and the IUVA Excel spreadsheets (jbolton@boltonuv.com).
- The UV sensitivity of the challenge microorganism and shape of each UV dose-response curve should be consistent with expected inactivation behavior for that challenge microorganism; accordingly, confidence bands developed for MS2 and other surrogates as a test of the quality of the UV dose response data should be used.
- Challenge Microorganisms with Shoulders or Tailing – In the case of a challenge microorganism with a shoulder or tailing in the UV dose response curve, the UV sensitivity should be defined as the sensitivity over the region of linear log inactivation that occurs between the shoulder and the onset of tailing. It is recommended that organisms with a shoulder not be used for this bioassay.

VALIDATION DATA ANALYSIS

Validation testing of UV reactors produces the following types of data for each experimental test:

- Concentration of the challenge microorganism in the influent and effluent sample.
- UVT of water.
- Flow rate.
- UV intensity as measured by the UV sensor.
- Lamp or lamp and ballast power.
- Status (on/off) for each lamp.

All experimental data should be documented, preferably in tabular format, and included in the validation report. The RED should be calculated for each experimental test using a combination of reactor testing data and collimated beam results. An additional analysis of RED data depends on the reactor's UV dose-monitoring strategy. For the UV intensity set point approach, RED results are averaged for each test condition and evaluated to identify the minimum value. For the calculated dose approach, all RED values and associated test conditions are used to create a UV dose-monitoring equation.

ADDITIONAL OPTIONAL ANALYSIS USING ADVANCED TOOLS AND EXISTING DATA

New and emerging technologies are being developed to aid in the validation of UV reactors. Computational Fluid Dynamics and Lagrangian Actinometry with Dyed Microspheres are two such technologies. They are presented here as optional analysis tools to provide greater flexibility and understanding of reactor design, not as alternatives to bioassays.

Computational Fluid Dynamics (CFD)

Computational fluid dynamics (CFD) is a branch of fluid mechanics that uses numerical methods and algorithms to solve and analyze problems that involve fluid flows. Computers are used to perform the millions of calculations required to simulate the interaction of fluids with light and microorganisms in a complex UV system.

The results from the CFD model must be calibrated with the results from the bioassay to validate the computer model. Once calibrated, the CFD model can be used to develop a CFD uncertainty factor and this can be used to predict the average and range of UV doses with a UV system

CFD can also be used to calculate the UV dose of a system at parameters (e.g. flow rate, and UV Transmittance), which have not been bioassayed.

For more information on modeling UV reactors with CFD see Ho et al. (2011)

Lagrangian Actinometry with Dyed Microspheres (DMS)

Lagrangian actinometry is a newly developed test method that uses dyed microspheres to determine the UV dose-distribution of a UV reactor (Blatchley III et al. 2008). Microspheres are modified by the attachment of a dye that allows measurement of the UV dose. When subjected to UV radiation, the dye undergoes a photochemical reaction to yield a stable, fluorescent compound that can be easily and accurately differentiated from the non-fluorescent parent compound. By measuring a large population of exposed microspheres, the UV dose-distribution with the reactor can be measured directly.

Grandfathered Protocols

It is our recommendation that a similar guideline to that described in the UVDGM is adopted for a uniform wastewater UV validation protocol, namely that UV equipment validated prior to the publication of a uniform wastewater protocol, be recognized within the following limitations:

- The microorganisms used in the previous validation were the same as those recommended in this document.
- QA/QC procedures that are generally inline with this document were followed.
- Data analysis was generally in-line with the methods outlined in this document.
- A qualified third party conducted and certified the results of the bioassay.

Related Equipment Verifications

It should be recognized that in addition to UV dose delivery performance validations, there are other related equipment tests that are used to verify operational performance. These include:

- Lamp output measurement.
- Lamp age factor testing.
- Cleaning mechanism testing (quartz sleeve fouling).

Whilst these verification tests can be completed separately to bioassay testing, it is recognized that they are used in the final equipment sizing and as such deserve attention.

The IUVA Manufacturers Council has published a protocol for the measurement of the UV output of low pressure lamps (Oliver et al. 2008), and it is our recommendation that this be adopted. A similar protocol for medium pressure lamps is pending.

Separate, updated, protocols for lamp aging and quartz fouling are required; however, in the short term it is recommended that the existing NWRI/AwwaRF guideline be followed.

REPORTING

A formal validation report is an important element of any validation testing. Both the National Water Research Institute and Awwa Research Foundation 2003 guideline and the USEPA UVDGM (US EPA 2006) include reporting guidelines. Since the UVDGM (US EPA 2006) details a more comprehensive outline of the key elements of a validation report, together with checklists helpful for review and approval, this guideline should be adopted for a uniform wastewater UV validation protocol.

SUMMARY

To ensure the objective of environmental protection is adequately met when UV light is being used to disinfect wastewater discharges, it is important to verify equipment performance. The widely accepted method for completing this validation is by determining the UV dose delivery performance using biodosimetry. The preceding proposed protocol takes elements of existing protocols for drinking water and reuse water and applies them to the specific application of wastewater as defined in this document.

However, unlike drinking water or reuse water, the wastewater regulatory community looks to effluent disinfection compliance as the sole target for UV disinfection performance and not system design or system testing processes. Therefore, the components of this document are not designed to propose a regulatory standard, but rather as a tool that allows for direct comparisons of UV systems during the design of such systems and to help to properly size a UV system.

REFERENCES

Blatchley III, E.R., Shen, C., Scheible, O.K., Robinson, J.P., Ragheb, K., Bergstrom, D.E., and Rokjer, D. (2008) Validation of large-scale, monochromatic UV disinfection systems for drinking water using dyed microspheres. *Wat. Res.* 42, 677-688.

Bolton, J. R. and Linden, K. G. (2003) Standardization of Methods for Fluence (UV Dose) Determination in Bench-scale UV Experiments. *J. Environ. Engr.* 129(3), 209–216.

Ho, C.K., Khalsa, S.S., Wicklein, E. and Wright, H. (2011) *Computational fluid dynamics based models for assessing UV reactor design and installation* Report of Water Research Foundation, Denver, Colorado, USA.

National Water Research Institute and Awwa Research Foundation (2003) *Ultraviolet Disinfection Guidelines for Drinking Water and Water Reuse Second Edition.* National Water Research Institute, Fountain Valley, California, USA.

Oliver, L., Dussert, B., Howarth, C., Platzer, K., Sasges, M., Muller, J., Whitby, E., Stowe, R., Adam, V., Witham, D., Engel, S., and Posy, P. (2008) Proposed method for measurement of the output of monochromatic (254) low pressure UV lamps. *IUVA News* 10 (1) 12-15.

Salveson, A. (2011) An NWRI review from a design and research perspective. In: *Workshop C: Disinfection of Reuse Water – a Presentation of Issues*. Disinfection 2011 Water Environment Association, Hyatt Regency Hotel, Cincinnati, Cincinnati, Ohio, USA April 10 to 12.

US EPA (2006) *Ultraviolet disinfection guidance manual for the final long term 2 enhanced surface water treatment rule.* EPA 815-R-06-007 Unites States Environmental Protection Agency, Office of Water, Washington, DC USA.

The Evolution of UV Disinfection in Indianapolis:
Balancing Between USEPA Guidance and Other Regulatory Frameworks

Bryan Townsend, Black & Veatch
Donnie Ginn, Black & Veatch
Dan Moran, Citizens Energy Group
Xi Zhao, Black & Veatch

Black & Veatch
10715 David Taylor Drive, Suite 240
Charlotte, NC 28262
(704) 510-8473

ABSTRACT

UV disinfection has been implemented as a disinfection barrier to provide *Cryptosporidium* and *Giardia* inactivation at four surface water treatment facilities serving the City of Indianapolis and surrounding communities. The first UV systems installed in Indianapolis were designed in accordance with the 10 State Standards having the disinfection capacity to provide a MS2 reduction equivalent dose of 40 mJ/cm^2, with operation based on the recommended approaches of the EPA's UV Disinfection Guidance Manual (UVDGM) to attain compliance with Federal regulations. As more UV systems were installed in Indianapolis and throughout Indiana, the increased experience and confidence in UV disinfection resulted in a shift in UV system design requirements, with the Indiana Department of Environmental Management moving away from 10 State Standards and towards the recommended procedures of the UVDGM. This evolution in UV system design and operation requirements in Indiana has resulted in significant savings in capital costs, energy usage and maintenance requirements at several facilities, all while operating with robust monitoring strategies ensuring the full protection of the public health.

KEYWORDS

Ultraviolet, UV, design, disinfection; drinking water, energy, regulations

INTRODUCTION

Following the release of the 2003 draft of the Ultraviolet Disinfection Guidance Manual (UVDGM: EPA, 2003), UV has experienced extensive applications in the primary disinfection of *Giardia* and inactivation of *Cryptosporidium* for compliance with the requirements of the US EPA's Long Term 2 Surface Water Treatment Rule (LT2ESWTR). The final version of the UVDGM (EPA, 2006), released in 2006, is the primary source of guidance for UV system design, validation and operation for surface water treatment facilities in the United States. Over the past few years, a wealth of knowledge and experience has been gained from the validation and operation of numerous UV systems for the disinfection of potable water. This experience has contributed to the advancement of UV system validation techniques in an effort to maximize UV system operating efficiency through the minimization of validation uncertainty and bias.

However, some utilities, including those that provide drinking water in the State of Indiana, are also obligated to meet additional policies of the Recommended Standards for Water Works (Great Lakes – Upper Mississippi River Board of State and Provincial Public Health and Environmental Managers, 2007), also known as the "10 State Standards". The 2007 Edition of the 10 State Standards includes a policy statement on UV light for treatment of public water supplies. Although the policy statement states that the UVDGM be used as the basis for validation, design and operation of all UV systems used for public water utilities, it also notes that the UV system must be capable of "providing a UV light dose of 40 millijoules per square centimeter (mJ/cm^2)".

In contrast with the UV policy statement, the UVDGM allows for sizing of a UV system based on a specific pathogen (*Cryptosporidium*, *Giardia*, virus) and targeted level of inactivation (0.5 to 4-log). In addition, the term "UV dose" by itself is meaningless in the UVDGM, as several UV dose terms are defined in the manual (required, validated, calculated, and reduction equivalent dose). The UVDGM dose terms vary greatly from one another in how they are calculated and applied to UV system design and operation. They are a function of the target pathogen and required level of inactivation, challenge microorganism, and the uncertainty of the validated performance models, none of which are addressed by the 40 mJ/cm^2 dose requirement of the UV policy statement.

As a result of these two conflicting approaches, there has been quite a bit of confusion regarding the design and operation of UV systems within the "10 States". Several different approaches have been utilized over the past years in various States in order to interpret and comply with the 10 State Standards in an effort to adequately size UV systems, while at the same time allowing for flexible and efficient UV system operation in accordance with the LT2ESWTR and UVDGM. Furthermore, with added experience and an increased level of confidence in UV disinfection, several State regulators have began to move away from the 10 State Standards UV Policy Statement and have embraced modern validation, sizing and operating techniques solely based on the recommendations of the UVDGM.

UV DISINFECTION OF DRINKING WATER IN INDIANAPOLIS

The majority of the drinking water for the City of Indianapolis and surrounding communities is supplied by four surface water treatment plants (WTP) which are owned and operated by Citizens Water, a division of Citizens Energy Group. These plants are the White River WTP, White River North WTP, Fall Creek WTP and T.W. Moses WTP. Citizens Water also has four groundwater stations (Geist Station, Harding Station, South Well Field and Ford Road Plant) that serve smaller portions of their service area, which is presented in Figure 1.

UV disinfection has been implemented at each of the surface water plants, with Black & Veatch designing the UV facilities at the White River, White River North and T.W. Moses WTPs. The White River WTP was the first UV facility installed for the disinfection of drinking water in Indianapolis and has been in operation since 2009. UV disinfection was then installed at the T.W. Moses WTP in 2011 followed by the White River North WTP in 2012.

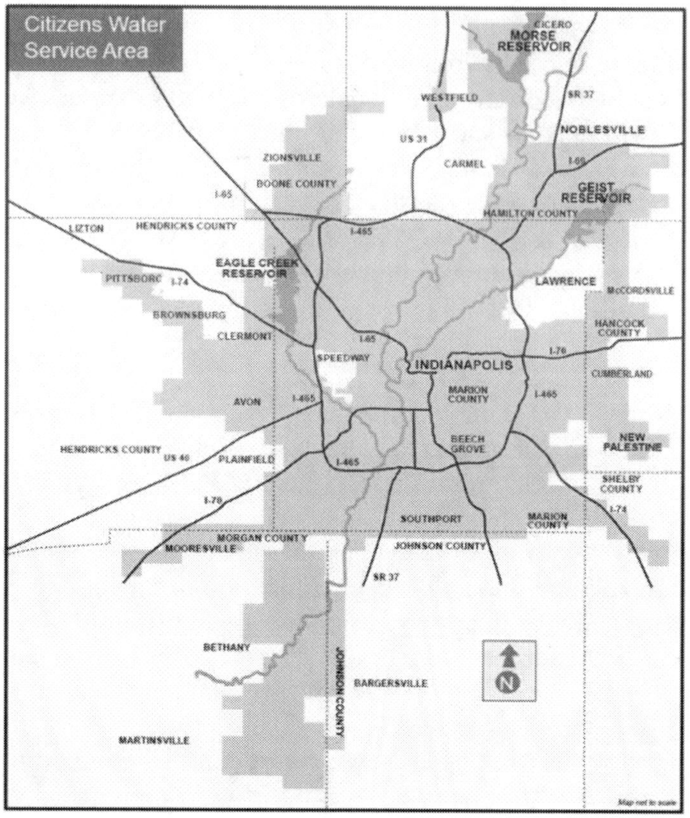

Figure 1 Citizens Water Service Area

White River Water Treatment Plant

The White River WTP is the largest treatment facility of the Citizens Water system. Portions of the facility date to the early 1900's. Numerous expansions and improvements have been made over the years, with the last major expansion constructed in 2009 with the addition of the UV disinfection facility. The White River WTP is a conventional treatment facility utilizing alum with flocculation and sedimentation; dual media filtration; UV disinfection of *Giardia* and *Cryptosporidium*; hypochlorite for primary disinfection of viruses; chloramines for a distribution system residual; and onsite storage for finished water prior to pumping into the distribution system.

Approximately 85 to 90 percent of the raw water supply to the facility is from the White River with the remainder of the supply originating from two well fields with a total of 19 wells. Water production at the White River WTP varies throughout the year, meeting a low average demand of approximately 65 mgd in winter months and a high demand of 120 mgd during peak summer days.

Prior to the addition of the UV facility, the White River WTP received pathogen disinfection credits through full conventional treatment followed by chemical disinfection with free chlorine. Source water monitoring for the LT2ESWTR resulted in a Bin 2 classification. The UV facility was added to the WRWTP to comply with the requirements of the LT2ESWTR, providing a minimum 1-log inactivation of *Cryptosporidium*. In addition, installation of the UV system assisted with compliance with the Stage 2 Disinfectants / Disinfection Byproducts Rule (Stage 2 DBPR), decreasing the dependency on chlorine for *Giardia* disinfection, thus reducing the potential for production of associated byproducts. Photographs of the White River UV disinfection facility are presented in Figure 2.

Figure 2 White River UV Disinfection Facility

Since the White River WTP is located in Indiana, the design of the UV disinfection facility required that the 10 State Standards UV Policy Statement be considered in the development of the disinfection requirements for the UV system design and operation. Discussions with the Indiana Department of Environmental Management (IDEM) resulted in a dual design/operating approach to satisfy the requirements of the UV Policy Statement, providing a design have a capacity to provide a MS2 RED of 40 mJ/cm^2, while operating the UV system using a UVDGM-compliant calculated dose monitoring strategy. The design requirements for the WRWTP UV Facility are presented in Table 1.

TABLE 1 WHITE RIVER WTP UV FACILITY DESIGN REQUIREMENTS	
Disinfection Requirements	
Regulatory Drivers	LT2ESWTR (Bin 2) Stage 2 DBPR
Disinfection Requirements	1-log *Crypto* (D_{VAL}: 2.5 mJ/cm^2) 0.5-log *Giardia* (D_{VAL}: 1.5 mJ/cm^2)
Design Basis	10 State Standards
Design Requirement	40 mJ/cm^2 MS2 RED
Operating Approach	UVDGM
Disinfection Target	3-log *Crypto* (D_{VAL}: 12 mJ/cm^2)
Water Quality / Hydraulics	
Influent Source	Combined Filter Effluent

TABLE 1 WHITE RIVER WTP UV FACILITY DESIGN REQUIREMENTS	
Flow rate (MGD)	
Design / Average	120 / 75 (454 / 284 MLD)
Range	50 – 120 (189 – 454 MLD)
UVT (%)	
Design / Average	86.3 / 87.4
Range	86.0 - 90.0

Details regarding the UV system are presented in Table 2. The White River UV facility consists of 5 (4 duty and 1 standby) 48-inch 6x20 kW Staggered Sentinel UV reactors, manufactured by Calgon Carbon.

TABLE 2 WHITE RIVER WTP UV SYSTEM	
UV System Design	
Manufacturer	Calgon Carbon Corporation
Reactor Model	Sentinel
Flange Diameter (in)	48 (122 cm)
No. of Reactors	4 duty + 1 standby
Reactor Components	
Lamp Type	Medium Pressure
No. of Lamps per Reactor / Total	6 / 30
Ballast Type	Electromagnetic
No. of Lamps Run per Ballast	1
No. of Ballasts per Reactor / Total	6 / 30
UV Sensor Type	Dry, germicidal
No. of Sensors per Lamp	1
No. of Sensors per Reactor / Total	6 / 30
Cleaning System	On-line Mechanical Wipers
Reactor Validation	
Validation Facility / Year	Portland, OR / 2009
Challenge Microbe	MS2, T1 and T7
Lamp Configurations	3, 6 or 9 operating lamps
Dose-Monitoring Approach	UVDGM Calculated LogI

Each reactor consists of six 20 kW medium pressure (MP) UV lamps, operated in two banks of 3 lamps (i.e. either 3 or 6 lamps can be operated in each reactor). Each lamp is monitored by a UV sensor and is controlled by a 20 kW electromagnetic ballast, which can be varied from 100 to 40 percent power. The UV system was validated at the Portland, OR UV Validation Facility in accordance with the UVDGM. Validation testing was conducted with three challenge microbes: MS2, T1 and T7, resulting in the development of a UV system control algorithm that calculates log inactivation of a target microbe as a function of flow rate, UVT, UV intensity, microbial UV sensitivity and number of operating banks of lamps.

The original operating strategy for the White River WTP UV system employed a *Cryptosporidium* inactivation target of 3-log, requiring a validated dose (D_{VAL}) of 12 mJ/cm^2. The monitoring algorithm employed by the reactor uses the UV sensitivity (D_L) of *Cryptosporidium* for direct the calculation of the *Cryptosporidium* log inactivation and associated validated dose. The estimated annual operation and maintenance (O&M) savings associated with the original 3-log *Cryptosporidium* operating strategy as compared to providing a MS2 RED of 40 mJ/cm^2, are presented in Figure 3. Annual maintenance costs include the expected labor and equipment costs associated with replacement of lamps, ballasts, sensors and sleeves; Annual energy costs include the power consumed by the UV system while providing the indicated disinfection target at the average annual flow and UVT for the White River WTP; and the total annual O&M cost is the sum of the annual maintenance and energy costs. In early 2012, IDEM approved the use of a reduced disinfection target, allowing the White River UV Facility to normally operate to achieve a 2-log inactivation of *Cryptosporidium*. Figure 3 presents the savings that are currently realized by the White River WTP, operating the UV system to provide a 2-log inactivation of *Cryptosporidium* (D_{VAL} of 5.8 mJ/cm^2) as compared to an MS2 RED of 40 mJ/cm^2.

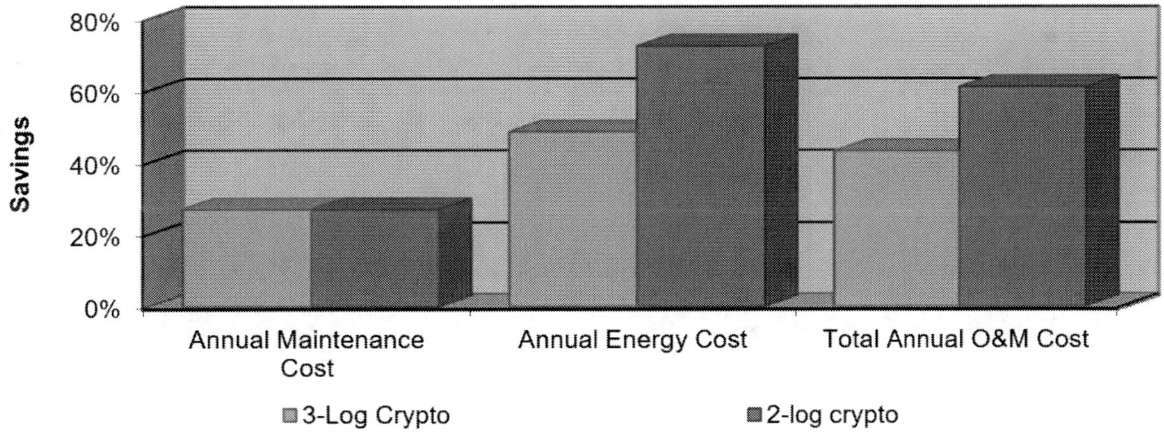

Figure 3 White River UV Facility Annual O&M Savings

The use of a UVDGM complaint dose-monitoring algorithm to achieve a 3-log inactivation of *Cryptosporidium* resulted in a 27 percent reduction in annual maintenance costs, 49 percent reduction in annual energy costs and 43 percent reduction in total O&M costs. When the operating target was further reduced to a 2-log inactivation of *Cryptosporidium*, the same general number of operating lamps were required, resulting in similar maintenance costs as compared to the 3-log *Cryptosporidium* operating target. However, the lamp power can be further reduced, allowing for a 73 percent reduction in energy costs and 61 percent reduction in the total annual O&M costs. This equates to a 20 year present worth savings of approximately $1,030,000 and

$1,460,000 while operating the UV facility to provide a 3 and 2-log inactivation of *Cryptosporidium*, respectively.

T.W. Moses Water Treatment Plant

The T.W. Moses Water Treatment Plant and raw water intake were constructed from 1974 to 1976. The facility is a conventional water treatment plant originally consisting of rapid mixing, flocculation/sedimentation, dual media filtration, finished water storage and a high service pumping station. Eagle Creek Reservoir is the source of water for this facility, which is located in the northwest part of the City of Indianapolis. Unlike the other water treatment plants serving Indianapolis, T.W. Moses WTP has only a surface water supply with no viable groundwater wells to provide a supplemental source of water.

As part of a recent technical upgrades project, a new UV disinfection facility was constructed, which went on-line in 2011. Similar to the White River WTP, the T.W. Moses WTP installed UV disinfection in order to comply with the LT2ESWTR following source water monitoring and classification in Bin 2, requiring a 1-log inactivation of *Cryptosporidium*. The facility also uses UV for *Giardia* inactivation in order to minimize the dependence on chlorine for primary disinfection in an effort to minimize byproduct formation and attain compliance with the Stage 2 DBPR. Photographs of the T.W. Moses UV facility are presented in Figure 4.

Figure 4 T.W. Moses UV Disinfection Facility

The identical dual design/operating approach used for the White River WTP was also utilized for the T.W. Moses UV disinfection facility to satisfy the requirements of the 10 State Standards UV Policy Statement while optimizing the operating efficiency of the UV system through the implementation of a UVDGM-compliant calculated dose monitoring strategy. The design requirements for the T.W. Moses UV Facility are presented in Table 3.

TABLE 3 T.W. MOSES WTP UV FACILITY DESIGN REQUIREMENTS	
Disinfection Requirements	
Regulatory Driver	LT2ESWTR (Bin 2) Stage 2 DBPR
Disinfection Requirement	1-log *Crypto* (D_{VAL}: 2.5 mJ/cm^2) 0.5-log *Giardia* (D_{VAL}: 1.5 mJ/cm^2)
Design Basis	10 State Standards
Design Requirement	40 mJ/cm^2 MS2 RED
Operating Approach	UVDGM
Disinfection Target	3-log *Crypto* (D_{VAL}: 12 mJ/cm^2)
Water Quality / Hydraulics	
Influent Source	Combined Filter Effluent
Flow rate (MGD)	
Design / Average	24 / 12 (91 / 45 MLD)
Range	6 – 24 (23 – 91 MLD)
UVT (%)	
Design / Average	88.0 / 90.0
Range	86.0 - 93.0

The T.W. Moses UV facility consists of 3 (2 duty and 1 standby) 24-inch 5x10 kW Sentinel UV reactors, manufactured by Calgon Carbon. Details regarding the UV system are presented in Table 4.

TABLE 4 T.W. MOSES WTP UV SYSTEM	
UV System Design	
Manufacturer	Calgon Carbon Corporation
Reactor Model	Sentinel
Flange Diameter (in)	24 (61 cm)
No. of Reactors	2 duty + 1 standby
Reactor Components	
Lamp Type	Medium Pressure
No. of Lamps per Reactor / Total	5 / 15
Ballast Type	Electromagnetic
No. of Lamps Run per Ballast	10 kW – 1 lamp, 20 kW – 2 lamps
No. of Ballasts per Reactor / Total	3 (1 10 kW, 2 20 kW) / 9
UV Sensor Type	Dry, germicidal
No. of Sensors per Lamp	1

TABLE 4 T.W. MOSES WTP UV SYSTEM	
No. of Sensors per Reactor / Total	5 / 15
Cleaning System	On-line Mechanical Wipers
Reactor Validation	
Validation Facility / Year	Portland, OR / 2007
Challenge Microbe	MS2, T1 & T7m
Lamp Configurations	1, 2, 3, 4 or 5 operating lamps
Dose-Monitoring Approach	UVDGM Calculated LogI

Each of three 12-inch Sentinel reactors contains five 10 kW MP UV lamps. The manner in which the system was validated allows for the ability to operate 1, 2, 3, 4 or 5 lamps in a single reactor. Three ballasts operate the lamps in each reactor, consisting of two 20 kW ballasts (running 2 lamps each) and a single 10 kW ballast, both allowing for variable lamp power between 100 and 40 percent. Each lamp is monitored by its own UV sensor. The system was validated at the Portland, OR UV Validation Facility in 2007 in accordance with the UVDGM. Validation testing was conducted with three challenge microbes: MS2, T1 and T7m, resulting in the development of UV system control algorithm that calculates log inactivation of a target microbe as a function of flow rate, UVT, UV intensity and microbial UV sensitivity for each combination of operating lamps.

The original operating strategy for the T.W. Moses WTP UV system employed a *Cryptosporidium* inactivation target of 3-log, requiring a validated dose (D_{VAL}) of 12 mJ/cm^2. The monitoring algorithm employed by the reactor uses the UV sensitivity (D_L) of *Cryptosporidium* for direct the calculation of the *Cryptosporidium* log inactivation and associated validated dose. The estimated annual O&M savings associated with operating the UV system to provide a 3-log inactivation of *Cryptosporidium* as compared to providing a MS2 RED of 40 mJ/cm^2, are presented in Figure 5.

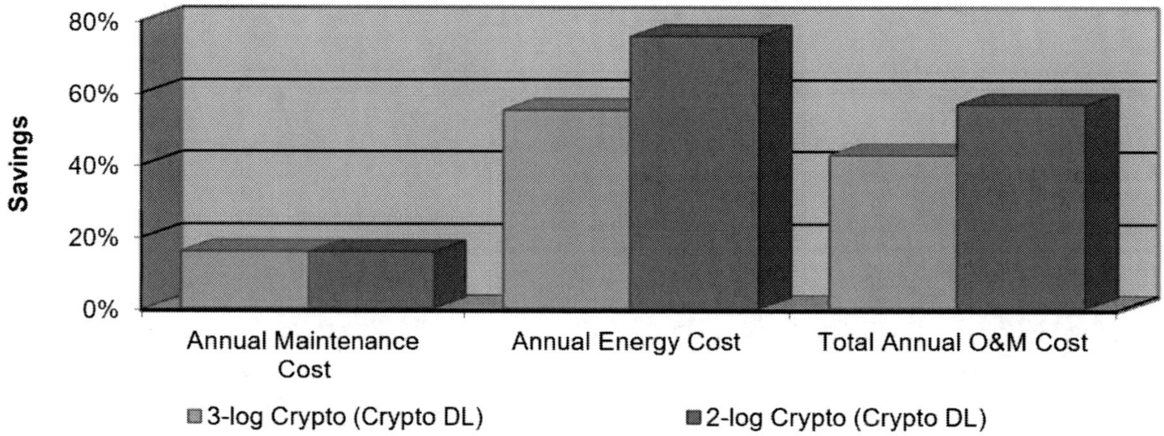

Figure 5 T.W. Moses UV Facility Annual O&M Savings

O&M costs were calculated in the identical manner as described for the evaluation conduced for the White River UV disinfection facility. In early 2012, IDEM approved the use of a reduced disinfection target, allowing the T.W. Moses UV Facility to normally operate to achieve a 2-log inactivation of *Cryptosporidium*. Figure 5 presents the savings that are currently realized by the T.W. Moses WTP, operating the UV system to provide a 2-log inactivation of *Cryptosporidium* (D_{VAL} of 5.8 mJ/cm^2) as compared to an MS2 RED of 40 mJ/cm^2.

The use of a UVDGM complaint dose-monitoring algorithm with a 3-log *Cryptosporidium* operating target resulted in projected annual maintenance, energy and total O&M cost savings of approximately 16, 55 and 43 percent. If the operating target is reduced to a 2-log inactivation of *Cryptosporidium*, maintenance costs are not significantly impacted as compared to the 3-log operating target, however further energy and thus total annual O&M savings can be realized. Operating the UV system with a 2-log *Cryptosporidium* inactivation target is expected to result in a 75 percent reduction in annual energy usage and 57 percent reduction in total annual O&M costs as compared to operating the UV facility to achieve a MS2 RED of 40 mJ/cm^2. These reductions equate to an approximate 20 year present worth savings of approximately $215,000 and $290,000 while operating the UV facility to provide a 3 and 2-log inactivation of *Cryptosporidium*, respectively.

White River North Water Treatment Plant

The White River North WTP and intake facilities were constructed from 1989 to 1990, with several improvements projects completed at the plant since 2001 including the addition of the UV disinfection facility in 2012. The facility is a conventional water treatment plant consisting of rapid mixing, flocculation/sedimentation, dual media filtration, UV disinfection, chlorine contact, finished water storage and a high service pumping station. White River is the source of water for this facility, with both the river intake and the treatment plant located on a single site in the City of Carmel, Indiana, north of the City of Indianapolis. Similar to the White River WTP, groundwater is used to supplement the river water supply at the White River North WTP. Groundwater is supplied by a separate treatment plant, rated at 12 mgd and utilizing pressure filters, located at the White River North WTP site

The drivers for the construction of the UV disinfection facility were identical to those for the other facilities in Indianapolis, with a 1-log inactivation of *Cryptosporidium* required for compliance with the LT2ESWTR and a 0.5-log inactivation of Giardia in an effort to reduce chlorinated byproducts and maintain compliance with the Stage 2 DBPR. Photographs of the White River North UV facility are presented in Figure 6.

The approach used for the design of the UV disinfection facility at the White River North facility differed from that used in the prior Indianapolis installations, departing from the conservative design approach based 10 State Standards UV Policy Statement. Instead of designing the UV facility to provide MS2 RED of 40 mJ/cm^2, IDEM approved the design of the facility to be based on providing a D_{VAL} of 12 mJ/cm^2 to achieve a 3-log *Cryptosporidium* inactivation. In addition, IDEM approved the use of a 2-log *Cryptosporidium* inactivation operating target (D_{VAL} of 5.8 mJ/cm^2), as compared to the 3-log target previously required to be implemented at the other UV facilities in Indianapolis. The design requirements for the White River North UV Facility are presented in Table 5.

Figure 6 White River North UV Disinfection Facility

TABLE 5 WHITE RIVER NORTH WTP UV FACILITY DESIGN REQUIREMENTS	
Disinfection Requirements	
Regulatory Driver	LT2ESWTR (Bin 2) Stage 2 DBPR
Disinfection Requirement	1-log *Crypto* (D_{VAL}: 2.5 mJ/cm^2) 0.5-log *Giardia* (D_{VAL}: 1.5 mJ/cm^2)
Design Basis	UVDGM
Design Requirement	3-log *Crypto* (D_{VAL}: 12 mJ/cm^2)
Operating Approach	UVDGM
Disinfection Target	2-log *Crypto* (D_{VAL}: 5.8 mJ/cm^2)
Water Quality / Hydraulics	
Influent Source	Combined Filter Effluent
Flow rate (MGD)	
Design / Average	30 / 14 (114 / 53 MLD)
Range	8 – 30 (8 – 30 MLD)
UVT (%)	
Design / Average	85.0 / 88.0
Range	83.0 - 97.0

IDEM's departure from the 10 State Standards UV Policy Statement was the result of the confidence and increased level of comfort gained from the design and operating experience from the existing UV facilities in Indiana and a desire to not only provide requirements to fully protect public health, but also allow for the use of modern validation, design and operating approaches that would result improved UV system operation and increased energy efficiency. The net effect of this approach is a smaller, more efficient UV system which while operating at a lower disinfection target will still normally provide at least two times the required level pathogen inactivation and have the available capacity to increase that to a factor of three. Therefore, the UV facility will be able to operate with a significant operating buffer (i.e. 1-log) to account for process fluctuations that could temporarily decrease the UV dose below the operating target, but easily maintain the dose above the compliance target, thereby fully protecting public health. In addition, the extra capacity in the design of the UV system will allow for future flexibility should the source water quality characteristics or regulations change requiring the need for additional disinfection.

The White River North UV facility consists of 3 (2 duty and 1 standby) TrojanUVSwift™ 6L24 UV reactors, manufactured by Trojan Technologies. Details regarding the UV system are presented in Table 6.

TABLE 6 WHITE RIVER NORTH WTP UV SYSTEM	
UV System Design	
Manufacturer	Trojan Technologies
Reactor Model	TrojanUVSwift™
Flange Diameter (in)	24 (61 cm)
No. of Reactors	2 duty + 1 standby
Reactor Components	
Lamp Type	Medium Pressure
No. of Lamps per Reactor / Total	6 / 18
Ballast Type	Electronic
No. of Lamps Run per Ballast	1
No. of Ballasts per Reactor / Total	6 / 18
UV Sensor Type	Dry, germicidal
No. of Sensors per Lamp	1 dual sensor / 2 lamps
No. of Sensors per Reactor / Total	3 / 9
Cleaning System	On-line Chemical-Mechanical Wipers
Reactor Validation	
Validation Facility / Year	Portland, OR / 2005-2006
Challenge Microbe	MS2 & T1
Lamp Configurations	6 operating lamps
Dose-Monitoring Approach	UVDGM Calculated Dose

Each of reactors contains six MP UV lamps. The manner in which the system was validated limits operation to all six lamps, with the UV dose provided by a reactor varied by increasing or decreasing the ballast power setting. One ballast controls a single lamp, having a variable output between 100 and 30 percent. The "dual" sensors have two photodiodes, with each photodiode monitoring a single lamp, for a total of 3 dual sensors per reactor. The system was validated at the Portland, OR UV Validation Facility over 2005 and 2006 in accordance with the UVDGM.

Validation testing was conducted with MS2 and T1 phage, resulting in the development of two UV system control algorithms that calculate either T1 or MS2 RED as a function of flow rate, UVT and UV intensity.

Design and operation of the reactors at the White River North UV facility was based on the T1 calculated dose monitoring algorithm. The modified design approach for this facility had a dramatic impact on the capital costs and equipment required for the UV system design. If the White River North design capacity was based on providing a MS2 RED of 40 mJ/cm^2, an additional duty UV reactor would need to have been included and 2 additional lamps installed per reactor for a total of 4 TrojanUVSwift™ 8L24 UV reactors. This would have not only resulted in a 25 percent increase in UV equipment and facility footprint, but would require an approximate 80 percent increase in installed power and supporting facilities (backup generator and UV system uninterruptable power supply).

The TrojanUVSwift™ reactors at the White River North facility are operated to provide a 2-log inactivation of *Cryptosporidium*, requiring a D_{VAL} of 5.8 mJ/cm^2. The estimated annual O&M savings associated with operating the UV system to provide a 2-log inactivation of *Cryptosporidium* as compared to providing a MS2 RED of 40 mJ/cm^2, is presented in Figure 7. O&M costs were calculated in a similar manner as was conducted for the White River and T.W. Moses UV disinfection facilities, with the exception that costs were determined based on the monthly average flow rates and UVTs as opposed to the annual average values. Figure 7 also presents a comparison of the savings that would have been obtained if the UV system was operated to provide a 3-log inactivation of *Cryptosporidium* (D_{VAL} of 12 mJ/cm^2), as was the prior operating strategy originally approved by IDEM for operation of the White River and T.W. Moses UV disinfection facilities.

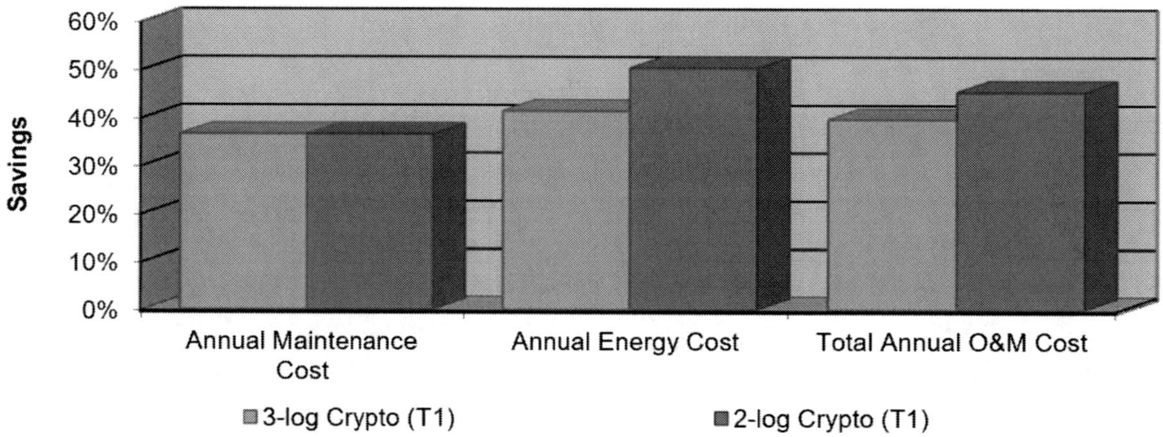

Figure 7 White River North UV Facility Annual O&M Savings

The use of a UVDGM complaint dose-monitoring algorithm with a 2-log *Cryptosporidium* operating target resulted in projected annual maintenance, energy and total O&M cost savings of approximately 36, 50 and 45 percent. The benefit of operating at a 2-log inactivation target is clear as the annual energy and total annual O&M savings are approximately 10 percent greater as compared to operating with a 3-log inactivation target. The increase in savings between the 2-

log and 3-log *Cryptosporidium* operating targets is less than that observed for the White River and T.W. Moses UV facilities partially due to the fact that during various months of the year the UV reactors at the White River North WTP are expected to operate at the minimum ballast power setting, thus limiting the system's ability to turn down further and recognize added energy savings. Operating the UV facility at a 2-log *Cryptosporidium* inactivation target equates to a 20 year present worth savings of approximately $320,000 as compared to an approximate present worth savings $280,000 associated with a 3-log operating target.

CONCLUSIONS

UV disinfection is a safe, effective and reliable microbial barrier that has been successfully implemented in numerous drinking water treatment plants for the inactivation of *Cryptosporidium* for compliance with the LT2ESWTR. Furthermore, implementation of UV disinfection allows drinking water facilities to reduce their dependency on chlorine for the primary disinfection of *Giardia*, resulting in the ability to decrease the formation potential of chlorinated byproducts and attain compliance with the Stage 2 DBPR.

The wealth of knowledge and experience gained from the validation and operation of numerous UV systems for the disinfection of potable water since the release of the UVDGM has resulted in the advancement of UV system validation, design and operation. Similarly, increased regulatory experience and confidence with UV disinfection has resulted in the migration of several States away from the overly conservative requirements of the 10 State Standards UV Policy Statement and towards modern approaches recommended by the UVDGM to attain the disinfection requirements of the LT2ESWTR. The surface water treatment plants serving the City of Indianapolis have benefited greatly from this experience, as the familiarity and an increased level of comfort gained by IDEM has resulted in the ability of recently designed UV facilities to benefit greatly from increasingly more efficient UV system design and operating requirements.

The evolution of the UV system design and operating requirements in Indiana has not only allowed for more recent UV facilities to capture significant reductions in capital costs, but has also allowed for older facilities to implement modern operating techniques, taking advantage of potential increases in operating efficiency and energy savings. While substantial savings will be recognized by the White River, T.W. Moses and White River North WTPs, the additional capacity included in the design and elevated operating target used in these facilities will also safely attain the required disinfection goals to fully protect public health today while allowing for the ability to meet the disinfection needs of tomorrow.

ACKNOWLEDGEMENTS

The authors of this manuscript would like to acknowledge the efforts of several individuals whose contributions were critical to the successful implementation UV disinfection at these facilities, including Karl Tanner, Jeff Miller, Dan Reidy and Ted Williams of Citizens Energy Group; and Amber McDonough, Jeff Neemann, Bob Hulsey and Bob Willet of Black & Veatch.

REFERENCES

EPA (2003) *Draft Ultraviolet Disinfection Guidance Manual, EPA 815-D-03-007.* U.S. Environmental Protection Agency, Office of Water, Washington DC.

EPA (2006) *Ultraviolet Disinfection Guidance Manual for the Final Long Term 2 Enhanced Surface Water Treatment Rule; EPA 815-R-06-007.* U.S. Environmental Protection Agency, Office of Water, Washington DC.

Great Lakes – Upper Mississippi River Board of State and Provincial Public Health and Environmental Managers (2007) *Recommended Standards for Water Works: Policies for the Review and Approval of Plans and Specifications for Public Water Supplies.* Health Research Inc., Health Education Services Division, Albany, NY

Alternate Indicator Organisms for Reclaimed Water in North Carolina

Patricia Drummey Stiegel, PE, Hazen and Sawyer
Melanie Mann, PE, Hazen and Sawyer
Marla Dalton, City of Raleigh
Hazen and Sawyer
4011 WestChase Blvd, Suite 500
Raleigh, NC 27607
(919) 833-7152

ABSTRACT

New reclaimed water rules that include limits for multiple microbial indicator organisms went into effect in North Carolina in 2011. One of the primary updates in these new rules is the establishment of two classes of reclaimed water: Type 1 is the less stringent class that is analogous to the older reclaimed water standard, and Type 2 is the higher quality reclaimed water, which requires the use of dual disinfection and which has additional permissible uses and stricter treatment requirements to protect public health. For Type 2 reclaimed water, the rules require treatment plants to provide a minimum 6 log reduction for *E. coli*, 5 log for coliphage, and 4 log for *Clostridium perfringens* in addition to single-digit concentration limits. To the author's knowledge, no other US state currently uses coliphage and *C. perfringens* as indicator organisms for reclaimed water regulatory compliance.

This paper reviews published data from various plants in the United States on the effectiveness of municipal wastewater treatment for removing or inactivating these alternate indicator organisms, as well as new data from one of the larger producers of reclaimed water in the state of North Carolina, the City of Raleigh. Data from the published studies suggest that relatively high disinfection doses may be required to meet Type 2 reclaimed water standards; however, the data from the City of Raleigh show that the Raleigh treatment plant would likely meet Type 2 microbial concentration standards with their current reuse regime of UV followed by sodium hypochlorite.

The published data suggest that an MS2 Reduction Equivalent Dose (RED) of 100 mJ/cm^2 may be required to meet the *C. perfringens* requirements for Type 2 reclaimed water, if the 4-log overall reduction requirement is met with 2-log inactivation by UV and 2-log removal by upstream processes. In this example, the energy required for Type 1 reclaimed water assuming basic UV disinfection with an MS2 RED of 30 mJ/cm^2, is less than one third the energy required to treat Type 2 reclaimed water with an MS2 RED of 100 mJ/cm^2.

KEYWORDS

Reclaimed water, reuse, coliphage, *Clostridium perfringens*, North Carolina, disinfection, UV, sodium hypochlorite

INTRODUCTION

Planned water reuse in North Carolina began in 1988 with the irrigation of several golf courses near the coast with treated wastewater (Safrit 2010). The rules initially established in the late 1980s underwent several iterations as interest in reclaimed water increased, especially during drought years. The most recent round of reclaimed water rules in North Carolina were adopted in 2011 and are referred to as the 2U rules, named after the North Carolina Administrative Code section 15A NCAC 2U that replaces previous reclaimed water regulation.

One of the key changes in the 2U rules was the establishment of two classes of reclaimed water. Type 1 is the lower quality class, which is equivalent to the previous reclaimed water standard in North Carolina. Typical uses of this class of water include irrigation of golf courses, ball fields, and crops such as corn for animal feed. Type 1 reclaimed water has microbial limits based either on *E. coli* or fecal coliform. Type 1 microbial standards are actually slightly less strict than the previous standard since *E. coli* is a subset of fecal coliform and the Type 1 limits have the same numerical criteria as the previous rules, which only included fecal coliform. Table 1 shows the microbial indicator concentration limits for both classes of reclaimed water. Each class also has limits for BOD_5, TSS, ammonia, and turbidity.

Table 1: North Carolina Reclaimed Water Microbial Indicator Requirements

Indicator	Type 1 Concentration (Monthly Geo. mean/daily max.)	Type 2 Concentration (Monthly Geo. mean/daily max.)	Type 2 Log Reduction Required
E. coli or fecal coliform	14 / 25 CFU/100mL	3 / 25 CFU/100mL	6 log
Coliphage	N/A	5 / 25 PFU/100mL	5 log
C. perfringens	N/A	5 / 25 CFU/100mL	4 log

Type 2 reclaimed water is the higher class of reclaimed water, and it has additional approved uses such as wetland augmentation and non-contact irrigation of food chain crops. This class of water must meet stricter microbial indicator organism limits for *E. coli* and also has strict microbial limits for the indicator organisms coliphage and *C. perfringens*.

While *E. coli* and fecal coliform are commonly used indicator organisms both in the United States and worldwide, coliphage and *C. perfringens* are not typically used for regulatory purposes in the United States. *E. coli* and fecal coliform are bacterial indicator organisms and are relatively easy to inactivate with chlorine and UV disinfection.

Coliphage has been used in academia for many years as a viral indicator. They are viruses that infect *E. coli* bacteria, and are thus relatively easy and safe to quantify. Coliphages are classified as either somatic or male specific (F+) depending on the type of *E. coli* they infect. Somatic coliphage attaches to the cell wall of *E. coli*, while male specific coliphage attacks "male" strains of *E. coli* with pili. Coliphages, and viruses in general, are much smaller than bacteria or protozoan pathogens, and this may affect what treatment processes in a wastewater plant are effective for removing or inactivating them.

Clostridium perfringens is a spore-forming anaerobic bacterium. It has also been suggested for use as an indicator organism for decades due to the fact that it forms extremely hardy spores that can survive both the environment and wastewater treatment processes. The spores may have some similarities to the cysts and oocysts forms of the protozoa *Giardia lamblia* and *Cryptosporidium*, although *C. perfringens* spores are somewhat smaller. One other difference is that *C. perfringens* spores are very resistant to UV disinfection, while cryptosporidium is inactivated relatively easily by UV.

Many utilities currently producing Type 1 reclaimed water in North Carolina already have advanced wastewater treatment to satisfy NPDES permits and often have secondary disinfection that meets the reclaimed water fecal coliform limits. The disinfection employed at plants that produce reclaimed water is often UV plus a form of chlorine, or chlorine disinfection followed by a second application point for chlorine. Little data from North Carolina is available regarding the ability of the plants to meet the Type 2 concentration and log reduction requirements for coliphage and *C. perfringens,* since these organisms have not previously been used for regulatory purposes in NC. In addition, few local laboratories are qualified and willing to do the analyses for these organisms.

The City of Raleigh currently produces Type 1 reclaimed water at their Neuse River wastewater treatment facility and recently commissioned testing of their reclaimed water to determine if they could meet the Type 2 alternate indicator standards. The Neuse River plant is a 227 mld (60-mgd) facility with primary clarifiers, 5-stage biological nutrient removal, and tertiary deep bed monomedia dentrification filters. The facility has a low pressure, high output horizontal UV system designed for basic disinfection of the entire plant flow, as well as supplemental sodium hypochlorite for the reclaimed water system.

COMPARISON OF PUBLISHED DATA TO TYPE 1 AND 2 REQUIREMENTS

Table 2 compares the North Carolina regulations to the results from a 2004 WERF report (Rose et al) examining six water reclamation facilities in Florida, California, and Arizona.
Only one of the six plants in this study (Plant F) met all of the microbial concentration limits associated with Type 2 reclaimed water in North Carolina. While all of the plants showed 6 log reduction of fecal coliform on average, no plants demonstrated 5 log reduction of coliphage, and only some of the plants could show 4 log reduction of *C. perfringens*. In general, these plants did not demonstrate 5 log removal or inactivation of coliphage because the influent concentration was not 5 logs greater than the detection limit for coliphage in this study. Plant E was the only plant with UV, and while specific dose data was not given, this plant achieved an average of 2-log reduction of *C. perfringens* through their medium pressure UV system.

Data from the WERF study showed that activated sludge and filters generally provided 2- to 3-log removal of both coliphage and *C. perfringens* and 3-to 5-log removal of fecal coliform at the plants in this study. By extension, dual disinfection for Type 2 reclaimed water in North

Carolina could potentially be designed to achieve a minimum of 3 log removal for fecal coliform and coliphage, and 2 log removal for *C. perfringens*. In practice, the design log removal or inactivation may need to be higher to account for higher influent concentrations and variations in water quality and plant operations. Designers should verify specific requirements with state regulators. While fecal coliform and coliphage are relatively easily inactivated using chlorine and/or UV, *C. perfringens* spores are extremely resistant to chlorine disinfection (Linden et al 2004) and require significantly higher UV doses than fecal coliform or coliphage. In addition, many reclaimed water providers use chlorine downstream of UV to provide a residual in the distribution system and do not currently rely on microbial inactivation from chlorine.

Published data show that the required dose to achieve 2 log reduction for *C. perfringens* is approximately 100 mJ/cm^2 (Hijnen et al 2006) while the dose required for 3 log reduction in somatic coliphage is only about 15 mJ/cm^2 (Linden et al 2004).

Table 2: Comparison of Facilities Producing Reclaimed Water From 2004 WERF Study (Rose et al) to North Carolina's Type 1 and Type 2 Reclaimed Water Standards

	NC Type 1 *E. Coli*/ Fecal Coliform Standard (25 CFU/ 100mL max)	NC Type 2 Coliphage Standard (25 CFU / 100 mL max)	NC Type 2 *C. perfringens* Standard (25 CFU / 100 mL max)	>6 log Fecal Coliform Removal	>5 log Coliphage Removal	>4 log *C. perfringens* Removal
Plant A: Activated Sludge, Cloth Filters, Chlorine	Pass	Fail	Pass*	Yes	No#	No
Plant B: Activated Sludge, Traveling Bridge Filters, Chlorine	Pass	Fail	Pass*	Yes	No#	Yes
Plant C: Activated Sludge, Anthracite Filters, Sodium Hypochlorite	Pass	Fail	Fail	Yes	No#	Yes
Plant D: Activated Sludge, Dual Media Filters, Chlorine	Fail	Pass**	Pass	Yes	No#	No
Plant E: Nitrification, Continuous Upflow Filters, UV	Fail	Pass**	Pass	Yes	No#	Yes
Plant F: BNR, Dual Media Filters, Chlorine	Pass	Pass*	Pass	Yes	No#	Yes

*Pass, but some values are over the monthly geometric mean limit
**Limit of detection for coliphage was 10 pfu/100mL in this study; proposed monthly limit is 5 pfu/100mL
>5 log removal coliphage could not be demonstrated since influent counts were generally 10^5 and detection limit was 10

(Adapted From Drummey Stiegel and Mann 2012)

RALEIGH DATA

The City of Raleigh staff at Neuse River collected samples of filter effluent (UV influent) and reclaimed water on several occasions in October, November and December 2012, and tested for total *C. perfringens*, spores of *C. perfringens*, somatic coliphages and F-specific coliphages, as listed in Table 3 and Table 4. On the final two sampling dates, the City also collected samples of UV effluent prior to the application of hypochorite for the reclaimed water, to show the separate effects of the UV disinfection process and hypochlorite on the alternate indicator organisms.

Table 3 – *Clostridium perfringens* in Pre-UV, Effluent, and Reuse Water at Neuse River WWTP

Sample Date	Total *Clostridium perfringens* (microbes per 100 mL)			Spores *Clostridium perfringens* (microbes per 100 mL)		
	Pre-UV	Effluent	Reuse	Pre-UV	Effluent	Reuse
09Oct2012	80	--	<0.3	93	--	<0.3
23Oct 2012	26	--	<0.3	13	--	<0.3
06Nov2012	87	--	<0.3	77	--	<0.3
13Nov2012	73	--	<0.3	97	--	<0.3
04Dec2012	26	21	<0.3	--	--	<0.3
18Dec2012	163	97	0.6	--	--	<0.3
Geo mean	62	45	<1	55	--	<1

Table 4 – Somatic Coliphage and F-Specific Coliphage in Pre-UV, Effluent, and Reuse Water at Neuse River WWTP

Sample Date	Somatic Coliphage (microbes per 100 mL)			F-Specific Coliphage (microbes per 100 mL)		
	Pre-UV	Effluent	Reuse	Pre-UV	Effluent	Reuse
09Oct2012	399	--	<1	2	--	<1
23Oct 2012	288	--	<1	2	--	<1
06Nov2012	602	--	<1	8	--	<1
13Nov2012	595	--	<1	<1	--	<1
04Dec2012	241	<1	<1	3	<1	<1
18Dec2012	496	12	<1	4	<1	<1
Geo mean	412	3	<1	3	<1	<1

The target UV dose at the Neuse River plant in this period was 35 mJ/cm^2, based on a point-source summation model. The target chlorine dose in the reclaimed water was 7 mg/L. The total chlorine residual in the reuse water storage tank varied from 0.10 mg/L to 0.54 mg/L.

In general, the Neuse River reclaimed water met the Type 2 reclaimed water microbial concentration standards. It is not known if the plant processes, taken together, provided the required log inactivations for Type 2 water because the plant influent was not tested at this time.

DISCUSSION

The data show that dual disinfection, as required for Type 2 reclaimed water and provided at Neuse River by UV disinfection and addition of hypochlorite to the reclaimed water, was necessary to achieve the *Clostridium perfringens* concentration limit. This result was expected given that the UV system at this plant is designed for Type 1 reclaimed water limits, and *C. perfringens* is quite resistant to UV disinfection. Higher UV doses may be required at some facilities to inactivate *C. perfringens*, especially those that have higher concentrations of *C. perfringens* entering the disinfection system. As expected, most of the *C. perfringens* that survived secondary treatment and filtration at the Neuse River plant were spores.

While the Neuse River UV system did not provide significant reduction of *C. perfringens*, the one plant with a UV system in the 2004 WERF study showed an average of 2 log reduction with UV. The delivered dose is not stated, but Florida and California both require MS2 REDs of 100 mJ/cm^2 for reclaimed water. With a higher UV dose, additional chlorine and contact time may not be required to meet log removal goals for Type 2 standards.

At Neuse River, coliphage concentrations were generally reduced to below the Type 2 microbial standard, with one data point higher than the allowable monthly geometric mean but lower than the allowable daily maximum.

CONCLUSIONS

The new data on coliphage and *C. perfringens* collected by the City of Raleigh at the Neuse River WWTP demonstrate that its existing UV and hypochlorite disinfection systems can meet the microbial concentration limits for North Carolina's new Type 2 reclaimed water. This data provides a valuable point of reference for North Carolina regulators and utilities.

REFERENCES

Blatchley, E. et al. 2007. Effects of Wastewater Disinfection on Waterborne Bacteria and Viruses. Water Environment Research 71, Number 181-92.

Burbano-Rosero, E. et al. 2011. Diversity of Somatic Coliphages in Coastal Regions with Different Levels of Anthropogenic Activity in Sao Paulo State, Brazil. Applied and Environmental Microbiology, Vol 77, No. 12, p. 4208-4216

Drummey Stiegel, P.N., Mann, M.A. Impact of North Carolina 2011 2U Reclaimed Water Rules on Design of UV Disinfection. Proceedings from NC AWWA/WEA 92nd Annual Conference, Raleigh, North Carolina, November 12-14, 2012

Hijnen, et al. 2006. Inactivation credit of UV radiation for viruses, bacteria and protozoan (oo)cysts in water: A review. Water Research 40 (1), 3–22.

Linden, K. et. al. Water Environment Research Federation, 2004. *Fate and Persistence of Pathogens Subjected to Ultraviolet Light and Chlorine Disinfection.* Report Number 98-HHE-2.

Novak, J. et. al. 2003. An ultrastructrual comparison of spores from various strains of *Clostridium perfringens* and correlations with heat resistance parameters. International Journal of Food Microbiology 86 (2003) 239-247

Rose, J, et. al. Water Environment Research Federation, 2004. *Reduction of Pathogens, Indicator Bacteria, and Alternative Indicators by Wastewater Treatment and Reclamation Processes.* Report Number 00-PUM-2T.

Safrit, D. The State of Reuse in North Carolina. Proceedings from NC AWWA/WEA 90th Annual Conference, Winston-Salem, North Carolina, November 15-17, 2010.

North Carolina Administrative Code 15A NCAC 2U. Published June 18 2011.

Spot-check Bioassay of a Full-Scale Gravity-Flow Open-Channel Vertical UV System At the Russian River Treatment Plant in Sonoma County, California

B. A. Ferran[1], Y. Borroum[2]

1. Ozonia North America, LLC.
2. Stantec Consulting Services, Inc.

ABSTRACT

The focus of this paper is to detail the findings of a "spot-check" bioassay that was conducted on a full-scale, gravity-flow, open-channel, vertical UV system at the Russian River Treatment Plant in Sonoma County, California. Field testing was performed by Stantec Consulting Services, Inc. in April 2012.

The new UV disinfection system is installed downstream of the tertiary filters and consists of two parallel reactor trains (UV channels). Each train contains eight banks (UV modules), each of which are equipped with forty (40) low-pressure high-output germicidal lamps.

Under the assumption that UV channel 2 would perform equivalently, the "spot-check" bioassay was limited to UV channel 1. The UV channel was tested at the nominal minimum flow rate of 0.25 mgd, the average flow per train of 0.7 mgd, and the design peak flow rate per train of 1.85 mgd.

In all eight tests the observed full-scale REDs are greater than the doses predicted using the model derived from the 2002 "off-site" bioassay. Placement of the UV modules did not affect performance of the UV disinfection system. De-rating is not needed for the UV system installed at the Russian River Treatment Plant.

Correcting the dose values for both reactor validation and full-scale "spot-check" bioassay per the new standard MS2 dose-response equation recommended in the third edition of the NWRI/WRF "Ultraviolet Disinfection Guidelines for Drinking Water and Water Reuse" did not impact the outcome of this "spot-check" bioassay.

KEY WORDS

Bioassay, Reactor Validation, Spot-check, Dose-response, Conformance

INTRODUCTION

Ultraviolet (UV) disinfection of wastewater effluents has become common practice in the United States over the past two decades in part due to increasing environmental awareness with a focus on protecting natural watersheds from microbial contamination. UV disinfection for wastewater reuse applications is a growing market as many communities seek alternative water supplies to help address sustained population growth and frequent droughts.

Due to an initial lack of regulatory requirements and a reproducible test method to assess the disinfection performance of UV reactors, the first and second editions of the NWRI/WRF "Ultraviolet Disinfection Guidelines for Drinking Water and Water Reuse" (thereafter referred to as the UV Guidelines) recommended reactor validation testing based on hydraulic testing (velocity profile testing) and biodosimetry. The first and second editions of the UV Guidelines also recommended that inlet and outlet velocity profile be maintained between the validated UV equipment and full-scale installation. The third edition of the UV Guidelines, released in August 2012, recommends that a UV reactor's "off-site" performance validation bioassay be verified at full scale at the reuse facility via a "spot-check" bioassay. This edition also recommends using a standard MS-2 dose-response equation for both reactor validation and "spot-check" bioassays.

The focus of this paper is to detail the findings of a "spot-check" bioassay that was conducted on a full-scale UV system at the Russian River Treatment Plant in Sonoma County, California. Construction of the UV system was completed in February 2012. Field testing was performed by Stantec Consulting Services in April 2012. The impact of the new standard MS-2 dose-response relationship from the UV Guidelines on the outcome of the "spot-check" bioassay is also evaluated.

RUSSIAN RIVER TREATMENT PLANT

The Russian River Treatment Plant is owned and operated by the Sonoma County Sanitation District. Location maps of the Russian River Treatment Plant are provided in Figures 1 and 2.

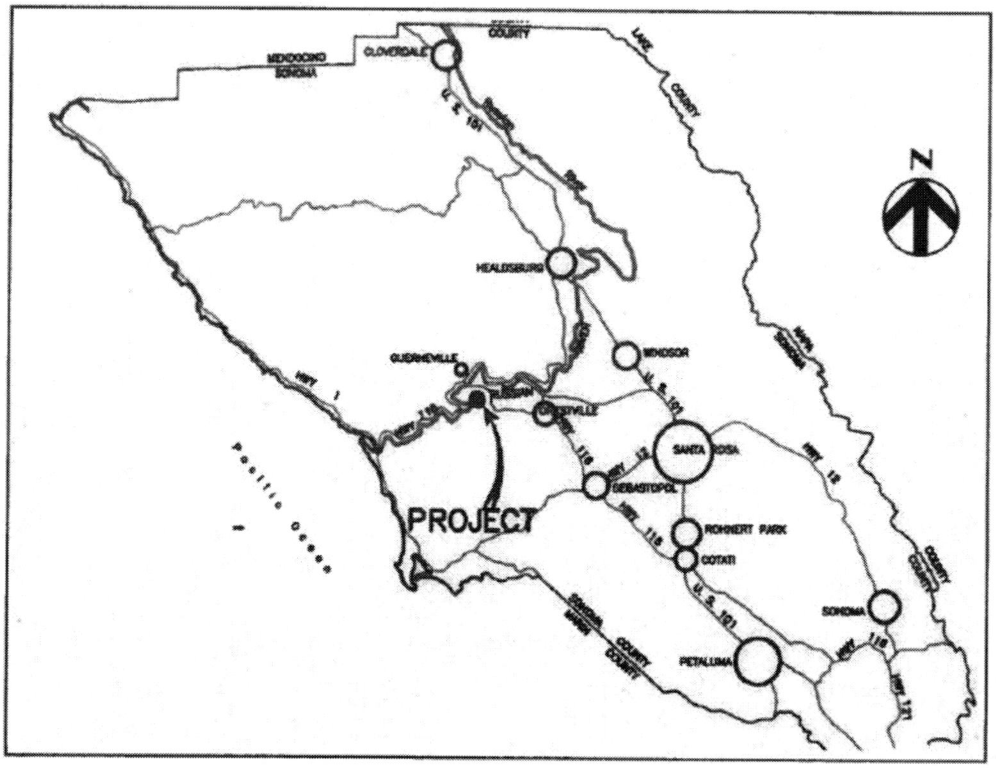

Figure 1 – Russian River Treatment Plant Vicinity Map

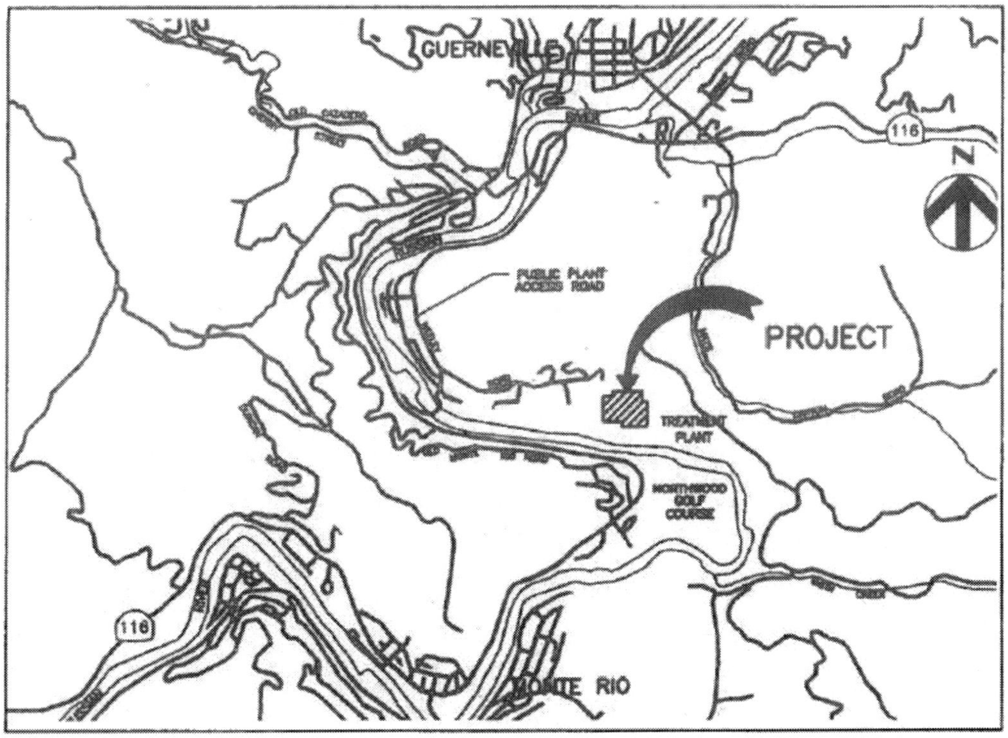

Figure 2 – Russian River Treatment Plant Location Map

The treatment facility is located near Guerneville, California and provides sewer services for approximately 3,000 residences and businesses in the area. The treatment facility currently treats an average dry weather flow of 0.71 mgd and a peak wet weather flow of 3.5 mgd.

Wastewater treatment at the Russian River Treatment Plant is accomplished by coarse screening, aerated grit removal, three (3) extended aeration activated sludge basins, three (3) clarifiers and two (2) tertiary filters. The third aeration basin is currently used as an additional storage basin for influent flow during high flow events. The new UV system is designed for a peak flow capacity of 3.7 mgd.

Tertiary treated, disinfected wastewater is held in a 3.5 million gallon effluent storage pond prior to discharge into the Russian river (October 1 – May 14) or the recycled water disposal system. Tertiary treated wastewater is supplied to the Northwood Golf course, located south of the treatment facility, where an average of 0.085 mgd is applied to an area of 43 acres during the irrigation season. Treated wastewater not used by the Northwood Golf Course is spray irrigated on 17 acres of wooded property adjacent to the treatment facility (the Burch property). During irrigation season (May 15 to September 30) approximately 0.02 mgd and 0.23 mgd, respectively, are currently applied to the "upper" and "lower" areas of the Burch property.
The new UV systems installed at the Russian River Treatment Plant is pictured in Figure 3. The UV system is open-channel, gravity-flow and modular consisting of two parallel channels, each containing eight UV modules.

Figure 3 – Picture of the New UV System at the Russian River Treatment Plant

A UV module consists of a rectangular stainless steel frame equipped with 40 non-dimmable low-pressure high-output mercury vapor discharge lamps. Each lamp yields a minimum of 52.0 watts UVC. As depicted in Figure 4 the lamps of a UV module are oriented vertically in a staggered grid of 5 rows by 8 lamps. A triangular-shaped side deflector is present for each row of lamps and is located on the side of the module where the lamp-to-wall spacing is greatest.

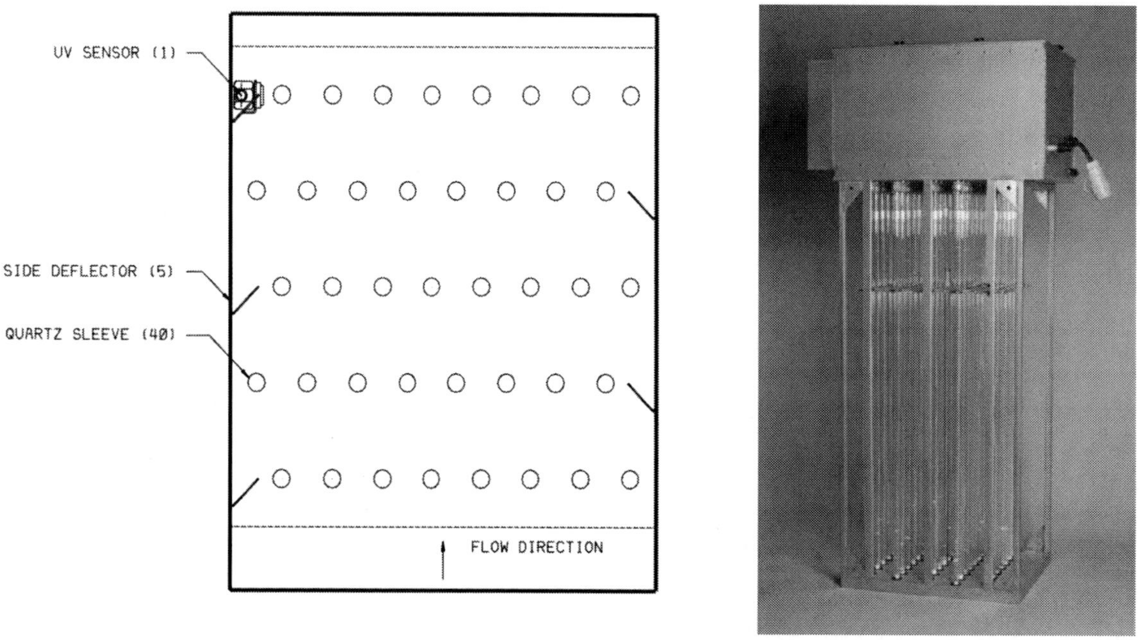

Figure 4 – UV Module for the Russian River UV System

Based on the performance results collected from reactor validation testing the Russian River UV disinfection system delivers a dose that exceeds the minimum of 100 mJ/cm^2 outlined in the UV Guidelines for media filtered secondary effluent wastewater. This performance is achieved with one excess UV module in each channel for redundancy.

REACTOR VALIDATION FOR THE RUSSIAN RIVER UV SYSTEM

Evaluation and validation of the Russian River UV system module was conducted in 2002 in accordance with the *Environmental Technology Verification (ETV) Protocol for Secondary Effluent and Water Reuse Disinfection Applications* (NSF, 2002). The development of this protocol was based on the first edition of the UV Guidelines (December 2000). HydroQual, Inc. was utilized as the third party engineering consultant responsible for all field testing and preparation of the final verification report.

Due to the amount of wastewater and extensive logistics that would have been required at the time of testing to bioassay full-scale 40-lamp modules, a decision was made to bioassay one-fourth-scale UV modules, consisting of 10-lamps. Figure 5 depicts the layout of the pilot-scale UV module with the lamps positioned in a staggered array consisting of 5 rows x 2 lamps.

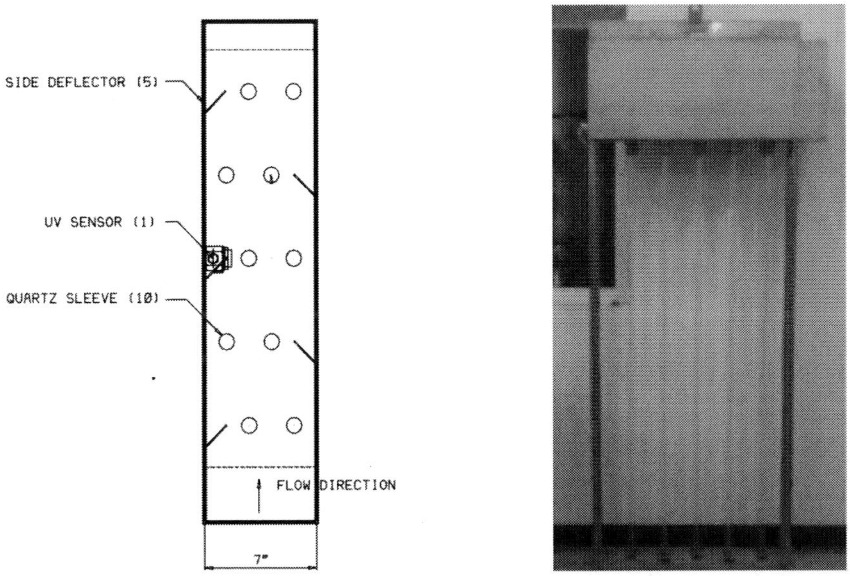

Figure 5 – Lamp Arrangement for the 10-lamp Test Module.

Ideally, the width of the 10-lamp module should have been one-fourth that of the full-scale 40-lamp module. However, the test modules fabricated had a width of 7.0 inches as compared to the 40-lamp module with a width of 24.0 inches, giving a ratio of 3.43. This discrepancy originates to satisfy engineering needs to avoid impractical wall to lamp spacing for the 10-lamp modules. Therefore 0.5 inch was added to both sides of the test modules, resulting in a final width of 7.0 inches. With the objective to keep the cross sectional velocity identical between full-scale and pilot-scale modules the width of the side deflectors (W_B) was reduced to satisfy the equation below.

$$\frac{4.Q_{10}}{(W_{CH40} - W_B - 8.d).H} = \frac{Q_{10}}{(W_{CH10} - W_B - 2.d).H}$$

Where:

- Q_{10} = Pilot-scale module test flow
- W_{CH} = Width of module
- W_B = Width of side deflector
- d = Outside diameter of sleeve
- H = Height of water column

Due to the larger width of 7.0 inch the average fluence rate of the pilot-scale test module remained 16% to 20% lower than that of the 40-lamp full-scale module as illustrated in Figure 6. This discrepancy between pilot-scale and full-scale modules provided for some inherent conservatism with the reactor validation.

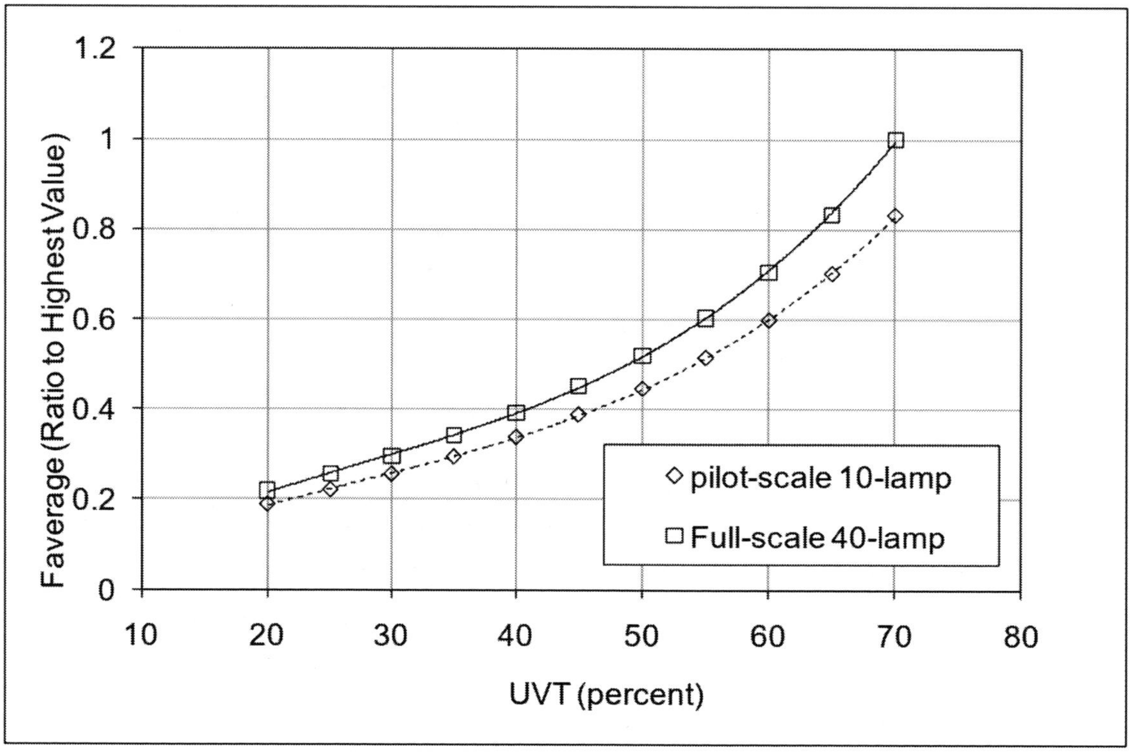

Figure 6 – Average Fluence Rate for Pilot-Scale and Full-Scale UV Modules.

As depicted in Figure 7 three 10-lamp test modules were installed in series inside a stainless steel channel under conditions identical to commercial applications with respect to lamp array geometry and lamp submergence. The challenge water was pumped to the inlet of the channel where a spreader baffle was used to provide even flow over the submerged cross sectional area of the channel. The liquid level in the channel was maintained at 60 inches using an adjustable-height weir located downstream of the three test modules.

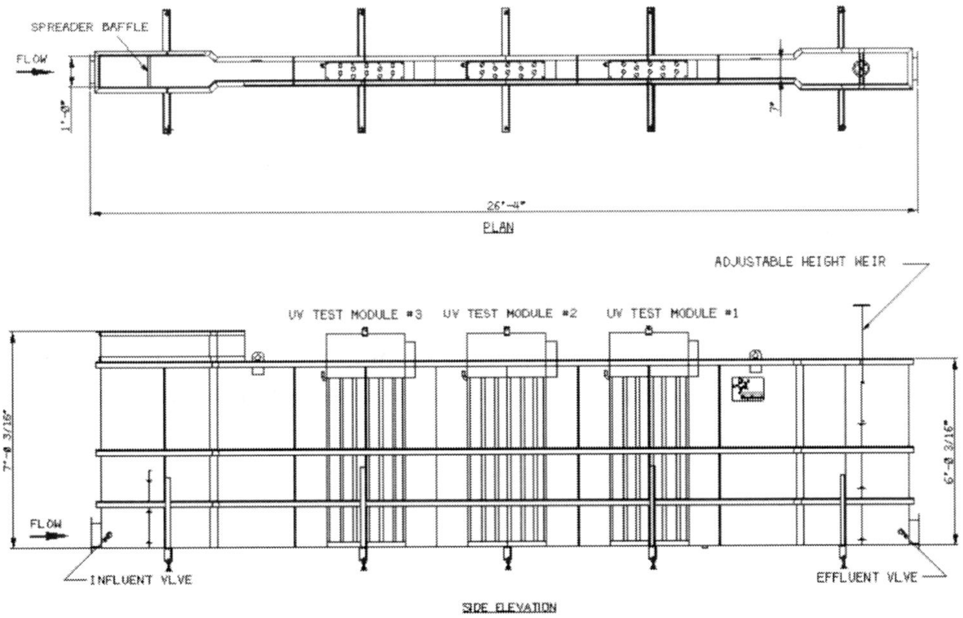

Figure 7 – Channel for Three 10-lamp Test Modules

The test site for this reactor validation program was the Parsippany-Troy Hills (PTRH) wastewater treatment plant in New Jersey. This plant was built to process 16.0 mgd of sewage with secondary treatment, nitrification and granular media filtration. The three 10-lamp test modules were fed with challenge water prepared in a batch tank that was pumped to the influent side of the channel. Before dose delivery verification began, the lamps were aged for 100 hours to allow the lamp intensity to stabilize.

The first water source used to simulate membrane-filtered wastewater was from the potable water supply, tapped off a local hydrant at the wastewater treatment plant. This was first dechlorinated with sodium thiosulfate in a batch tank and warmed to 15.0 °C using a 30 kW heater. The second water source was taken from the effluent side of the plant granular media filtering process. With both water sources, soluble coffee was added to adjust the UV transmittance of the water to prescribed levels. And, in both cases, the batches of water were spiked with a stock culture of MS-2 bacteriophage and thoroughly mixed. The MS-2 bacteriophage was ATCC 15597-B1 and the host *E. coli* was ATCC 23631. Assay and enumeration of the MS-2 bacteriophage were in accordance with ISO 10705.

Since the UV lamps from the Russian River UV system are not dimmable further lowering of the transmittance to reduce the fluence rate field in the module was used during reactor validation to simulate the specific test conditions. For the 65% nominal transmittance condition, which simulates membrane-filtered secondary wastewater effluent per the UV Guidelines the actual operating transmittance during testing was 56%. For the 55% nominal transmittance, which simulates media filtered secondary effluent wastewater per the UV Guidelines the actual operating transmittance was 46%. This approach of simulating a loss of lamp output power with a reduced UV transmittance is known to be conservative as it creates local areas of depressed fluence rates inside the UV reactor that would not be as pronounced had the lamp output been reduced via dimming.

The MS-2 dose-response relationship used for the dose flow assays was based on 13 valid collimated beam sequences conducted on challenge water adjusted to the transmittances used for testing. The MS-2 dose-response calibration curve is presented in Figure 8, with the dose as a function of log inactivation. This allows the computation of a calibration curve for the MS-2 bacteriophage stock, by fitting a second-order polynomial to the data points as follows.

$$\text{Dose (mJ/cm}^2\text{)} = 1.4822*(\text{-Log N/N0})^2 - 15.063*(\text{-Log N/N0}) - 0.1633$$

Figure 8 shows that 97.8% of the data points lie within the boundary limits referenced in the Verification Protocol (NSF 2002). This is well above the minimum requirement of 80%.

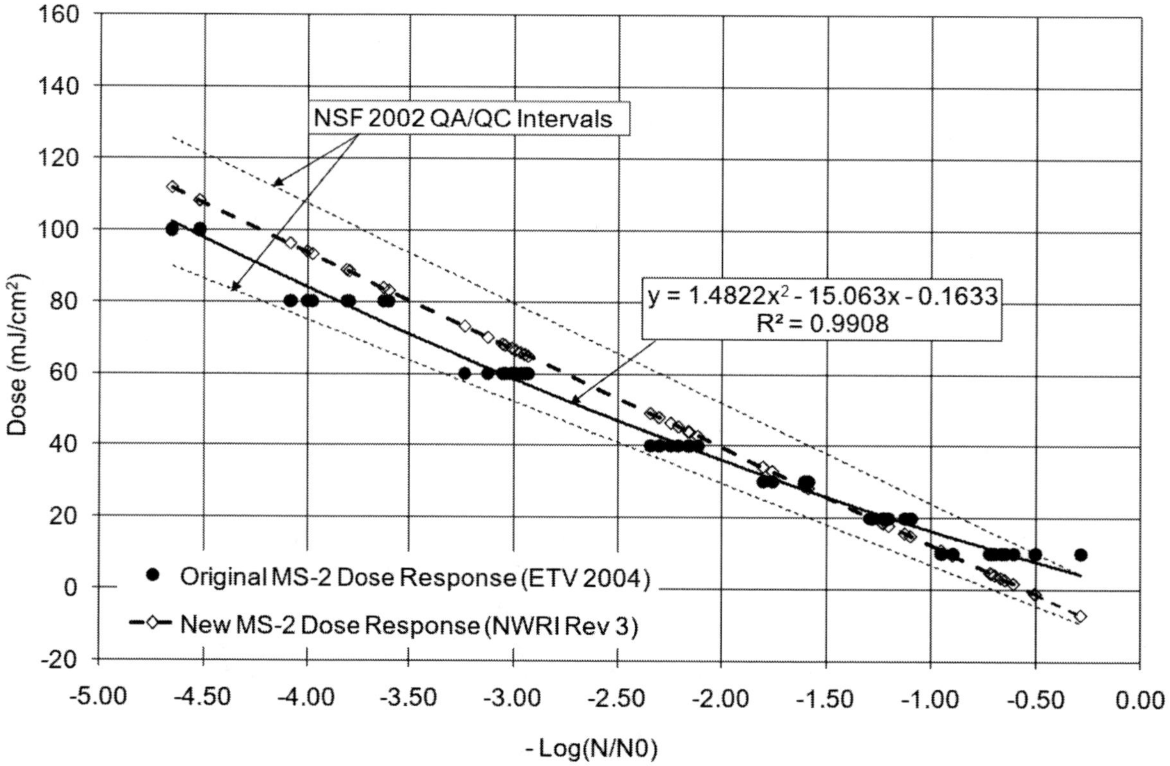

Figure 8 –MS2 Dose-Response Curve for Reactor Validation

Figure 9 displays the dose per module as a function of the full-scale flow per train for both the membrane-filtered effluent (65% nominal transmittance) and the granular media-filtered effluent

(55% nominal transmittance). Full-scale flow per train refers to four (4) times the flow of the 10-lamp test module. A train is defined as a row of single UV modules in series. A channel having "x" modules across the width comprises "x" parallel trains.

The dose values from Figure 9 are the lower 75% C.I. of the average values point by point based on a two-tailed student t distribution (n-1). The dose data was best correlated with two separate second-order polynomials, one for each nominal transmittance.

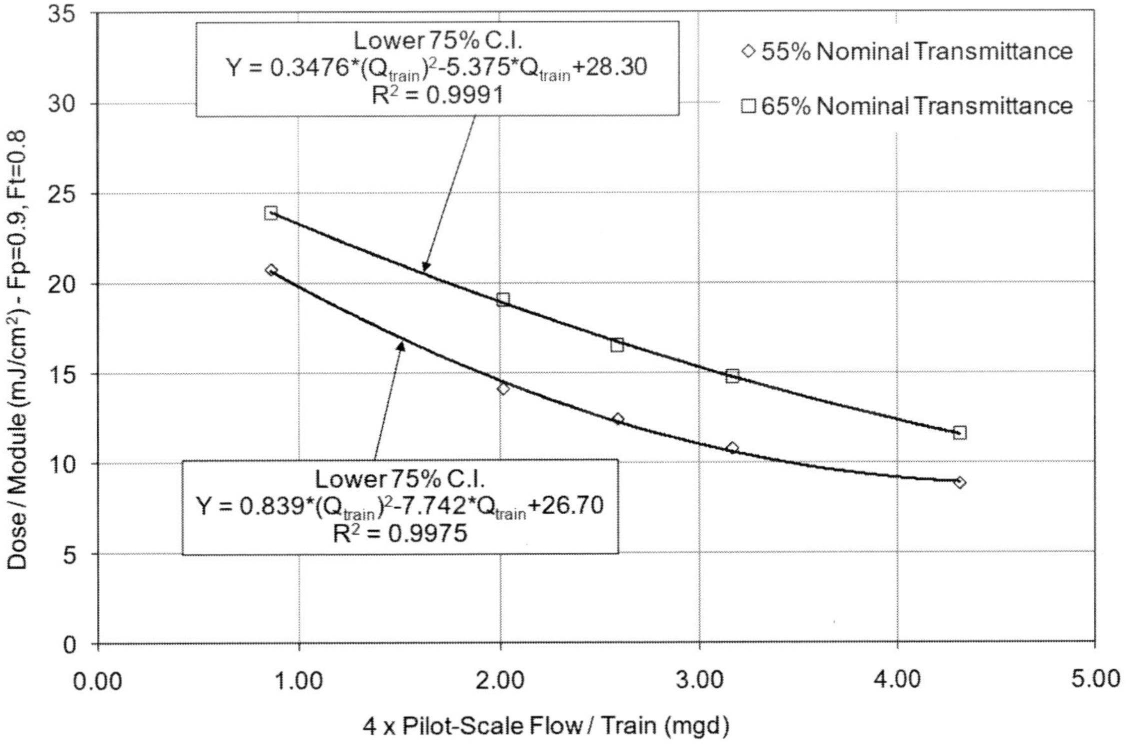

Figure 9 – Original Dose Delivery per Module (NWRI Revision 1&2)

Based on the third edition of the UV Guidelines the following new standard MS-2 dose-response relationship is to be used for both reactor validation and "spot-check" bioassay to convert the log inactivation rates to Reduced Equivalent Dose (RED) values.

$$Dose(mJ/cm^2) = \frac{-Log(N/N_0) - 0.5464}{0.0368} = -27.1739 \cdot \left(-LogN/N_0\right) - 14.8478$$

The dose-response curve that results from the equation above is included in Figure 8 and is labeled "New MS-2 Dose Response (NWRI Rev 3)". As can be seen from Figure 8 the new standard dose-response relationship is quite different from that of the various MS-2 batches that were actually used for reactor validation. For a log inactivation rate smaller than -1.50 the new standard dose-response curve corresponds to MS-2 bacteriophage that are more UV resistant. For a log inactivation rate greater than -1.50 the new standard dose response curve corresponds to MS-2 bacteriophage that are more UV sensitive.

Figure 10 compares the reactor validation original dose-flow curves from Figure 9 with those obtained based on the new standard MS-2 dose response equation. The dose values displayed in Figure 10 are based on the same lower 75% C.I. analysis of the average dose value point by point. A second-order polynomial best fit is also used to correlate the dose-flow data for each nominal transmittance. At 65% nominal transmittance (membrane-filtered water) the dose values obtained from the new standard MS-2 dose-response equation are 9.4% to 14.2% higher throughout the entire test flow range.

At 55% nominal transmittance (granular media-filtered water) the doses obtained from the new standard MS-2 dose-response relationship are up to 12.8% higher for test flows between 20% and 70% of full capacity and up to 12.8% lower at higher flows.

With higher dose values from reactor validation the commissioning "spot-check" bioassay becomes more challenging since in principle higher dose values need to be verified.

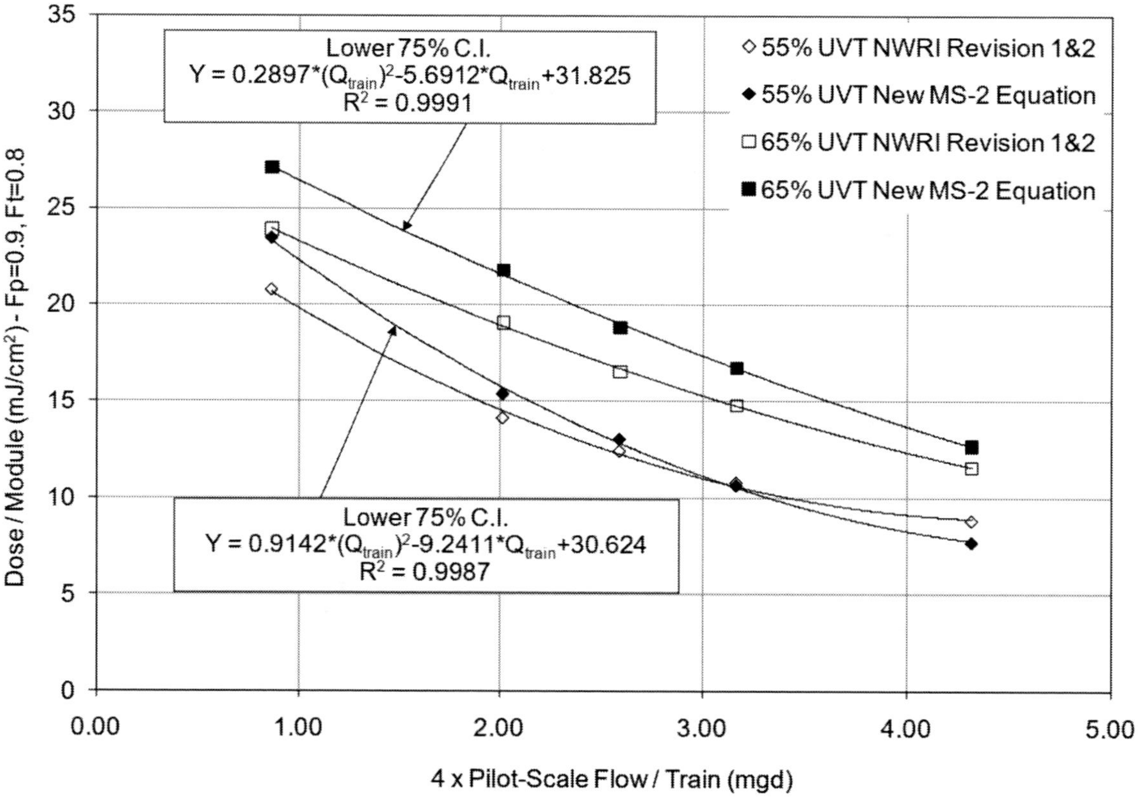

Figure 10 – Dose Delivery per Module with New MS-2 Dose-Response Equation

FULL-SCALE COMMISSIONING "SPOT-CHECK" BIOASSAY

The full-scale commissioning "spot-check" bioassay was intended only to verify that the Russian River new UV system applies a UV operational dose of 100 mJ/cm^2 per channel during peak, average, and minimum flow design conditions.

The validation procedure was for the full-scale UV modules in place, treating Russian River wastewater. The "spot-check" bioassay was limited to UV Channel 1 under the assumption that UV channel 2 would perform equivalently to UV Channel 1. The UV channel was tested at the nominal minimum flow rate of 0.25 mgd, the average flow per train of 0.7 mgd, and the design peak flow rate per train of 1.85 mgd. The inlet samples were collected at a depth in the middle of the effluent stream two feet upstream of the first module. All outlet samples were collected at the outlet structure to assure adequate mixing. Three replicate samples were collected at all sampling locations. Samples were immediately placed on ice upon collection. At the completion of the testing, samples were immediately shipped via express mail to Biovir Laboratories, Inc. in Benicia, California for analysis. Water used to conduct bioassays was free of chlorine as verified by conducting total and free chlorine sampling during each bioassay test.

The bioassay measurements were taken with a minimum lamp age of 100 hours. The UV channel was tested under the flow, transmittance, and power setting conditions highlighted in Table 1.

Table 1 – "Spot-check" Bioassay Test Conditions

Test #	Operating Modules[(a)]	Flow (mgd)	UVT (%)	Power Setting
1	1 and 2	0.68	55	100%
2	4 and 5	0.69	55	100%
3	7 and 8	0.67	56	100%
4	1 and 8	0.25	55	100%
5	1, 2 and 3	1.74	56	100%
6	3, 4 and 5	1.82	57	100%
7	5, 6 and 7	1.80	57	100%
8	1, 7 and 8	1.85	59	100%

MS-2 bacteriophage was used as the test organism, with assay and enumeration in accordance with standard protocols (US EPA 2000). The strain *E. coli* ATCC 15597 was substituted for *E. coli* (HS[pFamp]R) as the host bacterium for the enumeration of MS2 bacteriophage. All viral data including collimated beam testing was provided by Biovir Laboratories, Inc.

The MS2 stock was obtained from GAP EnviroMicrobial Services in London, Ontario, Canada. The inlet MS2 concentration was adjusted to be in the 5×10^8 plaque forming unit (PFU)/mL range. Total of 1.0 liters of 1×10^{11} concentrations virus was required to complete the tests.

Complete mixing of the MS-2 bacteriophage and UV absorbing solution was accomplished by injecting them on-line at a location upstream from the UV disinfection facility. The additives were injected at the filter outlet channel prior to the filter effluent weir. Verification of adequate mixing was established by comparing water transmittance at the inlet of the UV disinfection system to that measured at the outlet.

One collimated beam dose response curve was developed for the 24-hour of testing per the UV Guidelines. As depicted in Figure 11 the MS-2 dose-response curve obtained falls within the acceptable region described in the UV Guidelines. The data points can be fit with a straight linear function as follows.

$$Dose(mJ/cm^2) = \frac{-Log(N/N_0) - 0.2483}{0.0403} = -24.8139 \cdot (-LogN/N_0) - 6.1613$$

The new standard MS-2 dose-response curve proposed in the third edition of the UV Guidelines is also included in Figure 11 and is labeled "New MS-2 Dose Response (NWRI Rev 3)". By comparison to Figure 8 which pertains to the reactor validation actual and new MS-2 dose response curves are closer for the "spot-check" bioassay. The MS-2 bacteriophage stock used during the Russian River "spot-check" bioassay is more UV resistant than the various stocks used for reactor validation.

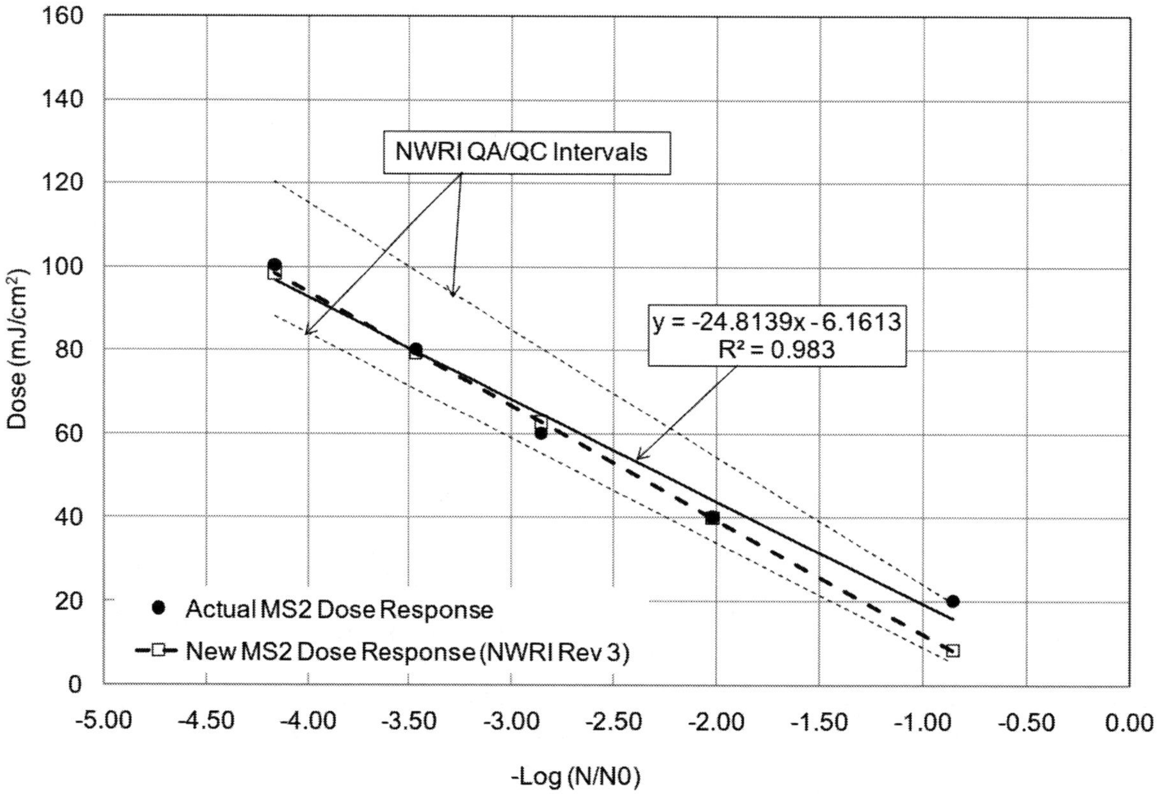

Figure 11- MS-2 Dose Response Curve for the Russian River "Spot-check" Bioassay

The viral data collected during the "spot-check" bioassay was analyzed in the following manner

1. Virus concentration (PFU/mL) for inlet and outlet samples reported by the laboratory were converted to logarithmic concentrations.
2. Virus inactivation (log based) for each pair of inlet and corresponding outlet conditions were calculated
3. Average virus inactivation (log based) achieved during each of the eight bioassay tests were calculated. The average was based on three pairs of inlet and outlet data collected for each test.
4. The log-inactivation results were converted to RED values via use of the actual MS-2 dose response curve highlighted in Figure 11.

The viral inactivation results obtained from the "spot-check" bioassay conducted at the Russian River Treatment Plant are enclosed in Table 2.

Table 2 – "Spot-scheck" Bioassay Results

Test #	Operating Modules[(a)]	Flow (mgd)	UVT (%)	Concentration (Phage/mL)		Average Log (concentration)		-Log (N/N_0)	Avg -Log (N/N_0)	RED (mJ/cm^2)
				Inlet	Outlet	Inlet	Outlet			
1	1 and 2	0.68	55	5.80E+04	2.00E+00	4.763	0.301	4.462	4.645	109.1
				8.60E+04	2.00E+00	4.934	0.301	4.633		
				6.90E+04	1.00E+00	4.839	0.000	4.839		
2	4 and 5	0.69	55	6.30E+04	1.00E+00	4.799	0.000	4.799	4.860	114.4
				7.00E+04	1.00E+00	4.845	0.000	4.845		
				8.60E+04	1.00E+00	4.934	0.000	4.934		
3	7 and 8	0.67	56	9.00E+04	1.00E+00	4.954	0.000	4.954	5.041	118.9
				9.20E+04	1.00E+00	4.964	0.000	4.964		
				1.60E+05	1.00E+00	5.204	0.000	5.204		
4	1 and 8	0.25	55	3.00E+05	1.00E+00	5.477	0.000	5.477	5.500	130.3
				3.30E+05	1.00E+00	5.519	0.000	5.519		
				3.20E+05	1.00E+00	5.505	0.000	5.505		
5	1, 2 and 3	1.74	56	1.30E+05	1.20E+01	5.114	1.079	4.035	3.818	88.6
				1.60E+05	4.70E+01	5.204	1.672	3.532		
				1.70E+05	2.20E+01	5.230	1.342	3.888		
6	3, 4 and 5	1.82	57	1.80E+05	1.40E+01	5.255	1.146	4.109	4.437	103.9
				1.40E+05	5.00E+00	5.146	0.699	4.447		
				1.70E+05	3.00E+00	5.230	0.477	4.753		
7	5, 6 and 7	1.8	57	1.70E+05	1.20E+01	5.230	1.079	4.151	4.086	95.2
				1.90E+05	1.70E+01	5.279	1.230	4.048		
				1.60E+05	1.40E+01	5.204	1.146	4.058		
8	1, 7 and 8	1.85	59	1.60E+05	1.50E+01	5.204	1.176	4.028	4.002	93.1
				1.80E+05	1.90E+01	5.255	1.279	3.977		
				1.50E+05	1.50E+01	5.176	1.176	4.000		

The REDs from Table 2 were compared to the doses predicted from reactor validation. The predicted doses are based on the dose-flow relationship from Figure 9 for a nominal transmittance of 55%. The resulting predicted dose values are corrected to reflect of actual test conditions during the "spot-check" bioassay in accordance with the equation below.

$$Dose(mJ/cm^2) = N_B \cdot \left[(0.839 \cdot Q^2 - 7.742 \cdot Q + 26.70) \cdot TF \cdot IF \cdot \frac{F_P}{0.9} \cdot \frac{F_t}{0.8} \right]$$

Where,

- N_B: Number of operating UV modules tested (2 or 3)
- Q: flow per train/channel in mgd
- TF: A temperature adjustment factor calculated for an actual wastewater temperature (T) as follows

$$TF = \frac{-0.0017 \cdot T^2 + 0.0776 \cdot T + 0.553}{1.438552}$$

- IF: A fluence rate adjustment factor calculated for an actual wastewater transmittance (UVT) as follows

$$IF = \frac{0.00003338 \cdot UVT^3 - 0.002668 \cdot T^2 + 0.137 \cdot UVT - 0.1832}{4.834698}$$

- F_p: lamp end-of-life factor that is equal to 1.0 for new lamps as tested during "spot-check" bioassay. During reactor validation F_p was 0.9 for lamp at end of life.
- F_t: Sleeve fouling factor that is equal to 1.0 for clean sleeves as tested during "spot-check" bioassay. During reactor validation F_t was 0.8 for fouled sleeves.

Predicted doses and REDs obtained for each of the eight tests conducted during "spot-check" bioassay are presented in Table 3.

Table 3 – "Spot-scheck" Bioassay REDs vs. Predicted Doses (UV Guidelines Revision 1&2)

Test #	Operating Modules[(a)]	Flow (mgd)	UVT (%)	T (°C)	TF	IF	Pred. Dose (mJ/cm^2)	RED (mJ/cm^2)	RED/Pred. Dose (Must be >1.00)
1	1 and 2	0.68	55	17.1	0.96	1.00	58.3	109.1	1.87
2	4 and 5	0.69	55	17.1	0.96	1.00	58.1	114.4	1.97
3	7 and 8	0.67	56	17.1	0.96	1.03	60.3	118.9	1.97
4	1 and 8	0.25	55	17.1	0.96	1.00	66.3	130.3	1.97
5	1, 2 and 3	1.74	56	17.1	0.96	1.03	65.1	88.6	1.36
6	3, 4 and 5	1.82	57	17.1	0.96	1.06	65.5	103.9	1.59
7	5, 6 and 7	1.80	57	16.9	0.96	1.06	65.7	95.2	1.45
8	1, 7 and 8	1.85	59	16.5	0.95	1.13	68.5	93.1	1.36

(a) The module numbering starts from the outlet weir end of the channel

For all eight tests the "spot-check" bioassay REDs are significantly higher than corresponding predicted doses. Placement of the UV modules inside the channel did not affect performance of the UV disinfection system.

Adjusting the content of Table 3 based on the new standard MS2 dose-response relationship from the UV Guidelines results in Table 4. The predicted doses are based on the dose-flow relationship from Figure 10 for a nominal transmittance of 55%. The resulting predicted dose values are corrected to reflect of actual test conditions during the Russian River "spot-check" bioassay in accordance with the equation below.

$$Dose(mJ/cm^2) = N_B \cdot \left[(0.9142 \cdot Q^2 - 9.2411 \cdot Q + 30.624) \cdot TF \cdot IF \cdot \frac{F_P}{0.9} \cdot \frac{F_t}{0.8}\right]$$

Where,

N_B: Number of operating UV modules tested (2 or 3)
Q: flow per train/channel in mgd
TF, IF, F_p and F_t as previously detailed

The REDs contained in Table 4 are obtained from the same average log inactivation rates listed in Table 2 however using the new standard MS2 dose-response equation from the UV Guidelines as opposed to the actual dose-response relationship from Figure 11.

Table 4 - "Spot-scheck" Bioassay REDs vs. Predicted Doses
(New MS-2 Dose Response Equation from UV Guidelines)

Test #	Operating Modules[(a)]	Flow (mgd)	UVT (%)	T (°C)	TF	IF	Pred. Dose (mJ/cm^2)	RED (mJ/cm^2)	RED/Pred. Dose (Must be >1.00)
1	1 and 2	0.68	55	17.1	0.96	1.00	66.1	111.4	1.68
2	4 and 5	0.69	55	17.1	0.96	1.00	65.9	117.2	1.78
3	7 and 8	0.67	56	17.1	0.96	1.03	68.4	122.1	1.79
4	1 and 8	0.25	55	17.1	0.96	1.00	75.8	134.6	1.78
5	1, 2 and 3	1.74	56	17.1	0.96	1.03	71.5	88.9	1.24
6	3, 4 and 5	1.82	57	17.1	0.96	1.06	71.7	105.7	1.47
7	5, 6 and 7	1.80	57	16.9	0.96	1.06	72.0	96.2	1.34
8	1, 7 and 8	1.85	59	16.5	0.95	1.13	74.8	93.9	1.26

(a) The module numbering starts from the outlet weir end of the channel

All the predicted doses and REDs from Table 4 increased compared to Table 2. However, the rate of increase is much more significant for the predicted doses. This finding makes sense based on the fact that actual and new standard MS-2 dose response curves are closer for the "spot-check" bioassay (Figure 11) compared to the reactor validation (Figure 8).

Once again, the "spot-check" bioassay REDs from all eight tests are significantly higher than the corresponding predicted doses. However, the safety gap between the two dose values is consistently smaller as a result of implementing the new standard MS2 dose response relationship from the UV Guidelines.

CONCLUSION

For all eight tests of the "spot-check" bioassay conducted at the Russian River Treatment Plant the observed REDs were greater than predicted based on the 2002 reactor validation. Therefore, the performance equation that derives from the 2002 validation test is considered appropriate for use with the Russian River UV system. De-rating is not needed for this UV system.

Implementation of the new standard MS-2 dose-response curve from the third edition of the UV Guidelines for both reactor validation and "spot-check" bioassay narrowed the safety gaps between predicted and actual full-scale dose values. However, the outcome of the "spot-check" bioassay conducted at the Russian River Treatment Plant as meeting overall conformance with the "spot-check" bioassay concept and UV Guidelines remains unchanged.

REFERENCES

International Standards Organization (ISO). 1995, Water Quality-Detection and Enumeration of Bacteriophage, Part I: Enumeration of F-Specific RNA Bacteriophage." Switzerland: International Standards Organization, ISO 10705-1.

National Water Research Institute (NWRI) and American Waste Water Association Research Foundation (AWWARF), 1st edition, 2000. *Ultraviolet Disinfection Guidelines for Drinking Water and Water Reuse.* (Fountain Valley, California).

NSF (2002) *Verification Protocol for Secondary Effluent and Water Reuse Disinfection Applications.* Prepared by HydroQual, Inc. for NSF International and the U.S. Environmental Protection Agency under the Environmental Technology Verification Program.

National Water Research Institute (NWRI) and American Waste Water Association Research Foundation (AWWARF), 2nd edition, 2003. *Ultraviolet Disinfection Guidelines for Drinking Water and Water Reuse.* (Fountain Valley, California).

NSFI/US EPA Environmental Technology Verification Report (2004). *Ondeo Degremont, Inc. Aquaray® 40 HO VLS Disinfection System.*

Stantec Consulting Services, Inc. (June 2012). *Russian River UV Disinfection System. Field Commissioning Test Results.*

National Water Research Institute (NWRI) and Water Research Foundation (WRF), 3rd edition, 2012. *Ultraviolet Disinfection Guidelines for Drinking Water and Water Reuse.* (Fountain Valley, California).

Oxidation and Disinfection Processes as Barriers to the Dissemination of Bacterial Antibiotic Resistance Genes via Water and Wastewater

Peiran Zhou, Shu Li, Michael C. Dodd*

University of Washington, Department of Civil and Environmental Engineering
*Author to whom correspondence should be addressed, at: doddm@uw.edu

ABSTRACT

Disinfection processes in drinking water and wastewater treatment facilities may be an important barrier for combating antibiotic resistance from antibiotic resistant bacteria (ARB) and extracellular/intracellular antibiotic resistant genes (ARGs). However, conventional water and wastewater treatment plants are designed to minimize the dissemination of ARB, not ARGs. The intact extracellular and intracellular ARGs that survived disinfection could still propagate antibiotic resistance by means of natural transformation and/or transduction - horizontal gene transfer processes that do not require live donor cells. Thus far, little is known about the effects of common disinfectants on ARGs. This study systematically evaluated the likelihood of intact, biologically-active extracellular/intracellular ARGs surviving disinfection processes in buffered water matrixes. Our data suggests that ARGs can indeed survive disinfection processes intact.

KEYWORDS

Disinfection, oxidation, antibiotic resistance genes, DNA, transformation, deactivation

INTRODUCTION

Antibiotic resistance in bacterial communities is becoming a severe challenge to human health. Considering that the discovery of new drugs is slower than the development of antibiotic resistance, it is possible we could ultimately regress to a time in which our ability to treat bacterial infections is severely diminished (Carlet, Jarlier et al. 2012). In addition, recent findings suggest that the assumption that resistance genes will be lost from ARB when antibiotic pressure is removed to reduce cell fitness cost might not be valid (Domingues, Harms et al. 2012). Therefore, the deactivation of ARB and ARGs during disinfection processes in water and wastewater treatment becomes an important barrier for limiting the dissemination of antibiotic resistance. However, current disinfection processes target the inactivation of living cells, not DNA, and to the best of the authors' knowledge, no systematic studies were done to evaluate the effects of common disinfectants on DNA. Therefore, the DNA containing ARGs may survive intact from the disinfection processes and propagate the antibiotic resistance by means of natural transformation and/or transduction (horizontal gene transfer mechanisms that do not require live donor cells).

This research aims to systematically study the likelihood of intact, biologically-active extracellular/intracellular ARGs surviving the treatment by free available chlorine (FAC, where FAC = $HOCl + OCl^-$), monochloramine (NH_2Cl), ozone (O_3), chlorine dioxide (ClO_2), or ultraviolet (UV) light, with specific emphasis on measuring the changes in (i) ARG copy number (measured by qPCR) and (ii) ARG transforming activity (measured as the ARGs' ability to transform non-resistant bacterial strains to antibiotic resistant genotypes).

METHODOLOGY

Chemical Reagents and Materials. FAC was dosed to solutions as aqueous sodium hypochlorite (NaOCl). NH_2Cl was dosed as a pre-formed aqueous stock solution, where NH_2Cl stocks were prepared by mixing NH_3 with FAC at a 2:1 molar ratio at pH 8.5. O_3 was dosed into solutions from a pre-dissolved stock of aqueous O_3, prepared by bubbling gaseous O_3 through a chilled flask of deionized water. ClO_2 was synthesized through the oxidation of chlorite by potassium peroxodisulfate. All other chemical reagents and growth media were purchased from commercial suppliers, certified-DNase free, and of at least reagent-grade purity. Monochromatic UV light (λ = 254 nm) was generated using a 15-W, low-pressure Hg lamp contained within a "merry-go-round" photoreactor apparatus. Residual oxidant concentrations were monitored colorimetrically using DPD for FAC and NH_2Cl, indigo for O_3, or ABTS for ClO_2. UV fluence was quantified using potassium iodide as a chemical actinometer.

Bacterial Strains and ARGs. ARG donor and recipient strains (*B. subtilis* 1A189 and 1A1, respectively) were obtained from the Bacillus Genetic Stock Center (BGSC) at Ohio State University. Chromosomal DNA of donor *B. subtilis* 1A189 exhibits an A-T base-pair deletion from the 2767 bp-long wild-type *bltR-blt-bltD* genome segment (i.e., that of *B. subtilis* 1A1). The mutant genome segment encodes efflux-mediated resistance to a wide variety of antibiotics (including fluoroquinolones, doxorubicin, and acriflavine), whereas the wild-type segment does not (Ahmed, Lyass et al. 1995). Chromosomal DNA was isolated by phenol-chloroform extraction (Saito and Miura 1963).

Deactivation of ARGs. For extracellular ARG deactivation experiments sterile, DNase-free, 10-mM phosphate buffer (PB) solutions dosed with 1 or 20 mg/L of ARG-containing DNA isolated from *B. subtilis* 1A189 were treated with FAC, NH_2Cl, O_3, ClO_2, or UV. Experiments were conducted by either: (a) varying exposure time for a fixed excess concentration of oxidant (using $Na_2S_2O_3$ to quench residual oxidant at pre-defined time intervals) – to measure kinetics of DNA oxidation and to simulate *Ct* or *It* values expected during treatment of municipal drinking water and wastewater, or (b) dosing the DNA samples with increasing concentrations of each oxidant and allowing them to react to complete oxidant consumption. For intracellular ARG deactivation experiments, 1L solutions of sterile, DNase-free, 10-mM phosphate buffer (PB) containing 10^6 CFU/ml *B. subtilis* 1A189 were treated with fixed excess concentration of FAC, NH_2Cl, O_3, ClO_2 for varying times, or 25-mL solutions of sterile, DNase-free, 10-mM phosphate buffer (PB) containing 10^8 CFU/ml *B. subtilis* 1A189 were exposed to UV for varying times. Following disinfectant exposures, chromosomal DNA was isolated from treated cells.

DNA Analyses. qPCR reactions were conducted with commercially-available real-time MasterMix from 5Prime for primers targeting a 193-bp amplicon region of the *B. subtilis* 1A189 *bltR-blt-bltD* segment. Purified chromosomal DNA was used as qPCR standards. The transforming activity of control and disinfectant-treated chromosomal DNA isolated from *B. subtilis* 1A189 was quantified by natural transformation of non-resistant *B. subtilis* 1A1 (Bott and Wilson 1967), with acriflavine as the selective agent for enumeration of antibiotic-resistant transformant cells.

RESULTS

Deactivation of ARGs and ARB. Table 1 summarizes the Ct_{90} and Ct_{99} values of deactivation of extracellular ARGs, intracellular ARGs, and ARB by FAC, NH_2Cl, O_3, and ClO_2, and the It_{90} and It_{99} values of deactivation of extracellular ARGs, intracellular ARGs, and ARB by UV, where Ct_{90} and It_{90}, and Ct_{99} and It_{99} represent the Ct and It values required for 1-log and 2-log of deactivation, respectively. Table 1 also summarizes Ct and It ranges typically applied for water disinfection (1999).

Table 1. Summary of disinfectant exposure required for 1-log and 2-log transforming activity deactivation of extracellular DNA, intracellular DNA , and cells of *B. subtilis* 1A189 in 10 mM PB at pH 7

Disinfectant	Deactivation Target	Ct_{90} (mg/L*min)	Ct_{99} (mg/L*min)
FAC	Extracellular ARGs	~10	~20
	Intracellular ARGs	~45	~100
	ARB	<<1	<<1
	Practical *Ct* range	10^1-10^2	
NH_2Cl	Extracellular ARGs	>>1500	>>1500
	Intracellular ARGs	>>5000	>>5000
	ARB	<<60	<<60
	Practical *Ct* range	10^2-10^3	
O_3	Extracellular ARGs	~0.009	~0.02
	Intracellular ARGs	~0.2	~0.5
	ARB	<<0.02	<<0.02
	Practical *Ct* range	10^{-1}-10^0	
ClO_2	Extracellular ARGs	>>120	>>120
	Intracellular ARGs	>>20	>>20
	ARB	<<1	<<1
	Practical *Ct* range	10^0-10^1	
		It_{90} (mW/cm^2*s)	It_{99} (mW/cm^2*s)
UV	Extracellular ARGs	~20	~100
	Intracellular ARGs	~60	~120
	ARB	<<1	<<1
	Practical *It* range	10^0-$10^{2.3}$	

qPCR Results. A reasonably strong correlation was found between extracellular 1A189 ARG transforming activity and qPCR measurements of the 193-bp ARG amplicon copy number following disinfectant treatment, as shown in Figure 1. However, qPCR results consistently underestimated measured losses in ARG transforming activity.

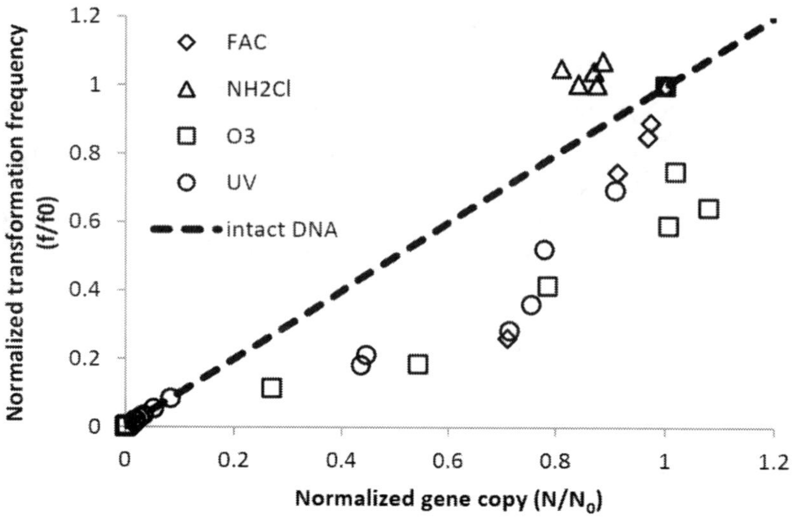

Figure 1. Correlations between normalized residual transforming activities of FAC-, NH_2Cl-, O_3-, and UV-treated *B. subtiils* 1A189 DNA (1 mg/L) versus normalized residual 193-bp *bltR-blt-bltD* amplicon copy number (N/N_0) as measured by qPCR.

DISCUSSION

The ARGs and ARB deactivation data showed that FAC and O_3 could achieve extensive deactivation (i.e. >3-log) of extracellular DNA's transforming activity at the typical exposure levels applied in practice, while significantly higher *Ct* values were required to achieve the same level of deactivation for intracellular ARGs, which is likely due to either an inherently lower reactivity of intracellular DNA (e.g., due to three-dimensional conformation) or the presence of other oxidant-consuming substances in the bacterial cytoplasm, such as RNA, enzymes, and lipids. For NH_2Cl and ClO_2, only minimal deactivation was observed for both extracellular and intracellular ARG transforming activity even at *Ct* values far beyond the practical range, which are more than 20× the *Ct* value needed for 6-log inactivation of the *B. subtilis* cells. UV light can achieve ~2-log deactivation in DNA's transforming activity in the typical exposure range. The UV deactivation trends are consistent with the previous observation by Mckinney et al. (McKinney and Pruden 2012). In addition, there appears to be no significant difference between the extracellular and intracellular ARG deactivation trends for UV. This is presumably due to the fact that UV directly targets DNA, and does not interact to a significant extent with other cellular constituents.

Although reasonable correlations between qPCR and ARG transforming activity data were found using the 193-bp amplicon described above. qPCR data could likely be improved by utilizing a primer set covering a longer segment of the ARG locus, in accord with previous observations for correlations between qPCR assay of viral nucleic acid damage and loss of infectivity (Pecson, Ackermann et al. 2011). With improved correlations, qPCR might in turn be utilized as a simpler, faster, and less expensive analytical surrogate for monitoring loss of ARG transforming activity during disinfection processes.

CONCLUSIONS

The findings confirm that all disinfectants investigated can sufficiently eliminate the model ARB. More than 6-log inactivation of cells can be achieved at the lower end of typical practical *Ct* or *It* ranges for each disinfectant. However, higher exposures are needed to achieve the same levels of deactivation of ARGs. While FAC, O_3, and UV light can effectively deactivate ARG transforming activity within typical disinfectant *Ct* ranges, NH_2Cl and ClO_2 yield minimal loss in ARG transforming activity at practical *Ct* values. These results demonstrate that even though the ARB are effectively inactivated during the disinfection processes, their ARG-containing DNA could retain its transforming activity and propagate the encoded antibiotic resistance traits to downstream bacteria by means of natural transformation. In addition, qPCR was found to be a potentially useful surrogate for ARG transforming activity. Future work will be directed toward investigating the behavior of ARGs relevant to clinically-relevant human pathogens during treatment with the disinfectants utilized here, as well as with disinfectants and antiseptics commonly utilized in health care. In addition, the use of primer sets specific to longer amplicons spanning the full length of target ARGs will be examined as a means of achieving optimal correlations between qPCR and transforming activity measurements.

REFERENCES

(1999). *Disinfection profiling and benchmarking guidance manual* United States Environmental Protection Agency.

Ahmed, M., L. Lyass, et al. (1995). "*Two highly similar multidrug transporters of Bacillus subtilis whose expression is differentially regulated.*" J. Bacteriol. **177**(14): 3904-3910.

Bott, K. F. and G. A. Wilson (1967). "*Development of competence in Bacillus subtilis transformation system.*" J. Bacteriol. **94**(3): 562-570.

Carlet, J., V. Jarlier, et al. (2012). "*Ready for a world without antibiotics? The Pensières antibiotic resistance call to action.*" Antimicrob Resist Infect Control. **1**(11). DOI: 10.1186/2047-2994-1-11.

Domingues, S., K. Harms, et al. (2012). "*Natural transformation facilitates transfer of transposons, integrons and gene cassettes between bacterial species.*" PLoS Pathogens **8**(8): e1002837. doi:10.1371/journal.ppat.1002837.

McKinney, C. W. and A. Pruden (2012). "*Ultraviolet disinfection of antibiotic resistant bacteria and their antibiotic resistance genes in water and wastewater.*" Environ. Sci. Technol **46**: 13393-13400.

Pecson, B. M., M. Ackermann, et al. (2011). "*Framework for using quantitative PCR as a nonculture based method to estimate virus infectivity.*" Environ. Sci. & Technol **45**(6): 2257-2263.

Saito, H. and K. Miura (1963). "*Preparation of transforming deoxyribonucleic acid by phenol treatment.*" Biochim Biophys Acta **72**(4): 619-629.

UV or PAA for Wastewater Disinfection: a Comparison of the Impact on Virulence Genes

Basanta Kumar Biswal[1], Ronald Gehr[1,#], Ramzi Khairallah[1], Kareem Bibi[1], Alberto Mazza[2], Luke Masson[2], Dominic Frigon[1]

[1]Dept of Civil Engineering and Applied Mechanics, McGill University, Montréal, Québec, H3A 0C3, Canada

[2]National Research Council of Canada, Montréal, Québec, H4P 2R2, Canada

[#]corresponding author. ronald.gehr@mcgill.ca

ABSTRACT

Ultraviolet (UV) and peracetic acid (PAA) disinfection have been used in municipal wastewater treatment plants (WWTPs) for their biocidal effect on microorganisms. The current study used PCR-Bioplex and DNA microarray genotyping techniques to investigate the impact of UV and PAA on the change in frequency of virulence genes in *E. coli* isolates and on the prevalence of uropathogenic *E. coli* (UPEC) isolates in wastewater effluents. Effluents from four WWTPs (activated sludge [AS], biofiltration [BF] and physicochemical [PC1 and PC2]) located in Québec, Canada, were sampled before disinfection and exposed to UV or PAA doses in the lab to reach a target count of approximately 200 CFU/100 mL. To achieve this, the required UV fluences ranged between 7 – 30 mJ/cm^2, depending on the plant, while the PAA dose varied between 0.9 – 2.0 mg/L. *E. coli* isolates totaling 1,766 were extracted from the samples pre- and post- disinfection, then screened by PCR/Bioplex to detect those likely to be UPECs, using three virulence genes (*hlyA*, *papC* and *cnf1*). The UPEC pathotypes of the positively screened isolates were confirmed by microarrays. The proportion of UPEC isolates decreased in all samples after disinfection, with that due to UV varying between 22% – 80%, and the reduction due to PAA ranging between 11% – 100%. The average reductions by UV (55%) and PAA (52%) were statistically significant ($P<0.05$). Gene frequency analysis revealed that the decline in the population of UPEC pathotypes by UV or PAA was not linked to specific virulence factors as most virulence genes were lost, suggesting that entire pathogenicity islands (PAIs), carrying clusters of virulence genes, were lost through disinfection. Thus, this study showed that both UV and PAA disinfection appear to significantly reduce the proportions of UPECs in the surviving *E. coli* populations in wastewater effluents.

KEYWORDS

UV, PAA, wastewater disinfection, uropathogenic *E. coli*, virulence genes, genotyping

INTRODUCTION

Ultraviolet (UV) and peracetic acid (PAA) have been used for disinfection in municipal wastewater treatment plants (WWTPs) due to their strong biocidal effect at low doses on a wide range of microbes including *Escherichia coli*, and their low production levels of disinfection byproducts (Kitis 2004, Oliver and Cosgrove 1975). Wastewater disinfection is carried out to meet indicator microorganism discharge limits set by local authorities, such as < 200 colony forming units (CFU/100 mL) (Health Canada 2012)

E. coli are normally considered as nonpathogens, however, some specific strains (e.g. *E. coli* O157:H7) are pathogenic and can cause disease outbreaks associated with recreational waters (Hlavsa et al. 2011); thus *E. coli* are a good model species to study the dynamics between pathogenic and non-pathogenic strains. Based on their distinct virulence properties and clinical symptoms of the host, *E. coli* strains are classified into three main categories: commensal (non-pathogenic), intestinal pathogenic, and extraintestinal pathogenic. Furthermore, seven human pathotypes are included in the intestinal pathogenic *E. coli* (IPEC) category, while three human pathotypes comprise the extraintestinal pathogenic *E. coli* (ExPEC) category (Table 1) (Kaper et al. 2004).

Table 1. Pathogenic *E. coli* pathotypes based on the host's clinical symptoms

Pathotype classification based on the location of pathogenic attacks	
Intestinal pathogenic *E. coli* (IPEC)	Extraintestinal pathogenic *E. coli* (ExPEC)
Enteropathogenic *E. coli* (EPEC)	Uropathogenic *E. coli* (UPEC)
Enterotoxigenic *E. coli* (ETEC)	Septicemia-causing pathogenic *E. coli* (SEPEC)
Shiga toxin-producing *E. coli* (STEC)	Meningitis-associated *E. coli* (MNEC)
Enterohemorrhagic *E. coli* (EHEC)	
Enteroaggregative *E. coli* (EAEC)	
Enteroinvasive *E. coli* (EIEC)	
Diffusely adherent *E. coli* (DAEC)	

E. coli pathogenesis is based on the assembly of a suitable number and combination of virulence genes (VGs) that encode one or multiple virulence factors (VFs) (Bruant et al. 2006). According to the recent virulence factor data base (VFDB) report (Chen et al. 2012), UPECs contain a total of 88 VGs encoding the following 5 VFs: adherence, invasion, autotransporter, iron uptake, and toxins.

Previous studies have used partial genotyping methods, including PCR (Lee 1996, Mokracka et al. 2011), colony hybridization (Hamilton et al. 2010) and phylotyping (A, B1, B2, C, D and E) (Clermont et al. 2000, Escobar-Páramo et al. 2004) for determination of a few VGs in order to identify pathogenic *E. coli* of environmental origin. For a more comprehensive analysis of *E. coli* populations, oligonucleotide microarrays containing 189 –195 virulence or virulence-related genes and 30–96 antimicrobial resistance genes have been developed to study surface waters and municipal wastewater effluent (Bruant et al. 2006, Frigon et al. 2012, Hamelin et al. 2006, Hamelin et al. 2007). These studies found that in natural waters, ExPECs (9–50% isolates) were the predominant pathotypes, while IPECs (0 –11%) were present either sporadically or most often at low levels. In WWTP effluents, UPECs were the pathotypes with the highest frequency within the ExPECs (58 – 93%) (Frigon et al. 2012).

The "neutralization" of a pathogenic strain is not necessarily the result of complete inactivation of this strain, but it could also be due to deletion of a number of VGs, thus rendering the strain non-pathogenic (Frigon et al. 2012). Most UPEC VGs are clustered in mobile genetic elements called pathogenicity islands (PAIs) (Middendorf et al. 2001a). It is known that some UPEC PAIs are relatively unstable and they can be lost upon exposure to environmental stresses such as lower temperatures (20°C instead of 37°C) (Middendorf et al. 2004) and the presence of antimicrobial agents such as quinolones at sub-inhibitory concentrations (Soto et al. 2006). Similarly, UV or PAA may be stressors which cause the loss of PAIs. Alternatively, disinfectants inactivate microorganisms by reacting with different biomolecules such as amino acids, lipids and nucleic acids, which result in the arrest of the various cellular or molecular functions (Dodd 2012). In our study, the two selected disinfectants differ in their inactivation mechanisms. PAA is a chemical disinfectant (oxidizing agent) and it mostly interacts with sulfhydryl and sulfur bonds in proteins and enzymes (Kitis 2004), whereas UV is a photo-irradiation agent and it excites and dimerizes pyrimidine bases in the DNA (Dodd 2012). Hence, if the two disinfectants interact in different ways with different cellular components of *E. coli*, the types of VGs lost due to UV or PAA may also be different. Therefore, the following question is warranted: are the losses of VGs dependent on the type of function or VFs encoded, or can the VGs loss be predicted from the loss of PAIs?

It is well known that inactivation of *E. coli* by UV or PAA is affected by wastewater quality, including suspended solids which cause shielding of *E. coli* from disinfectants (Kollu and Örmeci 2012). At this point, it is hypothesized that the impact of the disinfectants on the proportion of UPEC isolates in the free-swimming vs. particle-associated *E. coli* populations would be different. In the case of reduction of the proportion of UPECs caused by the disinfectants, it is to be expected that this reduction would be more pronounced for the free-swimming population.

The effectiveness of UV and PAA disinfection for municipal wastewaters has been amply demonstrated (Caretti and Lubello 2003, Gehr et al. 2003, Koivunen and Heinonen-Tanski 2005a), however there is no information on the differential inactivation or selection of pathogenic and/or non-pathogenic proportions of the microbial populations due to these disinfectants. Hence, the goal of the current study is to determine the dynamics between pathogenic and non-pathogenic strains following disinfection. *E. coli* was chosen as the model species. To the best of our knowledge, this is the first study of the impact of UV or PAA disinfection on changes in the sub-population of UPEC pathotypes and on the occurrence of VGs in *E. coli* populations in wastewaters.

METHODOLOGY

Sampling of Wastewater Effluents and Analysis of Wastewater Characteristics

For this study, four municipal WWTPs, located near Montréal, Québec, Canada, were chosen. Two plants used biological treatment: conventional activated sludge (AS), and biological filtration (BF). Two plants, designated as PC1 and PC2, used physicochemical treatment processes. Effluent samples were collected prior to any disinfection in two 1-liter sterile plastic bottles. For the UV study, samples were obtained between May 02 and August 22, 2012 for the AS, BF and PC1 plants, and between July 30 and August 08, 2008, for the PC2 plant. For the PAA study involving only non-filtered samples, the samples were taken between August 10 and October 15, 2011 from the AS, BF and PC1 plants. For the PAA study involving non-filtered vs. filtered samples, samples were collected from the BF plant (designated as BFa) on October 9, 2012. All samples were collected as grab samples from the center of the clarifier effluent channel. After collection, samples were immediately stored in a closed cooler containing ice and shipped to the laboratory where they were stored at 4 °C and analyzed within 24 h. Key physicochemical parameters including pH (method 4500-H^+ B), % UV transmittance (253.7 nm, method 5910 B), suspended solids (SS, method 2540 D) and chemical oxygen demand (COD, method 5220 D) were determined following Standard Methods (APHA et al. 2005).

Disinfection Experiments

UV disinfection was carried out using non-filtered (AS, BF, PC1 and PC2), as well as centrifuged/filtered effluent samples (PC2) to determine the effects of particles. In parallel with the non-filtered portion, a portion of the PC2 sample was centrifuged at 3,500 rpm (1,233 x g) for 5 min, then filtered using a 20 µm pore-size filter paper to collect only the free-swimming part of the *E. coli* population. For UV irradiation, 50-mL sub-samples were placed in a 75-mL crystallization dish and exposed to a collimated UV beam from a low- pressure mercury lamp emitting light at 253.7 nm. Exposure time was varied to establish the inactivation curve and achieve a target *E. coli* level close to 200 CFU/100 mL. The incidence UV intensity was measured using a radiometer (International Light, Newburyport, Massachusetts) and the average UV intensity was calculated using the Morowitz equation (Morowitz 1950). UV dose was calculated by multiplying the average intensity with the irradiation time. Immediately after UV exposure, the crystallization dishes were covered with aluminum foil to avoid photoreactivation, and filtration for *E. coli* enumeration was performed within one minute.

PAA disinfection was performed on non-filtered samples from the AS, BF and PC1 plants and on a 20-µm filtered sample from the BF plant. The PAA solution (12% w/w, Solvay Interox Inc., Houston, Texas) was added to a 100-mL sub-sample according to the methods described by Wagner et al. (2002). The volume of PAA solution was varied to obtain a dose-response curve for a contact time of 30 min for the AS and PC1 plants, and 30 min (non-filtered) or 60 min (non-filtered or filtered) for the BF plant. Plates corresponding to *E. coli* counts of ~ 200 CFU/100 mL after disinfection were used for isolation. After the each experiment, the PAA residual was measured using the DPD (N, N-diethly-p phenylenediamine) colorimetric method.

Enumeration and Isolation of *E. coli*

E. coli levels were enumerated using 0.45-µm membrane filters, and incubation at 44.5°C on mFC (modified fecal coliform) agar plates with 5-bromo-6-chloro-3-indolyl glucuronide (BCIG) (CM1111; Oxoid Ltd., England) for 18 – 24 h (method 9221 F, APHA et al. 2005). *E. coli* colonies which appeared blue were picked at random and streaked for isolation onto Luria-Bertani (LB) agar plates consisting of 1% tryptone, 0.5% yeast extract, 1% NaCl, and 1.5% agar to ensure pure isolates. Confirmation was performed by growing colonies on Chromocult agar plates (EMD Chemicals inc, Germany) and observing dark-blue/violet colonies, and further testing these colonies using Kovac's reagent (EMD Chemicals, Germany) for indole production (development of pink color upon testing). Between 60 and 149 *E. coli* isolates were randomly picked from each plate for a total of 1,766 isolates analyzed in this study. Non-disinfected and disinfected samples from AS, BF and PC1 effluents were each characterized by 91 to 130 *E. coli* isolates (a total of 1,597 isolates) whereas the samples from the PC2 plant were each characterized by 40 – 48 isolates (a total of 169 isolates).

Extraction of Genomic DNA and Genotyping

DNA extraction was carried out as described by Hamelin et al. (2006). Briefly, *E. coli* isolates were grown on LB media overnight and 1 mL of the liquid cultures was centrifuged at 13,000 × g for 1 min to obtain pellets. The pellets were washed with 1 mL of distilled deionized (ddH_2O) water then resuspended in 175 µL ddH_2O. The culture suspension was boiled at 95°C for 10 min and centrifuged at 13,000 × *g* for 3 min, and 150 µL of the supernatant containing the extracted DNA was taken for genotyping analysis.

Except for the PC2 samples, the isolates' libraries from each sample were initially screened for likely UPEC isolates using a quadruplex PCR amplification of three UPEC VGs: *papC* (P fimbriae), *cnf1* (cytotoxic necrotizing factor 1) and *hlyA* (α-hemolysin) and a positive control gene *uidA* (β-glucuronidase). The reverse primer of each gene was biotinylated at the 5-prime end to facilitate attachment to the SAPE (streptavidin–R-phycoerythrin, receptor) dye for detection in the Bioplex assay. PCR amplifications were performed in 96-well plates in a reaction mixture of 25 µL containing 12.5 µL of multiplex master mix (Qiagen, Canada), 10 µL of primer mixes, 1.5 µL of sterile distilled water and 1 µL of template DNA. The primers and their concentrations are shown in Table 2. PCR amplification was carried out in a thermal cycler (Mastercycler-epgradient, Eppendorf AG, Hamburg, Germany) under the following conditions: a predenaturation at 95°C for 15 min, followed by 35 cycles consisting of a denaturation at 94°C for 30 sec, annealing of primer at 61°C for 60 sec and primer extension at 72°C for 2 min. The final extension was held at 72°C for 10 min. The PCR amplified products were subsequently used in the Bioplex process for interpretation of the presence/absence of the four genes in the *E. coli* isolates using the following protocols.

Table 2. Oligonucleotide sequences used in the PCR assay

Target gene	Primer designation and sequences (5' - 3')	Concentrations of primers (pmol/µL)	Amplicon size (bp)
hlyA	Forward, ACGAGTTGGTTGATTAGC	0.6	575
	Reverse, ATGCCAACAATAACCACTGCAC	0.6	
papC	Forward, ATCAGAATACTCAAGCCAG	0.6	423
	Reverse, TTGTTGTCTCTGTTTTGTAACGC	0.6	
cnf1	Forward, CACATCTGCACTCTTTGTT	2.4	287
	Reverse, CAAATTTCCCTTCACCTG	2.4	
uidA	Forward, TGCCGGAATCCATCGCAGCGTAAT	1.2	126
	Reverse, ATT GGCCACCACCTGCCAGTC	1.2	

Bead Preparation and Probe Coupling

As per the recommended Luminex protocol and further developed at NRC Montreal, uniquely addressed magnetic coated beads (BIO-RAD, Canada) were selected for one specific bead type ("region") per probe. These were individually chemically coupled to their respective probes in $1x10^6$ bead lots. The beads were washed once in 200 µL of 0.1M MES (2N-morpholino-ethanesulfonic acid) buffer pH 4.5 (Sigma) and suspended in 8.5 µL (per bead lot) of 0.1M MES buffer pH 4.5. C_6 probe DNA oligonucleotides (2.0 µL at 0.1 nmol/µL) were added and the reaction initiated with the addition of 2.0 µL of EDC (1-ethyl-3-(diethylaminopropyl) carbodiimidehydrochloride) (Pierce). This was followed by 30 min incubation at room temperature in the dark. A second round of EDC treatment and incubation was performed using conditions described above. Coupled beads were washed once in 1.0 mL 0.02% Tween-20, once in 1.0 mL 0.1% (w/v) SDS and finally once in 1.0 mL tris-EDTA (TE) pH 8.0. Final re-suspension for storage and use was in 50 µL TE pH 8.0. Beads were stored in the dark at 4°C at all times. Bead concentration was determined using a TC10 automated cell counter (BIO-RAD, Canada). Final bead recovery was generally in the 60-80% range. Validation for successful coupling was performed with biotinylated RevComplement oligonucleotides using the hybridization procedure described below.

Bead Hybridization and Bioplex Assay

All hybridization assays were performed in a final volume of 50 µL. Appropriate bead types were mixed together in equal proportions and diluted in TE pH 8.0 to a target amount of 2,000 beads per assay. These were dispensed into individual wells of flat-bottomed (glass) plates in 33 µL aliquots. In a separate 96-well PCR plate, biotinylated target DNAs (amplicon or RevComplement) at desired concentrations were prepared in a per sample volume of 17.0 µL of TMAC hybridization solution composed of 4.5M tetramethylammonium chloride (TMAC), 75 mM Tris-HCl pH 8.0, 6.0 mM EDTA, and 20% (w/v) Sarkosyl (sodium-lauryl-sarcosinate). Target DNA samples were heated to 95°C for 5 min and quickly dispensed to appropriate bead wells followed by 30 min incubation at 37°C with vigorous shaking. Post-hybridization washes were performed at 37°C: one wash with 0.1x SSC, 0.1% SDS with 10 sec mixing and one last wash in 0.1x SSC with 10 sec mixing. Removal of hybridization/wash solutions was performed by the magnetic collection of beads at the bottom of individual wells using a magnetic deck and pipette removal of bead-free supernatant. After that, hybridized beads were incubated at room temperature for 10 min with shaking in a 2.0 µg/mL solution of SA-PE (streptavidin-R-phycoerythrin) in TMAC. Beads were collected as described above and the SA-PE solution replaced with TE pH 8.0 and resuspended with vigorous shaking for immediate analysis on the Bioplex 200 system (BIO-RAD, Canada). Fluorescence signals were expressed as MFI (Mean Fluorescence Intensity) units and converted to ratios over background for interpretation purposes: the bead intensity above 3 times the background intensity was considered positive. *E. coli* isolates which showed the presence of one or more VGs (i.e., screen-positive) were further genotyped by DNA microarrays.

DNA Labeling and Microarray Experiments

The screen-positive isolates for the AS, BF and PC1 samples and all the isolates from the PC2 samples were genotyped by microarray. The extracted DNA was first labeled using the BioPrime DNA labeling kit (Invitrogen Life Technologies, Burlington, Ontario, Canada), according to the procedure described by Hamelin et al. (2006). Then, labeled DNA was purified using the Pure Link PCR purification kits (Invitrogen Life Technologies), according to the manufacturer's instructions. The amount of incorporated fluorescent Cy5 dye was quantified by scanning with a Nanodrop spectrophotometer (ThermoFischer Scientific, USA) at wavelengths from 200 to 700 nm and using the Internet-based Percent Incorporation Calculator (http://www.pangloss.com/seidel/Protocols/percent_inc.html).

DNA microarrays used in this study were developed and validated by Bruant et al. (2006) and contained 70-mer probes targeting 195 virulence and virulence-related genes and 96 antimicrobial resistance genes (Frigon et al. 2012). Briefly, microarray slides were prehybridized at 50°C for 1 h using 60 µL of DIG Easy Hyb Buffer (Roche Diagnostics, Laval, Quebec, Canada) which was supplemented with 5% (vol/vol) bovine serum albumin (1 mg/mL; New England Biolabs Inc., Beverly, MA). Following the prehybridization, the slides were washed by dipping in 0.1×SSC and air-dried. The hybridization process was performed overnight at 50°C in a SlideBooster using 1,000 ng of Cy5-labeled genomic DNA which was dried and resuspended with15 µL of DIG Easy Hyb Buffer supplemented with 0.1 µg/µL single-stranded salmon sperm DNA (Sigma-Aldrich, ON, Canada) after denaturing for 5 min. at 95°C. After hybridization, slides were washed four times at 37°C, i.e. three times with 0.1×SSC and 0.1% SDS for 5 min. each and once with 0.1×SSC for 5 min. Finally, the microarrays were air-dried. Microarray slides were scanned with a ScanArray Lite fluorescent microarray analysis system (Perkin-Elmer, Missasauga, Ontario, Canada) and the images were generated using the ScanArray Express software (Perkin-Elmer, Foster City, CA). Fluorescent spot intensities were quantified using ImaGene v8.0 (BioDiscovery, Inc., El Segundo, CA). Sample:background intensity ratios greater than 3 were considered positive.

Assignment of Pathotype Based on Microarray Genotyping Results

Pathotyping of *E. coli* isolates was performed based only on microarray results according to the number and combination of VGs/VFs which they carried. The full list of VGs with their probes and pathotype classification rules were presented by Frigon et al. (2012). Specifically for the UPEC pathotypes, the presence of a minimum of five VGs was necessary with a distribution among VFs as follows: 2 genes for adherence (*papA*, *papC*, *papG* or *pixA*), 1 gene for capsule (*kpsM(III)* or *kpsM(II)*), 1 gene for iron acquisition (*fyuA, iron, irp1, irp2, iucD, iutA, sitA,* or *sitD*), and 1 gene for haemolysins and other toxins (*hlyA*, *vat*, *cnf1*, *cnf2*, or *sat*). The UPEC isolates determined in this study are considered "potentially pathogenic" since these isolates were not tested for their capability to cause diseases on humans or animals. Isolates which did not correspond to any VG combinations for defining pathotypes were deemed non-pathogenic *E. coli*.

Statistical Analysis

Statistical analyses of the variation of the frequencies of UPEC isolates or of specific VGs due to disinfection were performed using the log-linear model (Sokal and Rohlf 1969). Calculations were carried out using the CATMOD procedure of the SAS/STAT statistical software (SAS version 9.2, SAS Institute, Inc., Cary, NC). For the statistical tests, each plant was considered as an independent sample and all tests were considered two-tailed.

RESULTS

Disinfection Experiments

Average operational conditions and effluent qualities at the wastewater treatment plants at the time of sampling are summarized in Table 3.

Table 3. Treatment processes and effluent qualities of the four municipal WWTPs

Description/Parameter[a]	Wastewater treatment plants				
(unit)	AS	BF	BFa	PC1	PC2
Treatment processes	Conventional activated sludge	Biological filtration		Physicochemical using alum and organic polymer as coagulants	Physicochemical using ferric chloride or alum and organic polymer as coagulants
Population served ($\times 10^3$)	64. 4		40.0	59.0	1,862
Mean flow ($\times 10^3$ m^3/d)	65.3		38.3	44.0	7,600
HRT (h)	20		2.1	3.0	est. 3
SRT (d)	7		NA[b]	NA	NA
pH	7.1	7.7	–	7.2	7.1
UV transmittance (%)	67.2	63.6	–	54.5	42.6
SS (mg/L)	10	5.0	14	15	18
COD (mg/L)	37.6	46	62	45.1	92.2
E. coli (CFU/100 mL)	7.8×10^3	1.1×10^5	3.0×10^5	1.6×10^5	9.9×10^5

[a]HRT: Hydraulic retention time, SRT: Solids retention time
[b]NA: Not applicable

Dose-response curves for UV (Fig. 1a) and PAA (Fig.1b) disinfection were determined by varying the irradiation times and the initial concentrations, respectively. The target *E. coli* level after disinfection was set to 200 CFU/100 mL (~ 2.30 logs). Depending on the sample, this target corresponded to CFU reductions of 1.6 – 3.7 logs. Before tailing, the inactivation rates were similar in all samples, but tailing started to occur around 10^3 CFU/100 mL with UV for the BF, PC1 and non-filtered PC2 samples (Fig. 1a) and around 5×10^2 CFU/100 mL with PAA for all samples. The UV dose needed to reach the target *E. coli* counts for the AS sample was 7 mJ/cm^2, but it was much higher (15 – 30 mJ/cm^2) for the BF, PC1 and PC2 samples, mainly because of the tailing effect. As expected, a lower UV dose was needed for the biological effluents (7 – 15 mJ/cm^2) than for the physicochemical effluents (15 – 30 mJ/cm^2), which corresponds to the lower concentration of particles in the former case, as the SS levels are lower (5.0 – 14 mg/L) in the biological effluents than in the physicochemical effluents (15 – 18 mg/L) (Table 3). Filtration of the PC2 sample removed the tailing effect (Fig. 1a), but the high initial *E. coli* counts still required a dose of 15 mJ/cm^2 to reach the target disinfection level.

The PAA doses to achieve the *E. coli* target level with a 30-min treatment time were 0.9, 2.0 and 1.75 mg/L for the AS, BF and PC1, respectively (corresponding to CT values 30.9, 60 and 52.6 mg-min/L, respectively; Fig. 1b). These doses were similar to the range of doses obtained by other researchers (2 – 7 mg/L) for indicator bacteria in municipal secondary wastewater effluents (Koivunen and Heinonen-Tanski 2005b). Filtration did not remove the tailing effect observed with the BF samples for contact times of 60 min; the tailing appeared to correlate with the level of COD in the sample (Table 3), which, according to Koivunen and Heinonen-Tanski (2005b), causes a rapid consumption of the PAA, and this is seen in Fig. 2.

Our study also investigated the difference in the rate of inactivation of *E. coli* over a period of 120 min by applying 0.5 mg/L PAA immediately, then 0.5 mg/L every 30 min for 120 min (a total 2 mg/L, data not shown). It was presumed that *E. coli* could develop resistance to PAA on application of low doses (0.5 mg/L) over a period of time; however the rate of inactivation was the same for both cases.

Fig. 2 shows PAA residual concentrations immediately after dosing, and after 30 or 60 min. contact time. The initial concentrations depended on the initial effluent *E. coli* counts. At the end of the contact time, the residual levels reflect the consumption of PAA, mainly by reduced organics, as suggested by the COD.

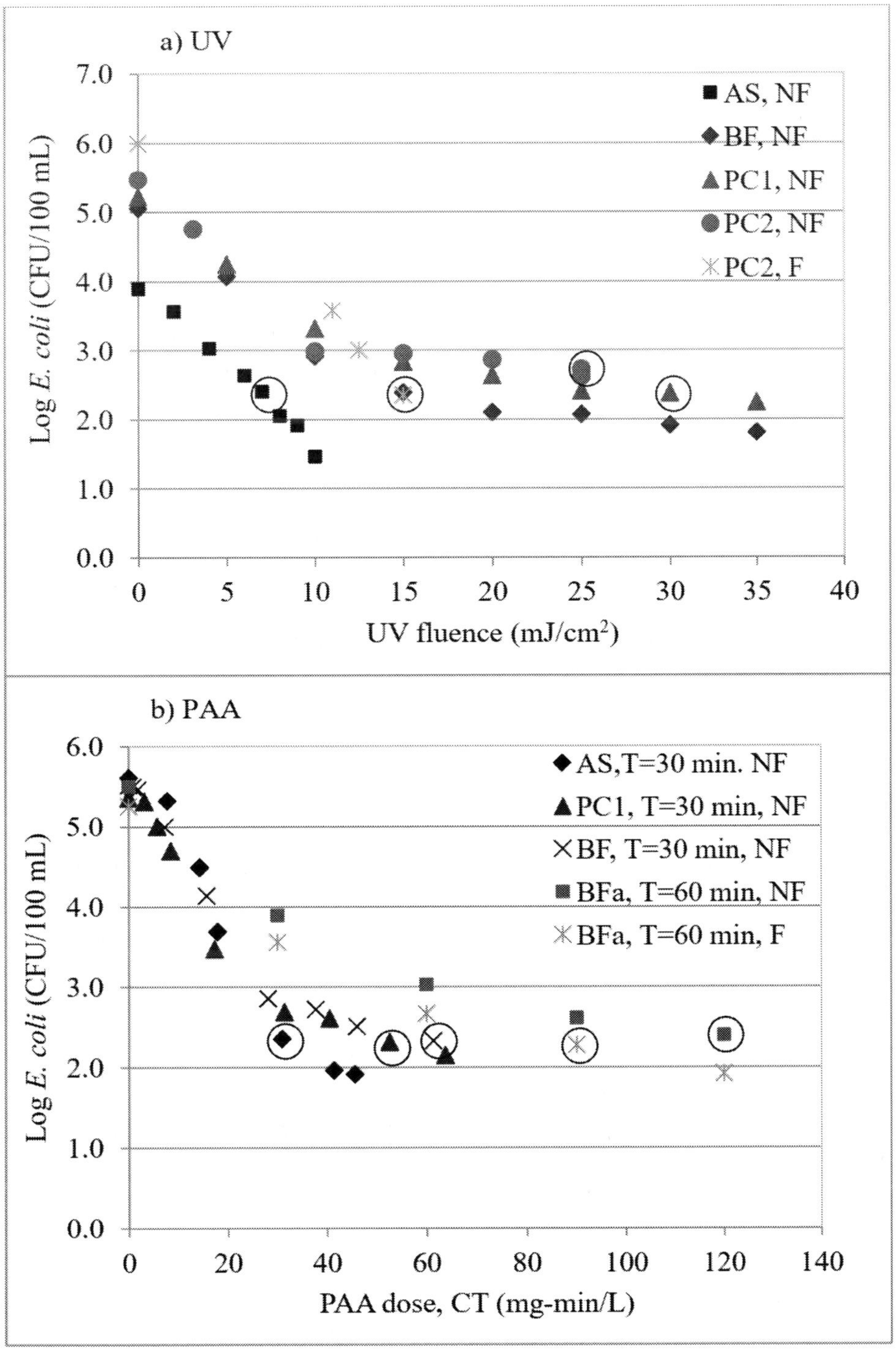

Figure 1. UV (a) and PAA (b) *E. coli* inactivation curves for non-filtered (NF) and filtered (F) samples. PAA contact times (T) were 30 or 60 min. CT: Concentration × Time (Data points circled are those conditions for which *E. coli* isolation was done)

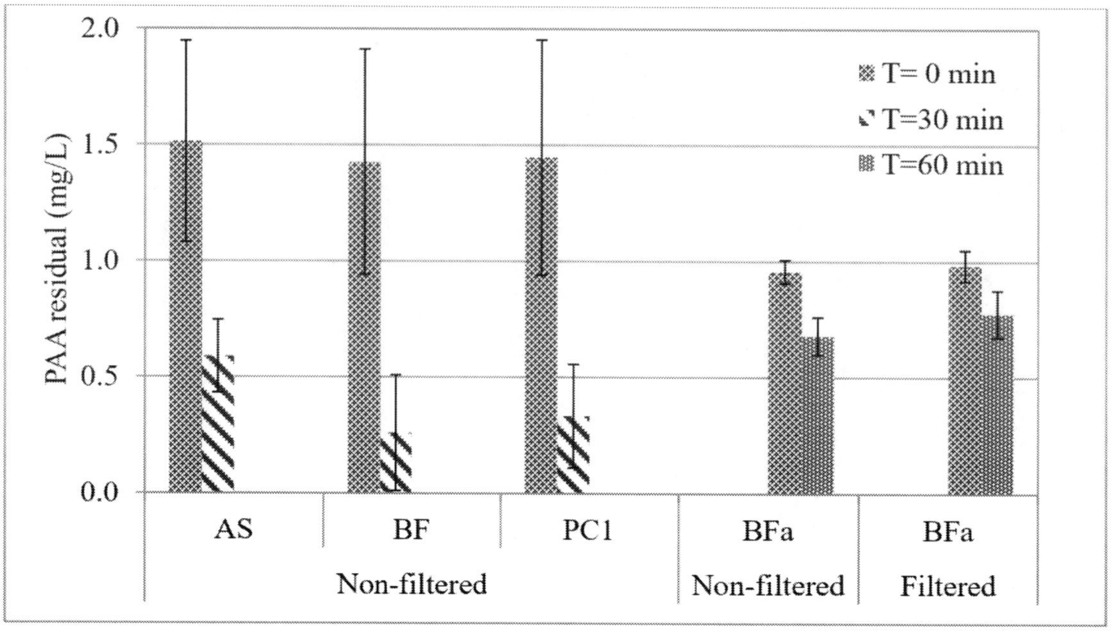

Figure 2. Residual PAA concentrations at time zero and after 30 and 60 min for non-filtered (NF) and filtered (F) samples. Error bars represent the standard error among 3 replicates

Prevalence of UPEC in Non-disinfected and Disinfected Effluents

In the non-disinfected samples analyzed, the proportion of UPEC ranged between 6 – 18% (Fig. 3). Following disinfection, the proportion of UPEC decreased independently of the disinfectant used. The reduction of UPEC pathotypes by UV disinfection of non-filtered samples varied between 22 – 80% (Fig. 3a), for a statistically significant ($P<0.05$) average of 55%. Similarly, PAA disinfection of non-filtered samples reduced the UPEC proportions by 12 – 100% (Fig. 3b), with a statistically significant ($P<0.05$) average decrease of 52%. Given that there was a significant reduction of UPEC pathotypes for all samples and both disinfectants, it seems that any mechanisms causing the reduction are very effective.

Impact of Particles

It is known that particles can shield *E. coli* from UV radiation (Kollu and Örmeci 2012). Consequently, it was decided to investigate whether UV irradiation had a different effect on free-swimming (filtered samples) and particle-associated (likely causing tailing in non-filtered samples) *E. coli* strains. The reduction in UPEC pathotypes for the PC2 plant was 61% for the non-filtered sample and 71% for the filtered sample. Though the reduction of UPECs in the non-filtered samples was lower than in the filtered samples, the difference was not significant ($P>0.05$). For this comparison, the isolates were not screened by the PCR procedure, but were all genotyped by the microarray. Thus, it is possible to compare directly the number of UPEC genes within the non-UPEC portion of the *E. coli* population. The average number of UPEC genes within the non-UPEC isolates increased from 1.7 to 2.3 (out of 26 possible genes) upon UV irradiation for the non-filtered sample, while it decreased from 2.3 to 2.0 for the filtered sample. The combined higher reduction of UPEC isolates and of UPEC genes frequency in non-UPEC isolates for the free-swimming population suggests that the impact of UV on the gene loss is more pronounced for the free swimming than for the particulate-associated populations.

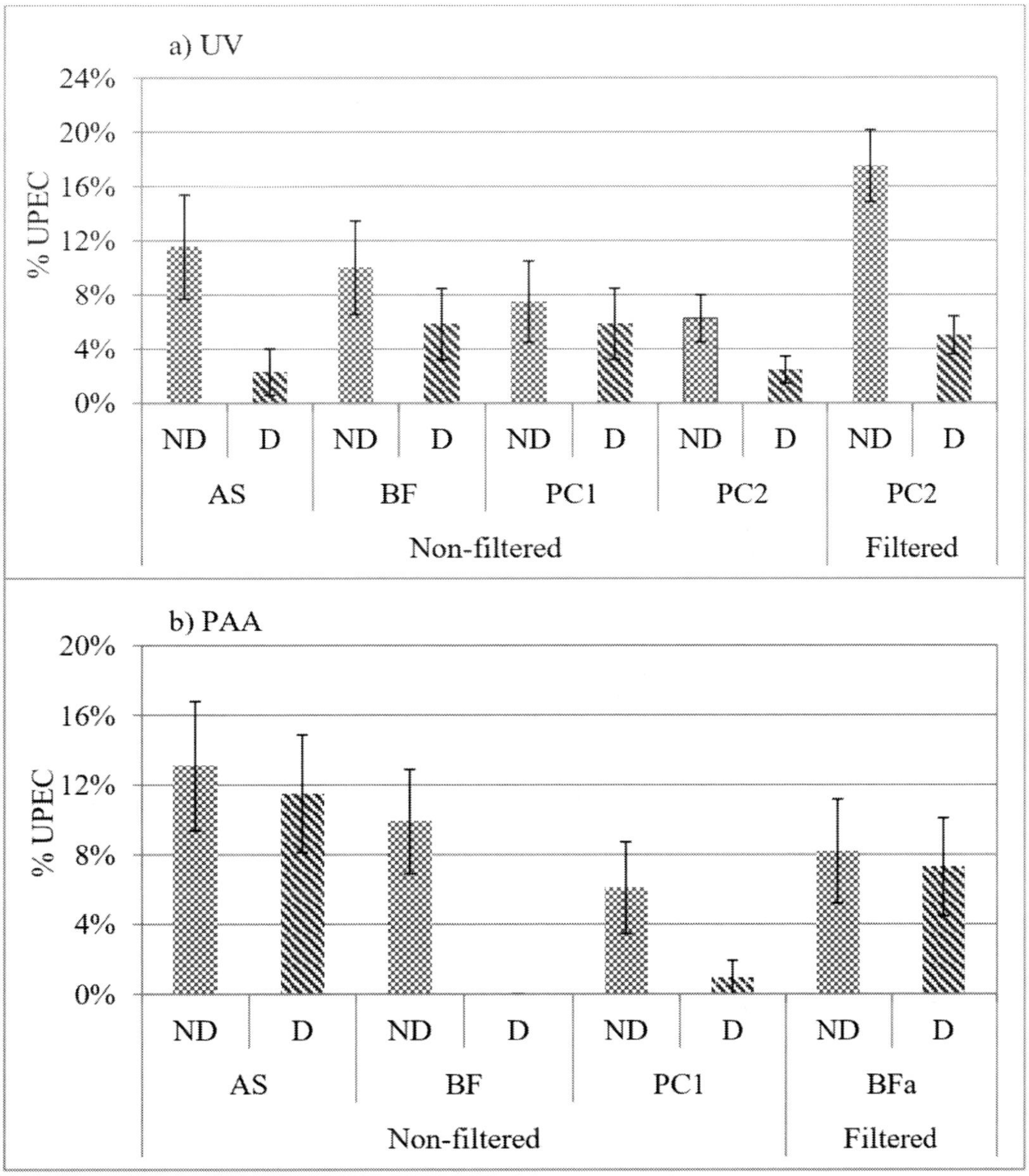

Figure 3. Proportion of UPEC in non-disinfected (ND) and disinfected (D) effluent samples by UV (a) and PAA (b). Error bars are the standard errors as calculated using the log-linear model

Change in the Frequency of UPEC Virulence Genes by UV or PAA

It is of interest to observe whether the two disinfectants (UV and PAA), which interact with different components of the *E. coli* cell, would affect the VG frequencies differently. The changes in the occurrence of UPEC VGs were analyzed between the non-disinfected and disinfected samples that were screened by PCR/Bioplex for likely UPEC pathotypes (Table 4).

Although the proportion of UPEC pathotypes decreased upon disinfection in all experiments (Fig. 3), the screening procedure yielded approximately 14% false-positives (i.e., truly non-UPEC that are screen-positive; data not shown), which resulted in an average of 65% of the screen-positive isolates that were not confirmed as UPEC by microarray analyses. Therefore, the likelihood of observing any UPEC VGs was likely lower after disinfection because of the decrease in the proportion of UPEC pathotypes, but it was unlikely to have been biased because the initial level of UPEC was always less than 18% (Fig. 3).

The microarray used in this study contained probes detecting 198 VGs, which were classified in ten categories representing nine VFs and one miscellaneous cellular function (Frigon et al. 2012). Of these VGs, 26 genes were either used for UPEC pathotyping (19 VGs among four VFs) or were monitored because of their common association with the UPEC pathotypes (7 VGs). On examining the change of the frequency of UPEC VGs in non-UPEC isolates, the average gene frequency decreased from 5.1 to 3.6 by UV irradiation while by PAA, it declined from 3.8 to 3.3. The decrease of VGs by UV and PAA was 29% and 13% respectively and these reductions were not statistically significant ($P>0.10$). However considering the change in the proportion of the VGs in both UPEC and non-UPEC isolates, the decreases were much higher (8 – 100%) (Table 4). Of the 26 UPEC genes, at least 17 (65%) had been reported to be present on various UPEC PAIs, including *papA*, *papC*, *papG* and *pixA* (adherence), *kpsM (II)* and *kapsM (III)* (capsule), *fyuA*, *iroN*, *irp(1)*, *irp(2)*, *iucD*, *iutA* (iron acquisition system), *hlyA*, *cnf1* and *sat* (toxins) and *malX* and *usp* (various cellular functions) (Yoon et al. 2007) and the majority of these VGS were reduced by both disinfectants. For the UV disinfection, a total of 21 (81%) VGs were reduced in the range of 7.8% – 68.6% with an average reduction of 35.6% and the decrease was significant ($P<0.10$) for 10 VGs taken individually. Furthermore, 13 VGs previously reported on PAIs were reduced by UV and the reductions were significant ($P<0.10$) in six cases. None of the increases were statistically significant ($P>0.10$). For the PAA disinfection, 16 (62%) UPEC VGs were reduced by 9.3% – 100% with a mean reduction of 32.5%, and the decrease was significant ($P<0.10$) for the six VGs. Again, none of the increases were statistically significant ($P>0.10$). Among the VGs that were lost by PAA, at least 9 VGs were reported present on UPEC PAIs. On comparing the types and proportion of reduction of VGs in each VF, it is noticed that the decrease was between 16% – 45% by UV and 24% – 64% by PAA, and the variation of loss of VGs between UV and PAA was not statistically significant ($P>0.10$). Our data indicate that, though UV and PAA are believed to interact with different cellular components of *E. coli* (Dodd 2012, Kitis 2004), the proportion and types of reduction of VGs were similar by both disinfectants.

Table 4. Change in the prevalence of UPEC virulence gene in non-disinfected (ND) and UV or PAA-disinfected (D) samples

	No. of isolates possessing the UPEC virulence gene									
	UV					PAA				
	ND	D	% change[a]	Odds ratio	Stat. sig.[b]	ND	D	% change	Odds ratio	Stat. sig.
ISOLATES										
Total number	370	370	NA[c]	NA	NA	423	434	NA	NA	NA
Screen +ve	83	66	-20.5	1.3	–	99	75	-26.2	1.5	++
UPEC	36	17	-52.8	2.2	++	39	22	-52.0	1.9	++
UPEC VFs/VGs[e]										
Adherence										
papA	60	44	-7.78	1.3	–	73	56	+1.26	1.0	–
papC	48	26	-31.9	2.1	++	43	22	-32.5	1.8	+
papG	37	19	-35.4	2.0	+	40	23	-24.1	1.5	–
pixA	4	1	-68.6	3.3	–	1	1	+32.0	0.8	–
Capsule										
kpsM(II)	52	27	-34.7	2.4	++	45	26	-23.7	1.6	–
kpsM(III)	1	1	+25.8	0.8	–	2	2	+32.0	0.8	–
Iron acquisition										
fyuA	17	19	+40.6	0.6	–	45	44	+29.1	0.6	+
iroN	25	14	-29.6	1.6	–	17	10	-22.4	1.3	–
irp(1)	19	20	+32.4	0.7	–	45	43	+26.1	0.6	–
irp(2)	46	25	-31.7	2.0	++	54	49	+19.8	0.6	–
iucD	16	9	-29.3	1.5	–	19	11	-23.6	1.4	–
iutA	25	15	-24.5	1.5	–	25	10	-47.2	2.2	+
sitA	60	38	-20.4	1.9	+	55	30	-28.0	1.9	++
sitD	35	25	-10.2	1.2	–	51	27	-30.1	1.9	++
Toxins										
hlyA	33	11	-58.1	3.3	++	23	18	+3.3	1.0	–
vat	27	12	-44.1	2.2	+	33	18	-28.0	1.6	–
cnf(1)	13	6	-42.0	1.9	–	13	14	+42.2	0.7	–
cnf(2)	0	0	NC[d]	NC	NC	2	0	-100	NC	NC
sat	14	12	+7.8	0.9	–	20	21	+38.6	0.7	–
Associated functions										
b(1432)	14	6	-46.1	2.0	–	13	6	-39.1	1.7	–
deoK	30	15	-37.1	1.9	+	32	22	-9.3	1.2	–
gimB(orf1)	19	7	-53.7	2.5	+	12	12	+32.0	0.7	–
malX	53	23	-45.4	3.3	++	41	25	-19.5	1.4	–
pic	14	6	-46.1	2.0	–	10	5	-34.0	1.6	–
usp	27	15	-30.1	1.6	–	38	20	-30.5	1.7	–
chuA	40	25	-21.4	1.5	–	42	23	-27.7	1.7	–

[a] % change: +, increase; -, decrease

[b] Stat. sig. (statistical significance): $P<0.10$,+; $P<0.05$,++; $P>0.10$,–

[c] NA: Not applicable

[d] NC: Not computed due to the low frequency

[e] bolded virulence genes (17 of 26 VGs) were reported on various pathogenicity islands (PAIs) of UPEC

DISCUSSION

The frequency of UPEC isolates varied between 6 – 18% in the non-disinfected effluents of the treatment plants sampled (Figure 3). Although in the current study the isolates had been pre-screened for the likelihood of being UPEC by a PCR/Bioplex assay before microarray analysis, this variation is similar to the one observed in our previous study, in which all isolates were directly analyzed by microarray and where the proportions of UPECs in non-disinfected effluents were between 8% – 28% (Frigon et al. 2012). The variability between plants and samples may be due to differences in the quality of raw wastewaters and their sources, such as being only domestic, or a mixture of industrial and domestic (Anastasi et al. 2010). In our previous study, it was shown that activated sludge or physicochemical treatment processes both reduced the proportion of UPECs, and there was no evidence that the kind of treatment processes affected the reduction differently (Frigon et al. 2012), however, the presence of factors such as antimicrobial agents (quinolones) may affect the reduction. The current study shows that a small fraction (5%) of isolates contained the *qnrs1* gene which encode the quinolone antibiotics and decreased by UV and PAA. The presence of a significant population of UPECs after biological or physicochemical treatment, as demonstrated by the current study and previous ones (Anastasi et al. 2010, Frigon et al. 2012) may cause health problems to humans and animals upon discharge of such non-disinfected effluents to the environment.

The pathotyping of *E. coli* is based on the appropriate number and suitable combinations of VFs/VGs, hence the loss of one VG can render a pathogenic *E. coli* non-pathogenic. The data presented above showed that UV or PAA reduced the UPEC population among surviving *E. coli* isolates on average by 52 and 55%, respectively. It is possible that the metabolic functions encoded by certain genes make the cell more susceptible to disinfectant attacks or that the stability of the genes themselves may be influenced by exposure to disinfectants. To generate hypotheses on the likely mechanisms leading to this reduction, it is important to identify the kind of lesions or adaptations to which the cells likely respond. In the experiments presented here, the time of contact with the disinfectant was less than 2.5 min for UV and 30 – 60 min for PAA. As the relaxation time of gene induction is in the order of a few hours (Stephanopoulos et al. 1998), the experiments are not likely to detect cells that adapted their gene expression during the course of disinfection. Nonetheless, the cells that survived the disinfectant may have been injured by the disinfectant, and the defense mechanisms triggered may take effect during the growth period on the selective media.

UV and PAA were studied in parallel, which provided an opportunity to test whether the cellular functions are the basis for the reduction of UPECs in the *E. coli* population. As noted earlier, UV is a photo-irradiation agent acting mainly on pyrimidine bases and subsequently inhibiting DNA synthesis or causing a high mutation rate (Dodd 2012), whereas PAA is a chemical disinfectant which inactivates microorganisms mainly by oxidizing sulfhydryl and sulfur bonds in proteins and enzymes on the cell surface (Kitis 2004). Given that the inactivation modes of UV and PAA are different, they would generate different gene loss profiles if the cellular functions are the basis for UPEC reduction. A comparison of the frequency of UPEC virulence genes among the screen-positive isolates between non-disinfected and disinfected samples showed that both UV and PAA caused the loss of VGs to nearly the same extent (35.6% and 32.5%, respectively), which correlates with a similar reduction in UPEC levels (52 and 55%, respectively). The profiles of gene frequency reduction among the 26 UPEC VGs encoding the five VFs were similar for both UV and PAA (Table 4). Fifteen of the 25 VG tested (one was not tested due to its low frequency) showed overall reductions with UV and PAA disinfection, although these reductions were not always significant. These 15 VGs were distributed among the 5 VF categories, suggesting that no single VF was specifically affected. These results suggest that the mechanisms leading to the gene loss were likely to be the same despite the differences in the disinfectants' physical state and mode of attack.

A possible mechanism that could lead to this loss of VGs could be the deletion of UPEC PAIs following UV or PAA treatment. In our data, the majority (13 out of 17) of VGs that had been reported previously to be present in various UPEC PAIs (Yoon et al. 2007) were reduced by UV or PAA (Table 4). It has been shown that a number of UPEC PAIs are unstable and can be lost at a higher rate upon exposure to a temperature of 20°C (rate: 10.4×10^{-5}) instead of the normal 37°C (rate: 0.54×10^{-5}) (Middendorf et al. 2001b) or the presence of antimicrobial agents such as quinolones in the growth media (Soto et al. 2006). It is notable that quinolones, which are inhibitors of DNA gyrase, cause DNA lesions and induce the DNA repair systems such as the SOS response (Piddock et al. 1990, Soto et al. 2006). This effect is also well documented from gene expression studies for the exposure of *E. coli* to UV (Tornier et al. 2001). Similar gene expression studies of *Bacillus cereus* have found induction of DNA repair systems by hydrogen peroxide and PAA, but not by a quaternary ammonium compound (QAC) or by hypochlorous acid (Ceragioli et al. 2010). Together, our data and the results from these other studies suggest that upon exposure to the disinfectants, DNA lesions may promote the loss of UPEC PAIs, which leads to a reduction in the prevalence of UPECs in the *E. coli* population.

CONCLUSIONS

The UV fluence needed to achieve the target 200 CFU/100 mL (~2.30 log) *E. coli* concentration in the effluents from two biological treatment plants (AS and BF) was in the range of 7 – 15 mJ/cm^2 whereas for the two physicochemical treatment plants (PC1 and PC2), the doses were higher at 15 – 30 mJ/cm^2. The PAA doses to reach the target *E. coli* level were in a narrower range of 0.9 – 2.0 mg/L for all plants. Centrifugation and filtration eliminated the tailing effects and reduced the required UV dose but not the PAA dose to attain the target *E. coli* levels. The proportion of UPECs in the non-disinfected effluent samples used in both disinfection studies ranged from 6 % – 18%, while in the UV and PAA disinfected effluent samples UPECs constituted only 2% – 6% and 0% – 12% respectively. The reduction in the proportion of the UPEC sub-population was observed in all disinfection experiments; average reductions were 55% and 52% for UV and PAA, respectively. The UPEC pathotypes and genes among the free-swimming *E. coli* sub-population appeared to be more sensitive to the effects of disinfection by UV than the particle-associated ones as filtration enhanced (from 61% to 71%) the reduction of UPEC isolates and promoted the loss of UPEC genes among the non-UPEC isolates. Both UV and PAA caused a loss of UPEC VGs with average reductions of 35.6% and 32.5%, respectively. Irrespective of differences in the reaction mechanisms of UV and PAA in the *E. coli* cell, the types of VGs lost and their proportions were similar for both disinfectants.

ACKNOWLEDGEMENTS

This work was funded by the Natural Sciences and Engineering Research Council of Canada (NSERC) (grant no. STPGP 35117-07) in collaboration with Trojan Technologies Inc. We thank the plant operators of the four municipal wastewater treatment plants for assisting with effluent sample collection.

REFERENCES

Anastasi, E.M.; Matthews, B.; Gundogdu, A.; Vollmerhausen, T.L.; Ramos, N.L.; Stratton, H.; Ahmed, W.; Katouli, M. (2010) *Prevalence and persistence of Escherichia coli strains with uropathogenic virulence characteristics in sewage treatment plants*; Applied and Environmental Microbiology 76(17), 5882-5886.

APHA; AWWA; WEF. (2005) *Standard methods for the examination of water and wastewater*; 21st ed. American Public Health Association, Washington, DC.

Bruant, G.; Maynard, C.; Bekal, S.; Gaucher, I.; Masson, L.; Brousseau, R.; Harel, J. (2006) *Development and validation of an oligonucleotide microarray for detection of multiple virulence and antimicrobial resistance genes in Escherichia coli*; Applied and Environmental Microbiology 72(5), 3780-3784.

Caretti, C.; Lubello, C. (2003) *Wastewater disinfection with PAA and UV combined treatment: A pilot plant study*; Water Research 37(10), 2365-2371.

Ceragioli, M.; Mols, M.; Moezelaar, R.; Ghelardi, E.; Senesi, S.; Abee, T. (2010) *Comparative transcriptomic and phenotypic analysis of the responses of bacillus cereus to various disinfectant treatments*; Applied and Environmental Microbiology 76(10), 3352-3360.

Chen, L.; Xiong, Z.; Sun, L.; Yang, J.; Jin, Q. (2012) *VFDB 2012 update: Toward the genetic diversity and molecular evolution of bacterial virulence factors*; Nucleic Acids Research 40(D1), D641-D645.

Clermont, O.; Bonacorsi, S.; Bingen, E. (2000) *Rapid and simple determination of the Escherichia coli phylogenetic group*; Applied and Environmental Microbiology 66(10), 4555-4558.

Dodd, M.C. (2012) *Potential impacts of disinfection processes on elimination and deactivation of antibiotic resistance genes during water and wastewater treatment*; Journal of Environmental Monitoring 14(7), 1754-1771.

Escobar-Páramo, P.; Clermont, O.; Blanc-Potard, A.-B.; Bui, H.; Le Bouguénec, C.; Denamur, E. (2004) *A Specific Genetic Background Is Required for Acquisition and Expression of Virulence Factors in Escherichia coli*; Molecular Biology and Evolution 21(6), 1085-1094.

Frigon, D.; Biswal, B. K.; Mazza, A.; Masson, L.; Gehr, R. (203) *Biological and Physicochemical Wastewater Treatment Processes Reduce the Prevalence of Virulent Escherichia coli*; Applied and Environmental Microbiology. 79 (3), 835- 844.

Gehr, R.; Wagner, M.; Veerasubramanian, P.; Payment, P. (2003) *Disinfection efficiency of peracetic acid, UV and ozone after enhanced primary treatment of municipal wastewater*; Water Research 37(19), 4573-4586.

Hamelin, K.; Bruant, G.; El-Shaarawi, A.; Hill, S.; Edge, T.A.; Bekal, S.; Fairbrother, J.M.; Harel, J.; Maynard, C.; Masson, L.; Brousseau, R. (2006) *A virulence and antimicrobial resistance DNA microarray detects a high frequency of virulence genes in Escherichia coli*

Isolates from Great Lakes recreational waters; Applied and Environmental Microbiology 72(6), 4200-4206.
Hamelin, K.; Bruant, G.; El-Shaarawi, A.; Hill, S.; Edge, T.A.; Fairbrother, J.; Harel, J.; Maynard, C.; Masson, L.; Brousseau, R. (2007) *Occurrence of virulence and antimicrobial resistance genes in Escherichia coli isolates from different aquatic ecosystems within the St. Clair River and Detroit River areas*; Applied and Environmental Microbiology 73(2), 477-484.
Hamilton, M.J.; Hadi, A.Z.; Griffith, J.F.; Ishii, S.; Sadowsky, M.J. (2010) *Large scale analysis of virulence genes in Escherichia coli strains isolated from Avalon Bay, CA*; Water Research 44(18), 5463-5473.
Health Canada (2012) *Guidelines for Canadian Recreational Water Quality, Third Edition. Water, Air and Climate Change Bureau, Healthy Environments and Consumer Safety Branch, Health Canada, Ottawa, Ontario. (Catalogue No H129-15/2012E).*
Hlavsa, M.C.; Roberts, V.A.; Anderson, A.R.; Hill, V.R.; Kahler, A.M.; Orr, M.; Garrison, L.E.; Hicks, L.A.; Newton, A.; Hilborn, E.D.; Wade, T.J.; Beach, M.J.; Yoder, J.S. (2011) *Surveillance for waterborne disease outbreaks and other health events associated with recreational water - United States, 2007-2008*; Morbidity and Mortality Weekly Report 60(SS-12), 1-32.
Kaper, J.B.; Nataro, J.P.; Mobley, H.L.T. (2004) *Pathogenic Escherichia coli*; Nature Reviews Microbiology 2(2), 123-140.
Kitis, M. (2004) *Disinfection of wastewater with peracetic acid: a review*; Environment International 30(1), 47-55.
Koivunen, J.; Heinonen-Tanski, H. (2005a) *Inactivation of enteric microorganisms with chemical disinfectants, UV irradiation and combined chemical/UV treatments*; Water Research 39(8), 1519-1526.
Koivunen, J.; Heinonen-Tanski, H. (2005b) *Peracetic acid (PAA) disinfection of primary, secondary and tertiary treated municipal wastewaters*; Water Research 39(18), 4445-4453.
Kollu, K.; Örmeci, B. (2012) *Effect of particles and bioflocculation on ultraviolet disinfection of Escherichia coli*; Water Research 46(3), 750-760.
Lee, C.A. (1996) *Pathogenicity islands and the evolution of bacterial pathogens*; Infectious Agents and Disease 5(1), 1-7.
Middendorf, B.; Blum-Oehler, G.; Dobrindt, U.; Mühldorfer, I.; Salge, S.; Hacker, J. (2001a) *The Pathogenicity Islands (PAIs) of the Uropathogenic Escherichia coli Strain 536: Island Probing of PAI [image]*; The Journal of Infectious Diseases 183(ArticleType: research-article / Issue Title: Supplement 1. The Millenium Symposium on Pyelonephritis and Urinary Tract Infection / Full publication date: Mar. 1, 2001 / Copyright © 2001 Oxford University Press), s17-s20.
Middendorf, B.; Blum-Oehler, G.; Dobrindt, U.; Mühldorfer, I.; Salge, S.; Hacker, J. (2001b) *The Pathogenicity Islands (PAIs) of the Uropathogenic Escherichia coli Strain 536: Island Probing of PAI II536*; Journal of Infectious Diseases 183(Supplement 1), S17-S20.
Middendorf, B.; Hochhut, B.; Leipold, K.; Dobrindt, U.; Blum-Oehler, G.; Hacker, J. (2004) *Instability of pathogenicity islands in uropathogenic Escherichia coli 536*; Journal of Bacteriology 186(10), 3086-3096.
Mokracka, J.; Koczura, R.; Jabłońska, L.; Kaznowski, A. (2011) *Phylogenetic groups, virulence genes and quinolone resistance of integron-bearing Escherichia coli strains isolated from a wastewater treatment plant*; Antonie van Leeuwenhoek, International Journal of General and Molecular Microbiology 99(4), 817-824.

Morowitz, H.J. (1950) *Absorption effects in volume irradiation of microorganisms*; Science 111(2879), 229-230.
Oliver, B.G.; Cosgrove, E.G. (1975) *The disinfection of sewage treatment plant effluents using ultraviolet light*; The Canadian Journal of Chemical Engineering 53(2), 170-174.
Piddock, L.J.V.; Walters, R.N.; Diver, J.M. (1990) *Correlation of quinolone MIC and inhibition of DNA, RNA, and protein synthesis and induction of the SOS response in Escherichia coli*; Antimicrobial Agents and Chemotherapy 34(12), 2331-2336.
Sokal, R.R.; Rohlf, F.J. (1969) *Biometry; the principles and practice of statistics in biological research;* W.H. Freeman, San Francisco.
Soto, S.M.; Jimenez De Anta, M.T.; Vila, J. (2006) *Quinolones induce partial or total loss of pathogenicity islands in uropathogenic Escherichia coli by SOS-dependent or -independent pathways, respectively*; Antimicrobial Agents and Chemotherapy 50(2), 649-653.
Stephanopoulos, G.; Aristidou, A.A.; Nielsen, J.H. (1998) *Metabolic engineering : principles and methodologies;* Academic Press, San Diego.
Tornier, C.; Bessone, S.; Varlet, I.; Rudolph, C.; Darmon, M.; Fleck, O. (2001) *Comparative gene expression profiles following UV exposure in wild-type and SOS-deficient Escherichia coli*; Genetics 158(1), 41-64.
Wagner, M.; Brumelis, D.; Gehr, R. (2002) *Disinfection of wastewater by hydrogen peroxide or peracetic acid: Development of procedures for measurement of residual disinfectant and application to a physicochemically treated municipal effluent*; Water Environment Research 74(1), 33-50.
Yoon, S.H.; Park, Y.K.; Lee, S.; Choi, D.; Oh, T.K.; Hur, C.G.; Kim, J.F. (2007) *Towards pathogenomics: A web-based resource for pathogenicity islands*; Nucleic Acids Research 35(SUPPL. 1), D395-D400.

An Assay for Estimating UV Disinfection Dose by Direct Quantification of DNA Damage in Indigenous Microorganisms in Drinking Water

James Blyth, Imperial College London, London, United Kingdom
Michael R. Templeton*, Imperial College London, London, United Kingdom
Bill Cairns, Trojan Technologies, London, Ontario, Canada
*Department of Civil and Environmental Engineering,
Imperial College London
London, United Kingdom SW7 2AZ
m.templeton@imperial.ac.uk, +44(0)2075946099

ABSTRACT

A new method was developed for estimating the ultraviolet (UV) dose that is delivered by operating UV disinfection reactors in drinking water treatment plants, through the quantification of pyrimidine (i.e. thymine-thymine) dimer formation in the DNA of indigenous microorganisms (i.e. no organisms spiked into the water). An enzyme-linked immuno-sorbent assay (ELISA) was applied to the indigenous DNA that was extracted from water samples that were collected from drinking water treatment plants and from lab tap water and exposed to collimated beam UV doses, resulting in a near-linear relationship between the assay response and UV dose. The new method has the potential to provide a relatively quick, sampling-based means of monitoring the UV dose being delivered by operating UV reactors in drinking water treatment plants, without the need to spike a biodosimeter or actinometer into the water.

KEYWORDS

Ultraviolet disinfection, biodosimetry, dimer, molecular methods, DNA, drinking water.

INTRODUCTION

The current method for validating the drinking water disinfection dose delivered by UV reactors is biodosimetry, whereby the manufacturer of the reactor conducts a series of tests which involve passing a microorganism with known UV dose-inactivation characteristics through the reactor under different flow rates, reactor conditions (e.g. lamp ages), and water quality conditions (e.g. a range of UV transmittances); the so-derived UV disinfection doses are linked to sensor readings which are then used to estimate the online UV dose when the reactor is operating in treatment plants (USEPA, 2006). Online biodosimetry (i.e. such a test conducted on an already installed UV reactor) is not commonly done because of the challenge of injecting biodosimeter organisms into the flow through operating drinking water treatment plants. In this study it was considered whether the indigenous microorganisms that is already present in water could be used as biodosimeters by measuring the DNA damage (i.e. pyrimidine dimer formation) within the microorganisms as an indication of the delivered UV dose by the reactor. If successful, this would provide a water sampling-based method for checking the disinfection dose of operating UV reactors, to complement and check existing sensor-based estimates of UV dose. The goal was to achieve an assay that would be relatively quick (i.e. no longer than the microbiological steps used in biodosimetry) and use inexpensive and commonly available materials and techniques.

METHODOLOGY

In brief, the conceptual approach is that an enzyme-linked immuno-sorbent assay (ELISA), similar to that described by Roza et al. (1988), is applied to indigenous DNA that is extracted from water samples collected immediately upstream and downstream of the UV reactor and the assay response would be compared against the dose-response relationship obtained from UV collimated beam exposures of the same water in order to determine the UV dose applied by the reactor.

Samples were filtered onto nitrocellulose filters and DNA was extracted using a PowerWater® DNA isolation kit; the kit used a standard technique in which cells are lysed in a lysis buffer and proteins and lipids are removed by low pH salt precipitation. DNA at low pH was bound to a resin column and washed with ethanol prior to elution from the column. The concentration of DNA was determined using PicoGreen® (Invitrogen) with fluorescence response using a plate reader fitted with a 485nm 10-nm band-pass excitation filter and 520nm 10-nm band-pass emission filter. Ninety-six well polystyrene plates were coated with poly-L-lysine for 30 minutes at room temperature and were washed three times with tris-buffered-saline before 50 µl of extracted DNA was incubated in the coated wells for 16-20 hours at 37 °C. After binding DNA, wells were washed three times with 0.05% Tween 20 in phosphate buffered saline (PBST). A primary antibody, mouse anti-thymine dimer monoclonal antibody (H3 clone, Sigma), was diluted 1:1000 in PBST containing 1% heat inactivated fetal bovine serum (PBST-FCS). Fifty microlitres of the primary antibody dilution was incubated in wells for 45 minutes at 37 °C. Wells were washed three times with PBST-FCS and incubated for one hour at room temperature with 50 µl of a 1:1000 dilution in PBST-FCS of a polyclonal goat anti-mouse secondary antibody conjugated to horseradish peroxidase. Following aspiration of the secondary antibody, wells were again washed three times with PBST-FCS and then three times with citrate buffer at pH 5.0. A two times stock of development buffer was made in citrate buffer pH 5.0 and contained 2,2'-azino-bis-3-ethylbenzothiazoline-6-sulphonic acid (ABTS) at 4.8 mg/ml and hydrogen peroxide at 180 µg/ml. Twenty five microliters of this stock was mixed in equal volumes in citrate buffer pH5.0 in the wells and the response , a coloured product, was measured by absorbance at 405 nm in a multi-well plate reader. In this study, a low pressure high output UV collimated beam apparatus was used to expose water samples to known UV doses calculated using the method of Bolton and Linden (2003).

RESULTS AND DISCUSSION

An example of the relationship between the assay response (i.e. dimers formed, normalised per unit of DNA in the sample) and the UV doses applied by the collimated beam is shown in Figure 1. This figure was generated using water samples collected post-filtration from a typical water treatment plant in the UK (processes upstream of the sampling point were ozonation and granular activated carbon filtration), i.e. at the stage of treatment at which a UV disinfection reactor would typically be installed. As shown, the linear approximation is quite good but a curved fit achieved a slightly higher R^2 value, suggesting slight tailing of the relationship at higher UV doses. A similar good relationship between assay response and UV dose was observed when exposing a sample from a second water treatment plant.

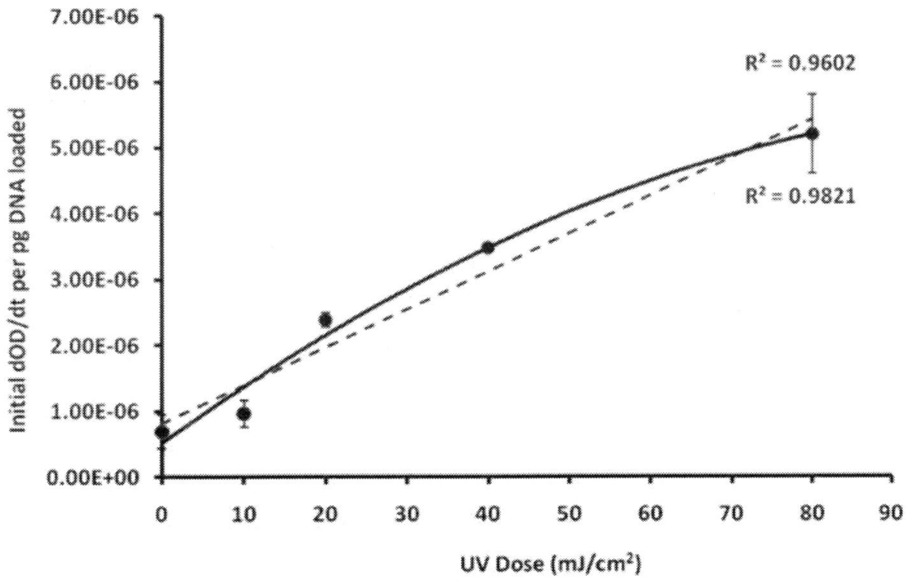

Figure 1 - The assay response versus UV dose from collimated beam exposures of indigenous microorganisms in water sampled post-filtration from a UK drinking water treatment plant. The units on the y-axis represent the fluorescence response normalised by the amount of DNA present in the sample.

The next step of the research is to compare the assay-estimated UV disinfection dose applied by online UV reactors against the sensor-based, biodosimetry-validated doses or doses estimated by computational fluid dynamics modelling. Also, this study only considered low pressure (monochromatic) UV light, so an extension would be to consider whether the dimer assay response relationship applies to medium pressure (polychromatic) UV doses as well.

CONCLUSIONS

Results to-date suggest that the developed dimer quantification assay produces a reproducible, consistent response to UV dose in real drinking water samples. The method provides the potential basis for estimating doses delivered by operating UV reactors without the need to spike a biodosimeter or actinometer into the flow.

REFERENCES

Bolton, J.R. and Linden, K. (2003). *Standardization of methods for fluence (UV dose) determination in bench-scale UV experiments.* Journal of Environmental Engineering-ASCE 129, 209-216.

Roza, L., Vanderwulp, K. J. M., Macfarlane, S. J., Lohman, P. H. M. and Baan, R. A. (1988). *Detection of cyclobutane thymine dimers in DNA of human cells with monoclonal-antibodies raised against a thymine dimer-containing tetranucleotide.* Photochemistry and Photobiology, 48, 627-633.

USEPA (2006). *Ultraviolet Disinfection Guidance Manual for the Final Long-Term 2 Enhanced Surface Water Treatment Rule.* Office of Water, EPA 815-R-06-007, Washington DC.

From systematic process optimization to alternative indicator development; How MSDGC has utilized innovative approach to optimize the chlorination disinfection efficacy

Ting Lu[1*], Ryan Welsh[2], and Donald Linn[2]

[1]Black & Veatch Corporation
[2]Metropolitan Sewer District of Greater Cincinnati, City of Cincinnati, Cincinnati
* lut@bv.com

Abstract The Metropolitan Sewer District of Greater Cincinnati (MSDGC) is responsible for conveying and treating an average of 192 million gallons per day (MGD) of wastewater and it operates seven major treatment plants in Hamilton County. The Mill Creek WWTP – the largest wastewater treatment facility operated by MSDGC – as part of this plan, is currently undergoing upgrades to provide secondary treatment (BOD and TSS removal only) for up to 910,000 m^3/d (240 MGD). This paper presents how MSDGC has used unit process optimization and a novel approach to indicator detection to control dosage and disinfection costs.

Key words: Disinfection, pathogen

Figure 1. Aerial view of the Mill Creek WWTP including on-going secondary treatment process and disinfection channel.

Introduction

The Metropolitan Sewer District of Greater Cincinnati (MSDGC) is responsible for conveying and treating an average of 192 million gallons per day (MGD) of wastewater and it operates seven major treatment plants in Hamilton County. Figure 1 shows an overview of plant configuration. Disinfection at the Mill Creek WWTP is achieved by sodium hypochlorite injection. Because this plant serves a substantial number of industrial users, the wastewater effluent color changes quite often. Due to the current limitation, stable operation is a challenge due to limited treatment capacity due to current long-term construction projects and wet weather flows. Consequently, chlorination dosage is not stable. During wet weather, the flow will bypass secondary treatment after it reaches 130 MGD. The primary flow and secondary flow combine to go through disinfection channel for final disinfection treatment.

Unit process optimization

The plant has established a performance based operation as follows: establishing key performance indicators with targets, monitoring/controlling indicators to targets, determining meeting target or not, and documenting the action. A five-step process has been developed to systematically improve operational performance (Figure 3). This process has been started with disinfection, as the last step in wastewater treatment process and worked all the way to the beginning of the wastewater treatment process. Figure 3 below shows the key elements that impact the disinfection efficacy. More importantly, this effort has identified the upstream processes and outputs that played key roles. The results presented in this study show significant chlorine dosage reduction since and cost control since 2009 (Figure 4).

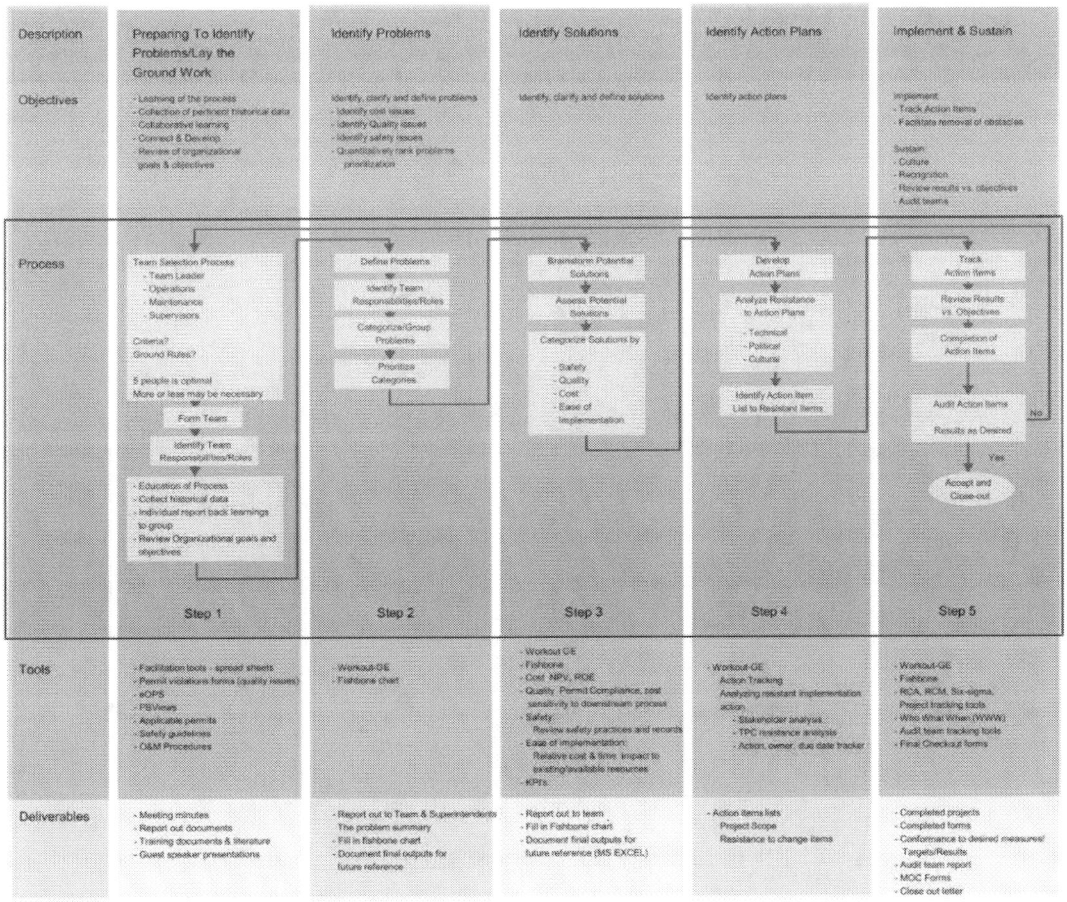

Figure 2. Five step process developed by the Mill Creek operation team.

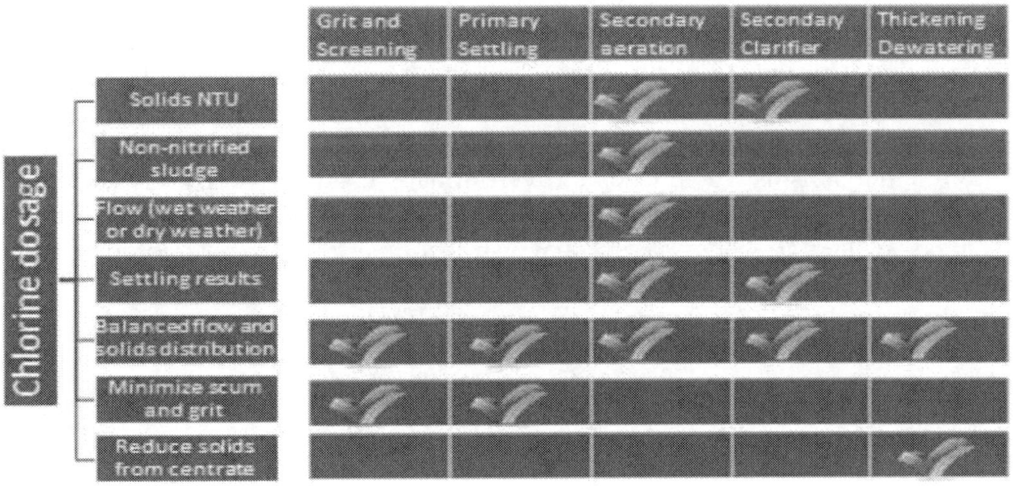

Figure 3. Key elements from the upstream processes that impact the chlorination usage and disinfection efficacy.

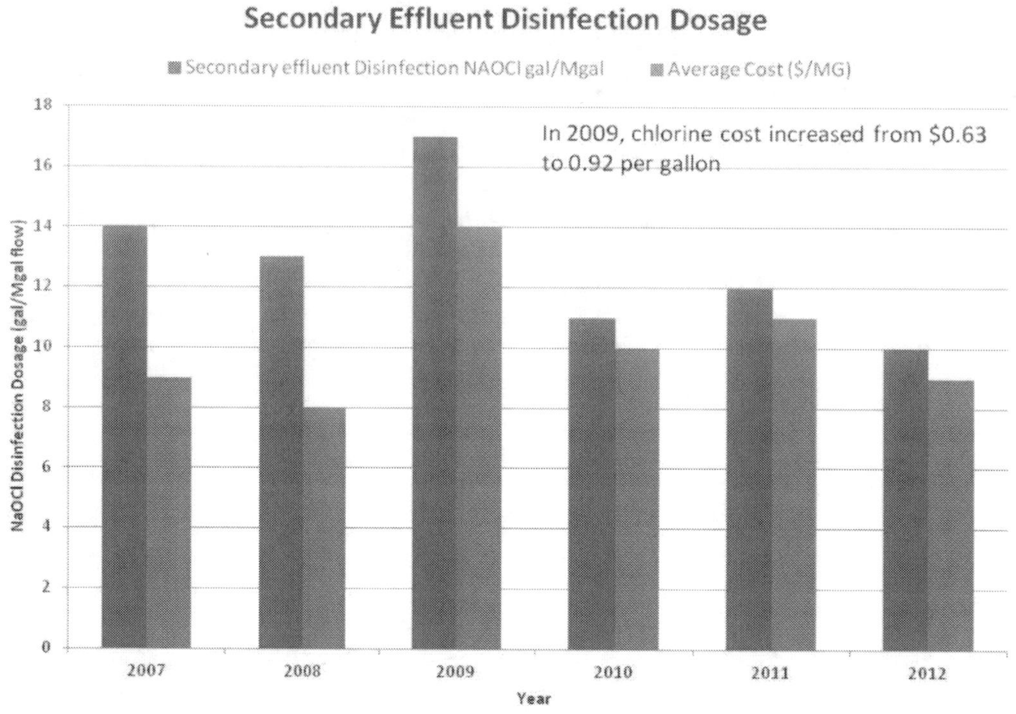

Figure 4. Secondary Effluent Disinfection dosage since the year of 2007.

Innovative molecular tools for pathogen indication

In this section, the authors present a cutting-edge technology that effectively monitors pathogen and disinfection efficacy for both wastewater and drinking water process.

Monitoring the true pathogens is as important as selecting the right disinfectant, dosage, and contact time. Currently the culture based methods are accepted as "gold standard" for *E.coli* accounts and evaluating the effectiveness of the disinfection process (Waters et al., 2006). The drawback of this method is that it excludes Viable But Non-Culturable (VBNC) microorganisms which leads to overestimate disinfection efficacy (Camper and McFeters, 1979). In order to protect public health, development of alternative procedures that can be applied to better determine disinfection efficacy for microorganisms is developed based on the "ribosome genesis" green fluorescent protein (GFP) reporter system. Ribosome genesis is a sensitive indicator to environmental change (Lu et al., 2009). This system was developed to target ribosome genesis, which is a direct indicator for cell growth. Lu et al. (2011) described the detailed method of constructing the GFP reporter system.

The reported system was first applied to *E. coli* cells during batch growth to monitor cellular activity, and it proved as an accurate, reliable and robust method for detection of active cells *in situ*. Then it was further applied to monitor real-time bacterial growth activity when cells were exposed to varying levels of chlorine disinfectant. Bacterial activity was successfully identified and quantified by this method.

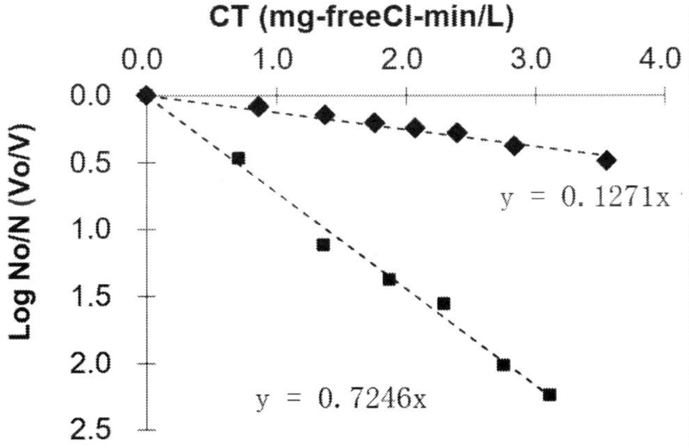

V_o, initial activity index value of transformed *E. coli* cells without exposure to chlorine; V, activity index of transformed *E. coli* cells at defined time points. N_o, initial concentration of *E. coli* K12 without exposure to chlorine (CFU/mL); N, concentration of *E. coli* K12 at defined time points.

Figure 5. Inactivation curve of the transformed *E. coli* based on the GFP-based reporter system (diamonds) and culture-based method (squares).

Figure 5 shows that in order to achieve the same log removal rate for bacteria, it requires a larger CT value calculated by the GFP reported system compared with that calculated by culture-based method. For instance, in order to inactivate 99% of E. coli, the CT value was 15.7 mg-free Cl-min/L based on the GFP-based reporter system, while this CT value was 2.8 mg-free-Cl-min/L based on culture-based method. This is because the quantification of the culture-based method is based on how many cells form into colonies, while the quantification of the GFP-reporter system is based on the gray value of fluorescent signals. This demonstrates that the method developed in this study is able to characterize bacterial activity, including dead, inactivated but not killed (IBNK), and live. This result shows that the disinfection efficacy results estimated by the GFP reported system was lower compared to that calculated by the culture based method in order to inactivate the same amount of the *E. coli* cells due to existence of viable but non culturable cells.

Conclusion

In summary, this study presents an integrated approach coupling practical process optimization and cutting edge research to increase chlorine disinfection efficacy and reduce human health risk. The process optimization efforts provide a common understanding to all the operation staff and enhance everyone's knowledge on wastewater treatment process. The results showed that GFP-reporter method is accurate and reliable to determine chlorination disinfection efficacy.

One of the more difficult challenges posed by National Pollutant Discharge Elimination System (NPDES) permitting is disinfecting either combined sewer or wet weather overflows. These flows are generally highly variable and infrequent, with highly variable influent quality that pose design and operational challenges. From a utility perspective, disinfection approach for wet weather facility needs to cost effectively supports highly variable wet weather flows. This study is **SIGNIFICANT** because it opened up a possibility to monitor real time cellular activity at a sing-cell level for environmental samples and it provides a more accurate measurement of disinfection process to protect public health. GFP reporter system is envisioned to provide an alternative indication of disinfection efficacy and it could provide important information for future regulations.

References

Camper, A.K.; McFeters, G.A. 1979. "Chlorine Injury and the Enumeration of waterborne coliform Bacteria" Applied and Environmental Microbiology 37:633-641.

Greater Cincinnati Water Works, 2012 "Chlorine Effects on Medium Pressure UV Reactor Sleeve Fouling", UV Chlorination Fouling Study Internal Report.

Heitz, Michael, George, B., and Welp, J., "Using an artificial neural network to predict chlorine consumption"

Lu, T., Stroot, P.G., and Oerther, D.B. (2009) Reverse transcription of 16S rRNA to monitor ribosome synthesizing bacterial populations in the environment. *Applied EnvironmentalMicrobiology* **75**: 4589-4598.

Lu, T., Chen, Y., Santo Domingo, J., and Oerther D., 2011. "Evaluation of Disinfection Efficacy by a Green Fluorescent Protein (GFP) Reporter System" WEF 2011, Disinfection Conference, 2011Proceedings of Research Symposium, Cincinnati OH, April 10-12, 2011 pg.243-248.

Waters, S.M., Murphy, R.A., and Power, R.F.G. (2006) Characterisation of prototype Nurmi cultures using culture-based microbiological techniques and PCR-DGGE. *International Journalof Food Microbiology* **110**: 268-277.

The Influence of Extracellular Polymeric Substance Composition on Bacteria Disinfection By Monochloramine

Kimberly Coburn[1], Zheng Xue[1], Youngwoo Seo[2]

[1]Civil Engineering, University of Toledo
[2]Associate Professor of Civil and Chemical Engineering, University of Toledo

3006 Nitschke Hall, Mail Stop 307
2801 W. Bancroft St.
Toledo, OH 43606-3390
Phone: 419-530-8113

Abstract

As a result of the implementation of the Stage 2 Disinfectants and Disinfection Byproduct Rules (DBPRs), many water treatment facilities in the United States began to utilize chloramines as their primary or secondary disinfectant. Although chloramines have been known to have better efficiency for biofilm control than chlorine, our understanding of the reaction, stability, transport and fate of chloramine in water distributions is unfortunately incomplete.

In this study, the impact of capsular extracellular polymeric substance (EPS) composition on bacteria disinfection by monochloramine was investigated using Matrix-Assisted Laser Desorption/Ionization Time-of-Flight/Time-of-Flight Mass Spectrometry (MALDI-TOF/TOF-MS) coupled with the multivariate statistical approaches of Canonical Correlation Analysis (CCA) and Partial Least Squares (PLS) regression for *Pseudomonas aeruginosa*. This research will not only provide fundamental information regarding the reactivity of monochloramine with bacterial EPS but it will illuminate some of the illusive reactions that bacteria undergo during the disinfection process.

Keywords

Monochloramine, Extracellular Polymeric Substance (EPS), MALDI-TOF/TOF-MS, CCA, PLS, and *Pseudomonas*.

Introduction

Monochloramine use in water distribution systems has dramatically increased in recent years due to stringent Stage 2 Disinfectants and Disinfection By-product Rules (DBPRs). Many water utilities seeking to form lower levels of disinfectant by-products have become compliant with DBPRs by switching from chlorine to monochloramine (Kirmeyer 2004). Other advantages of monochloramine include persistence in the water distribution system, effectiveness as a final disinfectant by being able to penetrate into biofilms better than chlorine, and minimized taste and odors (UnitedStatesEnvironmentalProtectionAgencyOfficeofWater 1999; Kirmeyer 2004). As a consequence of these advantages, monochloramine usage has increased from 68 million customers in 1998 to an estimated 100 million United States customers in 2012 (AmericanWaterWorksAssociation 2012).

Despite increasing reliance on monochloramine, the mechanisms behind monochloramine disinfection are still not completely understood despite its use in water treatment facilities since 1918 (Kirmeyer 2004). In order to enhance the overall efficacy of monochloramine disinfection, it is important to understand the disinfection mechanism. The mechanism is thought to involve the inhibition of protein associated processes; however, the detailed mechanism remains elusive (Jacangelo, Olivieri et al. 1991).

Moreover, microorganisms tend to aggregate together into what is known as biofilm, where the aggregates can not only adhere to each other but to surfaces. Biofilm formation does affect bacterial resistance to disinfectants where the amount of extracellular polymeric substances (EPS) in planktonic bacteria will also impact bacteria susceptibility (Xue, Hessler et al. 2012). EPS may account for up to 50 to 90% of total organic matter in a biofilm. This is of critical importance in water distribution systems because it has been found that at least 95% of total bacteria biomass is facilitated by the production of extracellular material (Flemming, Tamachkiarowa et al. 1998).

Thus, the primary objective of this study was to elucidate the role and influence of the chemical composition of EPS during the monochloramine disinfection process. In order to achieve this objective the bacteria *P. aeruginosa* was examined to determine its susceptibility, reactivity, and reaction sites to monochloramine. The changes that the components of EPS underwent during the disinfection process were determined using Matrix-Assisted Laser Desorption/Ionization Time-of-Flight/Time-of-Flight Mass Spectrometry (MALDI-TOF/TOF-MS). Multivariate statistical analyses were then used to determine which components of the EPS that contributed the most to the variability during the monochloramine disinfection process.

Methods

Bacteria Growth

This study utilized the well documented opportunistic pathogen *Pseudomonas aeruginosa*, which has been recovered in water distribution systems (Hardalo and Edberg 1997). For this study, the wild *P. aeruginosa* strain PAO1 was provided by Dr. Daniel Hassett of the University of Cincinnati (Maniatis, Fritsch et al. 1982). Bacteria were grown in one-tenth Luria-Bertani (LB) broth (2.5 g/L, Difco Laboratories, Detroit, MI) at 37°C. The bacteria were harvested in the late exponential phase, 16 hours, so consistent quality and quantity of capsular EPS could be harvested for each experiment (Wang, Hessler et al. 2012). The cells were harvested by centrifugation at 2,000 g for 10 minutes to minimize the removal of EPS (model 5804R centrifuge, Eppendorf, Hamburg, Germany) and then washed with Chlorine Demand Free (CDF) Buffer (0.54 g Na_2HPO_4 and 0.88 g KH_2PO_4 per liter) (Eboigbodin and Biggs 2008; Xue, Hessler et al. 2012). The optical density (OD) of the washed culture suspensions were measured and adjusted at 600 nm to obtain a concentration of 10^7 cells/mL as bacterial suspension for batch experiments. EPS extraction was done using a combination of centrifugation and cation exchange resin (CER) in order to minimize cell lysis (Wingender, Neu et al. 1999).

Monochloramine Preparation and Exposure

Stock monochloramine solution (2 mg/L) was prepared by combining solutions of sodium hypochlorite (5.65-6%, pH=8.3) and ammonium chloride (0.2 mM/L, pH=8.3) in a 4:1 chlorine-to-ammonia-nitrogen mass ratio so that monochloramine was the dominant species (Xue, Hessler et al. 2012). Monochloramine concentrations were measured using Monochlor F Reagent colorimetric methods using a DR/2700 spectrophotometer (HACH Company, Loveland, CO). Chloramine reactions were quenched using CDF buffer containing sodium thiosulfate (1 mM/L).

MALDI-TOF/TOF-MS

MALDI mass spectroscopy was employed in order to obtain more detailed information for monochloramine reaction sites in the EPS matrix. MALDI spectroscopy is a sensitive tool that has recently become a popular and versatile method to analyze biological macromolecules (Lay 2001; Marvin, Roberts et al. 2003). This is accomplished by embedding samples into analyte molecules known as a matrix that desorb laser energy so that analyte molecules can become ionized and traverse though the spectrometer. The two matrices, α-cyano-4-hydroxycinnamic acid (CHCA) and 2,5-dihydroxy benzoic acid (DHB), were mixed in 1:1 ratio in order to obtain the most robust spectra of glycoproteins (Gonnet, Lemaitre et al. 2003; Laugesen and Roepstorff 2003). Each matrix was saturated in 1:1 water–acetonitrile with 0.1% trifluoroacetic acid and 10 mM ammonium monobasic phosphate to minimize the formation of matrix adducts (Zhu and Papayannopoulos 2003). Samples were dried on a polished steel plate using a vacuum desiccator in the sandwich crystallization method. The MALDI-TOF/TOF-MS instrument UltrafleXtreme (Bruker Daltonics, Bremen, Germany) was utilized with the Bruker program WARP-LC to automatically acquire a random raster of samples. Bruker's FlexAnalysis software was used to smooth and subtract the baseline of all the spectra before multivariate analysis (Suckau, Resemann et al. 2003).

Multivariate Analysis

In this study, multivariate analysis was used to determine the relationship between the MALDI spectra and the inactivation of the bacteria (or time). Canonical correlation analysis (CCA) was performed to assess whether there was any significant correlation between the spectra and the inactivation data. Based on these results, the correlation was tested by constructing linear partial least squares (PLS) regression models for identification of individual components that were associated with the most variation from the disinfection treatment. These multivariate analyses were completed using the public domain available statistical software R 2.15.1 and several packages from the Comprehensive R Archive Network (CRAN) (RCoreTeam 2012). Raw spectra data were loaded into R using the readBrukerFlexData package (Gibb 2012). Preprocessing was done by using the MALDIquant package where spectra were smoothed, baseline-corrected, normalized, aligned, and approximated every 0.1 m/z (Gibb and Strimmer 2012). The preprocessing allowed at least three points on each peak to be represented for comprehensive data analysis. Canonical correlation analysis was performed using the R vegan package (Dixon 2009). Partial least squares regression analysis was done using the PLS package and VIP add-on, which automatically cross validates its models (Mevik, Wehrens et al. 2011).

Results

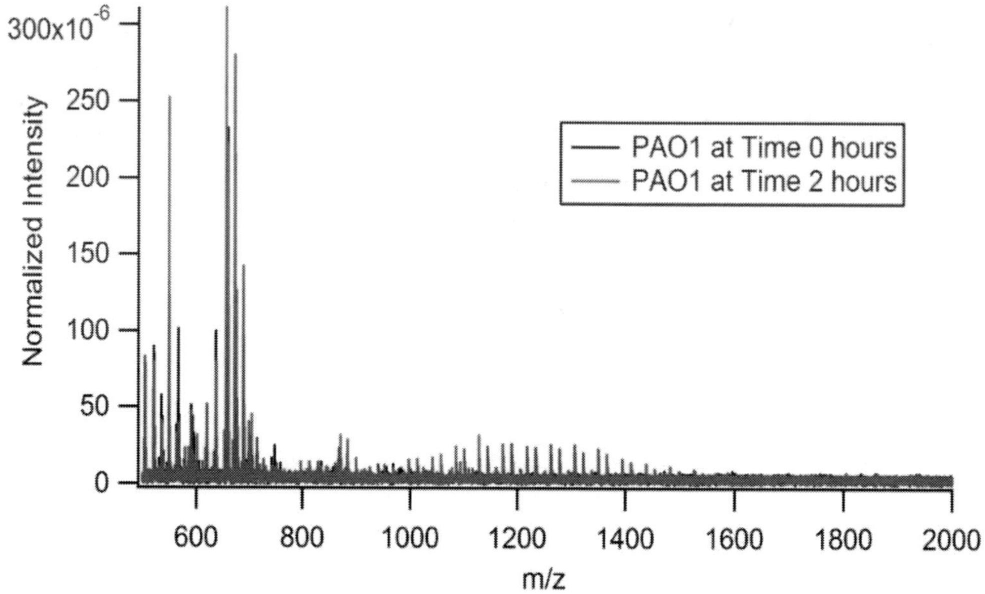

Figure 1: Preprocessed MALDI spectra for PAO1

MALDI spectra were preprocessed into forms similar to Figure 1. Due to the plethora of peaks and variability of peaks seen in the MALDI data, identifying the specific components that change and relating peaks to previously published research results was not possible. Since there were many potential reaction sites identified in the spectra with different magnitudes of reaction, multivariate analyses were employed to determine the fragments that contribute the most to the variability of the spectra during monochloramine disinfection.

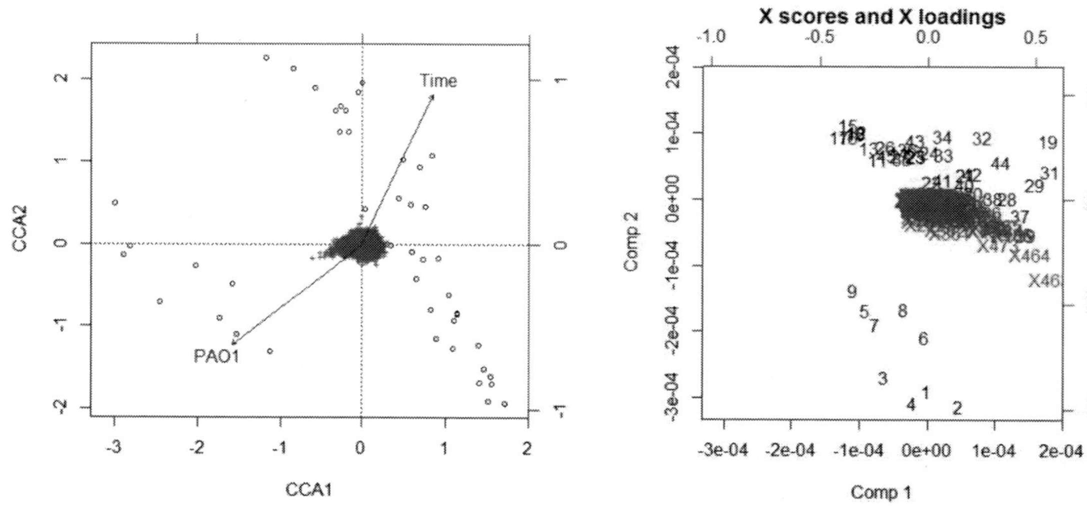

Figure 2: CCA and PLS bi-plots of extracted EPS

Canonical Correlation Analysis (CCA) was thereby employed to determine which variables are related, along with the strength and nature of relationship (Hair 1992; Dixon 2009). CCA was performed to assess whether there was any significant correlation between the data matrices. CCA works by maximizing the correlation between the two matrices through a set of linear correlations named canonical variables (Hair 1992; Dixon 2009; Nicolaou, Xu et al. 2012). The significance or degree of correlation of the data matrices is conveyed though the correlation coefficient, which means linear regression techniques such as PLS can be performed to elucidate the relationship between the variables.

Partial least squares (PLS) analysis is a predictive multivariate regression model used to quantitatively model complicated relationships that are notorious for being noisy, collinear, and even incomplete variables (Wold, Sjöström et al. 2001). In this study, PLS regression was used to predict a single target variable (PLS1) where the input spectra and target inactivation data were used for model generation (BrukerDaltronics ; Mevik, Wehrens et al. 2011; Nicolaou, Xu et al. 2012). One aspect of this modeling is known as the VIP (Variable of Importance in the Projection) score which allows one to quantify the variables that contribute the most to the overall variability of the model.

Biplots are a reliable way to visualize multivariate data because both observations and variable information can be displayed on the same graph (Bro 2011; Gower, Lubbe et al. 2011). As a result, outliers, groups, nonregularities, and previously hidden relationships between the matrices can be visually seen such as seen in Figure 2.

Conclusions

MALDI mass spectroscopy was used in conjunction with multivariate statistical analyses to reveal many protein sites that statistically undergo severe disruption. Future investigations into the reaction and role of these isolated proteins as well as bacteria response during the monochloramine disinfection process may further elucidate the molecular mechanisms for monochloramine inactivation. In summary, the results from this study provide fundamental insights into the function of capsular EPS against monochloramine inactivation as well as the inactivation mechanism.

Acknowledgements

This research was generously supported by the National Science Foundation Award Numbers: CBET-0933288 and TUES-1044823. Assistance was also provided by Dr. B. Leif Hanson for MALDI spectroscopy and Dr. Kevin Dong Zhang for Multivariate Analysis.

References

AmericanWaterWorksAssociation. (2012). "Monochloramines." from http://www.drinktap.org.

Bro, R. (2011). "Lesson in Chemometrics: Partial Least Squares Regression." from www.models.life.ku.dk.

BrukerDaltronics. "MALDI Mass Spectroscopy: An Introduction." from www.bdal.com.

Dixon, P. (2009). "VEGAN, a package of R functions for community ecology." Journal of Vegetation Science **14**(6): 927-930.

Eboigbodin, K. E. and C. A. Biggs (2008). "Characterization of the Extracellular Polymeric Substances Produced by Escherichia coli Using Infrared Spectroscopic, Proteomic, and Aggregation Studies." Biomacromolecules **9**(2): 686-695.

Flemming, H. C., A. Tamachkiarowa, et al. (1998). "Monitoring of fouling and biofouling in technical systems." Water science and technology **38**(8): 291-298.

Gibb, S. (2012). "readBrukerFlexData: Reads mass spectrometry data in Bruker flex format." R package version 1.3, from http://CRAN.R-project.org/package=readBrukerFlexData

Gibb, S. and K. Strimmer (2012). "MALDIquant: a versatile R package for the analysis of mass spectrometry data." Bioinformatics **28**(17): 2270-2271.

Gonnet, F., G. Lemaitre, et al. (2003). "MALDI/MS peptide mass fingerprinting for proteome analysis: identification of hydrophobic proteins attached to eucaryote keratinocyte cytoplasmic membrane using different matrices in concert." Proteome Science **1**(1): 2.

Gower, J., S. G. Lubbe, et al. (2011). Understanding Biplots, John Wiley & Sons.

Hair, J. F. (1992). Multivariate data analysis with readings, Macmillan.

Hardalo, C. and S. C. Edberg (1997). "*Pseudomonas aeruginosa*: Assessment of Risk from Drinking Water." Critical Reviews in Microbiology **23**(1): 47-75.

Jacangelo, J. G., V. P. Olivieri, et al. (1991). Investigating the mechanism of inactivation of *Escherichia coli B* by monochloramine. Denver, CO, ETATS-UNIS, American Water Works Association.

Kirmeyer, G. J. (2004). Optimizing Chloramine Treatment, American Water Works Association.

Laugesen, S. and P. Roepstorff (2003). "Combination of two matrices results in improved performance of MALDI MS for peptide mass mapping and protein analysis." Journal of The American Society for Mass Spectrometry **14**(9): 992-1002.

Lay, J. O. (2001). "MALDI-TOF mass spectrometry of bacteria." Mass Spectrometry Reviews **20**(4): 172-194.

Maniatis, T., E. F. Fritsch, et al. (1982). Molecular cloning: a laboratory manual, Cold Spring Harbor Laboratory.

Marvin, L. F., M. A. Roberts, et al. (2003). "Matrix-assisted laser desorption/ionization time-of-flight mass spectrometry in clinical chemistry." Clinica chimica acta **337**(1): 11-22.

Mevik, B. H., R. Wehrens, et al. (2011). "pls: Partial Least Squares and Principal Component regression." Vienna: R Foundation for Statistical Computing.

Nicolaou, N., Y. Xu, et al. (2012). "Detection and Quantification of Bacterial Spoilage in Milk and Pork Meat Using MALDI-TOF-MS and Multivariate Analysis." Analytical Chemistry **84**(14): 5951-5958.

RCoreTeam (2012). R: A language and environment for statistical computing. Vienna, Austria, R Foundation for statistical Computing.

Suckau, D., A. Resemann, et al. (2003). "A novel MALDI LIFT-TOF/TOF mass spectrometer for proteomics." Analytical and bioanalytical chemistry **376**(7): 952-965.

UnitedStatesEnvironmentalProtectionAgencyOfficeofWater (1999). Alternative disinfectants and oxidants guidance manual, U.S. Environmental Protection Agency, Office of Water.

Wang, Z., C. M. Hessler, et al. (2012). "The role of extracellular polymeric substances on the sorption of natural organic matter." Water Research **46**(4): 1052-1060.

Wingender, J., T. R. Neu, et al. (1999). Microbial Extracellular Polymeric Substances: Characterization, Structure, and Function, Springer.

Wold, S., M. Sjöström, et al. (2001). "PLS-regression: a basic tool of chemometrics." Chemometrics and intelligent laboratory systems **58**(2): 109-130.

Xue, Z., C. M. Hessler, et al. (2012). "Pseudomonas aeruginosa inactivation mechanism is affected by capsular extracellular polymeric substances reactivity with chlorine and monochloramine." FEMS Microbiology Ecology.

Zhu, X. and I. Papayannopoulos (2003). "Improvement in the detection of low concentration protein digests on a MALDI TOF/TOF workstation by reducing α-cyano-4-hydroxycinnamic acid adduct ions." Journal of biomolecular techniques **14**(4): 298.

Transforming Community Health with Simple Yet Effective Drinking Water Disinfection in Central Andes of Perú

Thomas R. Armitage, P.E., M.S. Environmental Engineer, Lieutenant Commander USPHS Commissioned Corps
Terry Jackson, BSEE, MBA, Missions Director, Immanuel Community Church, Fort Collins, CO

INTRODUCTION

Around the world, water quality greatly impacts public health. The United States and the western world have been treating water since the 1800s with filtration and chlorine or other disinfection techniques. Much of the developing world has untreated water. Generally, the more rural the area, the lower the quality of drinking water provided to the community. This situation leads to people living with a constant threat or perhaps a consistent state of poor health that hinders their ability to work and provide for their families. These water borne illnesses impact people's health in these areas from young to old.

Many organizations are seeking ways to address this fundamental health problem by improving water quality. Our process of improvement provides a low cost and tremendously effective treatment system that can be deployed to serve entire communities, quickly and easily. The water treatment process uses a design that addresses many of the complexities inherent to water borne microbial/pathogen disinfection yet is remarkably efficient. More importantly, the design does not require a great deal of technical understanding to implement or maintain. Unlike individual or household filters, this solution does not require cultural/household adaptation can thus immediately transform the health of a community. Additionally, this technology can be applied in many parts of the world. We discuss one such area's success here, how a simple chlorine tablet, gravity flow, disinfection system transformed the general health in Colca, Perú, a small community in the central Andes Mountains.

AREA BACKGROUND

The mountain village of Colca (Pueblo de Colca) is located in the Region of Junín south/southwest of Huancayo, Perú in the central Andes Mountains as shown in Figure 1. Colca is one of five district capitals located within the Canipaco River Valley, with several annexes or barrios associated to each district.

Figure 1: Project Site Location

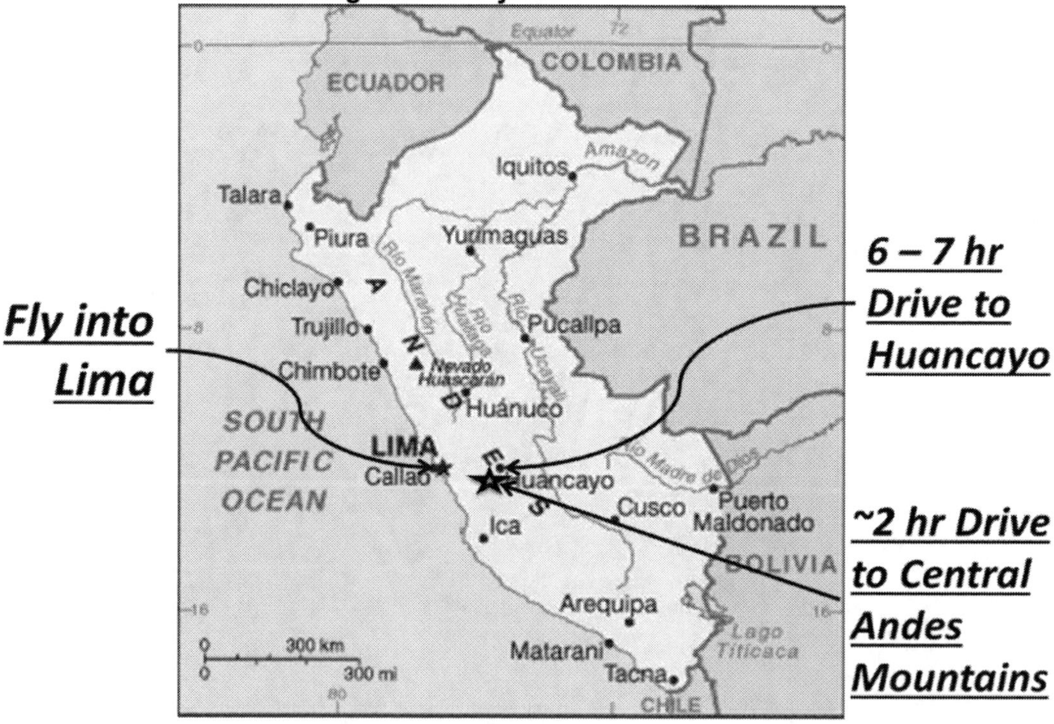

Population estimates for the area range around 9,500 people and Colca is one of the larger district capitals. All the districts serve the town that they are named for plus the surrounding countryside, since much of the population survives by agrarian and animal grazing activities. The populations are difficult to determine as estimates count families and some family members leave to find work in the mines or larger cities such as Huancayo or Lima for months, returning to the area for holidays and key activities like harvest. The District of Colca has an estimated population of around 1,600 based on 2005 census and the Puesto de Salud Colca (Colca Health Clinic) indicates that they serve the entire district. We assume that the Cocla Health Clinic serves an area consisting of ~1,600 people.

The primary sponsoring organization, Immanuel Community Church, Fort Collins Colorado, has been working in the area for nearly seven years. In March 2009, during one of their medical mission trips, an engineering evaluation of the area's water system was conducted to assess reoccurring diseases presumed to be water borne. The review found the water sources under the influence of surface activities. Follow-up included a study to find an effective disinfection process for the communities and a return trip in late 2009 into early 2010 to install treatment. Support for the project came from Immanuel, Applewood Baptist Church, Wheat Ridge, Colorado, and other donations from many people around the USA. In Perú, the Baptist International Mission Board (IMB) supplied translators and logistical support while a Baptist church in Huancayo offered support with personnel, logistics, and accommodations.

The engineering review revealed that the water sources are under the influence of surface water. Agricultural and livestock activities affected the captación de fuente (source water spring box) designed to capture flows along a drainage swale low point and direct them to the community via gravity flow through small diameter water distribution systems (1 to 4 inch pipes). These systems generally have one or more captacións directing water to a storage tank where the operator will occasionally add varying amounts of powder chlorine via a cylindrical dispenser with small drilled holes. As suspected, the perforated dispensers result in high chlorine residuals, for a short time (no one drinks the water), before returning to normal zero chlorine residuals. Such a process shocks the system and does not provide a consistent, controllable dose of chlorine disinfection.

Some of the more common water borne health related pathogens were suspected or found in the water by local health clinic personnel and through our review of composite patient health data. E-coli was verbally reported in the water and data from the Colca Health Clinic showed cases of intestinal infection (diseases and parasites), dysentery, gastroenteritis, and giardiasis. While there are several potential pathways to contract these illnesses, drinking contaminated water is one of the most likely scenarios for this community.

In addition to the powder chlorination activities, to help reduce the spread of diseases and illnesses, the local health officials regularly educated the communities on proper hygiene and sanitary practices. Boiling of drinking water is strongly recommended as a standard practice. The community may boil the water, however, with limited fire fuel, many or most will bring the water to a boil and then use it without permitting sufficient time to kill all the parasites and bacteria. Therefore, education and sanitation procedures provide inadequate defense against water borne disease and illnesses.

APPROACH AND RESULTS

The community of Colca has three water systems and we chose to work on the largest system, serving 75% of the population of the town and about 65% of the population seen by the Colca Health Clinic (~1000 of the ~1600 people in the District of Colca). The systems in Colca service different areas of the community based on date of installation and local terrain (i.e. a small ravine divides to the eastern and central systems and ridgelines may separate other parts of the systems). All systems provide piped water from a source(s) to a tank and on to residential taps for use. The focus system of this project provides piped water to residences in the central area of Colca (Sistema de Colca de Foncodes) as well as water to the Colca Health Clinic. Therefore, disinfecting this system's water would help the most number of people and assist the local health officials serve the community. The community of Colca and the three water systems are shown in Figure 2:

Figure 2: Colca Water Systems

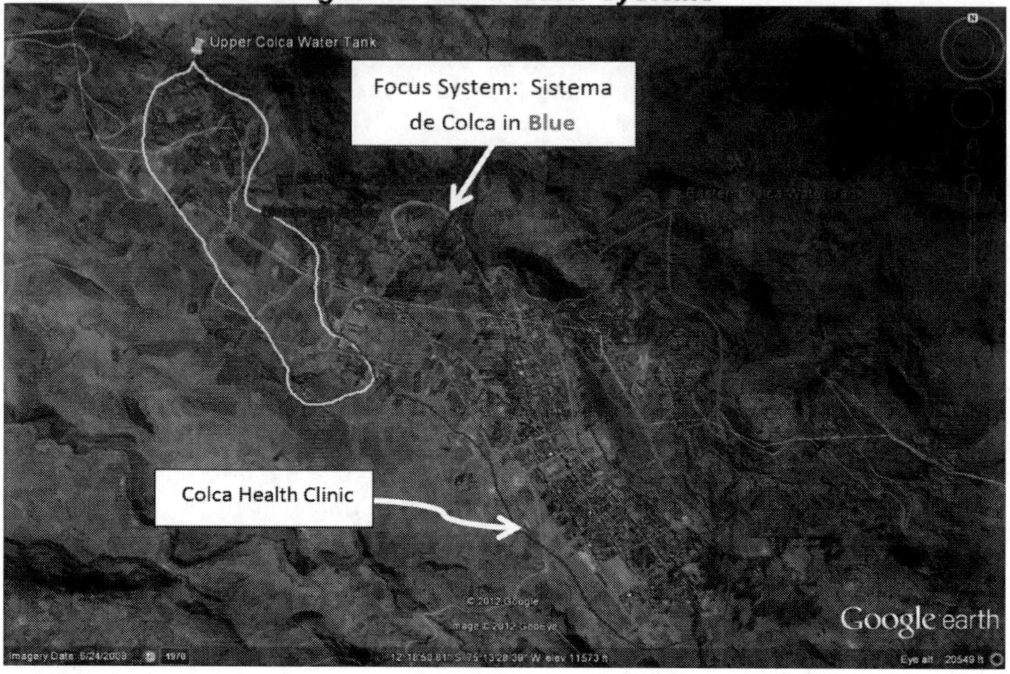

The solution to properly disinfecting the water in Colca required meeting several criteria. First, it had to effectively disinfect the drinking water on a consistent and controlled basis. Second, it had to be maintainable by the local health clinic staff and community water board/operators with locally available supplies (Huancayo, is a 2 to 3 hour drive). Third, it had to be acceptable to the local community and approved by the local health department and the department of sanitation.

After studying the area, treatment alternatives, and the local acceptance of current chlorination, the selected treatment process was the Norweco Bio-Dynamic® Model LF 3000 Tablet Feeder. Calcium hypochlorite tablets, 70% or 90% active, are available in Perú and in Huancayo on a broader basis each year. Santec Corporation (Castle Rock, Colorado) provided the equipment at cost and Hach Company (Fort Collins, CO) provided pocket chlorimeters as a donation to Immanuel Church. Piping, valves, and other appurtenances were obtained in Lima or Huancayo.

During the first visit in March 2009, the process began of informing and helping the local health department and community understand the benefits of watershed protection and consistent controllable disinfection. By August, the community had improved the watershed at the captación by removing all animal activities, adding three feet of soil to the area, and restricting agricultural practices to be all natural fertilizers (Figure 3 below).

Figure 3: Captación Watershed Before and After Comparison

March 2009: Shallow Spring Up-gradient Contaminants Include: animals wastes, fertilizers, insecticides, pesticides

Aug 2009: Watershed Improvements – Restricted Animal Grazing, Fertilizers are all natural, Limited Pesticides & Insecticides

Tributary watershed for this system required improvements. Implementation of protections such as removing all animal wastes and adding soil over the perforated intake pipes for this system provided needed contamination barriers and served to demonstrate concepts to the local health officials. By the end of December 2009, the health and sanitation departments for the Region of Junín in Huancayo had approved the tablet disinfection system. We presented the disinfection system plans with the support of the local health department at a community meeting of about sixty people. The community approved the plans and the location for the system and we began the installation.

The Colca system is gravity flow from the captación to the tank and throughout the distribution system. The terrain is mountainous with some very steep areas. Various pressure breaks are used to control pressure and limit rupturing pipes. In order to effectively disinfect the water system and provide contact time for chlorine disinfection, the main Colca water system line was intercepted down gradient of the captación and just above the storage tank with a bypass line where the Norweco tablet system was installed. Locally available 4-inch gate valves were used as flow control valves on the main and bypass lines to control flows to the Norweco tablet system and thereby control the chlorine dose to the system. The raw and super chlorinated flows combine at a second tee and a sample tap provides a monitoring point to check chlorine residuals/dosing levels.

As with a number of projects in rural Perú, this seemingly simple task did not go without hitches. In order to create the sample tap we pieced together the closest sized tee we could find, reducers, and a threaded adaptor to a sample tap. The most unusual problem was finding transition unions from a four (4) inch concrete (assumed asbestos concrete) pipe to 4-inch PVC piping system. We eventually found a fabricator in Huancayo that could create two transition unions out of PVC pipe couplings by expanding the 4-inch pipe to fit over the larger outside diameter of the concrete pipe. We cut the concrete pipe with a hacksaw and then, while looking for some replacement blades, we found some blades for sawing masonry block and concrete, which made the second pipe cut a little easier. Rainstorms made the going difficult on a couple occasions and stranded us once soaked and very cold. Of course, the day we finished the project we were given a rainbow that expanded across the water system service area, held for a few minutes, and then disappeared. Figure 4 shows the finished disinfection system, some of the locals, and the clinic doctor as we were adjusting the chlorine doses during system start-up:

Figure 4: Colca Tablet Disinfection System Installed

Shown (L – R): Doctor, Board President, neighbor, and water operator

For the system to operate properly (having the chlorine tablets dissolve uniformly), it must be level in both planes and have a free flowing discharge (no backwater into the outlet). This picture may not look like it, however the disinfection system is level and the flow drops down to the second tee where all flows combine prior to the long sample tap. From the start of operation, we were easily able to adjust the chlorine residual from 0.5 mg/L to 1.7 mg/L.

Initially, the local health official did not want a dose over 1 mg/L as Perú sets the highest permitted dose at 1 mg/L while requiring no less than 0.5 mg/L in the distribution system. After providing some process explanations and time to allow for distribution system cleansing by consistent chlorination, a dose of ~1.5 mg/L was applied to the raw water prior to the tank. The minimum residual of 0.5 mg/L generally became the controlling factor in the decision, as it was believed to be more important to maintain than to stay within the 1 mg/L upper limit. The dose of ~1.5 mg/L maintains about 1 mg/L at the first user in the distribution system and a residual of around 0.5 mg/L in the outer portions of the system. The public health outcome of this approach was dramatic.

The following data summary shows a significant reduction in water borne illnesses between 2010 and 2011. Colca and many from the surrounding communities have enthusiastically appreciated the overall reduction of 40% in water borne illnesses. The community lost the data from 2009 and prior due to a computer crash and unfortunately, we had not made copies of this data. The 2010 data includes impacts of the watershed improvements and about six months of an operational disinfection system (July – Dec). Hence, the numbers are skewed and are not fully reflective of the positive impact of the disinfection system. Therefore, the local health personnel believe the real impact is much greater. Regardless, the data between a partial year of disinfection and full year of disinfection shows a dramatic improvement in water quality.

In 2010, Colca reported 139 waterborne illnesses verses 83 in 2011. Based on 1600 people, normalized illnesses per 1000 people equate to 87 (2010) and 52 (2011). For comparison a local community, Chacapampa, with similar water systems and a clinic serving approximately 600 people, has reported around 230 illnesses per 1000 people. (In June 2012, we installed a tablet disinfection system on Chacapampa's main water system.) Key data from Colca is summarized in Figure 5 below showing the cases in 2010 verse the cases of the same constituents in 2011. Percent reductions are shown in green.

Figure 5: Health Improvements by Reductions in Water Borne Illnesses

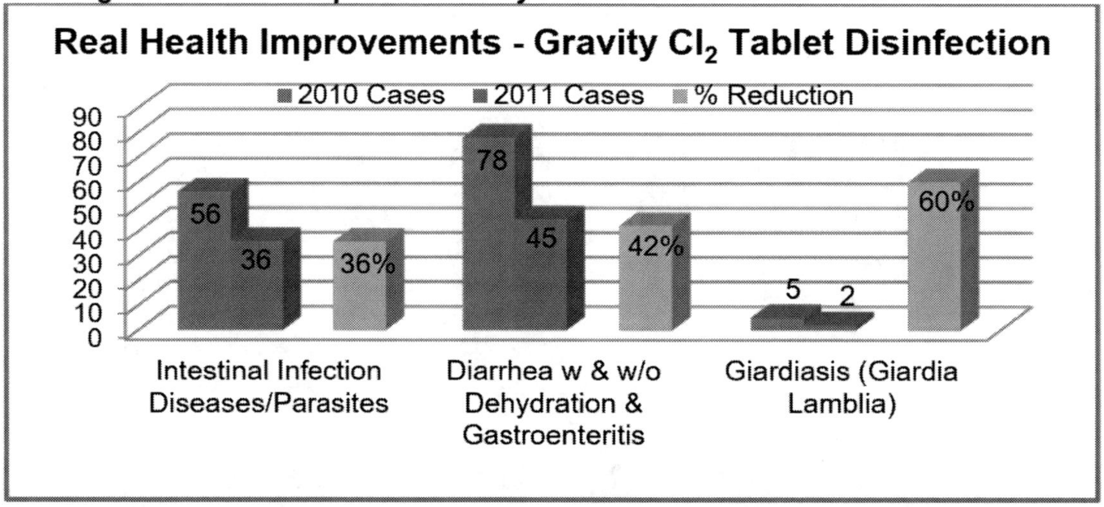

Important note: 35% of the population does not receive disinfected water. Therefore, the data from 2010/11 showing the positive impact on health may not accurately reflect the total impact of the disinfection system. If one assumes all the improvements in health are related to the disinfection system, the improvement in health could be around 60% for the central water system residents. The next project in Chacapampa will attempt to obtain better control of the data.

The system, equipment, travel, and sundry costs totaled around $4,250. The cost would be around $5,000 if one would pay full price for the Norweco tablet system and the Hach monitoring equipment. This cost is very reasonable to serve a population of about 1,000 people for the next 10 years or so. The reductions in water borne illnesses and accompanying increase in general public health make the project a great opportunity to improve the lives of many people.

The Chacapampa, Perú water system installation, June 2012, had great community interest and involvement, and we were able to reduce the installation time. As with Colca, the local health officials are working with us as we collect data and we expect results from the recently installed system in Chacapampa in late 2013. Other communities in the region, hearing of the success, are requesting the installation of systems in their communities. Plans are being made to address the current eleven requests based on limited funds and time constraints. Figure 6 shows the Chacapampa water facility before and after installation of the final building.

Figure 6: Chacapampa Water Treatment System June and July 2012

June 2012: Pictured (left to right): Enrique system operator, Tom Armitage, Water Board President, & Clinic Nurse

July 2012: Finished Water Treatment Facility Building with Terry Jackson on the right.

CONCLUSION

The use of the tablet chlorination equipment in a gravity flow through water system resulted in excellent reductions in water borne illnesses and improved general public health. Results indicated a 40% reduction in water borne illnesses in a small mountain community of Colca, Perú. Health officials and the communities express great appreciation of the system and results. The local health departments have requested additional water disinfection treatment processes for other water systems in the area. With this technically simple and effective process, many people can benefit in any location where a gravity flow through water disinfection process is applicable.

The Gift of Water, Inc.

Abstract for the Water Environment Federation – February 25, 2013

Gift of Water® (GoW) is a non-profit organization dedicated to providing filtered and clean drinking water to the third world communities that it serves, primarily in Haiti. Operational since 1995, GoW has a well proven water filtration system, along with an established operating model, for partnering with communities all around Haiti.

The Gift of Water filtration system incorporates both a chlorination step for disinfection and a physical filtration step for particle and protozoan removal. This dual approach leads to high quality treated water. The stacked, two bucket system uses a 1 micron polypropylene string-wound filter in the top bucket and a granulated activated carbon filter in the bottom bucket.

The operating model is as follows:

- A non-profit organization, often US based, with an established presence in a Haitian community, agrees to sponsor a water program. Sponsorship includes the purchase of the initial purification systems, the employment of Haitian technicians and, in some cases, financial assistance with the purchase of replacement filters.
- GoW manufactures, ships, and delivers units to the community.
- The sponsor organization works closely with a local leader, often a parish priest, clinic director or other community leader, to establish the process for selecting the receiving households.
- GoW sends a trained Country Liaison to that Haitian community to establish the program. GoW also trains a local Community Technician to manage the program locally, which includes household visits to provide maintenance, replacement parts, chlorine tablets, and ongoing training.

The Gift of Water organization witnessed a major change of management in January 2010. Since then, GoW has established or re-established programs in almost 50 communities, placed more than 10,000 new water purifiers, supplied replacement filters to maintain another 10,000, and distributed over 2 million chlorine tablets. Though the success stories are abundant, there remains challenges. Goals for 2013 include continuing to engage with local communities to strengthen already established programs by encouraging sustainability, increasing user compliance, supporting additional training programs, and working closely with the Haitian government and other non-government organizations to lay the ground work for continued growth.

It is Gift of Waters distinct and sincere pleasure to participate in the Water Environment Federation conference in Indianapolis in February, 2013.

Accuracy and Usability of Free Chlorine Residual Testing Methods

Anna Murray, Tufts University Medford, Massachusetts
Daniele Lantagne, PhD, PE, Tufts University Medford, Massachusetts

INTRODUCTION

Chlorination is a water treatment method proven to reduce the diarrheal disease burden in developing countires. Measuring free chlorine residual (FCR) after treatment indicates the potability of treated water. The several methods available for testing the presence of chlorine residual vary in terms of cost, measurement precision, and training required to perform the tests. This research aimed to evaluate and compare seven types of residual chlorine test kits and recommend which is most appropriate for potable water testing in developing countries. The study considers cost, accuracy at a variety of sodium hypochlorite (NaOCl) concentrations and light settings, and ease-of-use as determined by a focus group of volunteer participants.

METHODS

Laboratory Testing

Eleven concentrations of sodium hypochlorite (NaOCl) solution were prepared by adding bleach to deionized, chlorine demand-free water to create solutions at the following concentrations: 0.0 mg/L, 0.1 mg/L, 0.2 mg/L, 0.5 mg/L, 1.0 mg/L, 1.5 mg/L, 2.0 mg/L, 2.5 mg/L, 3.0 mg/L, 3.5 mg/L, 4.0 mg/L. Each of the seven test kits was used according to the manufacturer's instructions to measure the residual chlorine concentration in each solution dosage in the laboratory. To account for variations in lighting, the laboratory procedure was repeated in full outside in the sun. All measurements were repeated five times, and an arithmetic mean calculated. An error threshold of 10% was defined as our measure of accuracy.

Focus Group

A focus group was held with eight volunteer participants measuring FCR with the seven residual chlorine test kits. Participants were given written instructions on how to perform the tests, and they used each of seven kits to measure, in duplicate, the FCR concentrations of water samples at 0.2 mg/L, 0.5 mg/L, and 2.0 mg/L. Following the water testing, a survey was administered, asking participants to indicate the relative difficulty of the tests and their level of confidence in the results. They also were asked to choose which test type they would have the most confidence in using in a variety of situations.

RESULTS AND DISCUSSION

Laboratory Testing

The following figure and table summarize the results of the FCR laboratory testing. In Figure 1, the average FCR measurement values versus the actual values for each test and concentration with corresponding error bars are presented. In Table 1, the tests and chlorine concentrations with average measurement error exceeding the threshold limit of 10% are presented. The test kits in Table 1 are listed in decreasing order of accuracy.

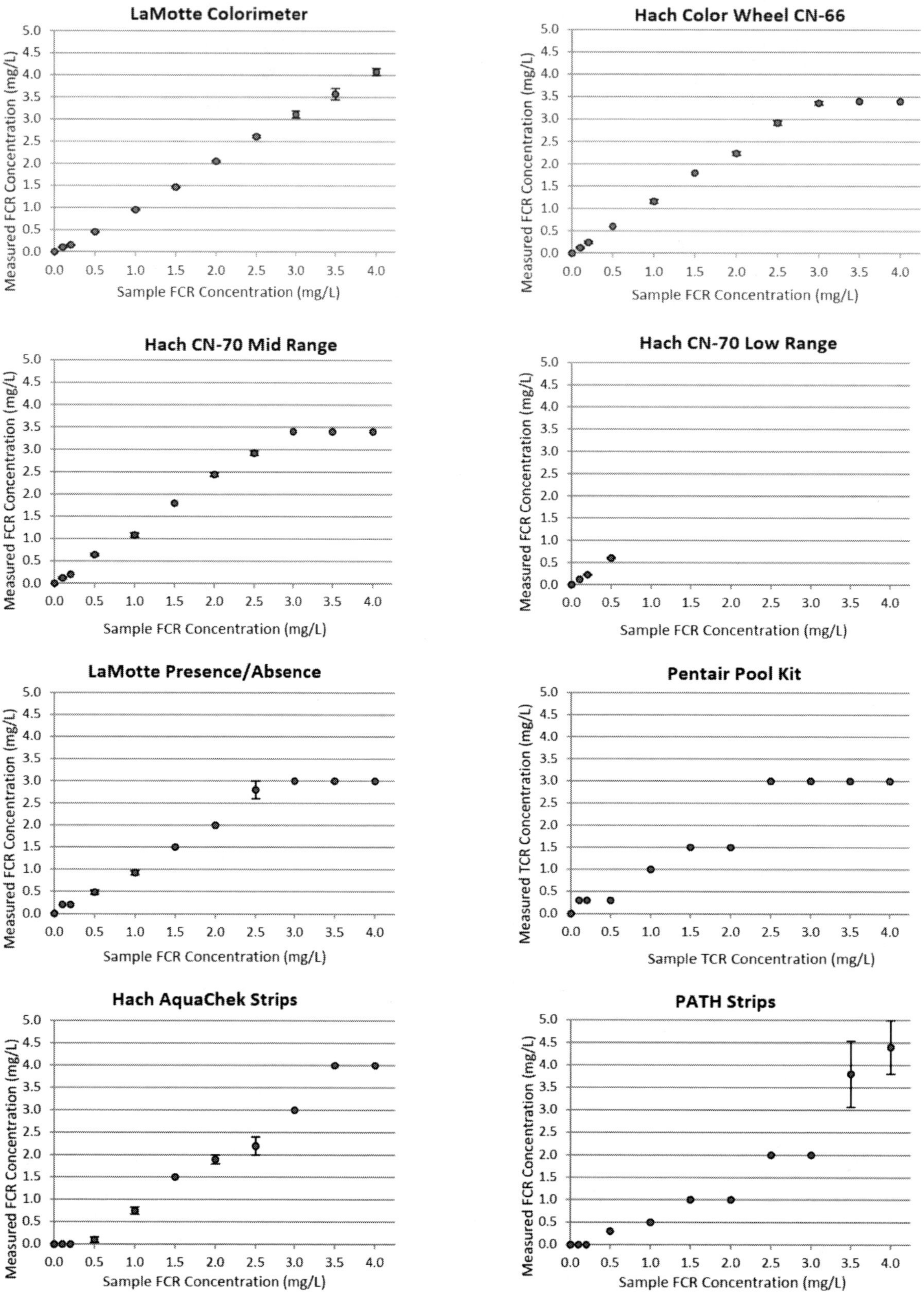

Figure 1: Laboratory measured results for free chlorine residual concentrations for each test kit and chlorine concentration. Results in FCR, except pool kit, which measures total chlorine residual.

Table 1: Measurement error in FCR laboratory tests for each test kit at varying free chlorine residual concentrations.

	FCR Concentration (mg/L)										
	0.0	**0.1**	**0.2**	**0.5**	**1.0**	**1.5**	**2.0**	**2.5**	**3.0**	**3.5**	**4.0**
LaMotte Colorimeter	<10%	<10%	>10%	<10%	<10%	<10%	<10%	<10%	<10%	<10%	<10%
LaMotte Presence / Absence Test Kit	<10%	>10%	<10%	<10%	<10%	<10%	<10%	>10%	<10%	>10%	>10%
Hach AquaChek Test Strips	<10%	>10%	>10%	>10%	>10%	<10%	<10%	>10%	<10%	>10%	<10%
Hach Color Wheel Test Kit (CN-70) Mid Range	<10%	>10%	<10%	>10%	<10%	>10%	>10%	>10%	>10%	<10%	>10%
Hach Color Wheel Test Kit (CN-70) Low Range	<10%	>10%	>10%	>10%							
Pentair Pool Chlorine Kit	<10%	>10%	>10%	>10%	<10%	<10%	>10%	>10%	<10%	>10%	>10%
PATH Chlorine Test Strips	<10%	>10%	>10%	>10%	>10%	>10%	>10%	>10%	>10%	<10%	<10%
Hach Color Wheel Test Kit (CN-66)	<10%	>10%	>10%	>10%	>10%	>10%	>10%	>10%	>10%	<10%	>10%

Lighting Conditions

With the exception of the Hach AquaChek Strips, chlorine residual readings were unaffected by taking measurements in sunlight versus fluorescent light. Figure 2 is a plot comparing measurements for the AquaChek strips. The concentrations for which measurements of FCR differed in varied lighting are as follows: 0.5 mg/L (100% measurement error), 1.0 mg/L (53%), 2.5 mg/L (36%), 3.5 mg/L (10%).

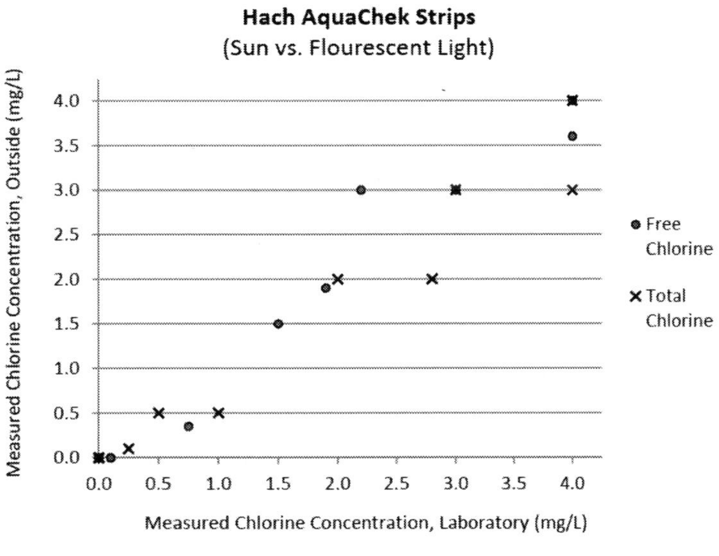

Figure 2: Chlorine residual measurement differences in differing lighting conditions, Hach AquaChek Strips.

Focus Group

All of the focus group participants (8/8) reported having some general lab experience, and 88% (7/8) reported prior experience testing for chlorine or other water quality parameters. Average measurement error across the three water samples for each test, in order from lowest to highest error, are as follows: LaMotte Colorimeter (4.5%), LaMotte Presence/Absence (6.6%), Hach CN-70 Low Range (14.9% on two concentrations only), Hach CN-66 (21.5%), Hach CN-70 Mid Range (28.8%), PATH Strips (28.8%), Pentair Pool Chlorine Test Kit (34.2%), Hach AquaChek Strips (54.0%).

Difficulty was indicated on a Likert scale 1-5, with 1 being the simplest and 5 being the most difficult. The average reported difficulty, from least difficult to most difficult, is as follows: PATH Strips (1.1), Hach AquaChek Strips (1.5), LaMotte Presence/Absence (2.4), Pentair Pool Test Kit (2.4), LaMotte Colorimeter (2.6), Hach CN-66 (3.3), Hach CN-70 (4.3). Users were most confident in the results from the LaMotte Colorimeter (ave. 5.0 on a scale of 1-5), and least confident in the Hach AquaChek Strips and PATH Strips (both ave. 2.3 on a scale of 1-5).

The comments on difficulty were divided into two subsets: ease of performing the test procedure, and ease of reading the results and comparing the color indicator to the standard colors. Half of the focus group participants commented that matching the colors or judging the difference between the standard color shades was difficult on each of the LaMotte Presence/Absence test, Hach Color Wheel, Pentair pool kit, and Hach AquaChek Strips, even though at least as many people also commented that these tests were each "simple," "easy to use," or "straight-forward." In contrast, only 25% (2/8) commented that they had difficulty distinguishing between colors with the PATH strips. The LaMotte Colorimeter is the only test kit that does not rely on a visual color comparison by the user. Several comments about the confidence in the LaMotte Colorimeter results indicated that the participants trust the colorimeter because it eliminates this human component of judging color ("the machine is designed to test differences in color intensity, so it can do it much better than I can").

The PATH Strips were the choice of most participants for both the easiest test to complete (8/8), and the test that they would choose for water testing in a developing country (3/8), but the results showed an average measurement error of almost 30%, and 53% measurement error at the low range (0.2 mg/L free chlorine). These results were only more accurate than the Pentair pool kit and the Hach AquaChek Strips in the focus group. The LaMotte Presence/Absence kit showed the second-lowest average measurement error, at 6.6%, and was the third easiest to use, but the average confidence level was third-to-worst (2.9 on a scale of 1-5).

Cost Comparison

Table 2: Comparison of initial and incremental costs of each FCR test kit.

	Equipment Cost, Initial	Incremental Test Cost
LaMotte Colorimeter	$435.00	$0.09
LaMotte Presence / Absence Test Kit	$5.85	$0.09
Hach Color Wheel Test Kit (CN-66)	$55.00	$0.16
Hach Color Wheel Test Kit (CN-70)	$72.00	$0.16
Hach AquaChek Test Strips	-	$0.31
PATH Chlorine Test Strips	-	$0.04
Pentair Pool Chlorine Test Kit	$14.00	$0.05

CONCLUSIONS

Considering accuracy by experienced and inexperienced users, ease-of-use, confidence in the results, and cost, we recommend that the LaMotte Presence/Absence test be used for measuring free residual chlorine concentrations in drinking water in developing countries. If cost is not a significant factor in the decision, the LaMotte Colorimeter is recommended. Table 3 presents the relative rankings of each test kit for each category considered, where 1 is the highest and 7 is the lowest.

Table 3: Relative rankings of each test over five categories.

	Cost	Accuracy, Lab	Accuracy, Focus Group	Difficulty, Focus Group	Confidence, Focus Group	Total
LaMotte Colorimeter	7	1	1	5	1	15
LaMotte Presence / Absence Kit	3	2	2	3	5	15
PATH Chlorine Test Strips	1	6	5	1	6	19
Pentair Pool Chlorine Test Kit	2	5	6	3	4	20
Hach Color Wheel Test Kit (CN-70)	5	4	3	7	2	21
Hach Color Wheel Test Kit (CN-66)	4	7	3	6	3	23
Hach AquaChek Test Strips	6	3	7	2	6	24

The Safe Global Water Institute: An integrated, collaborative approach for improving drinking water and sanitation globally

Aimee M. Gall [1,4], Vincent Madadi [3,4], Joanna L. Shisler [2,4], Benito J. Mariñas [1,4]

[1]Department of Civil and Environmental Engineering, University of Illinois at Urbana-Champaign, USA
[2]Department of Microbiology and College of Medicine, University of Illinois at Urbana-Champaign, USA
[3]Department of Chemistry, University of Nairobi, Kenya
[4]Safe Global Water Institute
4150 Newmark Civil Engineering Laboratory
205 N. Mathews Ave.
Urbana, IL 61801
(217) 333-6978

ABSTRACT

The World Health Organization (WHO) recently published a progress update for Millennium Development Goal 7, Target 10 dealing with sustainable access to safe drinking water and basic sanitation. As of 2010, 780 million people lack access to improved water sources globally, and many more drink unsafe water. Additionally, source waters can easily become contaminated from animal and human waste with 2.5 billion people worldwide lacking access to improved sanitation (WHO, 2012). Waterborne pathogens in contaminated drinking water sources are especially concerning because of their ability to cause high levels of morbidity and mortality. The recently established Safe Global Water Institute (SGWI) focuses on safe water and sanitation through fundamental research, the development of novel technologies, and building institutional and human capacity. This presentation will highlight some of the current trans-disciplinary efforts of the SGWI.

This research focuses on waterborne viruses because of their virulence and resistance to disinfectants, and also the limited knowledge available. Adenovirus, a human pathogenic virus present globally in drinking water sources, is highly resistant to monochloramine, low-pressure ultraviolet light, and solar disinfection. With a more fundamental understanding of how disinfectants inactivate viruses on a molecular level, more robust treatment technologies for highly resistant viruses like adenovirus can be designed and implemented globally. Additionally, this research studies the inactivation kinetics of a potential bacteriophage surrogate for adenovirus that will enable treatment technologies to be more easily tested in the laboratory and challenged in the field. There is currently no field method to test for infectious viruses in drinking water sources, and therefore the SGWI is engaging in research to develop a novel sensor to detect infectious adenovirus in drinking water.

Lastly, the SGWI strengthens collaborations between scientists, engineers, and sociologists in the US and developing regions. This presentation will emphasize SGWI partnerships in Eastern Africa including universities, NGOs, government agencies, and rural communities. These partnerships have involved surveying community members and analyzing source and household water quality with students from several international universities. At the University of Illinois, chemistry and social science graduate students from the University of Nairobi have been working with engineers on appropriate technologies for the region. The SGWI is working to improve drinking water and sanitation by engaging multidisciplinary international partners, developing novel technologies, and building capacity globally.

KEYWORDS

Safe Global Water Institute, safe drinking water, sanitation, adenovirus, viral sensors, capacity building.

METHODOLOGY

Adenovirus serotype 2 (ATCC VR-846) was propagated using human lung A549 carcinoma cells (ATCC CCL-185). This method was described previously (Sirikanchana et al., 2008). Similarly, after multiple rounds of infecting A549 cells, adenovirus stock solutions were purified using centrifugation, microfiltration, and ultrafiltration. Virus viability was determined using the plaque assay. This was accomplished by growing a monolayer of A549 cells and plating adenovirus onto the cells with a nutrient mixture and a soft agar overlay technique. Plaques were counted after ten days with 37 °C and 5% CO_2 incubation.

The bacteriophage PRD1 (ATCC BAA-769-B1) was propagated using a bacterial hosts *Escherichia coli* K-12 (ATCC BAA-769). After infecting the host bacteria, PRD1 was purified using centrifugation, microfiltration, and ultrafiltration. This yielded a high titer solution free from bacterial debris. The double agar layer plaque technique was used to quantify the bacteriophage (Adams, 1959). After incubation at 37°C, viral plaques were counted at 18-24 hours and concentrations calculated.

DISCUSSION

The Safe Global Water Institute (SGWI) is working to improve drinking water and sanitation globally through fundamental research and novel technology development. In the Department of Civil and Environmental Engineering at the University of Illinois at Urbana-Champaign (UIUC), the inactivation kinetics of adenovirus and a potential bacteriophage surrogate PRD1 are being studied after their exposure to drinking water disinfectants. When these viruses are exposed to certain disinfectants at specific pH and temperature conditions, their behavior can be remarkably similar. This has prompted further research with PRD1 as a surrogate for human adenovirus to test the efficacy of potential drinking water treatment technologies appropriate in East Africa. A visiting scholar from the University of Nairobi (UON), Kenya in the Department of Chemistry was involved in these studies in the summer of 2011. As our laboratory is trying to understand how viruses become inactivated on a molecular level after exposure to disinfectants, this UON researcher also collaborated with UIUC Department of Chemistry in the summer of 2012 to develop chemical techniques to elucidate these mechanisms. Within the Department of Microbiology at UIUC, molecular techniques are additionally being explored to elucidate these mechanisms. By studying PRD1 and adenovirus simultaneously in mechanistic studies, we hope to gain a more fundamental understanding of how disinfectants act on the molecular level.

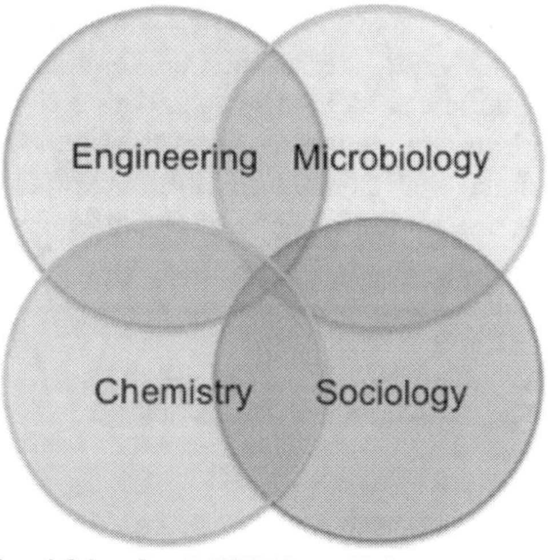

Figure 1. Research within the SGWI is collaborative and interdisciplinary

This research within the SGWI has encompassed many international partners as seen in Figure 2. After traveling to international rural communities with academic, NGO, and government partners, the team is able to assess drinking water quality and needs of the communities. Within this research for example, source water from Kenya, Ethiopia, and Mexico has been challenged with PRD1 to determine the efficacy of local drinking water treatment technologies on these waters. Highlighting the partnership within Kenya, several groups of undergraduate and graduate students have traveled to Kenya to test water quality onsite and collect samples for further analysis at UIUC. Working with the University of Nairobi in the Department of Chemistry and Department of Sociology, as well as the Department of Engineering at Bondo University College, and Bondo District Water, the team was able to assess the needs of rural communities in the Nyanza Province.

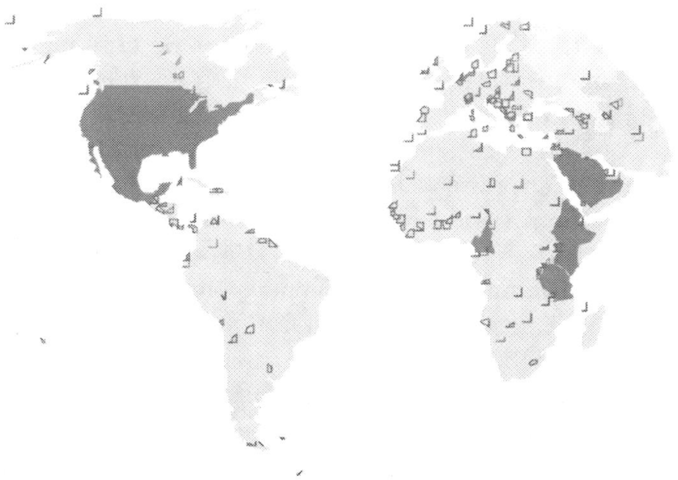

Figure 1. Some of the SGWI's global partners and target communities

The SGWI is also committed to developing sensors to detect infectious viruses in drinking water. Currently there is no method that exists to test for infectious viruses onsite. Additionally the isolation of infectious viruses remains a challenge in the laboratory because of the need to use cell culture. The Department of Civil and Environmental Engineering as well as the Department of Chemistry at UIUC are currently collaborating to develop a sensor for detecting infectious adenovirus in drinking water through the use of aptamers. An aptamer could then be applied in a simple 'dipstick' colorimetric assay to determine if water contains infectious adenovirus. This research could lead to a rapid, easy to use, and inexpensive way to test the microbial quality of drinking water in the field and laboratory, and could especially have an impact on drinking water quality in developing regions.

Overall, the Safe Global Water Institute is committed to improving the quality of drinking water and sanitation throughout the world by engaging in fundamental research, developing novel treatment techniques, building capacity, and developing sensors. Since the research needed to accomplish these challenges is so multidisciplinary, the SGWI consists of many international partners that span many disciplines and agencies with the goal of improving water and sanitation globally.

REFERENCES

Adams, M.H., 1959. Bacteriophages. Interscience Publishers, Inc., New York.

Sirikanchana, K., Shisler, J.L. Mariñas, B.J., 2008. Effect of exposure to UV-C irradiation and monochloramine on adenovirus serotype 2 early protein expression and DNA replication. Applied and Environmental Microbiology 74 (12), 3774-3782.

World Health Organization (WHO). *Progress on sanitation and drinking water: 2012 Update.* Geneva, 2012.

The role of sanitation and hygiene practices in contamination of household drinking water in rural households of South Africa

N Potgieter and TG Barnard*
Department of Microbiology, University of Venda, South Africa
Water and Health Research Unit, University of Johannesburg, South Africa

ABSTRACT

It has been shown that various household and cultural practices might be responsible for the recontamination of drinking water at the point-of-use. The aim of this study was to do a 6 month analysis of household practise by assessing the stored drinking water quality, the source water quality, the dishcloths used to clean the drinking water utensils and the kitchen floor. The pathogenic strains of *Escherichia coli* were used as the indicator of potential health risks in these households.

KEYWORDS

Escherichia coli, Sanitation, Water Quality, Contamination, Rural Households

INTRODUCTION

Diarrhoeal diseases are caused by protozoans such as *Giardia lamblia*, *Entamoeba histolytica, Cryptosporidium parvum*, *Isospora belli*, viruses such as Rotavirus, and Norwalk-like virus and frequently isolated bacterial diarrhoeagenic agents such as *Escherichia coli, Campylobacter jejuni/coli, Salmonella, Shigella,* and *Aeromonas* species (Obi et al., 1995, 1997; Lainson and Silva, 1999; Coker et al., 2002; Oyofo et al., 2002). *Escherichia coli (E. coli)* are, however, the focus of this study. The use of *E. coli* as an indicator organism is based on the fact that it is primarily associated with the gastrointestinal tract of humans and animals and therefore shows faecal specificity and it is unable to replicate and multiply in the environment outside of the host, due to the extreme changes in the environmental conditions (Brennan et al., 2010). Therefore, the concentration of *E. coli* in the secondary environment would be directly related to the constant faecal input of isolates from the primary host (Winfield and Groisman, 2003, Power et al., 2005). Gordon (2001) reported that most *E. coli* find themselves in an environment outside of their host at some stage and they may spend up to half their life in the environment outside of the host. Inside the host the environment have nutrients and carbon sources, constant temperature, microbial competition and protection from predation (Brennan et al., 2010). Outside the host, the environment might have lower temperatures, UV radiation exposure, limited available nutrients, limited moisture, environmental pollutants and predation which could lead to a decrease in the concentration of specific strains often to undetectable levels (Ishii and Sadowsky, 2008). Consequently, it is reported that the external environment does not actively support the growth of *E. coli*, forming the basis of its use as an indicator organism (Solo-Gabriele et al., 2000, Walk et al., 2007).

Although *E. coli* strains are part of the normal flora of animals and humans, some strains are pathogenic and infectious and may cause life-threatening infections such as urinary tract infections (Falagas and Gorbach, 1995), haemolytic colitis, neonatal meningitis, nosocomial septicaemia, haemolytic uremic syndrome and surgical site infections (Klein et al., 1986;

Thielman and Guerrant, 1999). These diseases account for a substantial degree of morbidity and mortality in different age groups (El-Sheikh and El-Assouli, 2001; Galane and Le Roux, 2001). Pathogenic *E. coli* strains include enteropathogenic *E. coli* (EPEC), enteroinvasive *E. coli* (EIEC), enterohaemorrhagic *E. coli* (EHEC), enteroaggregative *E. coli* (EAEC), and enterotoxigenic *E. coli* (ETEC) (Giron et al., 1991; Falbo et al., 1992; Blanco et al., 1992; Chan et al., 1994; Levine, 1987). Infection with pathogenic *E. coli* strains has been closely linked to poor sanitation and personal hygiene and therefore it was decided to use the identification of infectious *E. coli* strains as an indicator of hygiene and sanitation practices in rural households of South Arica.

METHODOLOGY

The study was carried out in rural communities in the Gauteng province in South Africa and the study sample consisted of 84 households. These communities did not have running water or flush toilets, many families have to share toilets and therefore, hygiene and sanitation were potential health risks for the people living there. The people had to rely on water that they collect from boreholes. The water was either from a central tap, large Jojo tanks or from a windmill and stored in 20 litre plastic containers in the homes and used for everything from cooking and drinking to bathing. Water (150 ml) which was used for household purposes was collected at the source [tap inside yard or communal tap shared] using published protocols. In addition, water samples (150 ml) from the household storage container were also taken to determine the storage conditions on the water quality of water used for drinking. Dishcloths were collected from the rural households and were placed individually in a sterile plastic ziplock bag and taken to the laboratory. In the laboratory, a piece of dishcloths were cut off from each of the selected dishcloths using sterile methods and then submerged into 100 ml sterile PBS, (pH 7.4) and vortexed for 1 minute. Toilet seats of each household with a toilet was swabbed separately using sterile swabs which were separately placed into sterile 100ml Phosphate Buffered Saline (PBS) solution (pH7.4) and transported to the laboratory for further analysis in 4oC cooler boxes. Samples were collected from the kitchen floors using sterile swabs to swab kitchen floors which were each added to sterile 100ml to sterile Phosphate Buffer Saline (pH 7.4) and transported to the laboratory in 4°C cooler boxes. Total coliforms and *E. coli* in all samples were enumerated in accordance with the Quanti-TrayTM method (IDEXX, Westbrook, ME, USA). Molecular characterisation of all E. coli isolates was done using a standardised adapted version of the guanidium thiocyanate/silica method (Omar et al., 2010).

RESULTS

The results indicated that more than 80% of the water sources and stored drinking water were positive for Total coliform counts and more than 50% were found to contain *E. coli* bacteria. Using molecular Polymerase Chain Reaction (m-PCR) technique to identify *E. coli* strains the results indicated the prevalence of several pathogenic strains of *E. coli* in both the source water and the stored water sample. In the dishcloth samples, more than 90% tested positive for Total coliforms and more than 80% positive for *E. coli.* Molecular Polymerase Chain Reaction (m-PCR) results showed that more than 20% of the samples contained at least one pathogenic strain of *E. coli,* more than 80% dishcloth samples contained a combination of more than one pathogenic *E. coli* strain. The results of the toilet seat samples showed that more than 90% samples were positive for TC and more than 70% were positive for *E. coli* of which more than 90% were pathogenic *E. coli* strains. The results of the kitchen floor samples indicated several *E. coli* pathogenic strains were prevalent on the kitchen floors.

DISCUSSION AND CONCLUSION

Several pathogenic strains of *E. coli* were detected in the source water and the stored water sample, the toilet seats, the dishcloth samples as well as the kitchen floor samples. The prevalence of pathogenic *Escherichia coli* found in samples collected from toilet seats indicate an important transmission route which needs to be investigated further to establish the health risk to children and vulnerable people in rural areas where hygiene practices is poor. Adequate cleaning and disinfection of toilet seats, which can be disease transfer points, is needed to reduce the potential of disease transmission (Potgieter et al., 2011). The results from this study clearly showed a potential risk to the health of households' members who fall into vulnerable group (children under 5 years, old people, pregnant women and immune compromised individuals). The prevalence of ETEC, EHEC, EIEC, EAEC and EPEC (especially the typical EPEC strains with bfp genes) showed the potential health risk in the transmission of diseases to the households' members.

REFERENCES

Blanco J, Gonzalez EA, Espinosa P, Blanco M, Garabal JI and Alanso MP (1992) Enterotoxigenic and necrotizing *Escherichia coli* in human diarrhoea in Spain. *Eur. J. Epidemiol* **8** 548-552.

Brennan, F. P., Abram, F., Chinalia, F. A., Richards, K. G and O'Flaherty, V. (2010). Characterisation of environmentally persistent *Escherichia coli* isolates leached from an Irish soil. *Applied and Environmental Microbiology*. **76** 2175-2180.

Chan KN, Phillips AD, Knutton S, Smith HR and Walker Smith JA (1994) Enteroaggregative *Escherichia coli*: Another cause of acute and chronic diarrhoea in England. *J. Pediatr. Gastroenterol. Nutr.* **18** 87-91.

Coker AO, Isokpehi RD, Bolaji NT, Amisu OA and Obi CL (2002) Human campylobacteriosis in developing countries. *Emerg. Inf. Dis.* **8** 237-243.

El-Sheikh SM and El-Assouli SM (2001) Prevalence of viral, bacterial and parastic enteropathogens among young children with acute diarrhoea in Jeddah, Saudi Arabia. *J. Health. Popul. Nutr.* **19** 25-30.

Falagas M and Gorbach S (1995) Practice guidelines: Urinary tract infections. *Infect. Dis. Clin. Pract.* **4** 241-257.

Falbo V, Famiglietti M and Caprioli A (1992) Gene block encoding production of cytotoxic necrotising factor 1 and hemolysin in *Escherichia coli*. *J. Infect. Immun.* **7** 873-880.

Galane PM and Le Roux M (2001) Molecular epidemiology of *Escherichia coli* isolated from young South African children with diarroeal diseases. *J. Health. Popul. Nutr.* **19** 31-37.

Giron JA, Asu HO and Shoonik GK (1991) An inducible bundleforming pilus of enteropathogenic *Escherichia coli*. *Sci.* **254** 710-713.

Gordon DM (2001). Geographical structure and host specificity in bacteria and the implications for tracing the source of coliform contamination. *Microbiology.* **147** 1079-1085.

Ishii S and Sadowsky MJ (2008). *Escherichia coli* in the environment: implications for water quality and human health. *Microbes Enivironment.* **23** 101-108.

Klein JO, Feigin RD and McCracken Jr GH (1986) Report of the task force on diagnosis and management of meningitis. *Pediatrics* **78** 959-982.

Levine MM (1987) *Escherichia coli* that cause diarrhoea: Enterotoxigenic, enteropathogenic, enteroinvasive, enterohaemorrhagic and enteroadherent. *J. Infect. Dis.* **164** 331-337.

Lainson R and Da Silva BAM (1999) Intestinal parasites of some diarrhoeic HIV-positive individuals in North Brazil, with particular reference to *Isospora belli* Wenyon, 1923 and *Dientamoeba fragilis* Jepps & Dobell, 1918. *Mem. Inst. Oswaldo Cruz* **94** 611-613.

Omar KB, Potgieter N and Barnard TG. (2010). Development of a rapid screening method for the detection of pathogenic *Escherichia coli* using a combination of Colilert Quanti-Trays/2000 and PCR. *Water Science and Technology: Water Supply* **10** 11-13

Obi CL, Coker AO, Epoke J and Ndip RN (1995) *Aeromonas* and *Plesiomonas* species as bacterial agents of diarrhoea in urban and rural areas of Nigeria: Antibiograms of isolates. *Cent. Afr. J. Med.* **4** 397-403.

Obi CL, Coker AO, Epoke J and Ndip RN (1997) Enteric bacterial pathogens in stools of residents of urban and rural regions in Nigeria: A comparison of patients with diarrhoea and controls without diarrhoea. *J. Diarrhoeal. Dis Res.* **15** 241-247.

Oyofo BA, Subekti D, Tjaniadi P, Machpud K, Komalarini S, Setiawan B, Simanjuntak C, Punjabi N, Corwin AL, Wasfy M, Cambell JR and Lesmana M (2002) Enteropathogens associated with acute diarroea in community and hospital patients in Jakarta, Indonesia. *FEMS Immunol. Med. Microbiol* **34** 139-146.

Potgieter N, Mpofu TB and Barnard TG. (2011). The Impact Water, Sanitation and Hygiene Infrastructures Have on People Living with HIV and AIDS in Zimbabwe, Microbes, Viruses and Parasites in AIDS process. Vladimir Zajac (Ed.). ISBN: 978-953-307-601-0. InTech. URL: http://www.intechopen.com/articles/show/title/the-impact-water-sanitation-and-hygieneinfrastructure-have-on-people-living-with-hiv-and-aids-in-z (Accessed: February 2012).

Power ML, Littlefield-Wyer J, Gordon DM, Veal DA and Slade MD (2005). Phenotypic and genotypic characterisation of encapsulated *Escherichia coli* isolated from blooms in two Australian lakes. *Environ. Microbiology.* **7** 631-640.

Solo-Gabriele HM, Wolfert MA, Desmarais TR and Palmer CJ (2000) Sources of *Escherichia coli* in a costal subtropical environment. *Applied and Environmental Microbiology.* **7** 230-237.

Thielman NM and Guerrant RL (1999) *Escherichia coli.* In: Yu VL, Merigan (Jr) TC and Barriere SL (eds). *Epidemiology and Pathogenesis of Escherichia Coli.* The Williams & Wilkins Company, Baltimore. 88-200.

Walk ST, Alm, EW, Calhoun LM, Mladonicky JM and Whittman TS (2007). Genetic diversity and population structure of *Escherichia coli* isolated from freshwater beaches. *Environ. Microbiology.* **9** 2274-2288.

Winfield MD and Groisman EA (2003). Role of nonhost environments in the lifestyle of *Salmonella* and *Escherichia coli. Applied and Environmental Microbiology.* **69** 3687-3694.

IMPACT OF WASTEWATER DISINFECTION ON REUSE APPLICATIONS

Mark LeChevallier, Zia Bukhari, Lauren Weinrich, Patrick Jjemba*
American Water, Voorhees, NJ
Email: mark.lechevallier@amwater.com

BACKGROUND

The United States does not have federal effluent quality or treatment standards for reclaimed water. Each state has adopted different regulations and guidelines, with California (under Title 22) recognized as one of the states with a comprehensive set of high effluent water quality treatment process requirements (California Department of Public Health, 2009). In terms of disinfection, the chlorine disinfection requirements in California for unrestricted reuse applications is specified as a CT value (the product of total chlorine residual and modal contact time measured at the same point) of not less than 450 mg-min/L with a modal contact time (time for highest concentration to pass through contact basin) of at least 90 minutes, based on peak dry weather design flow. Other states with well developed recycled water requirements include Florida, Washington, Arizona and Texas.

Membrane bioreactors (MBRs) are often the technology of choice for water recycling facilities. This technology has the capability to produce superior effluent water quality as compared to more conventional wastewater plants with tertiary treatment, particularly with respect to ammonia, organics, particles and microorganisms.

DISINFECTION OF MBR EFFLUENTS

MBR effluent water quality data was collected from a wide range of satellite facilities (38 facilities) to allow proper characterization of MBR effluents with respect to inorganic, organic, physical and microbial parameters (Jacangelo et al., in press). The MBR facilities sampled during the study utilized different process configurations (submerged and external), membrane geometries (hollow-fiber, flat-sheet and tubular), fouling control strategies (relaxation and backwash) and membranes of varying ages. The MBR facilities sampled were spread across six different states and three different USEPA regions; flow-rates at these facilities ranged from 0.001 to 1.8 MGD. Results from the reconnaissance survey demonstrated, in general, that satellite MBR facilities are capable of producing oxidized, nitrified effluents that have relatively lower concentration of particles and microbial indicators (Table 1). However, some plants experienced relatively poor water quality (for an MBR) and was attributed to sampling error (there was evidence of bacterial regrowth in MBR effluents, as evidence of relatively low bacteriophage levels in Figure 1) or no effluent requirement to meet certain parameters (e.g., the permit did not require treatment of TOC or ammonia in certain circumstances, Figure 2).

Bench-scale microbial inactivation experiments were conducted on the filtrate samples collected from the MBR system operating with a breached membrane to develop the residual-response curves for total coliform bacteria and seeded male specific bacteriophage. The objective of these experiments was to determine the free chlorine CTs necessary to achieve a desired level of microbial inactivation in MBR effluents produced from breached membranes. In these experiments, multiple filtrate samples were collected from the MBR system operating with breached membranes. The turbidity for these samples ranged from 2.8 to 6.9 NTU. To achieve the desired filtrate turbidity of 1.0 NTU, filtrate samples collected while the MBR system was operating with intact membranes were blended with those from the breached membrane.

Table 1. Summary of effluent water qualities produced by 38 satellite MBR facilities

Parameter	Concentrations			
	50th Percentile	90th Percentile	Minimum	Maximum
Turbidity (NTU)	0.21	0.71	0.09	8.58
Ammonia (mg/L-N)	<0.02	0.44	<0.02	3.41
TOC (mg/L)	4	8	2	27
Total Coliform Bacteria (CFU/100 mL)	1	100	<1	293
Indigenous MS-2 Bacteriophage (PFU/100 mL)	<1	21	<1	848
Particle Counts > 2 μm (Counts/100 mL)	26,715	145,840	2,648	2,044,564
UV-254 (cm^{-1}) / Corresponding UVT (%)	0.12 / 76	0.22 / 60	0.06 / 88	0.35 / 45

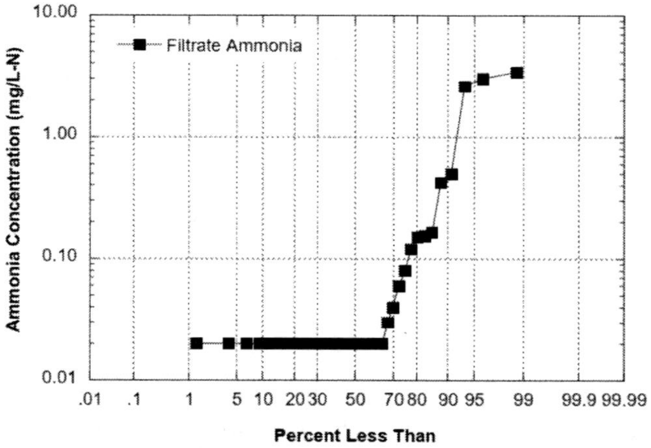

Figure 1. Cumulative probability distribution for the ammonia concentrations measured in the effluents of satellite facilities.

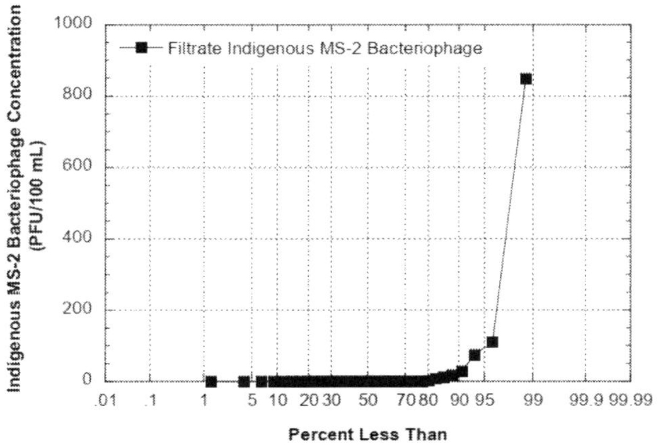

Figure 2. Cumulative probability distribution for indigenous male specific bacteriophage concentrations measured in the effluents of satellite facilities.

As shown in Figure 3, a CT of 30 mg-min/L was required to achieve greater than 5-log removal of seeded male specific bacteriophage and removal of total coliform bacteria at or below the method detection limit (2 CFU/100 mL) for samples with a filtrate turbidity of 1.0 NTU. These CT values were also sufficient to reduce the concentration of total coliform bacteria and male specific bacteriophage to below the method detection limit for the majority of the samples. Regulations in the state of California require MBR systems to shut down or divert the filtrate flow if the turbidity exceeds 0.5 NTU. As such, these experiments, being conducted on filtrate samples with turbidity of 1.0 NTU, represent a worst case MBR effluent scenario.

ON-LINE AMMONIA ANALYZER

In order to understand the impact of operating conditions and upsets in bioreactor basins on effluent ammonia, an online ammonia analyzer was installed at the CA-01 water recycling facility. Dissolved oxygen concentrations in the aeration basins and SRT of the facility were monitored to understand the impact of these parameters. Results from online ammonia monitoring showed that the CA-01 facility was able to consistently produce fully nitrified effluent for most of the test period although the effluent ammonia concentrations occasionally reached up to 3.4 mg/L-N during the first 700 hours of operation (Figure 4). Upon investigation, the DO concentrations in the aeration basins appeared to impact the ammonia concentration. Since the facility was not required to meet effluent ammonia limits, the DO set points for the aeration basins were set to a lower than desired value to achieve savings in operational costs with respect to power consumption by the process air blowers. After modifying the DO set point, the facility was able to consistently achieve low effluent ammonia concentrations. Since the SRT for the facility was always high enough (above 12 days) to retain the nitrifiers in the bioreactor basins, it is unlikely that SRT had any impact on effluent ammonia excursions.

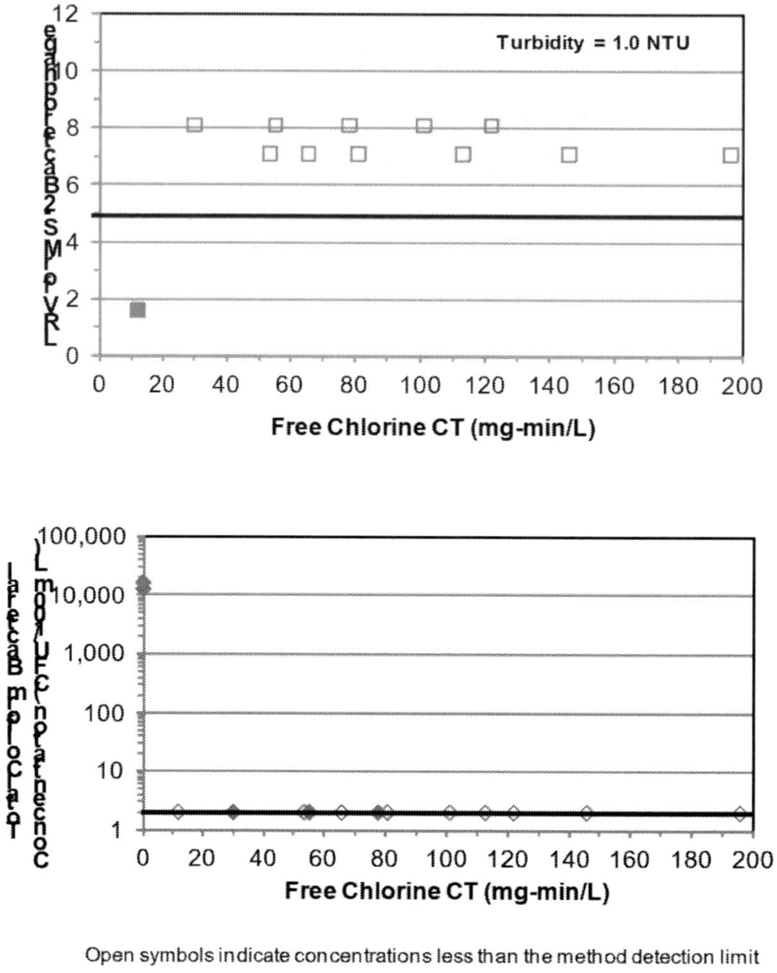

Figure 3. Inactivation of total coliform bacteria and seeded male specific bacteriophage in MBR filtrate at a turbidity of 1.0 NTU.

Figure 5 presents the dissolved oxygen concentrations measured in the aeration basin using an online DO analyzer and corresponding effluent ammonia concentrations. Elevated ammonia concentrations were observed when DO concentrations were measured below 1 mg/L in the aeration basin. It should be noted that the CA-01 facility does not have an ammonia discharge limit so the facility does not have to operate to continuously meet a low effluent ammonia concentration. On the contrary, to achieve savings in the operational cost, the dissolved oxygen set points in the aeration basins were lowered so the power consumption for the process air blowers could be reduced. Once the plant staff was made aware of the filtrate ammonia concentrations recorded by the online ammonia analyzer, DO set points were adjusted to achieve better nitrification.

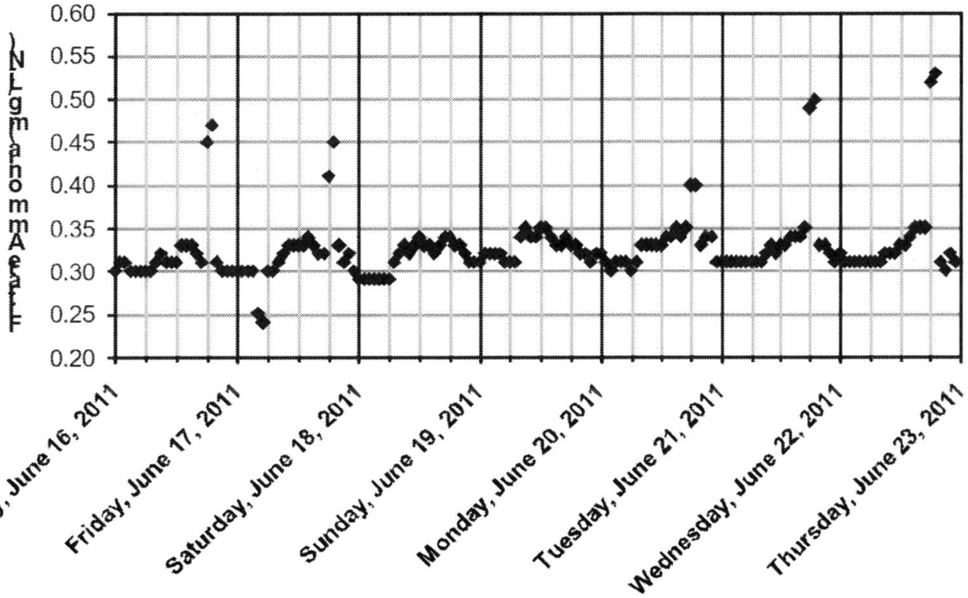

Figure 4. Diurnal variations in ammonia concentration in the MBR filtrate at the CA-01 facility measured using an online ammonia analyzer (week of June 16th, 2011).

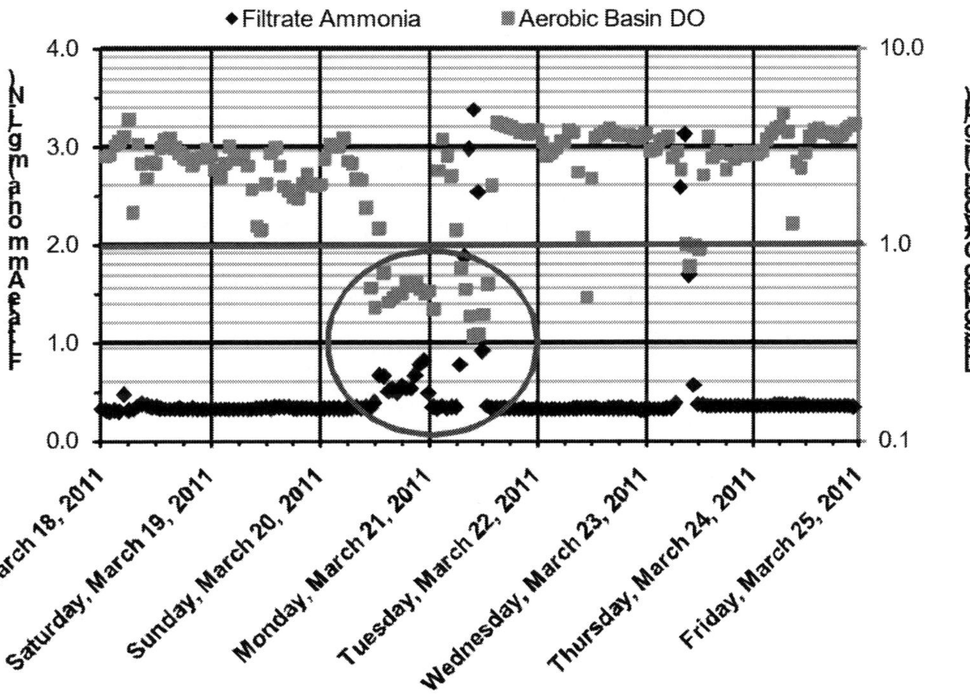

Figure 5. Dissolved oxygen concentrations in the aeration basin and corresponding effluent ammonia concentrations at the CA-01 facility (week of March 18th, 2011).

OCCURRENCE AND INACTIVATION OF *CRYPTOSPORIDIUM*

To monitor *Cryptosporidium*, an intensive sampling program was launched for 14 geographically dispersed wastewater utilities over a 12 month period, with 8 utilities employing secondary clarification, 2 using sand filtration, 3 employing MBRs, and 1 using cloth filtration. Sampling was designed to measure *Cryptosporidium* concentrations, infectivity, and genotype information after physical removal (i.e., pre-disinfection) and immediately post-disinfection. Method 1623 was used for sampling; with specific organism isolation using immunomagnetic separation (IMS) followed by immunofluorescence assay (IFA) based detection and enumeration.

Eight plants employing secondary treatment were frequently positive for *Cryptosporidium* (20–66.7%) and *Giardia* (83–100%). One of the two plants employing sand filtration demonstrated 8.3% of the post-disinfection samples were *Cryptosporidium* positive. *Giardia* cysts were detected at both sand filtration plants with one plant demonstrating a higher occurrence (i.e., pre-disinfection, 33.3%; post-disinfection, 50%) than the second plant (8.3% of samples positive). Of three MBR plants, only one plant demonstrated *Cryptosporidium*-positive samples (post-disinfection, 8.3%). In contrast two of the three MBR plants contained 58.3–83.3% *Giardia*-positive samples. Effluents from one plant employing cloth filtration were positive for both *Cryptosporidium* and *Giardia*. The pre-disinfection samples for this plant contained a substantially higher frequency of *Giardia*-positive samples (66.7%) than *Cryptosporidium*-positive samples (33.3%); however, 16.7% of the post-disinfection samples were positive for both organisms. Overall, 21.9% (73 of 333) of the samples were *Cryptosporidium* positive and 70.1% (234 of 333) of the samples were positive for *Giardia* by IFA. Comparing protozoan concentrations with various water quality parameters (i.e., turbidity, total suspended solids, ammonia, chlorine, fecal coliforms, dissolved oxygen, pH, and carbonaceous biochemical oxygen demand) indicated no obvious correlation.

A cell culture–IFA (CC-IFA) procedure was used to measure the infectivity of *Cryptosporidium* oocysts and indicated that 6.3% (18 of 285) of the samples were potentially infectious. The majority of the CC-IFA positive samples (n = 15) were derived from treatment plants employing secondary clarification. The infectivity data also demonstrated that infectious oocysts were detectable post chlorination; however, no infectious oocysts were detected following UV disinfection. When the data were expressed in terms of a risk of infection, the advantage of tertiary filtration, especially MBR filters, becomes apparent with risk levels that approach drinking water standards (Figure 6)

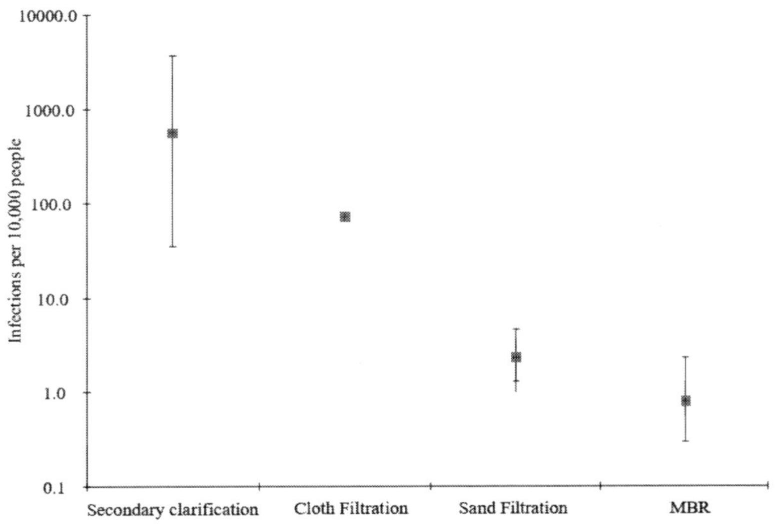

Figure 6. Estimate of *Cryptosporidium* infections per 10,000 exposed individuals.

The chlorine CT values recommended by California Title 22 are 450 mg-min/L with a modal contact time of 90 min for wastewater effluents (California Department of Public Health, 2009). Although chlorine can be effective for bacteria and viruses, it has little effect on protozoan parasites such as *Cryptosporidium*. Previous disinfection experiments have shown chlorine CT values between 6000 and 7200 mg·min L^{-1} yield marginal (i.e., 1.0–1.7 log) inactivation (Kroch et al., 1990; Vance et al., 1997). These results were verified in this study for CT values ranging from 90 to 600 gamin/L which showed 1 log or lower inactivation of oocysts, with the mean levels being <0.5 log (Figure 7). The impact of fresh and aged oocysts was also examined, but the differences were minimal, with aged oocysts showing marginally higher levels of inactivation.

Chlorination of organic compounds can also produce potentially carcinogenic disinfection byproducts (i.e., trihalomethanes and haloacetic acids), which can present short- and long-term human health impacts. As a result there is an increasing trend toward utilization of alternative disinfectants. It is estimated that 21% of POTWs are presently employing UV disinfection, with almost 40% of these systems switching between 2001 and 2005. Using a range of UV fluence, oocyst inactivation levels of >1.52 log were noted at 1 mJ/cm^2 and exceeded 4.5 log with a fluence of 20 mJ/cm^2 (Table 2). These data further support that CC-IFA data can reliably measure oocyst inactivation following UV disinfection. Also the disinfection data obtained with this cell culture assay were in agreement with data from previous UV disinfection studies using mouse infectivity assays (Bukhari et al., 1999).

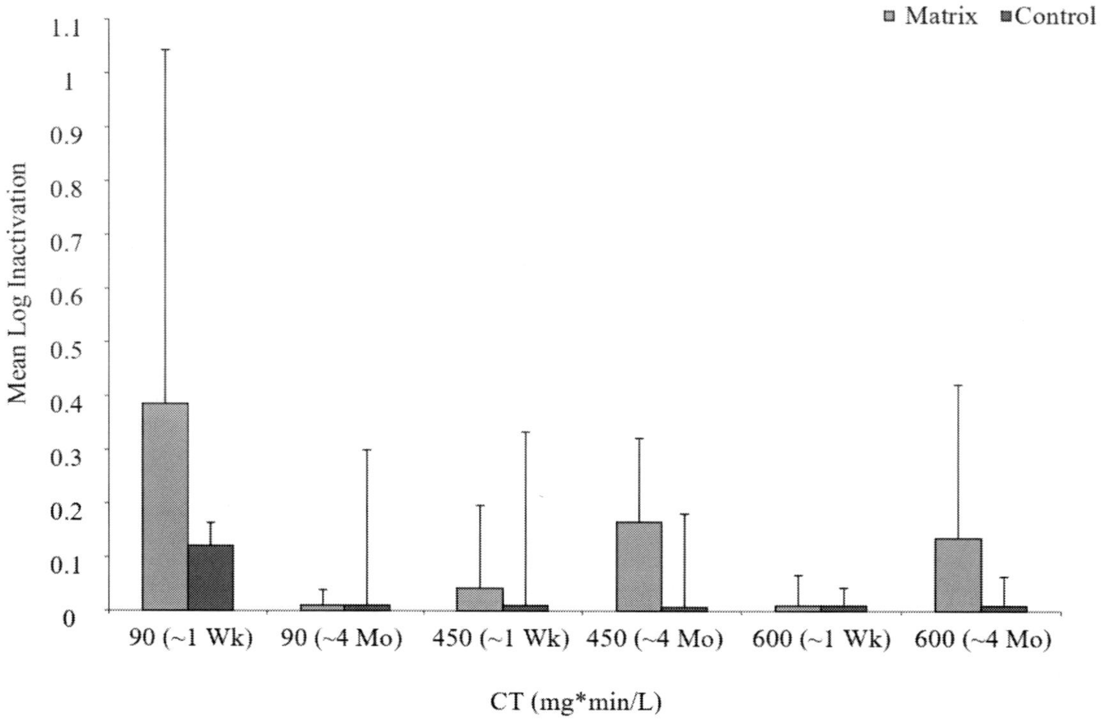

Figure 7. Inactivation of C. partum oocysts following chlorination. ***Note*:** **Wk = weeks; Mo = months.**

Table 2. Log Inactivation of *C. parvum* Oocysts Following Exposure to Various UV Doses

Target Inoculum	1 mJ · cm^{-2}	5 mJ · cm^{-2}	10 mJ · cm^{-2}	20 mJ · cm^{-2}
50,000	TNTC	>4.61	>4.69	>4.69
10,000	>1.67	>4.0	NA	NA
5000	>1.80	>3.69	NA	NA
5000	>1.92	>3.69	NA	NA
1000	>1.89	>3.0	NA	NA
500	>1.73	>2.69	NA	NA
50	>1.69	>1.69	NA	NA
Mean Log Inactivation	>1.78 ± 0.11	>4.61	>4.69	>4.69

Note: TNTC = too numerous to count; NA, not applicable; $n = 6$.

Ozonation experiment results for *C. parvum* and *C. hominis* oocysts suspended in deionized water and 30% matrix from a plant employing tertiary sand filtration have been presented in Figure 8. The original intent was to use undiluted effluent samples for the ozonation experiments; however, the matrix demand for ozone was high (>10 mg/L), making it impractical to meet these high ozone demands in the bench-scale experiments. Due to high matrix demands for ozone, effluents other than those from systems employing MBRs are unlikely to be candidates for ozonation during final disinfection. For MBR effluents that also use UV disinfection downstream, ozonation may be particularly beneficial in helping to improve UV transmittance, in turn enhancing the effectiveness of UV disinfection.

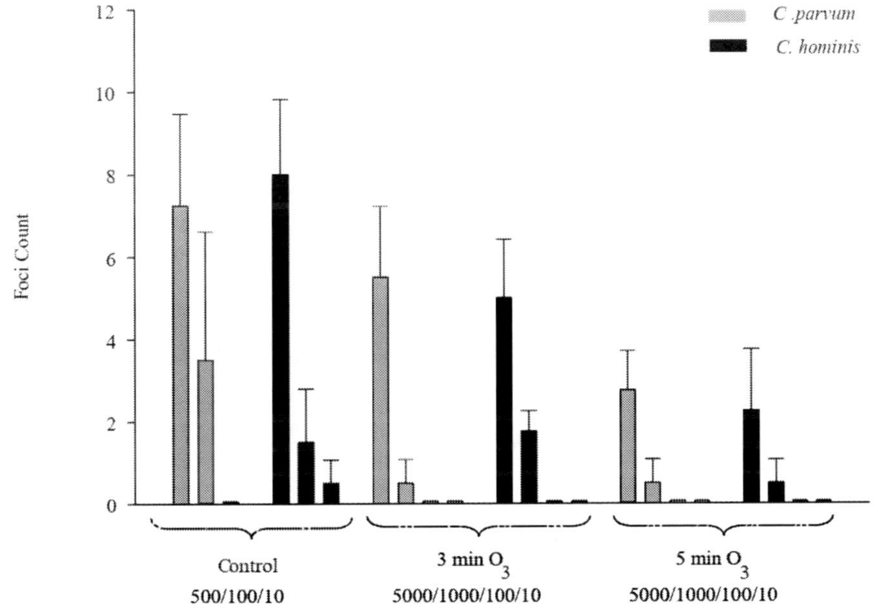

Figure 8. Comparing cell culture infectivity for C. *parvum* and C. *hominis* oocysts.

IMPACT OF DISINFECTION ON AOC LEVELS

Assimilable organic carbon (AOC) is the fraction of the total organic carbon that is easily utilized and has been related to the growth of opportunistic pathogens, heterotrophic plate count bacteria, and indicator organisms in water (van der Kooij et al., 1982). The levels of AOC in reclaimed water can be quite high. Jjemba et al. (2010) found average AOC levels ranged between 150 and 1400 μg/L and were in general, about 10 times higher than those found in drinking water systems. These levels are of concern because they could stimulate the growth of opportunistic pathogens including *Legionella* spp. and *Mycobacterium* spp. in reclaimed water distribution systems. Jjemba et al. (2010) found both *Legionella* spp. and *Mycobacterium* spp. present in the effluent and storage reservoir and throughout the reclaimed distribution system during the year (Table 3). In most instances, the density of *Mycobacterium* spp. was highest in the reclaimed water at the ends of the distribution system. *Legionella* spp. were also at least 10-fold higher in the distribution system than in the point of entry (or disinfection point).

Table 3. Geometric Mean Densities of *Mycobacterium* spp. and *Legionella* spp. in Four Reclaimed Water Systems[a]

Site and Species	Effluent	Storage	DS1[b]	DS2	DS3
Mycobacterium spp. (CFU/100 mL)					
CA	1 ± 1	5 ± 17	22 ± 15	35 ± 46	30 ± 120
FL	11 ± 20	65 ± 220	55 ± 390	73 ± 600	107 ± 800
MA	170 ± 190	2 ± 1[c]	57 ± 25	320 ± 130	120 ± 80
NY	6 ± 15	50 ± 80	42 ± 110	16 ± 14	31 ± 29
Legionella spp. (10^3 CFU/100 mL)					
CA	<0.3	2.2 ± 4.1	2.3 ± 7.1	0.9 ± 1.5	1.9 ± 1.6
FL	<0.3	3.0 ± 70	2.7 ± 13	3.5 ± 16	8 ± 52
MA	0.4 ± 0.1	<0.3[c]	1.6 ± 2.8	0.9 ± 2.0	0.6 ± 0.6
NY	0.6 ± 2.1	0.7 ± 0.6	0.5 ± 0.6	0.6 ± 0.6	0.6 ± 0.4

[a]Values are geometric means (±SE) based on aggregate densities over the yearlong monitoring.
[b]DS = distribution system.
[c]Disinfection point is at the storage tank for this location.

The impact of disinfection of AOC levels in reclaimed water systems was studied in a model pipe loop system (Weinrich et al. 2010). Data shown in Figure 9 illustrate large increases in AOC after disinfection with free chlorine relative to the un-disinfected control water. By comparison, chloramine disinfection did not produce a large increase in AOC. It is interesting to note that AOC levels declined in the free chlorine treated pipe loop presumably due to the growth of biofilm bacteria (at 100-150 ft, the residual had dissipated).

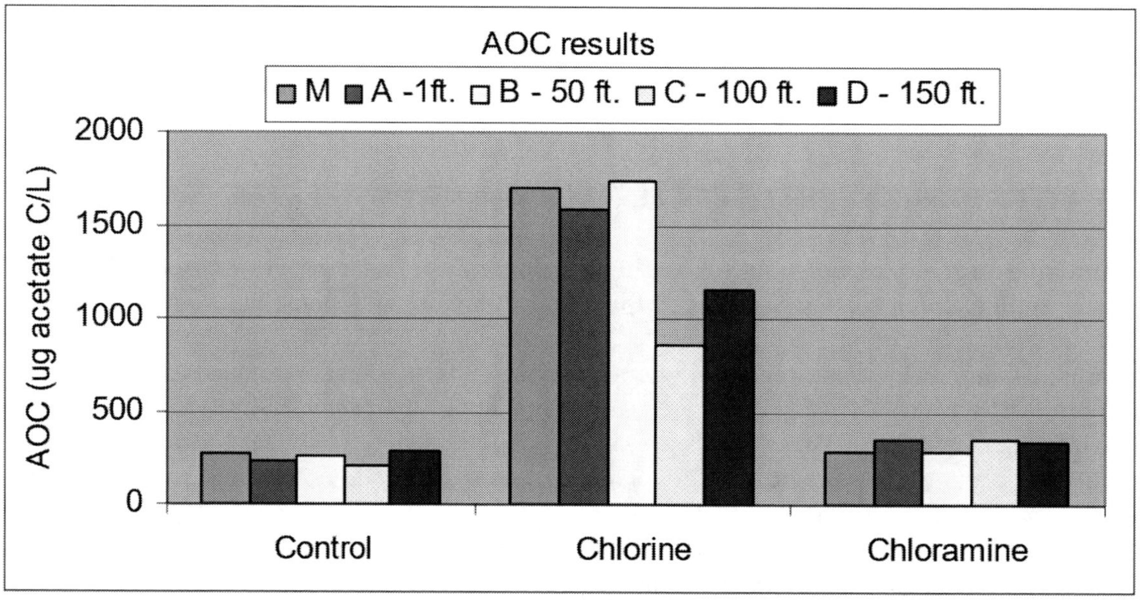

Figure 9. Impact of disinfection on the formation of AOC in a model pipe system. Abbreviations: M, mixed – the point at which disinfectants were added. Samples were collected after 1, 50, 100, and 150 ft travel within the pipe loop system.

Impact of Treatment and Disinfection on AOC Levels

A survey of wastewater facilities was conducted to examine the impact of various treatment processes on AOC levels in the finished effluent quality. The wastewater sources included the following categories, the number in parentheses corresponding to the number of utilities surveyed:

- domestic and municipal sources (7),
- domestic (6),
- domestic and industrial (5),
- municipal (2), and
- municipal and industrial (1).

The wastewater treatment technologies surveyed included activated sludge, RBCs, sequencing batch reactors, and MBRs. AOC levels ranged from 45 to 3200 μg/L, with a median value of 452 μg/L. All three of the MBR utilities surveyed had AOC levels below the respective median (Figure 10) although some activated sludge utilities produced effluent quality similar to those produced by MBRs based on AOC measurements. In general, the systems with the lowest AOC levels used UV for disinfection of the treated effluents.

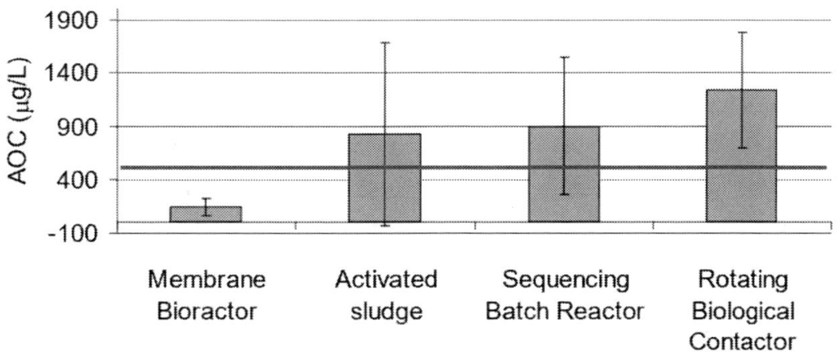

Figure 10. Impact of treatment process and disinfection on AOC levels in reclaimed water effluents. The median value for all systems was 452 μg/L.

CONCLUSIONS

Historically, disinfection requirements for reclaimed water have mandated extended disinfection contact times based on the assumption that ammonia would be present and that the prolonged exposure would be needed for virus and bacterial inactivation. However, fully denitrified systems can maintain a free chlorine and reliability achieve inactivation of bacteria and viruses with much shorter CT values (30 mg·min/L). Moreover, control of *Cryptosporidium* risks can only be managed using UV disinfection. Free chlorine disinfection can increase treated effluent AOC levels which can result in growth of biofilm bacteria including *Legionella* and *Mycobacterium* spp. MBR effluents treated with UV disinfection have some of the lowest AOC levels, although other treatment processes may be able to achieve similar results. The water operator must be aware of the interrelationship between treatment, disinfection, and reclaimed water AOC levels so that solutions can be found to provide reclaimed water with the lowest public health risk.

ACKNOWLEDGEMENTS

The studies presented here were funded by the WateReuse Research Foundation and the utility subsidiaries of American Water, Voorhees, NJ.

REFERENCES

Bukhari, Z.; Hargy, T. M.; Bolton, J. R.; Dussert, B.; Clancy, J. L. Medium-pressure UV for oocyst inactivation. *JAWWA* 1999, *91,* 86–94.

Bukhari, Z.; Holt, D. M.; Ware, M. W.; Schaefer, F. W., III. Blind trials evaluating in vitro infectivity of *Cryptosporidium* oocysts using cell culture immunofluorescence. *Can. J. Microbiol.* 2007, *53,* 656–663.

Bukhari, Z.; LeChevallier, M. Assessing UV reactor performance for treatment of finished water. *Water Sci. Technol.* 2003, *47,* 179–184.

California Department of Water Resources (2004) Water Facts No. 23 – Water Recycling. www.water.ca.gov (accessed Aug 2011).

Jacangelo, J., M. LeChevallier, J. Oppenheimer, Z. Hirani, and Z. Bukhari. In press. Disinfection Guidelines for Satellite Water Recycling Facilities. WRF-08-07. WateReuse Research Foundation, Alexandria, VA.

Jjemba, P.K., L.A. Weinrich, W. Cheng, E. Giraldo, and M.W. LeChevallier. 2010. Regrowth of Potential Opportunistic Pathogens and Algae in Reclaimed-Water Distribution Systems. *Appl. Environ. Microbiol.* 76(13): 4169–4178.

Korich, D. G.; Mead, J. R.; Madore, M. S.; Sinclair, N. A.; Sterling, C. A. Effects of ozone, chlorine dioxide, chlorine, and monochloramine on *Cryptosporidium parvum* oocyst viability. *Appl. Environ. Microbiol.* 1990, *56,* 1423–1428.

van der Kooij, D.; Visser, A.; Hijnen, W. A. M. Determining the concentration of easily assimilable organic carbon in drinking water. *JAWWA* 1982, *74* (10), 540–545.

Venczel, L.; Arrowood, M.; Hurd, M.; Sobsey, M. Inactivation of *Cryptosporidium parvum* oocysts and *Clostridium perfringens* spores by a mixed-oxidant disinfection and by free chlorine. *Appl. Environ. Microbiol.* 1997, *63* (4), 1598–1601.

Weinrich, L.A., P.K. Jjemba, E. Giraldo, and M. W. LeChevallier. 2010. Implications of Organic Carbon in the Deterioration of Water Quality in Reclaimed Water Distribution Systems. *Water Research* 44: 5367-5375.

Table 4. Wastewater Survey Sites, Process Details, and Water Quality Characteristics[a]

Coded Facility	Source	Process	Secondary Treatment	Disinfection	Tertiary Filtration	Nutrient Removal	TOC (mg/L)	BDOC (mg/L)	AOC (µg/L)
IA070108	Municipal/Industrial	AS	Extended Aeration	Chlorination	None	None	11.20	3.12	3203
NC070708	Domestic/Industrial	AS	Oxidation Ditch	Chlorination/ Dechlorination	None	N and P Removal	6.06	2.03	1292
PA070708	Domestic/Industrial	Attached Growth	Trickling Filter and AS	UV	None	N and P Removal	6.32	2.60	1313
GA070708	Domestic/Industrial	AS	A2O	UV	Yes	N and P Removal	4.43	1.50	351
OH070708	Domestic/Industrial	Attached Growth	RBC	Chlorination/ Dechlorination	Yes	P Removal	7.01	2.73	853
OKF70708	Domestic/Municipal	AS	Extended Aeration	Chlorination	Yes	None	5.73	1.65	452
OKE70708	Domestic/Municipal	AS	Extended Aeration	UV, Chlorination Dechlorination	Yes	None	5.41	1.17	432
AZA70808	Domestic	AS	MBR	UV	Yes[b]	N Removal	6.49	1.34	232
AZV70808	Domestic	AS	SBR	Chlorination	None	N Removal	5.49	1.36	645
AZN70908	Domestic/Municipal	AS	MLE	Chlorination/ Dechlorination	Yes	N Removal	7.62	3.07	1160
AZR70908	Domestic	AS	Extended Aeration	Chlorination/ Dechlorination		N Removal	5.07	0.24	45
CT071008	Municipal	AS	MBR	UV	Yes[b]	N Removal	1.78	0.17	128
CA072108	Domestic/Municipal	Attached Growth	RBC	Chlorination	Yes	None	5.50	1.39	1615
AL072108	Domestic/Municipal	AS	Extended Aeration			None	2.66	0.27	152
NJ072308	Domestic	AS	SBR	UV	Yes	N Removal	3.59	0.58	450
NJG72808	Domestic	AS	MLE	UV	Yes	N Removal	5.88	0.84	438
NJP72808	Municipal	AS	MBR	UV	Yes[b]	N Removal	0.70	0.03	81
NJB72808	Domestic/Municipal	AS	SBR	UV	Yes	N Removal	4.96	0.74	1859
NJH72808	Domestic/Municipal	AS	Extended Aeration	Chlorination/ Dechlorination		N Removal	5.42	1.45	664
NJR73008	Domestic	AS	SBR	UV	Yes	N removal	4.03	0.77	656
LA073008	Domestic/Industrial	AS	Aeration	Chlorination	None	None	9.60	0.83	438

[a]Median AOC = 452 µg/L.
[b]Plant treatment process is equivalent to tertiary filtration. AS = activated sludge; SBR = sequencing batch reactor.

Removals of Enteric Viruses and Indicator Organisms in Membrane Bioreactor and Conventional Secondary Activated Sludge Wastewater Plants

Ashley Williston, CT Consultants; Kimberly Riddell, City of Delphos, Ohio, City Council President and Smith Environmental; Terry Gellner

ABSTRACT

Membrane bioreactors (MBRs) are a relatively new wastewater-treatment technology in which a combination of suspended growth activated sludge biological treatment and membrane filtration equipment is used to perform the critical solids/liquid separation function that is traditionally accomplished using secondary clarifiers and filters. Information is lacking on the effectiveness of MBRs in removing human enteric viruses from wastewaters, particularly as compared to removal by conventional plants before and after disinfection. During three recreational seasons (May–October), when disinfection of effluents is required in Ohio, removals of bacterial indicators, coliphage (indicator viruses which are not harmful to humans), and human enteric viruses were studied throughout the treatment processes in three MBR and two conventional secondary activated sludge municipal wastewater treatment plants. Removals of all organisms were higher across secondary treatment for MBR plants than for conventional plants and ultraviolet disinfection after MBR secondary treatment provided little additional removal of any organism except for one type of coliphage. Reductions in capital cost, O&M cost, environmental impacts, and safety impacts are other considerations further supporting the argument that a disinfection process unit after MBR secondary treatment provides little significance.

KEYWORDS

Activated sludge, membrane bioreactor, MBR, viruses, microorganisms, disinfection, secondary treatment, post treatment, chlorination, ultraviolent light, UV

INTRODUCTION

The use of membrane bioreactor (MBR) technology for wastewater treatment is increasing in the United States because the quality of effluent is higher than with conventional secondary treatment and plant capacity can be increased by retrofitting the technology into existing conventional secondary-treatment activated-sludge units. By submerging the membranes in the activated sludge process the membranes provide the critical solids/liquid separation function that is traditionally accomplished using secondary clarifiers and filters.

An important component of wastewater treatment is to minimize exposure of the public to bacterial, viral, and protozoan pathogens. Viruses, which are considerably smaller than other types of pathogens, may be able to pass through some membranes based on size exclusion alone. Human enteric viruses include enteroviruses, adenoviruses, noroviruses, rotaviruses, and Hepatitis A virus, and they cause a wide range of diseases and symptoms. In spite of the potential public health risk, wastewater effluents are seldom monitored for enteric viruses. Instead, bacterial indicators, such as *Escherichia coli* (*E. coli*) and fecal coliforms, are the required microbial measures of effluents for wastewater discharge permits in the United States.

Enteric virus removal by membranes in wastewater matrices has been investigated in the past, but none of the MBR full-scale studies (Kuo et al., 2010, and Simmons et al., 2011) included a component in which removal of viruses was compared before and after chlorine or ultraviolet (UV) disinfection. In wastewater treatment plant pilot studies, MBR treatment removed different types of microorganisms, including enteric viruses in some studies, more effectively than did conventional secondary treatment (Ottoson et al., 2006; Zhang and Farahbakhsh, 2007).

Therefore state and federal regulators still require all plants, including those with MBRs, to have disinfection systems as part of the overall treatment process. It is expected that this requirement will remain in effect throughout the United States, until such time as data is produced for the regulators to consider otherwise.

The four objectives of the study included: 1) compare the reductions of enteric viruses in MBR systems to the reductions found for conventional secondary treatment before disinfection; 2) compare the additional reduction of enteric viruses achieved by disinfection of secondary wastewaters at conventional plants with those achieved by disinfection of MBR treated wastewaters; and 3) compare reductions of indicator bacteria and indicator viruses (coliphage) in MBR and conventional systems, and (4) identify which indicator(s) best represent the removal of enteric viruses.

METHODOLOGY

Samples were collected at three MBR activated-sludge and two conventional secondary activated-sludge municipal wastewater treatment plants in Ohio and analyzed for bacterial indicators, coliphage (indicator viruses), and enteric viruses. Sample results from one of the MBR are not reported because the plant was not operating properly during the study period. The plants selected for the study were small- to medium-sized municipal wastewater treatment plants in Ohio. Small plants are considered to have an average daily flow rate less than 1,000,000 gallons per day (gpd). Medium-sized MBR plants have average daily flow rates ranging between 1,000,000 and 5,000,000 gpd. The plants selected for the study were chosen because they are within the range of MBR plants currently being constructed. Other selection requirements for wastewater plants were as follows:

- The plants were to be located in Ohio for logistical purposes associated with the sampling.
- Existing conventional facilities with monthly discharge limitations not greater than 10 mg/L BOD/ 12 mg/L TSS/ 1 mg/L NH_3 that use chlorine or ultraviolet (UV) disinfection.
- MBR plants were in operation for at least one year.
- Facilities and processes within the plants needed to be well-defined with no other parallel treatment streams where the MBR process effluent could be mixed with conventional treatment effluent.

Generally there are several types of membrane systems used throughout the country in MBR activated sludge processes. The two most common types of membranes used in the activated sludge process today are hollow fiber and flat plate membranes. Different manufactures use different membrane pore sizes to perform the physical separation. For this study, the MBR plants that were selected utilize flat plate membranes manufactured by Kubota and distributed by Enviroquip a division of Ovivo USA. The flat plate membranes basically are composed of a membrane case and diffuser case. The membrane case consists of flat membrane sheets that are placed at fixed intervals of 0.4 to 0.7 mm. The membrane sheets by Kubota have a nominal pore size of 0.4 microns and are considered to be capable of microfiltration. Some other flat plate membrane manufacturers use membrane sheets having pour sizes (0.037-microns) similar to the hollow fiber which are considered ultrafiltration. Based on size exclusion alone, sampling results of the Kubota flat plate membrane plants should be considered applicable to other membranes since the flat plate studied has a pore size classified as microfiltration and the smaller pore sizes mentioned herein are classified as ultrafiltration.

Log removals of microorganisms were determined throughout treatment processes (post-preliminary, post-MBR, post-secondary, post-tertiary, and post-disinfection) during the 2008, 2009, and 2010 recreational seasons (May–October), when disinfection of effluents is required in Ohio.

Detailed information on methodology can be obtained from Francy et al. (2011) and Francy et al. (2012). Briefly, 6-L grab samples were collected for all bacterial indicators and coliphage. Analysis of samples for bacterial indicators was done by preparing serial dilutions for plating by standard membrane filtration methods. Analyses of samples for F-specific and somatic coliphage were done by use of the single agar layer (SAL) procedure. Viruses were concentrated by use of glass-wool fiber filters (Lambertini et al., 2008) from approximately 4-L of post-preliminary, 140-L of post-secondary, or 500-L of post-MBR, post-tertiary, and post-disinfection wastewater (Fig. 1). Viruses were eluted from the glass-wool filters and concentrated. Final concentrates were analyzed by quantitative polymerase chain reaction (qPCR) for adenovirus; by quantitative reverse transcriptase PCR (qRT-PCR) for enterovirus, norovirus GI and GII, rotavirus, and Hepatitis A virus; and for culturable viruses by the cell culture method. The qPCR and qRT-PCR methods are molecular methods that target viral DNA and RNA, respectively.

Figure 1. Sampling setup for concentrating viruses by glass-wool filtration.

Log removals for all organisms were calculated from the equation a = c–b, as follows:

Equation 1 (a) Log removal across conventional secondary or MBR treatment = (c) log10 concentration post-preliminary – (b) log10 concentration post-secondary or post-MBR

Equation 2 (a) Log removal across disinfection = (c) log10 concentration post-secondary or post-MBR – (b) log10 concentration post-disinfection

RESULTS

In post-preliminary wastewater samples, concentrations of *E. coli* (Fig. 2), fecal coliforms, and enterococci ranged from 10^6 to 10^7 CFU/100 mL, and concentrations of somatic and F-specific coliphage ranged from 10^4 to 10^7 CFU/100 mL. Median concentrations of *E. coli*, other bacterial indicators, and coliphage in post-secondary and post-tertiary samples in conventional plants were higher than those in post-MBR samples. In post-disinfection samples for the two MBR plants (Plants 1 and 4, Fig. 2) and the UV conventional plant (Plant 3, Fig. 2), concentrations of *E. coli* were near or below detection. For the chlorine conventional plant (Plant 5, Fig. 2), *E. coli* concentrations were in the single or double digits. The same types of patterns were seen for other bacterial indicators and coliphage. All of the plants met the NPDES required effluent limits on the days sampled. At the plants involved in this study the fecal coliform permit limits are/were approximately 1,000/100 mL monthly and 2,000/100 mL daily. The *E.coli* limits range from 126/100 mL to 161/100 mL monthly and 284/ 100 mL to 362/100 mL daily depending on the receiving stream.

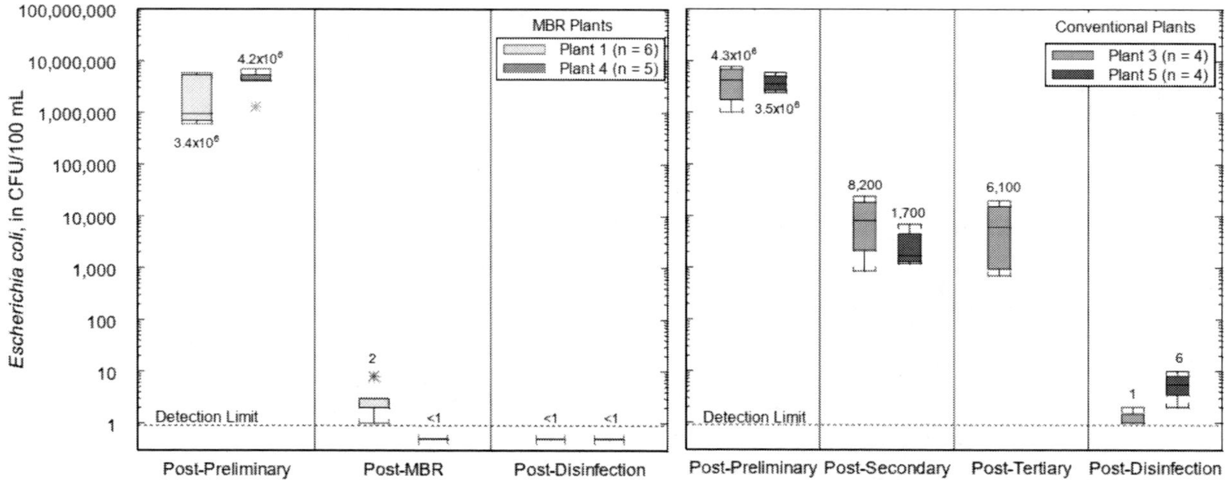

Figure 2. Concentrations of *E. coli* in wastewater samples throughout treatment processes at MBR and conventional plants, 2008-10. Median values are indicated for each boxplot.

Samples were analyzed for enteric viruses by both molecular and cell culture methods. By molecular methods, adenovirus was detected in 100%, enterovirus was detected in 91% and 63%, and norovirus GI was detected in 64% and 50% of post-preliminary samples from MBR and conventional plants, respectively (Fig. 3). Adenovirus was detected in 45–75%, enterovirus in 9-18%, and norovirus G1 in 0–25% of post-treatment samples. Norovirus GII and Hepatitis A virus were not detected in any samples, and rotavirus was detected in one sample. Although culturable viruses were found in 12 out of 19 post-preliminary samples, they were not detected in any post-secondary, post-MBR, post-ultraviolet, or post-chlorine samples.

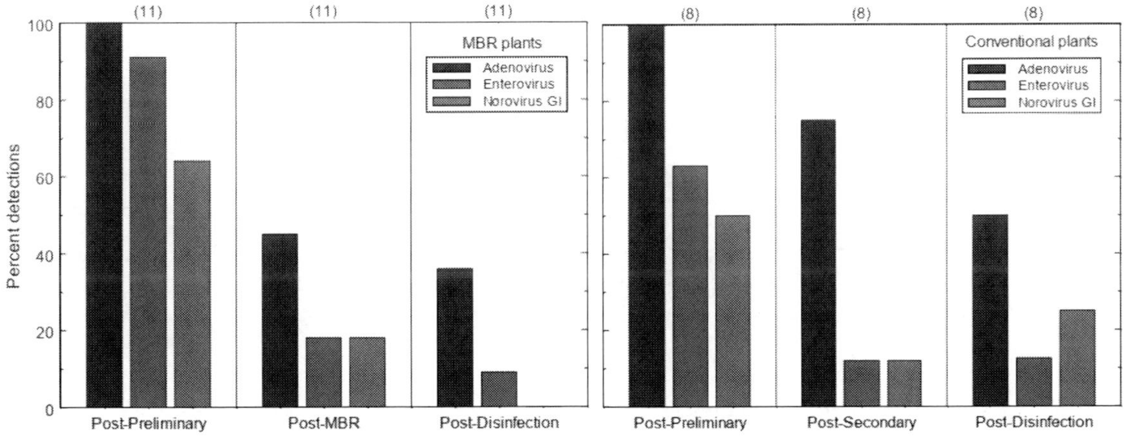

Figure 3. Percent detections of enteric viruses in wastewater throughout treatment processes in MBR and conventional plants, 2008-10. Numbers of samples are indicated in parentheses.

Median log removals for all organisms were higher for MBR secondary treatment (3.02 to >6.73) than for conventional secondary treatment (1.53 to 4.19). Ultraviolet disinfection after MBR treatment provided little additional log removal of any organism except for somatic coliphage (>2.18), whereas ultraviolet or chlorine disinfection after conventional secondary treatment provided significant log removals of all indicator organisms. Median log removals of adenovirus across disinfection were low in both MBR and conventional secondary plants (no removal detected and 0.24). *E. coli*, the indicator used to assess the microbial quality of wastewater, followed similar patterns to adenoviruses in two processes: (1) from post-MBR to post-disinfection in MBR plants and (2) from post-preliminary to post-secondary in conventional secondary plants (Fig. 4).

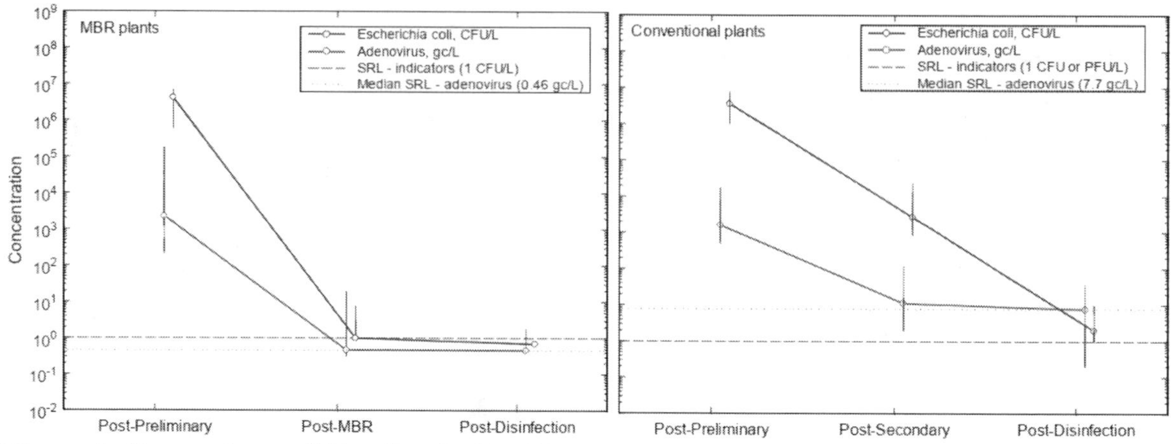

Figure 4. Comparisons of *E. coli* and adenovirus concentrations throughout treatment processes in MBR and conventional plants. The open circles indicate the median value, and the vertical lines are the range of detected values. SRL is sample reporting limit.

DISCUSSION AND CONCLUSIONS

The results from this study indicate clearly that the water quality prior to disinfection in MBR facilities exceeds the effluent quality from a conventional treatment plant and exceeds the current EPA regulations. Further, the data supports that disinfection after the MBR process does not provide any meaningful benefit to the health, safety, and well-being of the public. Reductions in capital cost, O&M cost, environmental impacts, and safety impacts are other considerations further supporting the argument that a disinfection process unit after MBR secondary treatment provides little significance. Additionally, if a permit violation occurs due to a failure in the membranes, the permit would require that compliance steps would have to be taken just like plants that currently have disinfection systems in place. Therefore, the permit compliance regulations would remain the same, so compliance would still be maintained. The following conclusions can be drawn from the study:

- Median log removals for all organisms were higher across secondary municipal wastewater treatment for MBR plants than for conventional secondary plants.
- Although culturable viruses were found in 63 percent of samples collected after primary treatment, they were not detected in any post-secondary, post-MBR, post-ultraviolet, or post-chlorine samples.
- Ultraviolet disinfection after MBR secondary treatment provided little additional removal of any organism except for somatic coliphage.

REFERENCES

Francy, D. S., Stelzer, E.A., Bushon, R.N., Brady, A.M.G., Mailot, B.E., Spencer, S.K., Borchardt, M.A., Elber, A.G., Riddell, K.R., Gellner, T.M. 2011. Quantifying Viruses and Bacteria in Wastewater—Results, Interpretation Methods, and Quality Control. Scientific Investigations Report 2011–5150, U.S. Geological Survey, Reston, VA.

Francy, D.S., Stelzer, E.A., Bushon, R.N., Brady, A.M.G., Williston, A.G., Riddell, K.R., Borchardt, M.A., Spencer, S.K., and Gellner, T.M. 2012. Comparative effectiveness of membrane bioreactors, conventional secondary treatment, and chlorine and UV disinfection to remove microorganisms from municipal wastewaters: Water Research 46, p. 4164-4178.

Kuo, D.H.-W., Simmons, F.J., Blair, S., Hart, E., Rose, J.B., Xagoraraki, I., 2010. Assessment of human adenovirus removal in a full-scale membrane bioreactor treating municipal wastewater. Water Res. 44, 1520–1530.

Lambertini, E., Spencer, S.K., Bertz, P.D., Loge, F.J., Kieke, B.A., Borchardt, M.A., 2008. Concentration of enteroviruses, adenoviruses, and noroviruses from drinking water by use of glass wool filters. Appl. Environ. Microbiol. 74 (10), 2990–2996.

Ottoson, J., Hansen, A., Westrell, T., Johansen, K., Norder, H., Stenström, T.A., 2006. Removal of noro- and enteroviruses, *Giardia* cysts, *Cryptosporidium* oocysts, and fecal indicators at four secondary wastewater treatment plants in Sweden. Water Environ. Res. 78 (8), 828–834.

Simmons, F.J., Kuo, D.H.-W., Xagoraraki, I., 2011. Removal of human enteric viruses by a full-scale membrane bioreactor during municipal wastewater processing. Water Res. 45, 2739–2750.

Zhang, K., Farahbakhsh, K., 2007. Removal of native coliphages and coliform bacteria from municipal wastewater by various wastewater treatment processes: implications to water reuse. Water Res. 41, 2816–2824.

Innocuous For Reuse: Scaling Sustainable Wastewater Treatment and Natural Disinfection Processes For Safe Water Recycling At Delhi, California USA and At Varanasi, Uttar Pradesh, India

Green, Franklin[1], Dawyot, Robert[2], Husby, Peter[3], McNaughton, Eugenia[3], Mishra, Shantanu[4], Mishra, Veer[4]; [1]GO2 WATER, INC., [2]Allied Engineers, Inc., [3]US EPA Region 9, [4]Sankat Mochan Foundation,

Access to clean water and sanitation is a basic human right. Safe drinking water and sanitation are essential to human and environmental health. Healthy communities and healthy cities depend upon safe drinking water and sanitation. Yet nearly one in five children born today will struggle to live without safe drinking water. More a larger number nearly three of five children born today will struggle to live without adequate sanitation. Life without safe drinking water is most fatal for children. Life without sanitation will inexorably cause the pollution and contamination with enteric pathogens of one's drinking water resource, be it surface water as in Varanasi, Uttar Pradesh, India or groundwater as in Delhi, California USA.
To deliver sustainable sanitation infrastructure and services to nearly three billion and to achieve the Millennium Development Goal for sanitation, municipal wastewater treatment technology must three times more efficient at one third the capital cost, one quarter the O&M cost, and one hundredth the life cycle cost as compared with conventional wastewater treatment technologies. One such technology, Advanced Integrated Wastewater Pond Systems, was developed at the University of California, Berkeley beginning in 1951 (Oswald, 1990). Oswald Green, LLC was established in 1996 to commercialize the AIWPS® Technology in the USA and in INDIA and other parts of the world where access to sanitation was either severely limited or non-existent. In 1996, the Delhi County Water District and its District Engineer selected the AIWPS Technology and retained Oswald Green, LLC to assist in the design of its new wastewater treatment facility (WWTF) that was commissioned in early 1998. The Delhi County Water District is a rural water and wastewater utility serving a small rural community of approximately 10,000 residents in northern Merced County near the Merced River in the San Joaquin River Valley of California.

Influent and effluent MPN data for the Delhi AIWPS WWTF indicates between one and two log E.Coli removal in each of six ponds in series without any chlorination and the very significant public health risk of disinfection byproducts. Initial MS Bacteriophage testing indicates a greater than four log removal. Based on earlier pathogen removal data from the renovated and improved, research-scale AIWPS WWTF at the Environmental Engineering & Health Sciences Laboratory of the University of California, Berkeley, the Delhi AIWPS WWTF was designed by Oswald Green, LLC as well as a much larger 200 million liter per day (MLD) AIWPS Sewage Treatment Plant (STP) for the City of Varanasi, Uttar Pradesh INDIA both in 1997. Based on more recent disinfection data from the Delhi AIWPS WWTF some of which was collected by the U.S. EPA Region 9 Water Quality Laboratory, GO2 Water, Inc., the successor of Oswald Green, LLC, designed in 2010 a second AIWPS STP for the City of Varanasi at an intermediate scale of 40 MLD that could be expanded with a fifth module to 50 MLD ADWF capacity as a demonstration of the AIWPS Technology under local conditions in India.

Sustainable municipal wastewater treatment technologies must utilize natural disinfection processes to remove human enteric pathogens including bacteria, viruses, protozoa, and parasites. Sustainable sanitation is essential for healthy cities and healthy communities in developing and developed communities alike. This paper discusses a sustainable wastewater treatment technology that provides disinfection through an integrated series of biological, chemical and physical processes occurring in each stage of treatment and in each particular pond or bioreactor. And the AIWPS Technology is compatible with and fully scalable to serve small rural communities up to large urban areas such as the City of Varanasi, Uttar Pradesh India.

Over four hundred million people in India depend on the River Ganges for drinking water. In Varanasi some 38 raw sewage outfalls and nallahs discharge untreated sewage into the Ganges every day along the seven kilometer city riverfront. The ancient city and its holy river are locked in a love-hate relationship, and to the pious Hindu, Ganga Ji is divine and above the profanity of four hundred million liters per day of untreated sewage discharging into the India's most holy river from it world heritage pilgrimage city containing all manner of waterborne disease pathogens at the city's river edge where more than one hundred thousand city residents and visiting pilgrims bathe at dawn each day. The Sankat Mochan Foundation has led a campaign for thirty years to prevent the sewage pollution of the holy river and to treat the municipal sewage to a degree that it becomes innocuous for reuse. The treatment and reclamation of municipal sewage proposed in the holy river city of Varanasi was developed at the University of California, Berkeley. Advanced Integrated Wastewater Pond Systems and the AIWPS Technology was pioneered by Professor William J. Oswald since 1951 (Oswald, 1957; Oswald, 1990). Two WWTFs using the first generation of the AIWPS Technology by small and medium sized cities in the Napa Valley of California USA since the mid-1960s were introduced to Dr. Veer Bhadra Mishra when in 1994 he visited Professor William J. Oswald and his senior research associate and doctoral student, Bailey Green, at the Environmental Engineering & Health Sciences Laboratory of the University of California, Berkeley, a collaborative field research laboratory of UC Berkeley's School of Public Health and College of Engineering located in Richmond, California from 1951 until 2001.

Advanced Integrated Wastewater Pond Systems integrate a number of physical, chemical, and biological processes to disinfect and remove and inactivate sewage pathogens. Helminth ova such as those from the roundworm parasite Ascaris Lumbricoides, as well as Giardia and Chryptosporidium oocysts are removed in the primary anaerobic in-pond digesters of Advanced Facultative Ponds. Enteric bacterial pathogens die off throughout the extended hydraulic residence time of an Advanced Integrated Wastewater Pond Systems Process that includes four types of ponds each uniquely designed to optimize one or more unit processes in sequence and by treatment stages using bacterial and algae symbiotically to provide sustainable primary, secondary, and tertiary treatment of municipal sewage, water reclamation and safe beneficial reuse. Influent and effluent data for E. Coli and coliform Most Probable Numbers (MPN), and colony forming units (CFU), per one hundred milliliters are presented for the AIWPS® Wastewater Treatment Facility (WWTF) at Delhi, California, USA provided by U.S. EPA Region 9 Water Quality Laboratory, and for the Ganges River at Varanasi provided by the Sankat Mochan Foundation and its own Water Quality Laboratory at Tulsi Ghat Varanasi that is part of its long campaign to clean the Ganges at Varanasi by promoting and providing sustainable sanitation technology with natural disinfection for water reclamation and reuse and to prevent waterborne disease and to protect human and environmental health at Varanasi.

Disinfecting processes integrated by the AIWPS Technology result in the elimination of helminth ova, protozoan oocysts, more than four-log removal of MS Bacteriophage used as an indicator of the presumed removal of specific human viral pathogens, and by design as many logs of fecal and total coliform MPN as required by local, state, and federal regulatory authorities. The AIWPS Sewage Treatment Plant (STP) designed for the City of Varanasi in 1997 was based on the final reclaimed water fecal coliform MPN of less than 50. Consequently, three Maturation Ponds to be operated in series were designed to safeguard the recycled water quality ("innocuous for reuse") for crop irrigation during the long dry ten-month dry season in Varanasi.

In California, under Title 22 reclaimed wastewater safe and permeable for unrestricted reuse requires non-detected on all four types of pathogens, and for this higher than WHO standard, we have used multimedia filtration followed by UV light disinfection for a final disinfection stage to meet Title 22 unrestricted reuse requirements of non-detect on all tested pathogens. By using UV disinfection when necessary to meet the most stringent reuse requirements in the world, the production of carcinogenic disinfection byproducts are avoided. Coliform tested performed by the U.S. EPA Region 9 Water Quality Laboratory for the Delhi County Water District and earlier coliform MPN data for other AIWPS WWTFs in California since the 1960s indicate on average a four- to six-log removal of total and fecal coliform MPN/100 ml when a minimum of four ponds--Advanced Facultative Pond, High Rate Pond, Algae Settling Pond, Maturation Pond are operated in series.

Low(est) Energy Disinfection Using Pasteurization – Results of Demonstration Scale Testing in Ventura California

Dan Pfeifer, Shana Epstein, and John Willis, City of Ventura
Greg Ryan, Pasteurization Technology Group
Andrew Salveson and Nicola Fontaine Carollo Engineers

Pasteurization has come a long way. Louis Pasteur might not recognize the tools, but the concept of pasteurization employed now in Ventura California is the same one that he perfected 150 years ago in France. This latest incarnation of pasteurization began in 2003 by a small think-tank, now called the Pasteurization Technology Group (PTG). In 2005, PTG brought there concept of pasteurization to our industry. By 2007, pasteurization had received the coveted Title 22 approval for reclaimed water disinfection, demonstrating that heating secondary effluent and filtered effluent to 155 to 180 degrees F resulted in robust kill of all studied pathogens. The next step was economics. Pasteurization works efficiently for two reasons. First, the heat source for pasteurization is waste heat. Second, the PTG pasteurization process captures and holds that waste heat. Simply put, cold water in and out, heat in the middle (Figure 1). For Figure 1, the waste heat is heat coming off of a gas turbine.

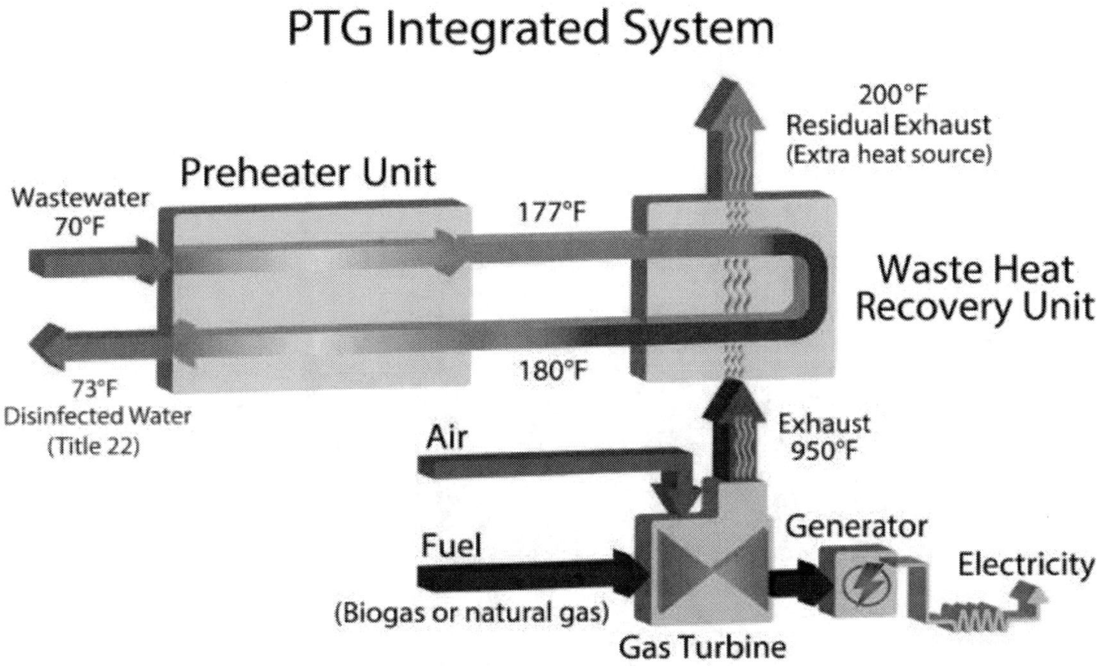

Figure 1 – Schematic of PTG Process (with example water temperatures)

PASTEURIZATION AT VENTURA

The City of Ventura (City) has previously investigated the cost and feasibility of pasteurization at the Ventura Water Reclamation Facility (VWRF). That study suggested that the use of pasteurization for disinfection of recycled water would be substantially less expensive than alternative disinfection technologies. Before implementation of this technology at a large scale, more detailed research of disinfection performance and energy efficiency would be required. In response to this need, PTG and the City designed and installed a 400 gpm demonstration pasteurization unit (Demo) at the (VWRF) (Figure 2). Carollo advised on the process design. The Demo consisted of full-scale parts and automated controls with the intent of simulating a future full-scale system. Successful test results demonstrated equal or better disinfection performance and disinfection efficiency then previously predicted. The testing also provided valuable information on the long-term operations and maintenance needs, based on the longer run times. The testing documented here bolsters confidence in pasteurization for the City and for other future potential users.

The testing at the VWRF consisted of short and long-term testing to determine the performance and economics of the pasteurization system. The specific objectives of Demo testing were to:

- Demonstrate "Title 22" total coliform and virus compliance over a range of operating temperatures;
- Demonstrate the efficiency of the heat exchangers, thus better defining the economics of full-scale operation;
- Measure the potential fouling (scaling) of the heat exchangers;
- Measure the head loss through the system, to better understand pumping needs and thus long-term operational cost impacts.

Figure 2 – 400 gpm PTG Reactor in Ventura California

DEMONSTRATION TESTING RESULTS

Long term total and fecal coliform results were robust, showing complete disinfection at 155 degrees F (Figures 3 and 4).

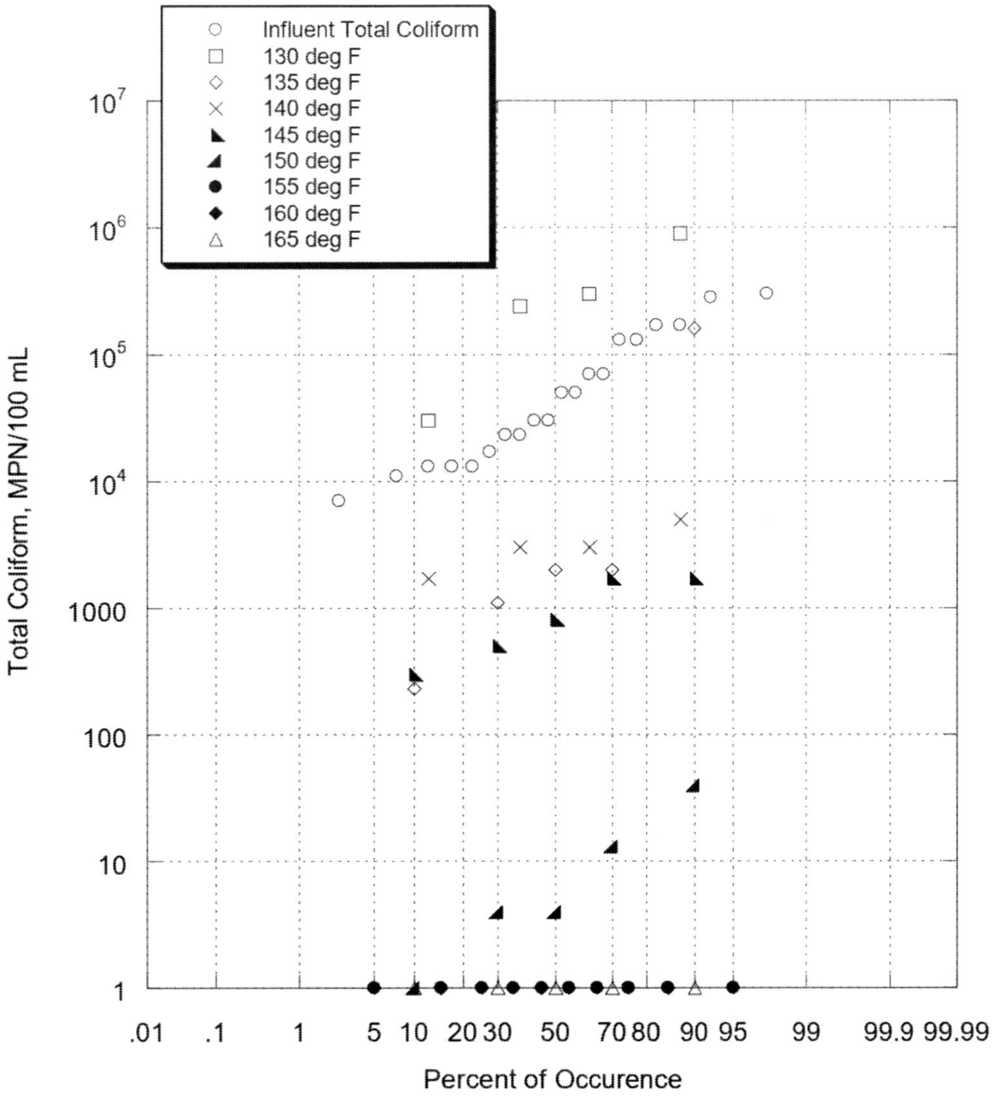

Figure 3 ***Probability Plot of Total Coliform at Various Temperatures (Port 5– after contact chamber)***

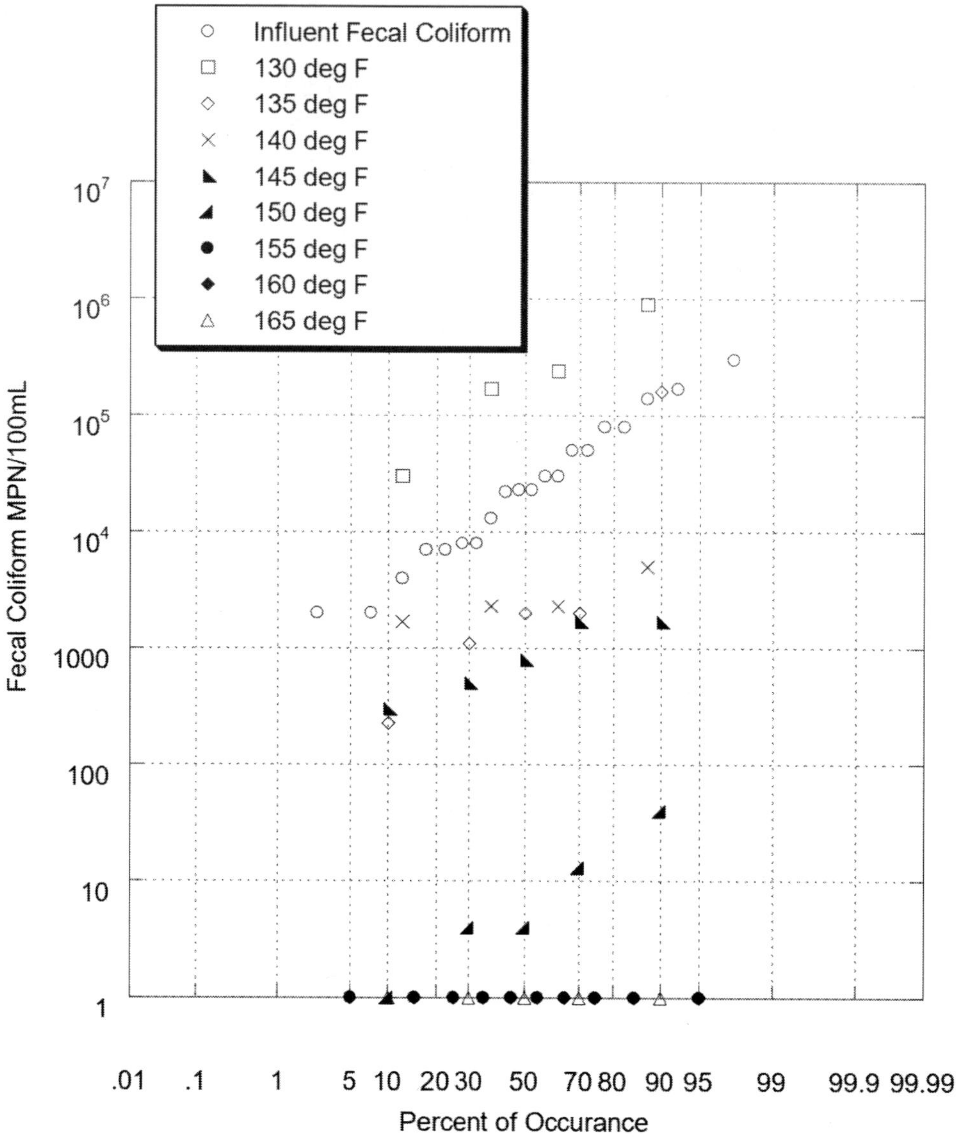

Figure 4 Probability Plot of Fecal Coliform at Various Temperatures (Port 5– after contact chamber)

The Demo was "stress-tested" to evaluate disinfection of MS2 at three temperatures, two contact times, and at several different monitoring locations throughout the Demo unit. The sample locations tested the disinfection effectiveness of system components as well as the Demo as a whole. ***As a point of reference, the target reduction of MS2 is 4 logs, which CDPH previously approved as equivalent to 5 logs of poliovirus kill.***

The starting MS2 "influent" concentration (N_0) was determined by sampling the effluent port with the heat off. Figures 5 through 8 show the part-by-part virus disinfection performance of the

Demo unit and Figure 9 shows the performance of the system as a whole. Key points from the testing include:

- The influent preheater alone achieved >5-logs reduction of MS2 at ~162 °F with a contact time of 84 seconds.
- The stack heater alone achieved almost 5-logs reduction of MS2 at ~165 °F with a contact time of 2 seconds.
- The first half of the contact chamber alone achieved >6-logs reduction of MS2 at ~164 °F with an estimated contact time of 9 seconds.
- The entire contact chamber alone achieved 7.0-logs reduction of MS2 at ~164 °F with a contact time of 18 seconds.
- The entire Demo achieved >6-logs reduction of MS2 at ~165 °F with a contact time of 195 seconds.

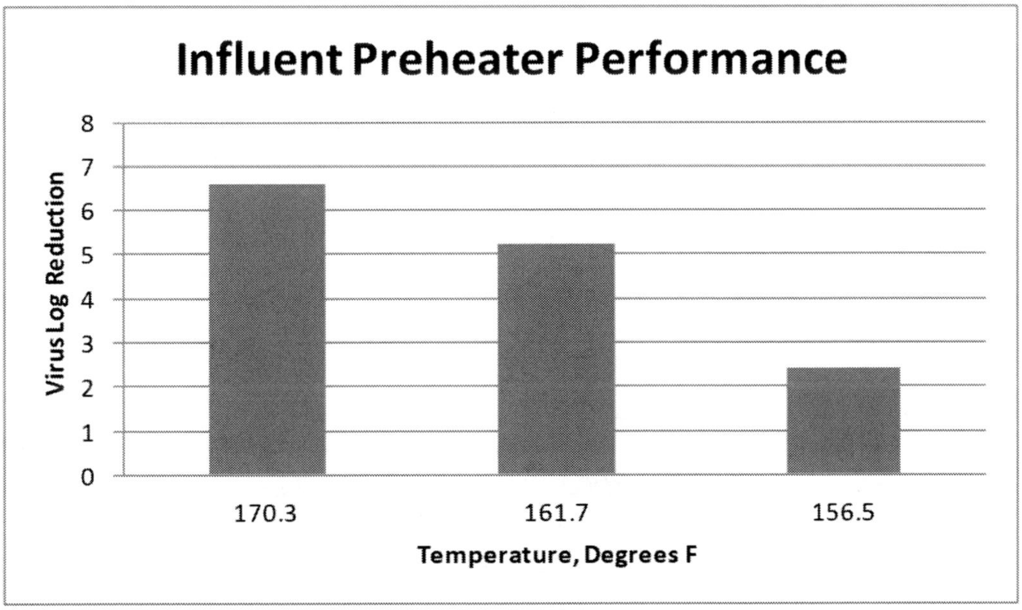

Figure 5 MS2 Inactivation – Influent Preheater Alone

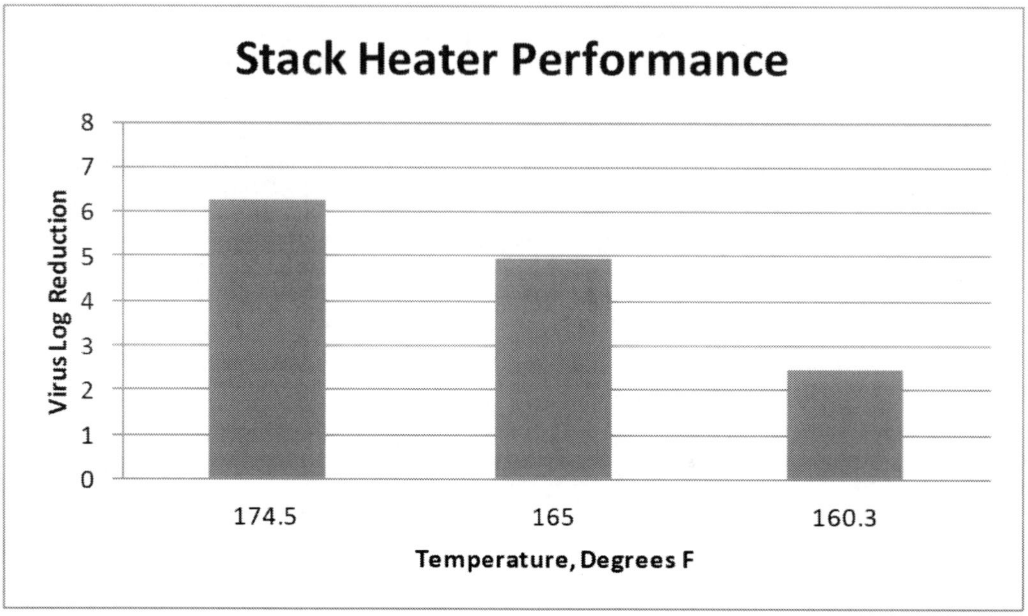

Figure 6 *MS2 Inactivation – Stack Heater Alone*

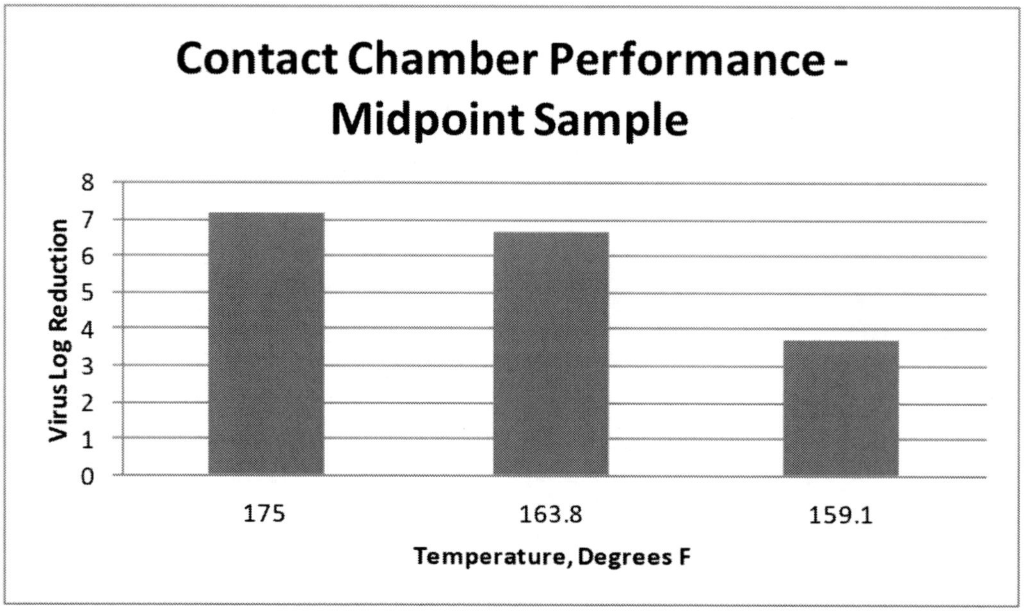

Figure 7 *MS2 Inactivation – First Half of Contact Chamber Alone*

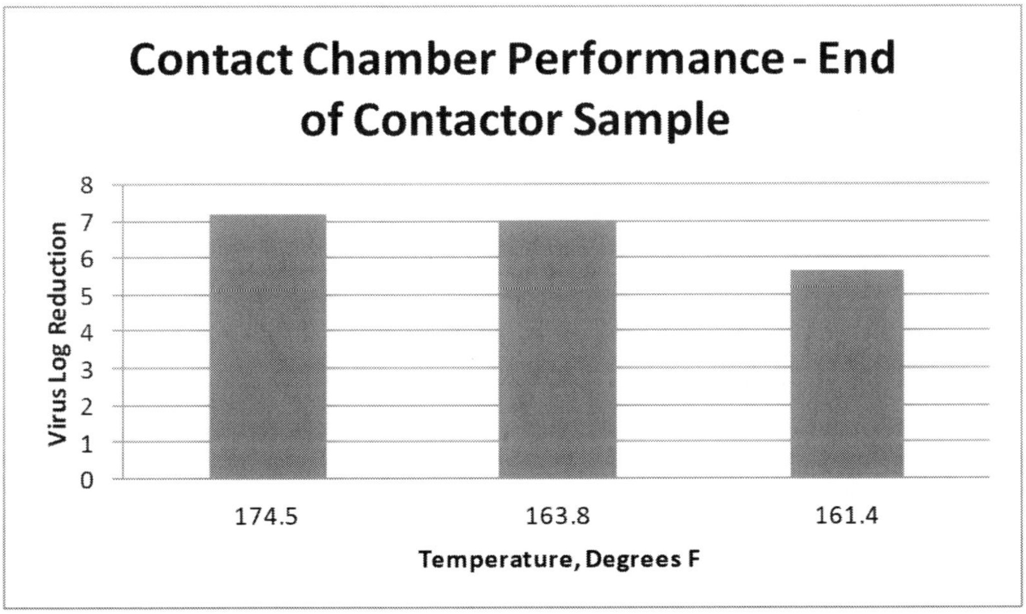

Figure 8 *MS2 Inactivation – Entire Contact Chamber Alone*

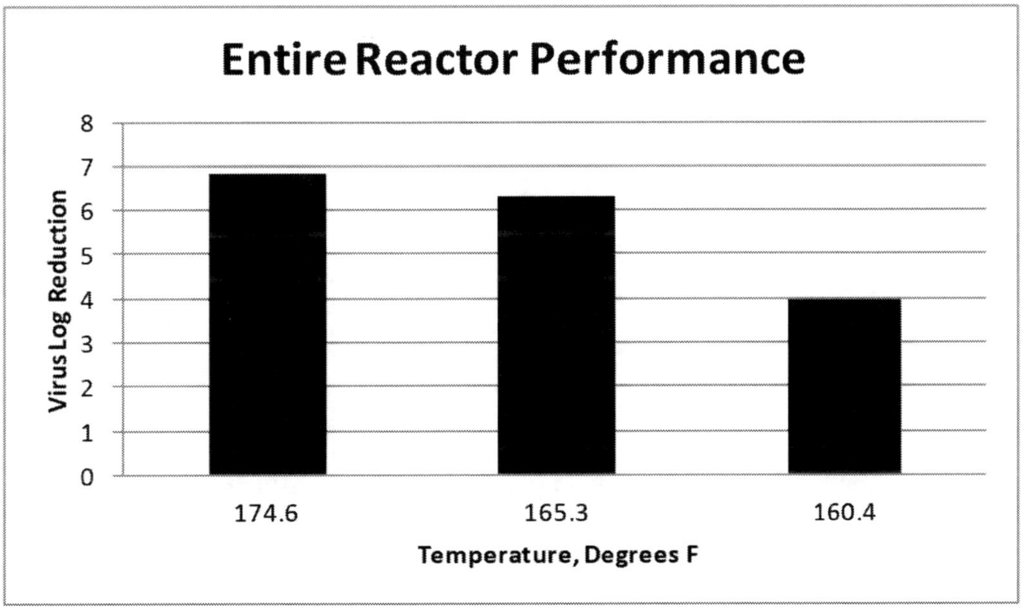

Figure 9 *MS2 Inactivation – Entire Demo Unit Performance*

Note that the log reduction calculations are limited by the ability to measure much more than 6 logs of reduction, so values in excess of 6-log are potentially higher than noted. ***While there is no intention to obtain >20-log disinfection credit, the component by component treatment analysis suggests that the total log reduction for this system at 165 °F is 22-logs (5 logs in***

Preheater (way in), 5 logs stack heater, 7 logs through contact chamber, 5 logs in Preheater (way out)).

The data also suggests that operation at 160 °F is sufficient to obtain the 4-log reduction of MS2, based upon >2 logs in the preheater (at ~157 °F), > 2 logs through the stack heater (~ 160 °F), >3-logs through the contact chamber (~159 °F, midpoint of the chamber), and again >2-logs through the preheater on the way out (at ~157 °F).

HEADLOSS AND FOULING THROUGH THE SYSTEM

Determination of headloss through the system is important for design considerations and determining pumping requirements. Fouling rates of the system (through mineral fouling, biofouling, or a combination of the two) are important in consideration of operation and maintenance requirements of the system. Figure 10 is a plot of the pressure increase in the preheater (on the way in) at 160 °F, which shows an approximate 4 percent increase in the pressure over the two-month testing period. Figure 11 is a similar plot of the preheater (on the way out) over the same time period and at the same temperature, which shows an approximate 13 percent increase in the pressure. Both figures demonstrate that fouling is occurring, but the fouling is not excessive. This data is valuable for predicting when a chemical cleaning of the system should take place in order to minimize energy loss through the system.

Figures 12 and 13 are plots of total system headloss both before and after cleaning. The results show that chemical cleaning resulted in recovery of the heat exchangers.

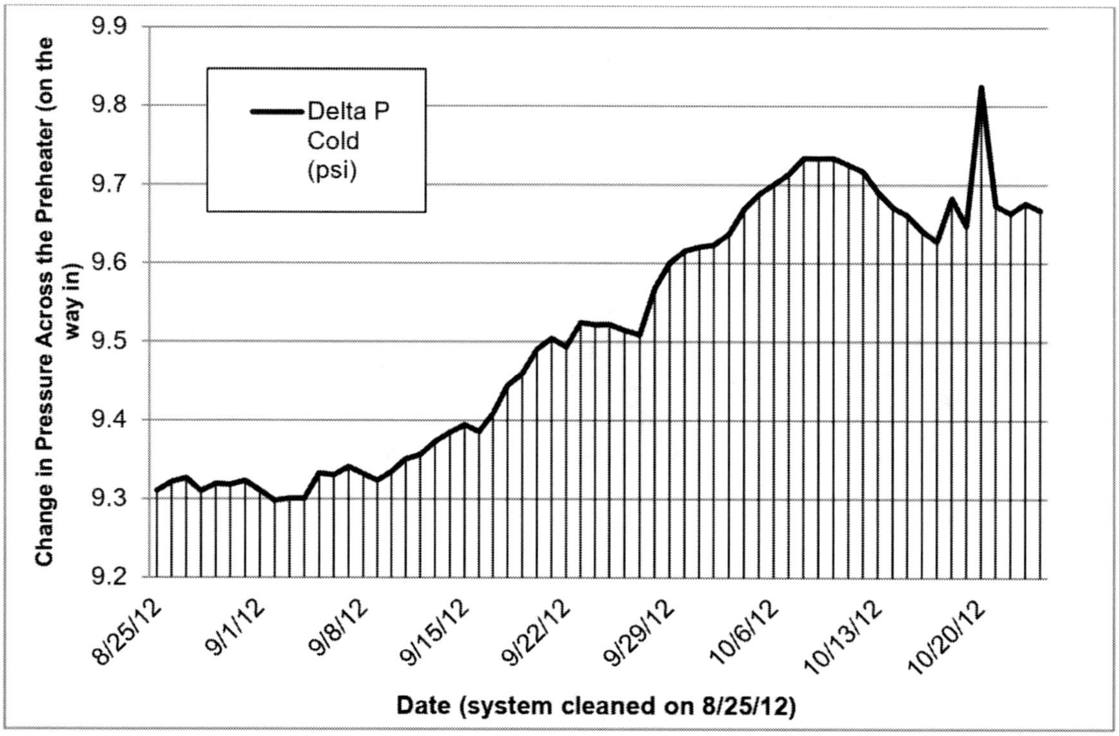

Figure 10 Headloss Through the Preheater (on the way in) with Time at 160 Degrees

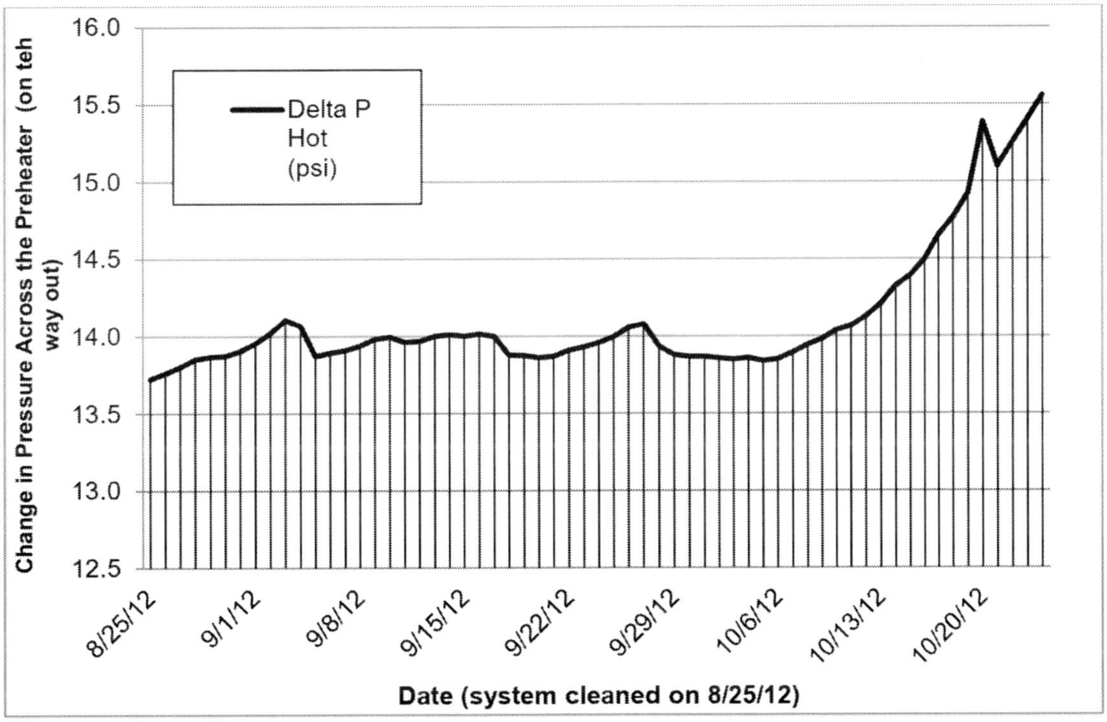

Figure 11 Headloss Through the Preheater (on the way out) with Time at 160 Degrees

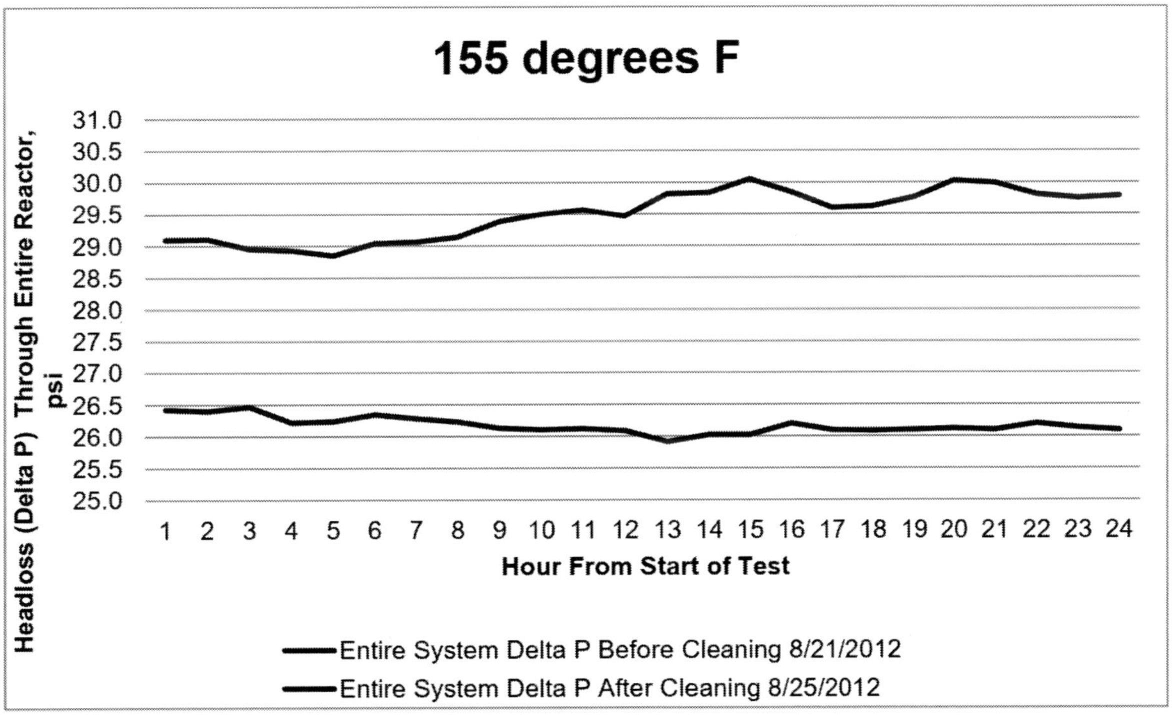

Figure 12 Total System Headloss Before and After Cleaning (155 degrees F)

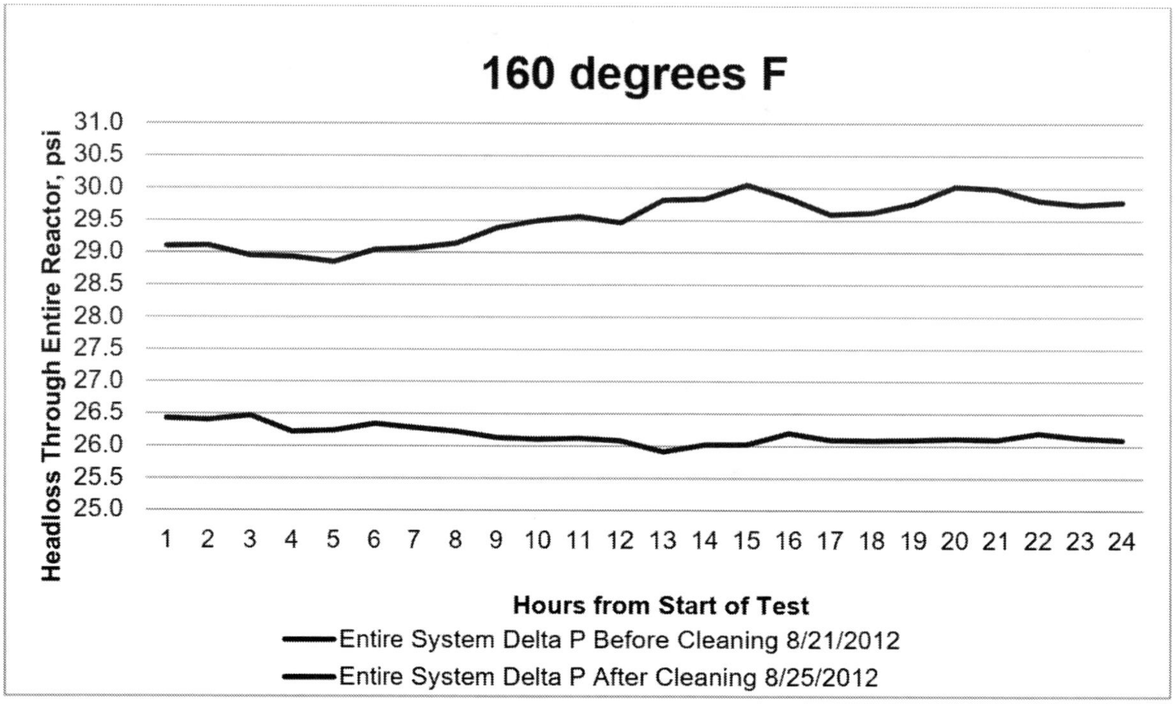

Figure 13 Total System Headloss Before and After Cleaning (160 degrees F)

HEAT LOSS THROUGH THE SYSTEM

The efficiency of the pasteurization system is based upon the maintenance of minimal heat loss. The original economic evaluation assumed a loss of 3 $^{\circ}$F. That heat loss is measured as the change in temperature between the influent water and the effluent water, also called delta T. Figure 14 contains a plot of the heat loss through the system from August 25, 2012 to October 25, 2012. The Demo unit was operating at 160 $^{\circ}$F during this period. The rise in delta T correlates with the increase in pressure differential across the Demo unit. After chemical cleaning, the delta T dropped as expected (Figure 15). Essentially, fouling of the preheater results in less efficient heat transfer.

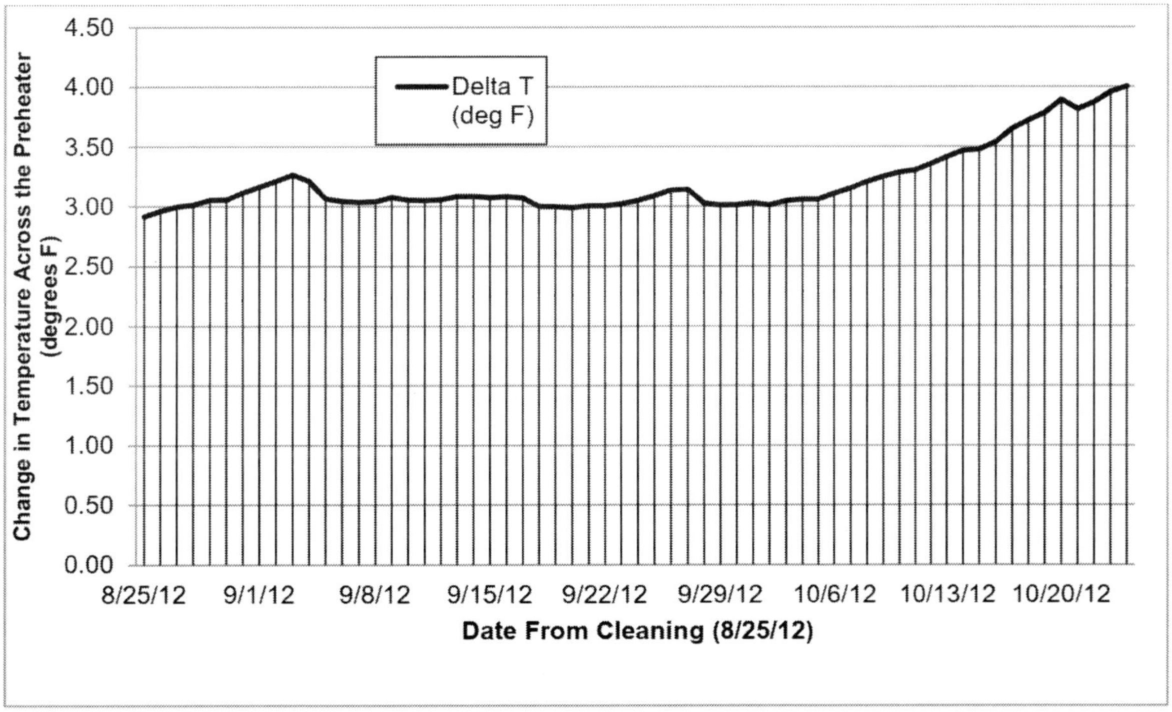

Figure 14 Heat Loss Through the Demo Unit Two Months After Cleaning

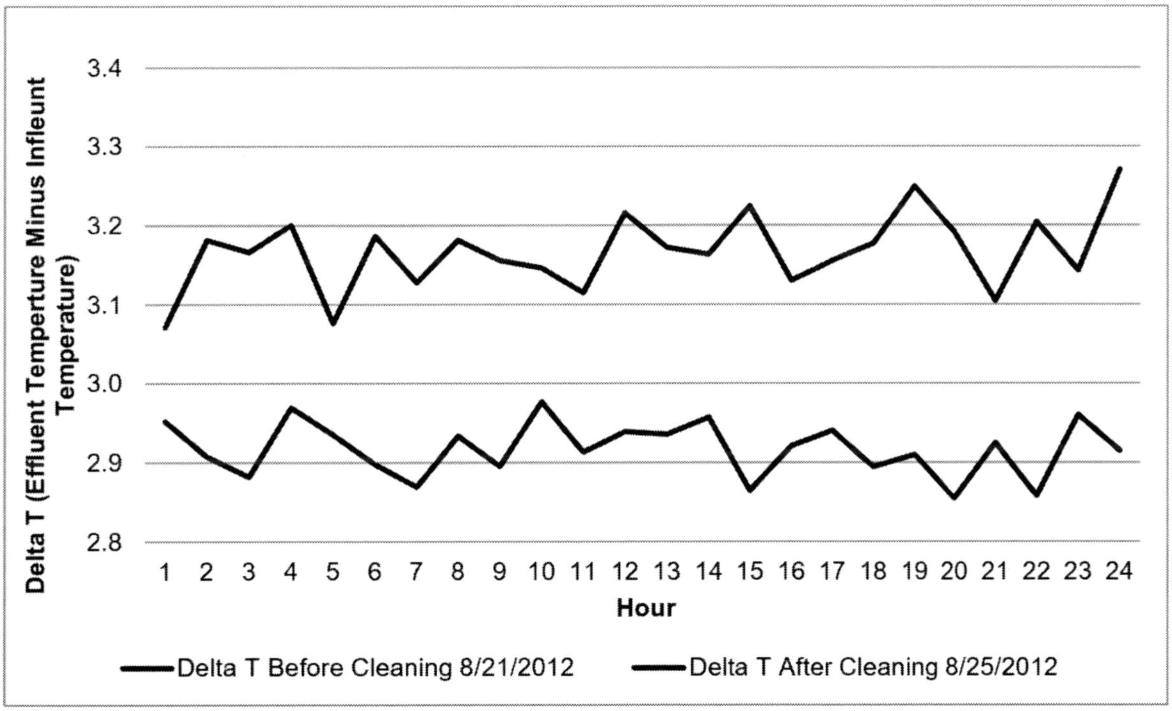

Figure 15 Improvement in Heat Exchanger Efficiency After Chemical Cleaning

CONCLUSIONS

The pasteurization Demo testing at the VWRF was highly successful. Initial economic evaluation work for the City was based upon 10 seconds of contact at a target temperature of 180 degrees F. The heat loss through the system was assumed to be 3 degrees F. The results indicate equal or better disinfection performance and efficiency. The fouling of the heat exchangers was not excessive and proved to be manageable with commonly used chemical solutions. In particular, the following can be concluded from this work:

- Coliform disinfection is robust. 155 degrees F is all that is needed at the VWRF to obtain non-detect levels of coliform.

- Coliform regrowth was not observed at disinfection temperatures of 155 degrees F and above.

- Virus disinfection is predictable and grossly redundant. 165 degrees F likely results in 22 log reduction of MS2, though only 4 log reduction is needed.

- Fouling is related to temperature. At 180 degrees F, fouling was rapid. At the targeted temperature of 165 degrees F, the fouling was substantially reduced and manageable. Chemical cleaning every 30 to 45 days may be required to maintain heat exchanger efficiency and minimize head loss.

- The overall change in temperature from the influent to the effluent is approximately 3 degrees F. The impact of this heat differential on downstream permit requirements must be examined.

- Given that the previous economic evaluation was conducted using more conservative assumptions than this Demo testing showed is necessary, the economics should be more favorable than previously estimated. The economic evaluation should be redone in light of these findings to more clearly define the costs and benefits of pasteurization compared to other alternative disinfection technologies.

Mixed Oxidants: Alternative disinfectant that can reduce operational costs and improve water quality

John Deogracias
Commercial Leader
Parkson Corporation
1401 W Cypress Creek Road
Fort Lauderdale, FL 33309

On site generation is a technology that has drastically improved in the last 10 years and is a safe, green and low cost alternative means of disinfection for a water supply systems. On site generation can be either a superior mixed oxidant or the industry standard 0.8% sodium hypochlorite. Both are generated by the simple feed stock of salt, water and power. Approved by EPA and follow the same standards of chlorine, mixed oxidant chemistry has provided water municipalities with chlorine residual enhancement, biofilm control, taste and odor improvement, DBP reduction, and chemical savings by microflocculation in conventional water treatment plants. Cedar Knox, NE and Oneonta, NY, Apple Valley, CA data and research from across the country has shown that mixed oxidants are able to penetrate the polysaccharide substrate that biofilm uses to attached to pipe distribution walls where standard chlorine and bleach chemicals could not. Additional studies by EPA have determined that ozone and chlorine dioxide is not present in mixed oxidants. Recent evidence from laboratory research indicate that mixed oxidants include H2O2 and other reactive oxygen species such strong oxidizing free radicals – hydroxyl radical, superoxide, and chlorine radicals. Research on the composition continues; but the evidence on the chemical and biocidal behavior continues to show, as it has for the past 20 years, that MOS is a superior oxidant to bleach alone. The presentation will discuss the water quality improvements and the operational savings mixed oxidants has been able to provide to water and wastewater utilities.

SUMMARY

On-site generators (OSGs) produce chlorine when a solution of sodium chloride is passed through an electrolytic cell and electricity is added. Many communities are turning to OSGs for their water distribution systems because of the benefits inherent in the process, including better safety, high quality disinfection, greener operations, and substantial economic savings. On-site generation of chlorine and other disinfectants is a technology that is based on decades-old scientific principles. In essence, OSGs take a solution of sodium chloride (salt) and water and apply electricity, which produces chlorine and other oxidant species. While OSGs have a number of applications in the industrial world, including providing disinfectants for swimming pools and cooling towers, the largest application of OSG technology is for municipal drinking water disinfection. Because of the benefits that OSGs provide, many water municipalities are switching to OSG systems as opposed to more traditional chlorine delivery systems such as chlorine gas, concentrated sodium hypochlorite, and bulk calcium hypochlorite.

HOW DO ON-SITE GENERATORS WORK?

While there are many different types of OSG systems available, there are also a number of components that almost every OSG uses (See Figure 1).

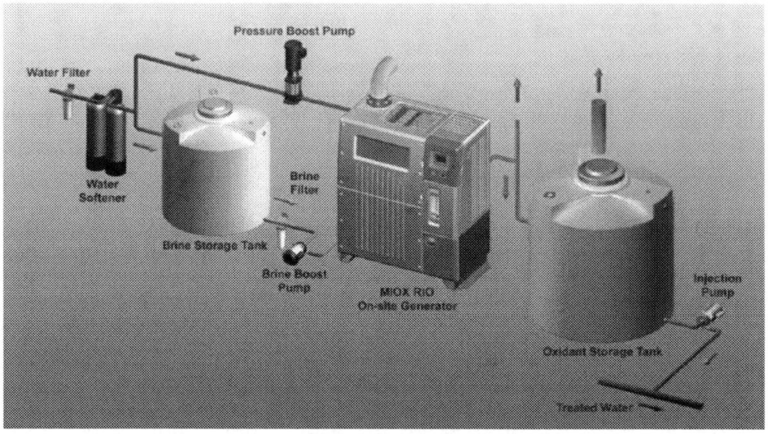

Figure 1: Process Schematic

Water coming into the OSG first goes through a softener, after which it is split into two lines. One line feeds directly into the electrolytic cell while the other is used to fill the brine tank. The brine tank stores a concentrated salt solution, prepared by having an excess of salt in the tank so that the solution is near-saturated brine, which is then injected into the softened water stream entering the electrolytic cell. When the dilute salt solution is inside the electrochemical cell, a current is passed through the cell, producing the oxidant (sodium hypochlorite or other oxidants) solution. After leaving the electrolytic cell, the oxidant solution is stored temporarily in the oxidant tank and is then metered into the water moving through the treatment process. Hydrogen gas is also produced inside the electrolytic cell, and the hydrogen is removed from the cell and oxidant storage tank through vents. The electrolytic cell, where the oxidants are actually produced, is central to the OSG (See Figure 2). Electrolytic cells consist of two electrodes, the anode and cathode, arranged so that both make contact with the mixed water and brine solution. When the OSG is activated, a voltage is applied to the cell so that current flows through the cell, causing chemical reactions to take place at the surfaces of both electrodes that eventually produce the disinfectants. The overall chemical equation for reaction of salt ($NaCl$) and water (H_2O) to form sodium hypochlorite ($NaOCl$) is:

[1] $NaCl + H_2O \rightarrow NaOCl + H_2$

Oxidation reactions are carried out at the anode where two chloride ions (Cl^-) are stripped of one electron each to produce chlorine gas:

[2] $2Cl^- \rightarrow Cl_2 + 2e^-$

Depending on the physical and working parameters of the cell (e.g., electrode to electrode spacing, cell applied potential, etc.), it is also possible to produce oxidants other than chlorine, which can provide enhanced removal of microbiological contaminants from water and other benefits. After it is produced, the chlorine gas dissolves in water to produce hypochlorous acid (HOCl) in the same way that bulk chlorine gas from cylinders acts:

[3] $Cl2 + H_2O \rightarrow HOCl + H^+ + Cl^-$

Chlorine production is balanced by the reduction reactions that occur at the cathode where water (H_2O) is converted into hydroxide ions (OH-) and hydrogen gas (H_2):

[4] $2\ H_2O + 2e^- \rightarrow 2\ OH^- + H_2$

During electrolysis, the hydrogen gas is produced as bubbles that must later be removed from the OSG and oxidant storage tanks to prevent buildup of the gas. The hydroxide ions produced at the cathode then react with the hypochlorous acid produced at the anode, producing the hypochlorite anion (OCl-), which is charge balanced with sodium cations (Na+) that originally came from the salt:

[5] $HOCl + OH^- \rightarrow H_2O + OCl^-$

Typically, the pH of oxidants that OSGs produce is around nine. The addition of these solutions often does not alter the pH of the water that is to be treated. Overall reaction:

[6] $NaCl + H_2O \rightarrow NaOCl + H_2$

WHAT ARE THE BENEFITS OF USING OSG?

There are four principal benefits associated with OSGs: (1) improved operator safety, (2) higher quality chemicals, (3) greener applications, and (4) cost savings. Improved Operator Safety Chlorine sources traditionally used in water disinfection pose a variety of hazards to the operator. Chlorine gas is probably the most hazardous source of chlorine used by water treatments plants; it is toxic and the use of chlorine gas cylinders also poses a pressure hazard. Industrial strength bleach used for water disinfection is a 12.5 percent-by-weight solution, which is caustic. OSG systems use only water and salt and produce nonhazardous oxidant solutions with a chlorine content that typically contains less than 0.8percent free available chlorine. Treatment plants that use OSG systems typically have to face less oversight from state health agencies, provide less safety training for operators, and have less of an insurance issue compared to those using traditional forms of chlorine. Higher Quality Chemicals Recent research has indicated that hypochlorite storage leads to chlorate (ClO_3^-), and perchlorate (ClO4-) production from hypochlorite anions. Additionally, factors such as time in storage, temperature at storage, and exposure to sunlight can cause hypochlorite loss through other chemical degradation pathways. These observations indicate that older hypochlorite will contain less and less free available chlorine and more degradation products. Storage issues mount in areas that are required to have 30-day or higher supplies of disinfectant chemicals on hand. OSG systems, on the other hand, typically produce only a two- to three-day supply of chlorine at a time, thus providing a potent disinfectant. Salt does not decompose, so that long-term requirements can be met by storing enough salt to comply with regulations.

OSGs mean greener operations compared to traditional chlorination methods. In addition to the reduction in use and potential accidental release of toxic chemicals, transportation of chemicals from factories to the water plant is reduced. For example, it takes one delivery of salt to produce the same amount of chlorine as more than three deliveries of 12.5 percent sodium hypochlorite solution. This, therefore, lessens the carbon footprint of the plant because less fossil fuel is needed to supply the plant with disinfectant. Cost Savings OSGs typically produce chlorine at a much lower cost than traditional delivery methods, primarily because there is no need to continuously purchase expensive chlorine chemicals. This is especially the case for systems using calcium hypochlorite. Additional savings come from decreased transportation and safety-related costs, and lower insurance premiums. Although OSG systems usually present a large, up-front capital equipment cost, most water plants realize a return on their investment in OSG equipment within two to three years.

Although OSGs are basically safe, there are a few items to consider.

Hydrogen safety
Hydrogen gas is colorless, odorless, and is flammable. All electrochemical systems that employ aqueous solutions—disinfectant OSGs included—produce hydrogen at the cathode as a byproduct of the electrochemical process. Hydrogen is more than 13 times lighter than air, so it will rapidly dissipate from an electrolytic cell or OSG system. Because OSGs are typically installed inside a building, the system and tanks need to be properly ventilated. Hydrogen safety concerns are mitigated by careful engineering of the OSG itself, as well as good planning when the OSG is installed. When considering an OSG system for a water treatment plant, it is important to ensure that the system meets standards set by groups such as Hydrogen Safety LLC.

Water quality and temperature
Water is the most common component of the salt solution that enters the electrolytic cell of an OSG and, thus, the composition of water is important. Potable water supplies can feed most OSG electrolytic cells, but it is very important that the water be softened so as to have the lowest possible hardness. If hard water is used to provide either the water or brine solutions for an electrochemical cell, scales will rapidly form on the surfaces of the electrodes, causing the electrolytic cell to fail. Similarly, the temperature of the water entering the electrolytic cell should be maintained within a range of 40 to 80 degrees Fahrenheit, to avoid damaging the electrolytic cell. If an OSG is installed in an area where water feeding the OSG will be outside of that temperature range a heater/chiller unit is typically added to the overall system.

Salt quality
Sodium chloride is the only chemical added to the water stream that is employed by OSGs to produce disinfectants so it is vital that the salt be of high purity. Some contaminants in salt can cause damage to the electrolytic cell, typically calcium and

magnesium salts found in sea salt. Another concern is that some salts contain other chemical species that are subject to oxidation, the most common being bromide (Br^-). In any electrochemical cell that produces chlorine, bromide will be oxidized to form bromates (BrO_3^-), which are regulated and have a MCL of 0.01 mg/L. Food quality salt is the most common form of salt recommended for OSGs.

Calcium carbonate comes from either the salt or the incoming water supply. The electrolytic cell is the most expensive part of an OSG and appropriate care should be taken. Flushing the cell with soft water after every usage helps to prevent salt-deposit buildup. Most OSG systems perform this action automatically upon system shutdown. Using appropriately softened water and high quality salt are the two most important factors of cell maintenance. Even under these conditions though, electrochemical cells will develop scales over time. This scale will impede the ability of the cell to generate chlorine and, if left unchecked, will eventually destroy the electrodes. Wash the cell periodically by flushing it with muriatic acid (hydrochloric acid) to remove the scale and clean the electrode surfaces. How much acid is needed and how often the electrolytic cell needs to be rinsed are factors that rely on variables such as how often the cell is in operation and the quality of water and salt that goes into the cell.

Figure 2: Electrolytic Cell

New OSG technology has become available at which the electrolytic cell performs a self cleaning feature automatically after 700 hours of operation which is about every 4 to 6 weeks. The self cleaning feature drastically reduces the amount of maintenance an operator will have to perform on the unit. Additionally, the two efficiencies that engineers often look at are salt and energy efficiencies and the self cleaning option insures that the cell is the most consistent in keeping the electrolytic cell clear of calcium carbonate deposits.

MIXED OXIDANTS

Despite years of field evidence demonstrating that mixed oxidant solution (MOS) is more effective than bleach in a variety of applications, identifying the specific oxidants in addition to free chlorine in MOS has been limited by the availability of analytical techniques capable of differentiating oxidant species. Now, however, recent laboratory data using a light-emitting analytical technique, chemiluminescence, strongly indicate the presence of both free available chlorine and trace amounts of hydrogen peroxide (H_2O_2) in freshly generated MOS. Mixed oxidant solution is made from exact same process and feed stock as 0.8% sodium hypochlorite as described above.

The mixed-oxidant solution (MOS) is produced by electrolysis of sodium chloride (NaCl) brine in an electrolytic cell that has been optimized for disinfection efficacy. This solution exhibits microbial inactivation properties that are superior to bleach alone. In MOS, free available chlorine (FAC) is the primary analyzable oxidant constituent. However, the chemical produced from MOS generators shows demonstrable differences from bleach in both field and laboratory studies, indicating the presence of other oxidant species beyond FAC alone. The additional oxidant species in MOS are responsible for enhanced biocidal efficacy (demonstrated on a variety of microorganisms), as well as enhanced behavior in several chemical processes important in water treatment. Figure 3 shows one example of the superior biocidal efficacy of MOS compared to sodium hypochlorite/bleach (NaOCl). There are several results published in the peer-reviewed literature that have demonstrated this superior efficacy.i Years of compiled field data also show differences.ii

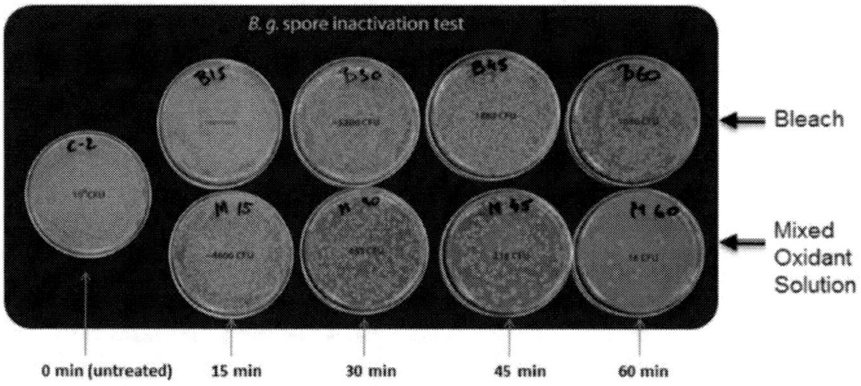

Figure 3: Superior Biocidal Efficacy

Analytical method limitations include FAC interference. A major obstacle to determining additional oxidant species present in MOS is the lack of analytical techniques for differentiating oxidants. Researchers continually improve oxidant analyses for FAC, ozone (O_3), hydrogen peroxide (H_2O_2) and chlorine dioxide (ClO_2).iii Almost invariably, however, each of the methods is based on the total oxidizing capacity of the solution being analyzed and is readily subject to interferences from the presence of other potential oxidizing agents or intermediates from associated chemical reactions.iv This is especially the case for analyzing additional oxidants in the presence of FAC.

Speciation research – ClO_2 and O_3 discounted and a growing body of evidence for H_2O_2 and/or other reactive oxygen species emerges. In 1994, Dowdv demonstrated the presence of a substantial systematic excess of oxidants over FAC alone in freshly-generated MOS produced at different brine flow rates. Methods included standard amperometric titration for FAC at pH 7 and a modification of the iodometric method at pH 2 in glacial acetic acid for total oxidants. While the nature and composition of the oxidants other than FAC was not identified, the method for total oxidants analysis suggested ClO_2, O_3, and H_2O_2 as possible components. The presence of ClO_2 and O_3 as analyzable components of the bulk produced MOS was discounted by subsequent research.vi

By the 2000s, a growing body of evidence for H_2O_2 and/or other reactive oxygen species in MOS had emerged. The chemical and biocidal behavior of MOS, coupled with potentials measured at the anode of the electrolytic cell, suggest the presence of other oxidants including reactive oxygen species (ROS). ROS are present in solutions of non-chloride brines electrolyzed under electrolytic conditions similar to those for producing MOS.vii More recent research on the chemical species present in MOS using non-chloride brines has provided evidence that stronger oxidants than chlorine (likely ROS) are present in the solution. O_3, H_2O_2, and hydroxyl free radicals (•OH) were detected at the point of generation.viii The longevity of each of these species in MOS is still not established, although that of O_3 and •OH is expected to be short – milliseconds to seconds, explaining why no O3 is detected in the bulk MOS. Literature data suggests that at higher pHs, such as those found in produced MOS, H_2O_2 has the potential to last hours to days in a chlorine matrix.

MIOX has applied this commonly-understood principle that is observed in nature to analyzing the mixed oxidant solution. Laboratory data using chemiluminescence analytical techniques indicate the presence of H_2O_2 in MOS. Certain chemiluminescent reagents are known to emit light in reaction with NaOCl, but not with H2O2. Very strong chemiluminescence is produced, however, in reactions when bleach and H2O2 are present simultaneously.10 The comparison of the reagent-mediated chemiluminescence of bleach and MOS in the pH range between 7.5 and 9 strongly suggests the presence of H2O2 and/or other reactive oxygen species in addition to FAC as hypochlorite (OCl-) in MOS.ix Upon aging, the reagent-mediated chemiluminescent properties of MOS become gradually similar to those of bleach, which is also reflected in decreased biocidal efficacy compared to freshly generated MOS.

Interestingly, the human immune defense system utilizes a similar strategy in which immune cells produce ROS at the sites of infection to kill bacteria. This is achieved when a chemical reaction occurs between NaOCl and H_2O_2.x

The benefits that customers have experienced with Mixed Oxidants over the last 15 years include:

- Superior Disinfection
- Biofilm Control
- Enhanced Residual
- Reduction and disinfection byproducts (Reduced TTHMs)

- Improved taste and odor (Improved T&O)
- Micro-flocculation
- Improved Oxidation. For example for Arsenic, Iron Manganese

Enhanced residual

Figure 4 provides evidence of municipalities using chlorine gas or bulk hypo and experiencing typical chlorine residuals seen at the end of the distribution system. These water municipalities made a switch over to mixed oxidants. Cedar Knox, NE, and North Table Mountain were able to reduce their dose and achieve a higher chlorine residual with mixed oxidants. In all these instances the municipalities were dealing with Biofilm issues.

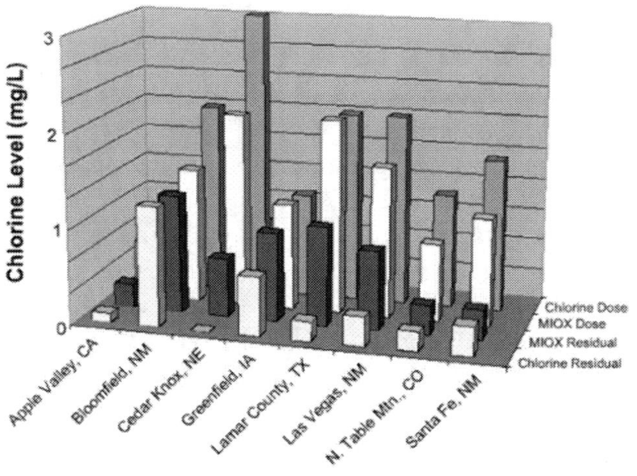

Figure 4: Chlorine versus Mixed Oxidants in the Distribution System

Montana State University study confirmed that mixed oxidant solution is able to penetrate the polysaccharide film that biofilm uses to attach to the distribution system walls. In all these instances there had to be a public awareness program as well as a flushing program to remove the particulate from the distribution system so it did not go into people's homes.

Reduction of disinfection byproducts

The MCL or maximum contaminate level for trihalomethanes is 80 parts per billion. Figure 5 shows several communities that were close or exceeding or very close to TTHM maximum contaminant limit while utilizing chlorine. When they made the switch over to mixed oxidants they were able to reduce their concentrations below or reduce it further from the 80 parts per billion MCL. This is all attributed to again to the removal of the biofilm. When they were able to remove the biofilm they were able to reduce the amount of chlorine that was being injected into the distribution system. With the reduction in chlorine then you have less of a reaction to the organic material that is in the water which creates your disinfection byproduct.

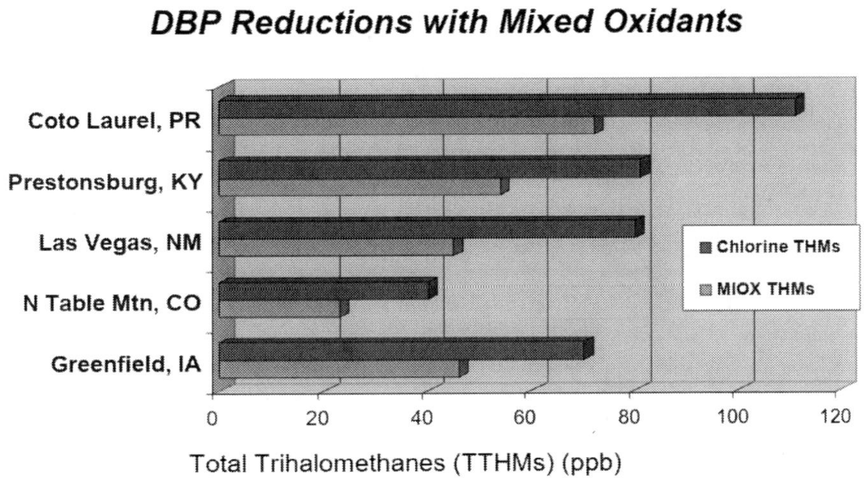

Figure 5: DBP Reductions with Mixed Oxidants

Often time's engineers ask how long it takes to see these types of reductions. It all depends on the complexity and size and severity of biofilm as well as the size of the distribution system. In Greenfield, Iowa, over a three month period of time from the time they started injecting the mixed oxidant that they were able to reduce the amount disinfection byproduct formation both at the treatment plant as well as the distribution system.

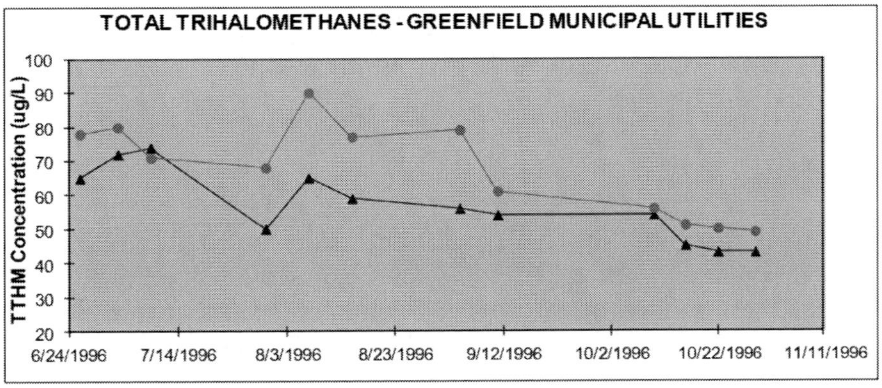

Figure 6: DBP Reduction Timeline

Biofilm

Diana WSC in Texas provided Figure 8 where they had simultaneous main breaks in an area where they were still using sodium hypochlorite and another area where they were using mixed oxidants. In the area where they were using sodium hypochlorite you can see the biofilm is still evident in the pipe distribution wall. This again reconfirms the study done by Montana State University with regards to biofilm.

Figure 7: Biofilm in the Distribution System with Chlorine

Micro-flocculation
Micro-flocculation is defined as a reduction in coagulant demand for the same final (settled water) turbidity, or a reduction in final turbidity at the same coagulant demand.

In this study at Dekalb Jackson, the use of MOS in pretreatment achieved both lower turbidity levels and reductions in alum required. Operators report faster floc formation and improved settling due to the quality of the formed floc. This translated to longer filter run times and fewer chemical cleaning events. Disinfection by-product (DBP) formation so far is also being reduced with the corresponding reduction in organics.

The Dekalb Jackson water treatment plant (WTP) is considered a conventional treatment plant, except that microfiltration (MF) is used instead of sand filtration. Surface water is subjected to flocculation, settling, MF, and final disinfection (Figure 8). Plant operators were interested in improving the flocculation and sedimentation steps due to the excessive filter backwashing and clean-in-place (CIP) procedures required for the MF membranes. Two parallel tube settler trains are used for sedimentation and bleach was used in pretreatment and final disinfection. The WTP has four banks of filters each containing 16 modules, all 0.1 μm pore-size polysulfone hollow-fiber membrane. Plant data suggested MOS would cause a microflocculation effect similar to that observed for ozone in pretreatment year-roundxi and jar tests were performed to confirm the prediction that MOS would improve settled turbidity values compared to the current operations. Jars using MOS produced larger, faster settling floc and a UV254 reduction in the supernatant. After the encouraging results demonstrating that MOS performs better than bleach, the manufacturer performed due diligence to determine that MOS would not cause premature degradation of the membrane.

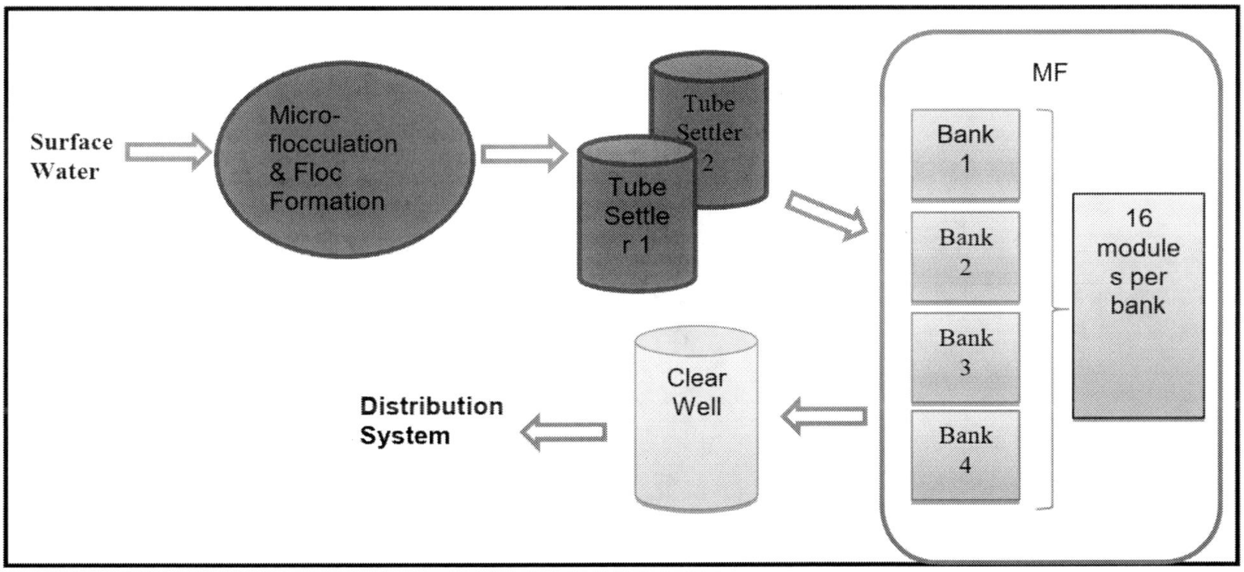

Figure 8: Dekalb Jackson Plant Schematic

The plant monitors UV254, turbidity, TOC and filter cleaning frequency. No changes to routine monitoring were required during the pilot. To minimize the potential formation of DBPs and deterioration of the membranes, FAC dose would be controlled to support microflocculation while still keeping the FAC residual low at the tube settlers.

The pilot test began 1 July 2011. By 22 July, preliminary results indicated the following:

- Rapid and good-quality floc formation before sedimentation in the tube settlers;
- Raw water turbidity spikes not seen in settled turbidity (normally such raw water spikes were seen in the settled turbidity);
- Settled turbidities were routinely <1 ntu – the target for this operation – values that had been difficult to achieve in prior operations using bleach for pre-treatment; and
- Settled water FAC easily controlled to <0.1 mg/L, well within requirements.

Tube settler performance data from the raw water turbidity spike in September, 2011 are compared with a similar event that occurred one year earlier in Figure 9.

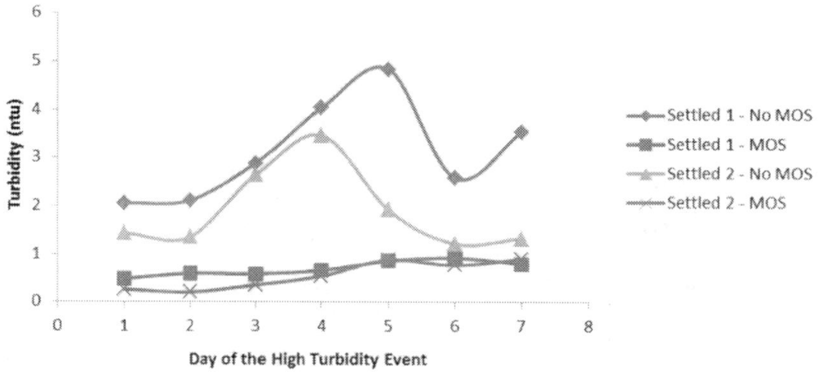

Figure 9: Performance Comparison in Tube Settler Trains

After MOS began to be used in pretreatment, engineers and plant operators experimented with a CIP protocol involving only a 3-hour soak. It is clear that filter run times between backwashing increased while full CIP cleaning frequencies declined. Plant operators attribute these improvements to reduced floc load due to microflocculation using MOS placing less emphasis on other operations such as chemically-enhanced backwash.

---

[i] Bajszar, G. and Dekonenko, A., "Stress-Induced Hsp70 Gene Expression and Inactivation of *Cryptosporidium parvum* Oocysts by Chlorine-Based Oxidants", *Appl. Environ. Microbiol.* 2010, 76(6):1732-1739; Venczel, L.V., Likirdopulos, C.A. Robinson, C.E. and Sobsey, M.D., "Inactivation of Enteric Microbes in Water by Electro-Chemical Oxidant from Brine (NaCl) and Free Chlorine" *Wat. Sci. Technol.*, 2004, 50(1):141-146; Son, H., Cho, M., Chung, H., Choi, S., and Yoon, J., "Bactericidal Activity of Mixed Oxidants: Comparison with Free Chlorine", *J. Ind. Eng. Chem.*, 2004, 14(5):705-709; Venczel, L.V., Arrowood, M., Hurd, M., and Sobsey, M.D., "Inactivation of Cryptosporidium parvum Oocysts and Clostridium perfringens Spores by a Mixed-Oxidant Disinfectant and by Free Chlorine", *Appl. Environ. Microbiol.*, 1997, 63(4):1598-1601.

[ii] Bradford, W.L., 2006, "The Differences between On-Site Generated Mixed-Oxidant Solution and Sodium Hypochlorite" MIOX Corporation, www.miox.com.

[iii] Standard Methods for the Examination of Water and Wastewater, 20th edition.

[iv] **"Disinfectant Residual Measurement Methods, Second Edition", AWWA Research Foundation;** Prepared by Gilbert Gordon, Miami University; William J. Cooper, Florida International University; Rip G. Rice, Rice, Incorporated; and Gilbert E. Pacey, Miami University. 1992.

[v] Dowd, M.T., 1994, "Assessment of THM Formation with MIOX", Master's Thesis, University of North Carolina, Department of Environmental Sciences and Engineering, School of Public Health, Chapel Hill, NC.

[vi] Gordon, G.L., 1998, "Electrochemical Mixed Oxidant Treatment: Chemical Detail of Electrolyzed Salt Brine Technology", prepared for the U.S. Environmental Protection Agency, National Risk Management Laboratory, Cincinnati, OH, May 1998.

[vii] Jeong, J., J.Y. Kim, and J. Yoon, 2006. "The Role of Reactive Oxygen Species in the Electrochemical Inactivation of Microorganisms", Environ. Sci. Technol., 40(19):6117-6122.

[viii] Weinberg, H., S. Rodriguez-Mozaz, and A. Sykes, 2008, "Characterization of the Chemical Constituents of Mixed Oxidant Disinfection", Final Project Report, presented to MIOX Corporation by the University of North Carolina, Department of Environmental Sciences and Engineering, Chapel Hill, NC, 23 July 2008.

[ix] Bajszar, G., 2008, "Co-Oxidation by Hypochlorite and Reactive Oxygen Species: Implications for Mixed Oxidant Solution Speciation", presented at MIOX Corporation, Albuquerque, NM 20 February 2008.

[x] Brestel, E. P. (1985) Biochem. Biophys. Res. Commun. 126:482-488.

[xi] Becker, W.C. and C.R. O'Melia, 1996, "Optimizing Ozonation for Turbidity and Organics (TOC) Removal by Coagulation and Filtration", American Water Works Research Foundation, Denver, CO, 209 pp.

Evaluation of Oxidants Products as Disinfectants Seeking Reuse of Refinery Wastewater

Andréa Azevedo Veiga, R & D Center of Petrobras, CENPES)
Ysrael Marrero Vera, Federal University of Rio de Janeiro
Vania Maria Junqueira Santiago, R & D Center of Petrobras, CENPES
Lídia Yokoyama, Federal University of Rio de Janeiro
Vera Lúcia dos Santos, Federal University of Minas Gerais
CENPES
Rua Horácio Macedo, 950 - Cidade Universitária
Rio de Janeiro, RJ – CEP 21941-915 – Brasil
55 (21) 9989-7052

ABSTRACT

Wastewater reuse from municipal and industrial effluents is a very emerging alternative to obtain water resources and is becoming increasingly important in many areas around the world. Water scarcity, local regulation and environmental impact are driving many industries to reuse the large volume of produced effluent. Disinfection is an essential component of many wastewater reuse treatment station (WRTS) of refinery that has as objective to inactivate or destroy pathogen organisms. Although the most common method of wastewater disinfection is chlorination, the potential toxicity of chlorination by-products generated in reaction with organic matter present in reuse water makes this process less attractive. The essential foundation for successful implementation of such WRTS is the capability of producing water of a desired quality that guarantee protecting worker health and the environment and the industry demands. In order to cope with the large fluctuations of wastewater quality generated by refinery and considering cost-effective operation of the system, studies based in the evaluating oxidizing disinfectants alternative to chlorine were conducted. The activities of commercial disinfectant with active bases in performic acid, peracetic acid, hydrogen peroxide and chlorine dioxide were evaluated against total aerobic heterotrophic bacteria (THB) and total coliform (TC) in laboratory scale to replace chlorine. Performic acid, chlorine dioxide and peracetic acid exhibited an excellent activity against TC. However, none of the products presented effectiveness for the reduction of THB in the conditions of contact time and concentrations used. The technical evaluation showed favorable economic application of performic acid. Currently, studies are being conducted with continuous application of performic acid for disinfection of effluent in the WRTS of refinery.

KEYWORDS:

Wastewater reuse, performic acid, disinfectants, pathogens, refineries

INTRODUTION

Recent Brazilian regulation on water resources management (Law 9433/97), which introduces the granting and collection and management tools, has led the productive sectors to seek ways to rationalize water consumption, once the water and its availability are limiting factors for urban, industrial and agricultural development. The refining industry is an important water consumer, both in terms of quantity and quality, treating the rational use of this resource as a priority in its strategic planning.

One of the Brazilian refineries lies in a region of high criticality in terms of water availability. The water demand in this region is greater than the available capacity. Moreover, high levels of river pollution are caused by irregular occupation in the areas of watershed protection.

The refinery takes water from a wellspring and uses it in the oil refining process. The effluents are treated in Industrial Wastewater Treatment Plant (IWTP) in order to achieve environmental standards specification for disposal in the same river (Figure 1). The treatment consists of water and oil separators (API), where there is a reduction in the wastewater oil content, and activated sludge reactors for degradation of contaminants by microorganisms.

Figure 1 - Industrial Wastewater Treatment Plant: API separators (left) and activated sludge reactors (right).

After assessing the scenarios of increasing collected water quality criticality; increasing demand for use of water for industrial purposes and the rising cost of water treatment, a process was selected in order to realize the complementary treatment of the final effluent aiming the reuse.

The technology selected for the effluent further treatment was a high-rate clarifier that uses the principle of sedimentation with lamellar recirculation of sludge, which increases the efficiency of clarifying and reduces chemicals consumption. Reusing 140 m^3/h of the treated effluent allows the refinery to eliminate the effluent disposal into wellsprings and also increase the water supply for the companies of a petrochemical complex in the region.

Figure 2 shows a schematic Water Reuse Station (WRS), as well as a picture of the high rate clarification unit installed in the refinery.

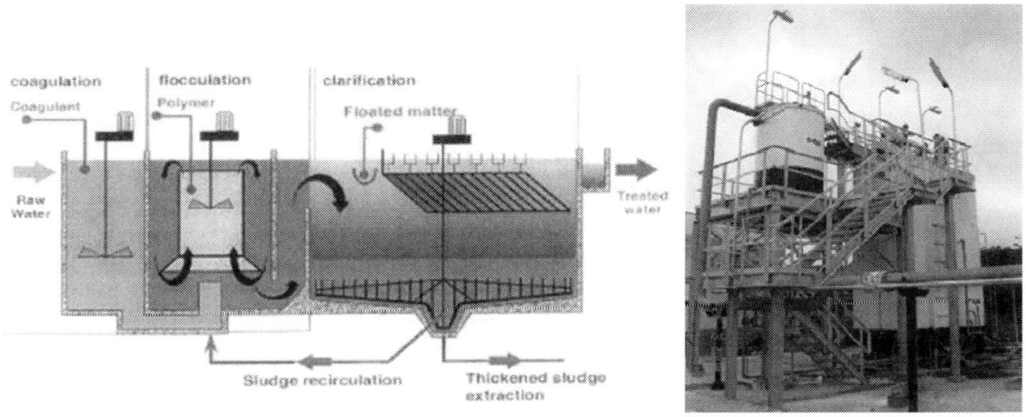

Figure 2 - Schematic figure of the photo and the high rate clarifier (WRS) installed in the refinery.

In general, the process of water disinfecting involves the use of a chlorine-based disinfectant products that have been applied due to its low cost associated with high microorganisms' removal efficiencies. However, in contact with higher organic waste loads, characteristics of reused water, the chlorine-based disinfectants can produce toxic (organochlorine), mutagenic, persistent, carcinogenic and persistent byproducts, generating adverse impacts for the environment.

Considering the limited treatment capacity of the wastewater clarification unit (WRS), fluctuations in the effluent quality generated by the refinery and local operators safety, an evaluation study of oxidizing disinfectants was proposed, in order to improve the treatment performance, ensuring the quality of water for reuse and removal of pathogenic micro-organisms.

The organisms of interest in the industrial wastewater recycling include total aerobic heterotrophic bacteria (THB) and total coliform (TC) due the fact that they are associated with human diseases.

Furthermore, the choice of a suitable disinfectant depends on the following criteria: absence of toxic products, operational safety, easy handling and also the ability to penetrate and destroy the agents of infection during the normal operation contact times. According to the requirements of the EPA (1992), waters that comes from the sanitary effluent treatment that are designated for reuse in urban application, without restrictions, should present no detectable concentrations (ND) of fecal coliforms per 100 mL of sample (ND/100mL) and <200/100 mL for the same organism. In order to achieve the required water quality parameters within the CONAMA 357/05 classification, the biggest restriction applies to effluents classified as class I, where the maximum fecal coliform is up to 200 CFU/100 mL.

Within limits of potability (Brazilian Regulation 518/04 updated to Ordinance 2914/11), the absence of coliforms (ND/100 mL) and the maximum limit of 500 CFU/mL for total heterotrophic bacteria (BHT) is required. Table 1 presents a summary of the requirements for disinfection in water for different environmental laws and their applications.

Table 1 - Environmental legislation for water use limits and concentrations of microorganisms according to the application.

Desinfection Legislation	Application	Thermotolerant Coliforms (per 100 mL)	Fecal Coliforms (E. Coli) (per 100 mL)	THB (per mL)
CONAMA 357/05	Class I Class II Class III	Max. 200 Max. 1000 Max. 2500	- - -	- - -
EPA 1992 (reuse from sanitary sewage)	Urban with out restrictions Industrial Reuse	- -	ND <200	- -
Regulation 518/04 & 2914/11	Potability		ND	<500

In this context, replacing chlorine in compliance with a demand that the refinery while seeking to ensure the operational safety of employees working in the area and also evaluate the potential for microbiological control in a water reuse, comparative studies were conducted on a bench scale by the Chemistry School in Federal University of Rio de Janeiro and the Research and Development Center of Petrobras - CENPES, Brazil, in order to evaluate the efficiency of disinfection from commercial disinfectant products, with active bases in performic acid, peracetic acid, hydrogen peroxide and chlorine dioxide, besides the chlorine. The products activities were evaluated against total coliform (TC) and total heterotrophic bacteria (THB) at concentrations and times pre-determined, based on results of preliminary studies.

EXPERIMENTAL METHODOLOGY

To obtain the inoculum, sterilized Erlenmeyer flasks with 250 mL effluent from WRTS clarified by filtration over a 0.22-µm filter (Millipore filter with nitrocellulose membrane) were added of 0.5g of biomass washed in sterile distilled water from secondary treatment of refinery and 0.25 g nutrient broth and tryptone soya (TSB). The flasks were incubated by 48 h with stirring to obtain a concentration of total heterotrophic bacteria (THB) and Total Coliform (TC) > 10^7 CFU/mL.

The assays of disinfection were performed in triplicate in sterilized Erlenmeyer flasks with 250 mL effluent from WRTS clarified inoculated with THB and TC to a final concentration of 10^6 CFU/ml, concentration similar to those found in the industrial effluent prevenient from WRTS. Commercial disinfectants with active bases in Cl_2, ClO_2, H_2O_2 and peracetic acid were added to the flasks to a final concentrations of 8.0 and 10.0 mg/L.

The performic acid was evaluated at concentrations between 0.5-2.0 mg/L. Controls were effluent added bacteria without disinfectant. Samples were taken at regular intervals from 15 to 30 minutes and evaluated as counting of BHT using NUTRILAB E slide culture kits (Laborclin) and CT count using 3M Petrifilm plates, according to manufacturer instructions.

RESULTS AND DISCUSSION

To facilitate the comparative evaluation of the effectiveness of disinfectants, was created a classification system for the same in three groups, considering their ability to reduce populations of THB and TC in log scale (Table 2). The class A includes the best results of removal efficiency of microorganisms in study and their removal rates in log scale, while C is the class that had the lowest removal efficiency.

Table 2 – Classification of commercial disinfectants based on the results of their efficiency in reducing populations of THB and TC in logarithmic scale.

Classes	Final microbial density (CFU/mL)	Rate Removal of Bacteria
A	$10^2<$ [THB] $<10^3$ [TC] $<10^1$	4 - 5 Log Max. 5 Log
B	$10^3<$ [THB] $<10^4$ [TC] $<10^3$	2 - 3 Log Max. 3 Log
C	$10^4<$ [THB] $<10^5$ [TC] $<10^4$	1 - 2 Log Max. 2 Log

Two dosages were tested simultaneously in a batch inactivation study with a contact time varying from 0 to 30 minutes for the ClO_2, H_2O_2, Cl_2 and peractic acid (Table 3). A dosage of 8mg/L ClO_2 to the clarified effluent with a contact time of 30 minutes ensured a TC count of less than 10^1 CFU/mL. This dosage succeeded in the bacteriological disinfection requirement as specified in the legislation (<150 CFU/100mL). However, the desinfectant showed low effectiveness in reducing THB at all concentrations an contact time evaluated, being classified in the C class (Table 2).

Table 3 - THB and TC densities in different periods of contact between commercial disinfectants and samples efluents.

Desinfectant	Concentration (mg/L)	Total Coliforms (CFU/mL)			Total Heterotrophic Bacteria (CFU/mL)		
		0 min	15 min	30 min	0 min	15 min	30 min
ClO_2	8	$1.5 X 10^6$	$4.0 X 10^2$	$5.0 X 10^0$	$10^5 – 10^6$	$10^4 – 10^5$	$10^4 – 10^5$
	10	$1.5 X 10^6$	$2.0 X 10^2$	$3.0 X 10^0$	$10^5 – 10^6$	$10^4 – 10^5$	$10^4 – 10^5$
H_2O_2	8	$2.5 X 10^6$	$5.0 X 10^6$	$1.6 X 10^5$	$10^4 – 10^5$	$10^4 – 10^5$	$10^4 – 10^5$
	10	$2.5 X 10^6$	$1.3 X 10^4$	$1.6 X 10^5$	$10^4 – 10^5$	$10^4 – 10^5$	$10^4 – 10^5$
Cl_2	8	$1.5 X 10^6$	$1.9 X 10^4$	$7.0 X 10^4$	$10^5 – 10^6$	$10^4 – 10^5$	$10^4 – 10^5$
	10	$1.5 X 10^6$	$8 X 10^3$	$5.2 X 10^3$	$10^5 – 10^6$	$10^4 – 10^5$	$10^4 – 10^5$
Peracetic Acid	8	$2.0 X 10^6$	$8.6 X 10^2$	$8.5 X 10^2$	$>10^6$	$10^4 – 10^5$	$10^4 – 10^5$
	10	$2.0 X 10^6$	$2.2 X 10^2$	$8.5 X 10^1$	$>10^6$	$10^4 – 10^5$	$10^4 – 10^5$

Hydrogen peroxide showed low effectiveness for removal of TC (C Class - Table 2), but was not effective in reducing the density of BHT in the evaluated conditions (Table 3). A reduction of the total coliforms density from 1.5 X 10^6 to 5.2 X 10^3 CFU/mL was observed by dosage of 10 mg/L Cl_2 and 30 minutes of contact time, being classified in the B (Table 2). The product showed low effectiveness in removing THB in all concentrations and contact times evaluated (Class C - Table 2).

At 10 mg/L, peracetic acid reduced concentrations of TC from 2.0 X 10^6 to 8.5 X 10^1 CFU/mL within 30 minutes of contact time, being classified as Class A (Table 2), but had limited efficacy in the THB control, reducing the density of this microbial group by only one order of magnitude (Class C - Table 2).

Performic acid showed low efficiency in reducing THB under the conditions tested (Table 4), but exhibited high specificity for removing CT. At concentrations higher than 1 mg/L and 15 min of contact time, removal rates exceeding values of 99.99% were obtained (class A - Table 2).

Table 4 - THB and TC values at baseline and after the contact time of 15 to 30 minutes of performic acid dosing.

[Performic acid] (mg/L)	Total Coliforms (CFU/mL)			Total Heterotrophic Bacteria (CFU/mL)		
	0 min	15 min	30 min	0 min	15 min	30 min
0.5	2,9 X 10^5	1,2 X 10^1	8,0 X 10^0	$10^5 – 10^6$	$10^4 – 10^5$	$10^4 – 10^5$
1.0	2,9 X 10^5	7,0 X 10^0	9,0 X 10^0	$10^5 – 10^6$	$10^4 – 10^5$	$10^4 – 10^5$
1.5	2,9 X 10^5	3,0 X 10^0	5,0 X 10^0	$10^5 – 10^6$	$10^4 – 10^5$	$10^4 – 10^5$
2.0	2,9 X 10^5	3,0 X 10^0	4,0 X 10^0	$10^5 – 10^6$	$10^4 – 10^5$	$10^4 – 10^5$

The comparative results of all the tests proved high efficiency of performic acid, chlorine dioxide and peracetic acid for removal of TC in the treated refinery effluent for reuse (Table 5). However, none of the products presented effectiveness for the reduction of THB in the conditions of contact time and concentrations evaluated.

Table 5 - Comparison between the technical and economical disinfectants evaluated.

Product	Product Features / Commercial	Applied Concentration (mg/L)	Removal Efficiency (THB and TC)	U.S. $ /m3 Treated Effluent	Note	Technical Evaluation
Performic Acid	Liquid/Desinfix®	1	C (THB)/A (TC)	0,074	High specificity for TC	Approved
ClO_2	Gas *in situ* - Generator 2 kg/h	8	C (THB)/A (TC)	0,222	-	Approved
Cl_2	Liquid/NaOCl	8	C (THB e TC)	0,003	Generation of toxic byproducts	Disapproved
		10	C (THB)/B (TC)	0,035		
Peracetic Acid	Liquid/Proxitane 1512®	10	C (THB)/A (TC)	0,203	-	Approved
H_2O_2	Liquid	8	C (THB e TC)	0,015	Low efficiency	Disapproved

Hydrogen peroxide (H_2O_2), despite having the lowest specific cost (U.S. $ 0.015/m^3 of treated effluent), showed low removal efficiency for microorganisms evaluated (Classification C - Table 2) in comparison with the other products tested (Table 5). For generate toxic byproducts, disinfection with Cl_2 was not considered a viable possibility. This step evaluation of cost per m^3 for the implementation of performic acid as disinfectant product in the treatment of reuse wastewater is essential to assess the technical feasibility of using the product on an industrial scale.

CONCLUSIONS

Performing technical evaluation showed favorable economic application of performic acid in conducting continuous studies in refinery wastewater for reuse in industrial scale. The continuous dosage of performic acid on an industrial scale for desinfection of clarified effluent from WRTS of refinery has validated of the results of the disinfectants efficacy in laboratory scale.

REFERENCES

CONAMA (2005) *Conselho Nacional do Meio Ambiente*, 17 de março de 2005. Resolução N° 357. Diário Oficial da União, Brasília. Brazilian Environmental National Council (CONAMA), Published in the Official Journal of the Union - D.O.U of March 17th, 2005. Resolution N° 357.

Environmental Protection Agency, EPA.*Edition of the drinkingwater standards and health advisories*, EPA 822-R-02-38, Washington, DC, 2002, - EPA (1992), Environmental Protection Agency.

Ministério da Saúde (2004). Portaria Nº 518. Diário Oficial da União, Brasília. Ordinance of Water Potability, MS 518, Published in Official Journal of the Union – D. O. U of March 25th.

Ministério da Saúde (2011). Portaria Nº 2914. Diário Oficial da União, Brasília. Ordinance of Water Potability, MS 2914, Published in Official Journal of the Union – D. O. U of December 12 th.

A Post-Implementation Assessment of Student Executed Clean Water Programs in Ecuador

Gabrielle String, Tufts University

ABSTRACT

The focus of this research program was to analyze the public health impacts, long-term sustainability, and post-construction management of recently implemented water projects in rural communities in Ecuador. Implemented projects include a slow-sand filter in El Cristal, and ceramic filters in La Margarita. Engineers Without Borders student chapters, the former at Tufts University and the latter at Clarkson University, implemented both of these projects during the summer of 2011. By critically evaluating the long-term impact of student-run service learning projects on local communities improvements to EWB projects can be suggested to maximize both student learning and community benefit.

KEYWORDS

Engineers Without Borders, Water, Monitoring, Sustainability, Slow-sand filter, Ceramic filter

INTRODUCTION

Many people take for granted reliable access to clean water; yet, according to the World Health Organization, 884 million people are without improved sources of drinking water [1]. The problem is exacerbated by the global trend towards urbanization that marginalizes the rural poor, resulting in a cycle of ill health and poverty within which children are typically the first to suffer from the burden of disease caused by dirty water and poor hygiene. A WHO 2010 report has determined that Ecuador has 89% access to an improved water supply in rural areas, with an <5 yrs. death due to diarrhea at 6% [2].

Engineers Without Borders (EWB-USA) is a non-profit, non-competitive, humanitarian organization established to partner with developing communities worldwide in order to improve their quality of life. This partnership involves the implementation of sustainable engineering projects, while involving and training internationally responsible professionals and students. EWB-USA currently has a loosely defined and limited post-construction management plan as a part of their project model [3]. Furthermore, only limited research has been conducted to determine the long-term effects of these projects in the communities that they serve.

BACKGROUND

Prior Research

Past epidemiological studies have examined how improvements in drinking water quality can have a positive impact on health. A meta-analysis by Fewtrell and Colford found that water quality improvements could reduce incidence of diarrhea by 39% [4]. The researchers looked at 12 studies that dealt with household water treatment, nine of which found that household treatment could reduce diarrhea illness by a statistically significant amount.

Brown and Sobsey studied pot-shaped ceramic water filters in Cambodia [5]. Resource International Development introduced approximately 1,000 filters in Kandal Province, and International Development Enterprises introduced over 1,000 in Kampong Chhang and Pursat Provinces. The study involved a cross-sectional examination of 506 households that received the filters to find the variables associated with filter uptake and use. The researchers also carried out a longitudinal prospective cohort study that looked at the microbiological effectiveness and health impacts for 80 households with the filters and 80 without.

Palmateer et al. tested the Manz intermittent slow sand filter for its ability to remove bacteria, parasites, and toxicants [6]. They determined that the filter could remove 83% of heterotrophic bacteria, 100% of Giardia cysts, 99.98% of Cryptosporidium oocysts, and 50-90% of toxicants. A study by Bellamy et al. found 1 to 2 log removal rates for total coliforms [7].

Background of Tufts EWB Project

The Tufts EWB group began working with the community of El Cristal in 2006. The agricultural community consists of 120 families in the rural and mountainous area of the Esmereldas. The community relies on water from a spring-fed mountain stream. This relatively pure water is polluted by pathogens from a cow pasture that is located on top of the source. Additionally, the turbidity of the water is too high for disinfection without filtration. Two slow-sand filters were constructed by the local government in the mid 1990's, but quickly fell into disuse and disrepair. With the help of the Tufts EWB team, the filters were brought back online as part of a piped water distribution to the community. In August of 2011 the team brought the filters up to their full operational capacity and added a chlorination system. In October 2011 the group officially closed their program with EWB national and created a new group on campus: Developing Rurally Optimized Projects for Sustainability (DROPS).

Background of Clarkson EWB Project

The Clarkson EWB group began working with the community of La Margarita in 2007. La Margarita is a small agrarian community located adjacent to the Los Tintos River, approximately 5 miles upstream of its confluence with the Guayas River in the Guayas Province of Ecuador. Like other similar communities in the coastal plains, La Margarita lacks access to clean drinking water and other basic services such as sewage treatment and refuse removal. With limited alternatives, villagers use Los Tintos River to remove untreated refuse and septic wastes from their community, as do other communities upstream. These activities, as well as runoff from livestock agriculture, have polluted the river with refuse and high levels of fecal pathogens. Unfortunately, the people of this and other nearby villages use water from the river for drinking purposes with little or no treatment. The result has been a high incidence of gastrointestinal illnesses in the community. The Clarkson University EWB chapter developed a clay pot filtration system capable of greater than 5-log reduction of fecal coliform bacteria to yield potable water from the untreated river water for the residents of La Margarita. The manufacturing and distribution center for the ceramic filters is an NGO called Hogar de Cristo that is located in Guayaquil. Filters were implemented in the community in May 2011.

METHODOLOGY

Public Surveys

An evaluation of the implementations was conducted, primarily focusing on the knowledge, attitudes and practices (KAPs) surrounding water treatment and usage. Confidential surveys acquired information from each willing family in the respective communities. The surveys were conducted verbally and required only 10-15 min of a participant's time. Each household-surveying plan involved questioning people from both partner communities as well as control communities in the surrounding areas that have not yet received aid from the project teams. For the Tufts EWB Project, 32 surveys were conducted in El Cristal, representing ~51% of the community. Approximately 30% of houses in the communities of Santa Rosa and Las Tolas, which are under the same governance as El Cristal, but not on the water system, were also surveyed. For the Clarkson EWB project, 44 surveys were conducted in La Margarita, accounting for ~59% of the community. As a control, 52 surveys were conducted at random in the Monte Sinai region, a slum of Guayaquil that Hogar de Cristo works in.

Water Quality Tests

Water quality analysis was conducted differently in each community. In order to compare post-implementation data with pre-implementation data, the water quality testing methods were based on prior methods in each community. For microbiological sampling, total coliform/E. coli tests were conducted using Millipore preparation equipment and growth mediums. In addition to microbiological testing, turbidity testing was also conducted in both communities. In El Cristal only, chlorine tests were carried out to determine the effectiveness of their chlorination system.

Water quality testing involved sampling of treated drinking water in each community of all households that have answered the surveys and agreed to have their water sampled. Additionally, samples in each community were collected of the source water for each household. For El Cristal, this meant sampling source water before it enters the slow-sand filter, before it is chlorinated, and before it is disseminated via pipes to the houses in the community. For La Margarita, this involved identifying where the household collects its drinking water from prior to use of the ceramic filters.

Water quality data will be used to determine the effectiveness of the implemented solutions in improving water quality. Prior data analysis conducted on slow-sand filters indicates a bacteria removal rate of 95-99% [6]. Field-testing of ceramic filters has indicated bacteria removal rates of 90-99% depending upon the pore composition [5]. This data will also be used to examine relationships between KAPs and water quality.

Operator Focus Group

A focus group was conducted to determine in better detail the practices of the El Cristal filter operators. The filter operators are volunteers from within the community who donate time to maintenance of the filters. The questions asked during the session were designed to have the operators explain different parts of their activities and respond to how well prepared they felt they were for the job. The group setting allowed the operators to discuss individual practices and see how they compared to each other. This allowed the group to bring to light differences in operation pertaining to what they say they were doing, what they were actually doing, and how they were trained to do it. Additionally, this session allowed the researcher to observe group dynamics and potential organizational issues that otherwise may not have been brought to light.

ANALYSIS

To understand the connections between exposure and outcome a relative risk analysis will be conducted using the water quality data sets. An odds ratio will be calculated for each exposure factor to determine its relationship to diarrheal illness. Additionally, a chi-square test will be used to determine the statistical significance of each of the two factors analyzed for relative risk. This information will be correlated with verbal responses from the survey to determine how KAPs have an impact on relative risk.

CITATIONS

[1] WHO and UNICEF. "Progress on Sanitation and Drinking-water: 2010 Update." WHO Press: Geneva, Switzerland. 2010.

[2] WHO. Country Statistics: Ecuador. Global Health Observatory Data Repository. Accessed 5 Mar 2012. < http://apps.who.int/ghodata/?vid=7900&theme=country>

[3] Engineers Without Borders, USA. Document 531: Post-Monitoring Report. Sourcebook Downloads, EWB-USA. Accessed 2 Mar 2012. < http://www.ewb-usa.org/myewb-usa/chapter-resources/sourcebook-downloads>

[4] Fewtrell, L. and Colford, J. "Water, Sanitation and Hygiene: Interventions and Diarrhoea: A Systematic Review and MetaAnalysis." Health, Nutrition and Population Discussion Paper. World Bank. July 2004. Accessed 8 Feb 2012. <http://www.bvsde.ops-oms.org/texcom/nutricion/ref7.pdf>

[5] Brown, J. and Sobsey, M. Independent Appraisal of Ceramic Water Filtration Interventions in Cambodia: Final Report. University of North Carolina School of Public Health and Department of Environmental Sciences and Engineering. Submitted to UNICEF May 2006.

[6] Palmateer, G., et al. "Toxicant and Parasite Challenge of Manz Intermittent Slow Sand Filter." Environmental Toxicology. 14.2 (1999): 217-225.

[7] Bellamy, W. D., Hendricks, D. W., and Logsdon, G. S. 1985. "Slow Sand Filtration: Influences of Selected Process Variables." Journal AWWA. 78:12:62-67.

Fighting Failure: Lessons Learned in Pueblo Nuevo, Honduras

Cassandra Dutt, Lehigh University Engineers Without Borders
Joshua Bryant, Lehigh University Engineers Without Borders
Alexa Baker, Lehigh University Engineers Without Borders
Kevin Hoogenboom, Lehigh University Engineers Without Borders

ABSTRACT

Introduction

Despite the efforts made in recent years, access to clean water is one of the most pressing issues facing the globe. In the Central American country of Honduras alone half of the population does not have access to clean water. Eighty percent of the illnesses within the country are water borne and 50,000 children under the age of 5 die as a result of these illnesses each year.

In the early 2000s, Lehigh University professor of archeology, David Small, visited the town of Pueblo Nuevo, Honduras to conduct social and cultural field research. Professor Small returned to Lehigh with a complete social assessment of Pueblo Nuevo, as well as a comprehensive understanding of the impoverished life led by the 1,000 members of this rural Honduran town. In 1998 Hurricane Mitch had ravished the town, most notably leaving their water distribution system in shambles. With no source of clean and reliable water, about 80% of the residents reported crippling water borne illnesses and routinely spent hours each day collecting the necessary water to carry out everyday tasks. In 2006, a group of dedicated students and faculty made the pledge to find a solution to the water issues of Pueblo Nuevo, Honduras, forming the Lehigh University chapter of Engineers Without Borders.

Throughout the next several years, the members of Lehigh University Engineers Without Borders (EWB-LU) made several assessment trips to Pueblo Nuevo. After much social and technical research, it was decided that the existing piping system should be used to distribute water to town after traveling from the spring to a storage tank and chlorination system. The town leaders agreed to repair the current water distribution system in order to save valuable resources and to avoid water losses through the distribution process. The technical designs were approved by EWB-USA and the town of Pueblo Nuevo and implementation was finished in January2010. However, the first system test brought attention to a major problem. The tank drained faster than it could be refilled, and the town was left with two choices: use this new system and have an unreliable water supply, or go back to the old one, which provided plenty of water. With a focus on quantity over quality, Pueblo Nuevo chose the ladder. The town was again left without a disinfected source of water. EWB-LU spent the next year attempting to determine why the system that had been so carefully designed failed to provide Pueblo Nuevo with a sufficient amount of water.

Project Turning Points

With a system that could not meet an unexpectedly high water demand , the chapter explored several solutions to combat the technical failures.

After two years of exploration and attempts at a resolution, EWB-LU made the tough decision to abandon the project, both for social and technical reasons that are still being analyzed by chapter members. Socially, the town did not have an understanding of our purpose, goals, and timeline for the implementation of the water distribution and sanitation system. There was no relationship between the early members of EWB-LU and the members of the town or local leaders and preliminary research was compiled with no explanation of what it was for. Because this relationship lacked, the people did not feel any ownership of the project and took no steps to help find the necessary technical solutions. Education was not a priority in the initial phases of the project and measures should have been taken to explain why EWB was there, why disinfected water was important, why water conservation would play a crucial role, and how the town could support the chapter. With this lack of trust and cultural understanding, the project failed to be relevant in the eyes of the town.

After a review of the technical aspects of the project, it became clear that pipe leaks in the previous distribution system were causing a large portion of the disinfected water to drain before reaching the intended destination. In conjunction with unplanned losses, the estimations of daily water consumption and town growth were not thoroughly researched so the tank could not supply enough water to keep up with the excessive demand. Because the tank size could not be adjusted and the people would not change their cultural water usage habits, the project would never supply an adequate amount of water for Pueblo Nuevo.

Ultimately, the project was abandoned because of a poor relationship with the *junta de agua* and unrealized expectations of the townspeople.

Lessons Learned in Pueblo Nuevo, Honduras

As a chapter, we have learned from our failures in Pueblo Nuevo. The chapter has been analyzing the relationships built with the town, as well as the technical implementation of the project in order to determine how to proceed with future endeavors.

Before beginning a new project it is vital that chapter members educate themselves on the town's culture and needs. Members must take the time to travel to the town, meet with the community and study their lifestyle patterns. Installing a new water system which will provide the community with a necessity that they have never had—purified water—will undoubtedly improve community lives and perhaps even the economy of the town. However, it cannot be expected that community members will change their habits so that they may benefit from the system. Once a thorough assessment of the town has been completed it is important for a chapter to agree on overall project goals that will take into consideration the unique social and cultural aspects of town life.

Beyond an understanding of the needs of those you are serving, it is imperative to develop a relationship with the community that is built upon trust. Moreover, a similar relationship should be maintained between the chapter and a Non-Governmental Organization (NGO) or in-county partner. Though it will take a lot of time and effort on the part of chapter members, it is essential in completing a successful project. Begin building such relationships by visiting each party and discussing with them your intentions. A contract or Memorandum of Understanding (MOU) will need to be drafted and signed by chapter members, community leaders and NGO partners. We have found that when creating such contracts between the chapter and the community, it is advantageous to require the community to contribute to the project in some way. This could be a monetary or labor-based contribution. Whatever their contribution, it is important that the community feel a sense of ownership for the project, so that when we have completed our work they will be invested in the maintenance and use of the system.

As the project progresses, continue to share all work completed on the project with the community and NGO to gain both feedback and approval. Showing the community that we value their input and have made project adjustments based upon their stated needs will give the community a reason to trust our engineering judgment. When traveling is not an option, maintain communication through phone calls or emails if at all possible.

Moving On: La Fragosa, Honduras

Since the halt of our work in Pueblo Nuevo, the chapter has focused on a new project in La Fragosa, Honduras. After receiving a letter from the town that expressed their need for a water distribution system, the chapter has responded by beginning the steps to a water sanitation and distribution design. We plan to use a multidisciplinary approach to design that incorporates social development, community education and technical implementation. Though the system is still in the early developmental phases, it is anticipated that the design will threefold: (1) A water storage tank with a hypo chlorinator or other means of water decontamination, (2) a piping and distribution system that will connect with a sanitation station at each home, and (3) educational programming that will promote the virtues of hygiene, the importance of clean water and the treatment of water borne illness. We have not yet completed the assessment phases of the project, but have been traveling to collect necessary data for the creation of finalized designs and hope to begin the implementation of the system in the spring of 2014. As we continue to work in the community, we hope to establish a sense of trust and understanding between our group and the people. By developing this relationship, we hope that the community will feel a true sense of ownership for the project upon completion. Because we have built such a strong relationship with the community, we believe that there is a higher potential for project success in La Fragosa.

Learning from our mistakes in Pueblo Nuevo, we solidified relationships with members of the La Fragosa's *junta de agua* (water board) and the Honduran NGO Agua y Desarrollo Comunitario (ADEC) before completing any work on a water system. With both parties we have clearly stated and agreed to contracts and good relationships which we maintain by communicating on a monthly basis. As a chapter we have determined that our goal for this project is to create an entirely new water distribution system that will send chlorinated water to a tap at each house. We have explained to the community that in order for them to be benefitted from the system water must only be used for potable purposes which they understand and agree to. We will continue to educate them on such water use.

Upon implementation, community members will be responsible for installing the system with the help of qualified personnel provided by ADEC. They will also be expected to maintain the system for which they will pay a small water tax. We trust that the community will hold true to these commitments because we have already witnessed their efforts to ensure that the system is completed. For the past year various community members have been farming land that belongs to the owner of the water source being used for the system as he requested this be done before he would relinquish ownership.

KEYWORDS

International development projects, project assessment, lessons learned, social and cultural understanding

A Self-Contained, PV-Powered Toilet and Domestic Waste Water Disinfection System

Yan Qu, Asghar Aryanfar, Clement Cid, Hao Zhang, Michael R. Hoffmann
California Institute of Technology, 1200 E. California Blvd., MC 131-24, Pasadena, CA 91125

A comprehensive design review, development, construction, and performance testing of a sustainable, traditional infrastructure-free approach to the handling and processing of human sanitary waste including human fecal solids control, urine processing, and liquid wastewater treatment. The treatment scheme incorporates sludge disinfection, treatment, and volume reduction. This solar toilet system has at its core unit process photovoltaic-powered (PV) electrochemical chemical reactors that generate H_2 as a potentially useful byproduct obtained during anoxic wastewater and fecal matter treatment including the complete disinfection of the total cultural bacteria initially present. The system has been designed to be free of an electrical grid or from subsurface urban infrastructure. Our prototype treatment scheme can be adapted for single family use as a sanitary toilet facility or upon scale-up, the design can be enlarged in size to handle the daily wastes of 500 people with a predicted break-even operating cost when powered by a PV array in which energy is stored for use throughout 24 hours of continuous operation. The fundamental concept has been tested at the bench-scale and also at the prototype scale with synthetic feces, with urine, with domestic wastewater, and with human feces. Based on both our bench-top laboratory experiments and on our larger-scale reactor systems (20 L and 40L process volumes), and in the prototype unit testing unit. Our general concept, specific design elements, and treatment approach has proven to be viable for the treatment of raw domestic wastewater, human urine, human feces, and synthetic human waste analogues. After several hours of PV-powered electrochemical treatment, the turbid, black water influent can be clarified with the elimination of the suspended particles along with the reduction or total elimination of the chemical oxygen demand (COD), total enteric coliform disinfection via in situ reactive chlorine species generation, and the elimination of measurable protein after 3 to 4 hours of PV-powered treatment.

In our PV-powered electrochemical concept, a semiconductor anode generates oxidizing and disinfection radical species (e.g., $OH^{\bullet}$, $Cl^{\bullet}$) (R1), which subsequently react with aqueous phase components and microorganism, while the cathode splits water into hydrogen (R2). Oxygen evolution via water oxidation (R3) is normally the complementary reaction to H_2 production (R2) and thus the generation of radical species (R1) at the anode results in non-stoichiometric water splitting (i.e., $H_2/O_2 > 2$).

$H_2O \rightarrow OH^{\bullet} + H+ + e^-$ ($E° = 2.74$ V vs. NHE) (R1)

$2\ H_2O + 2e^- \rightarrow H_2 + 2\ OH^-$ ($E° = 0$ V) (R2)

$2\ H_2O \rightarrow O_2 + 4H+ + 4e^-$ ($E° = 1.23$ V) (R3)

$C_xH_yO_z + (x + 0.25y - 0.5z)\ O_2 \rightarrow x\ CO_2 + 0.5\ yH_2O$ (R4)

The sodium chloride (NaCl) concentrations in domestic wastes are sufficient to serve as a supporting electrolyte in electrochemical water treatment.[4-7] In some cases, chlorinated substrates enhance anodic efficiencies due to the liberation of chlorine during the course of electrolysis1. Active chlorine species such as $Cl^{\bullet}$, $Cl_2^{\bullet-}$, and ClO^- are generated at the anode surface and act as direct/indirect disinfectant and oxidants for organic/inorganic reductants and microorganism.

The floor plan layout for the self-contained and PV-powered prototype toilet room design and its integrated treatment sequence is shown in Figure 1 along with a 3-dimensonal perspective view of the upgraded prototype test facility. To minimize the energy consumption for this system, gravity is the main driving power of wastewater flow in the design.

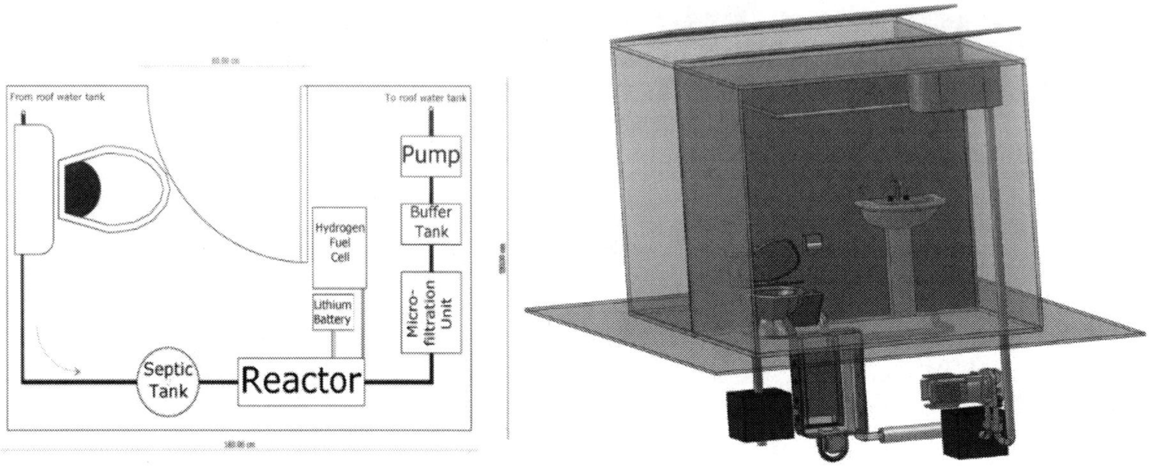

a) ***b)***

Figure 1. ***a)*** *The floor plan layout for the first generation prototype showing the various treatment units and devices linked in sequence.* ***b)*** *A three-dimensional view of the Phase 1 design will be expanded into a fully portable housing unit for Phase 2 test bed prototype integrated toilet facility where the treatment system is placed below ground level and the treated water is pumped back to the flush-water holding tank on the ceiling of 'advanced water closet (i.e., WC).'*

The core component of our solar-toilet and coupled waste treatment system is an electrochemical reactor that is powered by GPS-guided tracking solar panel coupled with battery storage of excess solar energy that is regulated. Photos of our 40-L reactor electrochemical reactor system which uses semiconductor electrodes (i.e., a sequential coating of semiconductor particles annealed on to conductive metal plates) are shown in Figure 2. The overall treatment scheme involves a sequence for unit processes including a septic waste holding tank design for the initial sedimentation of solids and partial anaerobic digestion, a hetero-junction composite semiconductor electrode array, a flow regulated reactor for continuous operation, the tracking solar pane as the sole source of energy, an alternative power source comprised of a thin flat panel multi-junction organic laminate photovoltaic power source, collector, a post-reactor microfiltration unit, a post-treatment holding tank, and a recycle system for water reuse for water flushing.

a) *b)* *c)* *d)* *e)*

Figure 2: ***a)*** *the core electrochemical reactor,* ***b)*** *a 10-anode electrode array,* ***c)*** *a GPS tracking photovoltaic panel with the back-up silicon triple-junction laminated solar collector, which can be integrated into the toilet facility housing,* ***d)*** *the core reactor filled with black wastewater,* ***e)*** *the appearance of the wastewater after 4 h of solar PV-powered electrochemical treatment.*

After assemble the main parts of this system according the design, Figure 3 below shows the layout of the first demonstration prototype. The toilet options include at typical western toilet, an Asian squat toilet and a waterless urinal. The demonstration treatment units are housed below the upper toilet platform. The waterless urinal uses a mechanical trap to prevent the back diffusion of ammonia and other gases such as H_2S and CH_4. In the future, it will be possible in an alternative configuration to separately treat the urine in order to optimize the recovery of useful by-products.

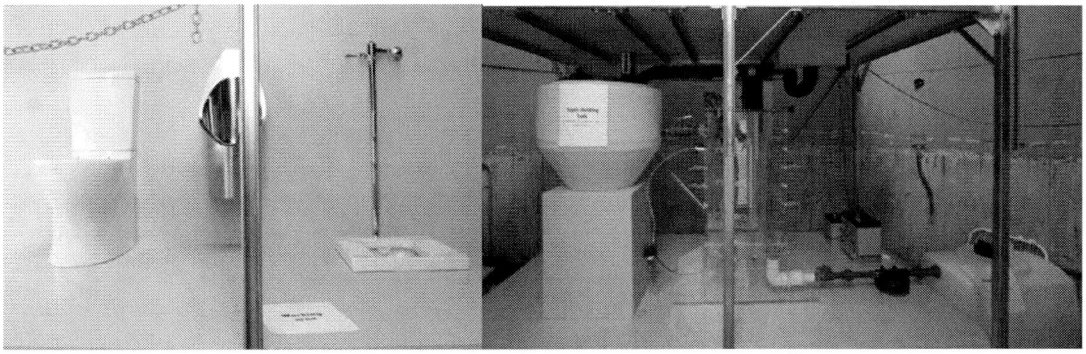

a) ***b)***

Figure 3: *a) The top portion of demonstration platform showcases the toilets and waterless urinal. b) The bottom view of the prototype system showcases the four unit processes and the battery pack and associated charge controller that stores and regulates the electrical output of the solar PV collectors.*

Domestic wastewater from the Los Angeles Country Sanitation Districts WWTP was used for disinfection performance test. Prototype scale experiments were did by 20 liters reactor with PV panel power source. The open cycle potential of solar panel was 30V. Since the output potential and current is changed by the electrochemical reaction and solution conductivity, different electrolyte concentrations in tap water were tested when we connect solar panel with reactor directly. The result was showed in table 1. As electrolyte concentration increase, the solar panel output potential and reactor potential decrease. However, the system current increased slightly.

C_{NaCl}/mM	Solar panel output voltage/V	Reactor voltage/V	Current/A
0	4.86	3.98	7.15
10	3.70	2.78	7.27
20	3.62	2.69	7.26
30	3.56	2.65	7.29
40	3.72	2.63	7.31
50	3.62	2.62	7.30

***Table 1:** Electricity characters of solar panel and reactor in prototype scale electrochemical treatment system (volume=20L, anode surface area=1280 cm^2)*

Domestic wastewater from WWTP was tested by the prototype scale electrochemical treatment system. When PV panel was used as direct power source, the output potential changed as the reaction. In this system, to regular the output potential of PV panel, a charge controller was used before the reactor. The charge controller keep the output potential stable at 12V. At the same time, to reach the ideal potential for electrochemical reaction, a 1Ω resistor was used to split 12V to 3V for electrochemical reaction. The reactive chlorine species generation mechanism of this system is shown in figure 4 along with the top-down view of hydrogen generation during the prototype working under sunlight.

An example of the time-dependent degree of treatment of actual domestic (human) wastewater can be seen in Figure 5. The test wastewater was obtained from a conventional wastewater treatment plant located in Los Angeles County. The total PV-powered reaction time is 4 hours. During this timeframe, the wastewater was completely oxidized due to the action of hydroxyl radical generated at the anode surfaces; the key material at the electrode-water interface is bismuth-doped titanium dioxide. Generation of reactive chlorine species such as hypochlorous acid, HOCl, via the oxidation of chloride by hydroxyl radical leads to the total disinfection of *E. coli* bacteria, total coliform bacteria, and fecal coliform bacteria as shown in Figure 5.

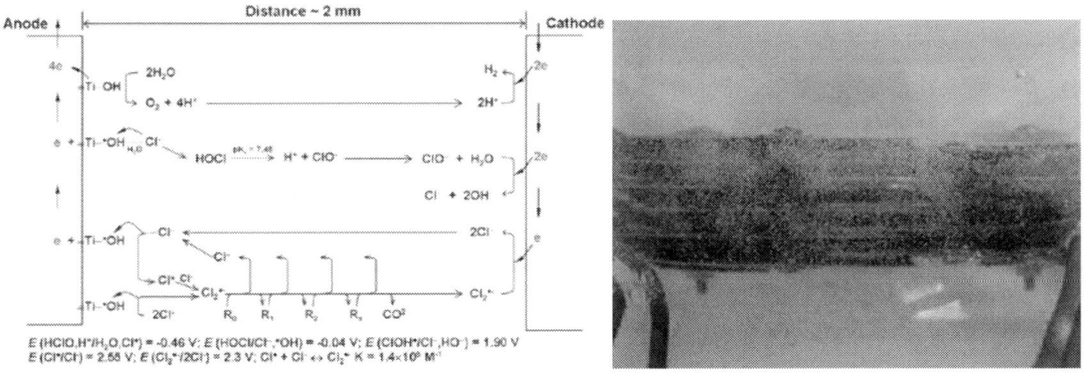

Figure 4: *Left) Reactive chlorine species are produced between the anode and cathodes due to interfacial chloride oxidation. Reactive chlorine serves to disinfect the water and wastewater during treatment. Chloride is recycled via reduction of the reactive chlorine species at the bare metal anodes. Right) A top-down view of the sandwiched electrodes (anode-cathode pairs) that shows the evolution of H_2 bubble during oxidation.*

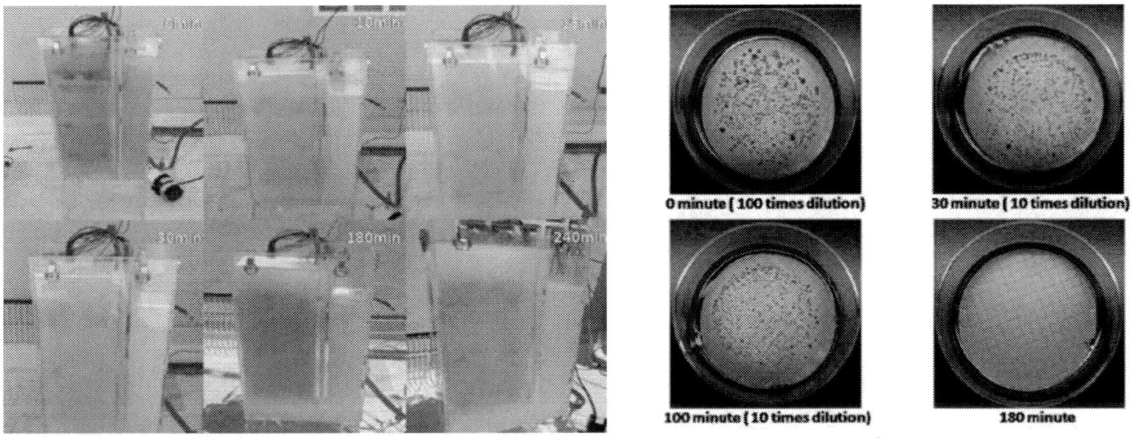

a) ***b)***

d) ***e)***

Figure 5: *a) A sequence of photos showing visible changes in color during solar wastewater treatment under ambient sunlight conditions over 240 minutes. b) A matched sequence of photos of the total fecal coliform colony forming units vs. time. After 180 minutes of treatment, no viable bacterial colonies can be cultured and thus total disinfection was achieved. A short video of the operating treatment system under full sunlight conditions can be viewed at http://www.youtube.com/watch?v=9Vuk6QVQklw&feature=plcp along with other videos* *illustrating the underlying electrochemical concepts show the generation of molecular hydrogen as a potentially useful by-product at* www.youtube.com/SOLARH2.

Based on our experimental results obtained at the bench-reactor scale in the laboratory and with the prototype reactor operating under ambient solar conditions, we are convinced that a properly engineered system using standard chemical engineering process technology will lead to the complete transformation of human waste into potentially useful by-products, while, at the same time, achieving the complete disinfection the wastewater. The fully treated water should be of sufficient quality to recycle back through the system for use as flushing water. Modest amounts of make-up water may be required due to small evaporative water losses during treatment. We have identified and quantified several useful by-products of treatment that include H_2 gas that can be compressed to serve either as a gaseous fuel for cooking or water heating, along with nitrate, urea, ammonia and phosphate for use as fertilizer. The residual organic solids are also completely disinfected for potential use as an organic soil amendment for agriculture.

FROM AMAZON TO THE ANDES: WASH PROJECTS IN RURAL PERU

J. Milliken, G. Stonebraker; George Mason University

ABSTRACT

Engineers for International Development (EfID) at George Mason University (GMU) have worked on two projects in Peru over two years, stretching from the Andes Mountains to the Amazon Rainforest Basin.

EfID's program started with the "Agua Para San Isidro" project in San Isidro, Peru in 2010. The San Isidro project required the demolition of an existing 15,000 liter ferrocement tank and installation of two 10,000 liter polyethylene tanks for water storage.

In August 2011, the program expanded to work with the Maijuna peoples near the Napo and Putumayo Rivers in the Amazon River Basin. Work with the Maijuna was established through a 4-year Memorandum of Understanding with the local community for future projects regarding water, sanitation, and hygiene, also known as WASH. The first phase of this relationship addressed drinking water quality concerns by paying for and teaching the community how to construct 25 slow sand filters, also known as a biosand filter. In addition, a hygiene promotion program was set up for education regarding sanitation and hygiene. The second phase of the relationship will address various sanitation solutions, further hygiene education in the community, and a possible sustainable biogas solution for energy or heating.

STARTING EFID

Since May 2010, students and faculty at George Mason University in Fairfax, VA have been teaming up with engineering professionals in the Northern Virginia area to explore, design, and build sustainable solutions for water, sanitation, and hygiene, also known as WASH, in developing countries. A group of 9 dedicated students first started Engineers for International Development (EfID) at GMU to follow the Engineers Without Borders format in order to educate themselves on how to help those in need using their engineering talents. Over the organization's two and a half year lifetime, EfID has involved over 60 students on 4 projects and 8 trips overseas to aid 7 communities totaling approximately 700 people.

Students involved in EfID have seen 2 themes of contrasts in how EfID has shaped their development as engineers and humanitarians. The first theme reflected in this paper is the balance between education and production. This theme is largely present because of the organization's dual goal to educate students and assist communities with their WASH needs. The second theme is the balance between technical skills and business skills. While the WASH projects implemented by EfID reinforce technical engineering knowledge acquired through students' coursework, most WASH projects are relatively simple engineering problems. Just like the real world, half of the work for the projects boils down to fundraising and the politics of pleasing both the client and investors.

Water for San Isidro project, as EfID's first project, is a good example of balancing business skills with technical skills. EfID chose the project in coordination with the University of Maryland's Engineers Without Borders chapter as a simple project to establish organizational processes such as funding sources, university relations, and meeting frequency.

It was decided to take a mixed bag approach to funding projects. As an official student organization at a state university, EfID was able to become part of the George Mason Foundation 501(c)3 for tax-deductible donations. Principal financial backers from the beginning included the university Provost Office and the engineering department's Civil Engineering Institute, which constitutes a number of presidents and vice presidents of local engineering companies that financially back the civil engineering department. These two principal investors therefore represented an occasional conflict in interest. The provost, as an overseer of the university at large, preferred EfID as a multi-department organization that would benefit all aspects of the university. The Civil Engineering Institute, however, is mainly concerned with the development of local engineers, so preferred EfID to focus on recruiting and developing engineering minds specifically. Thus, funding requests were handled with care and tact.

In addition, students were expected to raise money for the projects, both on their own and through organization-sponsored fundraising events. EfID has made it a priority to keep multiple fundraisers per year as fun activities for all. Keeping an element of fun is an essential part of the EfID's method of team building. In addition, the fun activities make for an excellent outreach to the rest of the student body at George Mason, promoting awareness and involvement for the chapter, killing three birds with one stone.

AGUA PARA SAN ISIDRO

San Isidro, Ayllu Tumbibamba sector, is next to Centro Compone, province and district of Anta (3,435 meters above sea level) counts with a population of approximately 100 people and it is about 36 kilometers from the city of Cusco 30 minutes away in car passing Izcuchaca town on 'Carretera Cusco Abancay.'

The population in San Isidro was found to lack water storage for their consumption and farming needs. One ferrocement tank was built in August 2009, but wasn't enough to sustain all community's needs. One side of the main highway was supplied with enough water while the other half had difficulty receiving decent flow. Additionally, some overflow was being misused from the CT tank due to the lack of storage.

EfID considered 2 alternatives. The first was to repair the damaged ferrocement tank walls, which would require complete reconstruction. Only one wire mesh was applied within the construction of the ferrocement walls, causing strength failure, rupture at some points and rusting of the rebar. Therefore EfID decided to choose the second alternative of purchasing a Polyethylene tank from a local supplier which facilitated the delivery to the site and provided a longer life span of the product (30 years per tank). Additionally, each tank was certified by ISO 14001 and approved by FDA (Food and Drug Administration) and 5 years warranty.

Community is Essential to Success

The huge effort of the San Isidro community in the implementation of the water initiative was formed out of the basic relationships built between the GMU members and the community leaders and people. Upon these relationships the project was overwhelmingly successful both in the involvement of the community and the completion of the goals set by the trip.

The EfID team met with the leaders of San Isidro, first thing in the morning the day after arrival at San Isidro. The team was taken on a short tour of the existing water system for the community. The initial plan was to demolish the entire existing ferrocement tank. However, at the request of the community leaders, the plan was changed to demolish only the roof and interior baffles of the existing tank. The village leaders requested that the walls of the ferrocement tank be left erect to provide protection for the new tank, which would fit within the walls of the existing tank.

GMU members reached out to the community in a variety of ways. Most importantly, the EfID team listened to the concerns of the community and made adjustments to the plans based on suggestions by the locals. In other words, GMU realized it is imperative that the concerns of the people of San Isidro are heard and taken into account with the implementation of the project. This not only insures a sustainable solution but gains community support because they are a part of the project. Furthermore, it was decided early on that the project would be completed with only locally-available materials to ensure that maintenance, upkeep, and future replacement could all be implemented by the community.

The team made special time to visit community members in their homes and eat meals under their roofs. This built a strong relationship between the EfID team and the Peruvians. Another way the relationship was built was simply with the foresight of bringing a soccer ball. (It is suggested to bring more soccer balls to give to the community in the future). With only a soccer ball, GMU members played soccer each day meeting new community members and building stronger relationships.

In response to reaching out to the community, the people of San Isidro gave overwhelming support during the duration of the project. Over thirty community members joined the GMU team for two consecutive days in order to help expedite the project and prepare for the second tank. On the second day of support, the tank was placed into position and the community blessed the tank with a cultural blessing where each member of the team and leaders of the community drank the special Inca corn wine, chicha, which was then thrown onto the tank to christen it.

The list of community involvement goes on. A local plumber was instrumental in plumbing the new tank and shopping for materials. A local cab driver drove members the GMU members to pick up supplies and acted as a guide in finding store locations. Finally, breakfast and dinner were provided by a local cook who ate with members of the EfID team in her dining room.

The first project of EfID in the Andes Mountains of Peru helped a great community with their water shortage issues, and provided Mason students quality experience for moving forward to help other communities in the future.

MAIJUNA PROJECT

In the summer of 2011, Mike Gilmore, an ethnobiologist at the George Mason University New Century College, read a news article published on EfID's work, and contacted the organization with interest in working on a WASH project in the Amazon of Peru. Mike had worked with a people called the Maijuna for 13 years, and they were reaching out through him to find an organization capable and interested in assisting with various WASH needs. Due to Mike Gilmore's involvement, EfID established a four year Memorandum of Understanding with the community during the assessment trip that was conducted in January, 2012.

Community Overview

The Maijuna, also known as the Orejón, are an endangered and marginalized indigenous group found in the northeastern Peruvian Amazon. Today, there are only 400 Maijuna individuals left living in four villages in a large area between the Napo and Putumayo Rivers, tributaries of the Amazon River. The intact nature of Maijuna ancestral lands and the biological diversity present within them is a testament to the past and present environmental stewardship of the Maijuna and the sustainability of their traditional resource use and management strategies. Unfortunately, because Maijuna ancestral lands are rich in resources, they are at times under siege by illegal incursions from poachers and resource extractors from outside their communities. The Peruvian government is considering building a road directly through the heart of Maijuna ancestral territory that would not only devastate the livelihood of the Maijuna, but destroy countless acres of the Amazon rainforest surrounding the road. The Maijuna are adamantly against this and are asking the Peruvian government to create a protected area that would formally protect over 336,000 hectares of their ancestral lands and the critically important biological and cultural diversity found there. The Maijuna, once nomadic in nature, are now a settler society that has congregated into four groups of about 100 people with roaming livestock per community. Water, sanitation and hygiene (WASH) is a big problem for the Maijuna as they consolidate into the four villages.

Community Meetings

Community meetings were held in each community visited. These meetings were imperative to assess the needs and desires of each community for planning future WASH improvements. The agenda of the meeting generally consisted the following:

- Introductions of community leaders
- Introductions of EfID team
- Overview of Maijuna historical requests for water and sanitation projects
- Participatory Mapping of community, where households get water, bathe and relieve themselves
- Requests for type of project the community would like to see
- Discussion of potential alternatives for future interventions
- Discussion of activities immediately available to community

After the community meetings, the EfID team walked each community and conducted sanitary surveys of most of the houses in the communities. This data was necessary for future decision making about what kind of WASH solutions EfID could use for the Maijuna and which would be most effective.

Common themes from all meetings

The Maijuna who chose to speak during the community meeting were all very excited that GMU-EfID was here to help them find solutions for clean water and sanitation issues. They said that these water and sanitation projects are absolutely critical and important for the health of the community and the children. One of the more noticeable issues is the animal and human waste that covers each community. In all of the communities, each household reported that they would go to the bathroom in the rainforest during the day and in the river at night. One of the reported reasons for using the river was the ability to use water for anal washing instead of leaves from the forest. Another reason for using the river was the danger of being in the forest at night due to the difficulty in seeing predators. They also reported bathing and retrieving drinking water from the river. Of the individuals who collected rainwater, none used the rainwater for drinking water. It was often used for cleaning dishes.

After information was gathered and the assessment team collaborated, several alternatives for clean drinking water were presented to and discussed with the community, including Solar disinfection, rainwater collection, chemical disinfection, three-pot filtration, ceramic water filters and bio-slowsand filters (biosand). The Maijuna and team agreed that biosand filtration would meet the Maijuna's needs in a sustainable manner.

A Simple Yet Effective Solution

A biosand filter is a very low-tech device that can be made with industrial and natural materials that are readily available to the Maijuna. A 30-gallon food grade plastic drum is used as housing for the filter medium. Pipes and fittings are placed inside the housing and medium and exit near the top of the plastic drum below the upper lip where the spout becomes visible. The first step in the assembly of the filter is attaching the pipe fittings together and placing them into the drum. Second, 10-20 centimeter layer of gravel is placed inside the container covering the bottom pieces of the pipes.Third, a 50-100 centimeter layer of sand is placed on top of the gravel layer that stabilizes the remainder of the pipes. Once the assembly is completed, water is placed inside of the barrel until the water level reaches just below the tap fitting of the pipe system. Pumps or machines are not needed to pump the water out of the filter because the pressure from the water head is used to push the water through the filter and out of the spout.

For three days the water is left in the barrel undisturbed in order for a thin biological layer to form underneath the first millimeter of sand called schmutzdecke. This layer is responsible for clearing the water mostly of the bacteria that causes diarrheal diseases. Maintaining the biosand filter is very easy. Once the container, gravel, pipes and fittings are purchased, the Maijuna have access to a natural sandbar located in a local river. This sand has been thoroughly tested and was found to be optimal for the biosand filter. Roughly once every six months, the schmutzdecke layer dies and a new layer will need to be grown. The top layer of sand, approximately one centimeter, needs to be scraped off and removed and the schmutzdecke layer is ready to be grown using the same technique previously stated. Once the sand layer has been removed to about 30 centimeter, all of the remaining sand needs to be removed and replaced with new sand for the local sandbar.

Excellent Results

In June, 2012 the implementation team traveled back to the Maijuna lands to the Sucusari community on the Sucusari River. It was decided that a test trial for one community would be conducted with the biosand filters in order to see if these were indeed the right solution for the Maijuna.

The Implementation Team was split into a Biosand Team and Hygiene Team to emphasize to the Maijuna the complementary importance of clean drinking water and hygiene practices. The Hygiene Team focused on teaching the Maijuna about sanitation and hygienic practices that would be useful for germ prevention. It was found that many of the Maijuna did not know what germs were and were told for the first time why washing their hands after using the bathroom or touching raw meat and other germ laden materials was crucial to their well-being. The team demonstrated how germs spread by putting lotion on all of the participants' hands. The individual teaching the Maijuna about germ prevention would then add glitter to their hands and proceed to shake the hand of one of the Maijuna. Each person after would shake at least on person's hand, spreading the glitter to each individual. After the germs were spread, each person involved in the demonstration washed their hands with soap and water, removing most or all of the glitter from their hands. This exercise really helped the Maijuna realize that hand washing was critical for germ and sickness prevention.

The Biosand Team focused on teaching the Maijuna how to build, use and maintain the Biosand filters. Unfortunately, just before the team travelled to the Maijuna lands, a one hundred year flood had hit the Amazon River and its tributaries. Due to the flooding, the Maijuna were uncomfortable retrieving the sand from the sandbar due to safety reasons. As an alternative, the Implementation Team purchasing sand from the nearby city of Iquitos. While constructing the first test biosand filter with the purchased sand, it was clear that the sand was not suitable for a biosand filter. Using a turbidimeter, the Sucusari river water - untreated - registered 35 Nephlometric Turbidity Units (NTU). The effluent out of the biosand filter during the first flow through read 800+ NTU. It was only after 6 gallons that the turbidimeter reached a level below 800 NTU for the biosand filter with the purchased sand. After 30 gallons of water, the purchased sand biosand filter was leveling out at around 50 NTU - a level 15 NTU higher than the untreated river water. This sand was clearly not suitable for a biosand filter. The purchased sand had high level of clay present in the sand that caused the water to flow more slowly and create a milky white effluent.

After explaining this to the Maijuna, the President and leader of the Sucusari community decided to retrieve sand. They obtained enough sand for the Team to build one biosand filter. After the biosand filter was made, turbidity tests were conducted. The very first water to pass through new biosand filter had a turbidity reading of 526.6 NTU, significantly better than the purchased sand biosand filter first effluent. After 30 gallons of water passed through the biosand filter made with river sand, the turbidity readings started to level out around 5 to 6 NTU with readings as low as 3.04 NTU. Bottled water that was purchased in Iquitos where the sand was originally purchased was also tested with the turbidimeter. The bottled water read 2.0 NTU. Unfortunately, no biological tests were conducted on the biosand filter due to time constraints, but after several days of preparing the biosand filter, schmutzdecke had begun to form.

One of the biggest challenges in all of this is balancing the engineering education of a first world country with real world production in a third world community. Biosand filtration, however effective at removing pathogens, is obviously not suitable to be a completely stand alone method of water treatment to United States standards, and should not be taught as anything more than just one part of the water treatment chain. Ideally, implementation of biosands should also be coupled with disinfection methods to kill all remaining pathogens. However, majority removal is still better than no removal. The reality is that children in the Maijuna communities are dying from untreated water every year.

What then is an American professor to do when faced with the task of teaching students to implement the optimal solution, while at the same time providing cleaner water to a community in need? This is a challenge that EfID now faces after the beginning of implementation of community wide biosand filters. American professional mentors are of course inclined to seek to implement a total solution including all aspects of safe source water, filtration, and disinfection. But reality is the optimal solution is not possible without time consuming cultural change. Future efforts of EfID will therefore continue to seek the optimal solution by introducing an element of disinfection or improved source water in combination with the biosand filters as a second phase of the Maijuna project. GMU-EfID is thrilled to continue their work with the Maijuna and improve their lives through simple yet effective WASH practices.

First Flush Disinfection: Challenges of chlorinating wet weather primary effluent to meet a stringent *E. coli* target

Matthew Scarborough[1*], Michael Arends[2], Samuel Jeyanayagam[1], Larry Schimmoller[1], Jason Assouline[1]

[1]CH2M HILL, Englewood, Colorado
[2]City of Omaha, Nebraska
* E-mail: Matthew.Scarborough@ch2m.com

ABSTRACT
A study to assist with the design of a chlorination facility for the Missouri River Wastewater Treatment Plant in Omaha, Nebraska is discussed. The facility will disinfect wet weather primary effluent flows to meet an effluent *E. coli* of 126 MPN/ 100mL as a geometric mean for an individual storm event. Samples were collected for seven events. Testing for samples collected during the first two wet weather events determined the presence, duration, and magnitude of a first flush. Subsequent wet weather samples were collected for three wet weather events. Paired primary influent and effluent samples were collected at the start of the event and at a time corresponding to the first flush. Samples were chlorinated at doses of 5, 15, 30, and 60 mg/L Cl_2 and tested at various batch reaction times. Testing indicated that a free chlorine residual was needed to achieve a reduction of *E. coli* to 126 MPN/100mL. Further testing was conducted using dry weather samples. Raw, bench settled, and bench chemically enhanced settled samples were disinfected with chlorine to ammonia ratios of 3, 5, 12, and 16:1 and various batch reaction times. Results of the dry weather sample testing confirm that a free chlorine residual is needed to meet the required disinfection; the chlorine dose required is directly related to the ammonia concentration of the wastewater. During a storm event, the ammonia concentration was determined to vary from approximately 15 mg-N/L (during the first three hours of an event) to 2 mg-N/L (once the first flush dissipates). The chlorination facility will be designed with a high turndown capacity to meet the high range of chlorine doses required to achieve a free chlorine residual. Options for storing the initial high-ammonia wastewater for later dilution and/or sending as much of the high-ammonia wastewater through secondary treatment will also be considered during final design of the chlorination facility.

KEYWORDS: Wet weather, disinfection, chlorine, monochloramine, ammonia, breakpoint

INTRODUCTION

The City of Omaha, Nebraska (City) has a consent decree to reduce the volume and frequency of combined sewage overflows (CSOs). The City developed a 15-year plan addressing water quality and quantity impacts to receiving streams. One major element is increasing overall plant treatment capacity for wet weather flows. With the upgrades, the Missouri River Wastewater Treatment Plant (MRWWTP) will receive 568 ML/d (150 mgd) during wet weather events, 185 ML/d (49 mgd) of which will undergo full secondary treatment. The remainder of the flow, 383 ML/d (101-mgd), will undergo preliminary and primary treatment followed by sodium hypochlorite disinfection to meet a discharge limit of 126 MPN/ 100mL of *E. coli*.

As chlorine is added to wastewater, it initially reacts with reduced inorganic substances, such as sulfide (S^{2-}), hydrogen sulfide (HS^{-}), sulfite (SO_3^{2-}), nitrite (NO_2^{-}), iron (II) (Fe^{2+}), and manganese (II) (Mn^{2+}). The byproducts of these reactions have no germicidal effect. These reactions generally occur quickly, often within the first minute after chlorine is added. As the reaction progresses, chlorine reacts with organic material and organic-nitrogen. After the initial chlorine demand is satisfied, a chloramine residual is formed due to the reaction between chlorine and ammonium. At a theoretical molar ratio of chlorine–to-NH_3.N of up to 1.0 (mass ratio of 5.1:1), monochloramine is predominantly formed causing TRC to increase. At molar ratios between 1.0 and 1.5, monochloramine is converted to dichloramine, but the relative amounts of the two species will depend on pH, temperature, and chlorine–to-NH_3-N ratio. For example, below pH 7, as the chlorine–to-NH_3-N molar ratio of 1.0 is exceeded, dichloramine will form more significantly. Trichloramine can also form, but typically only at very low pH (<6.5). As the theoretical molar ratio reaches 1.5 (mass ratio of 7.6), the destruction of chloramines is essentially complete (breakpoint) and all free ammonia is converted to nitrogen gas; any additional chlorine added will result in an increase in free chlorine. Free chlorine is the most powerful disinfectant of the chlorine species followed by monochloramine.

Many facilities throughout North America chlorinate wet weather flows as a method to reduce pathogens in CSO flows prior to discharging to receiving waters (WERF, 2005). The United States Environmental Protection Agency (EPA) CSO Chlorination Fact Sheet presents three different facilities in Philadelphia, Pennsylvania; Goose Point Woods, Michigan; and Lake Onodaga, New York. Chlorine doses at these facilities ranged from 0 to 25 mg/L and contact times ranged from 1 to 6 minutes. At the higher end of presented doses, fecal coliform log removals of 4 to 6 logs were achieved (EPA, 1999). The EPA conducted a study at a facility in Syracuse, New York, on both solids removal and disinfection alternatives, and dosed chlorine resulting in residual chlorine concentrations of 7.3 to 24 mg/L. This study found that, keeping the chlorine dose contant, the chlorine residual increased, as well as the log removal of fecal coliform, as the storm progressed (EPA, 1979). No studies were found investigating chlorination following conventional primary treatment during wet weather events.

The primary objective of this study is to assist with the design of the CSO 102 Chlorination/ Dechlorination facility, which is part of the Missouri River Wastewater Treatment Plant (MRWWTP) CSO Improvements Project (Project). Secondary objectives of this study are to (1) investigate the feasibility of disinfecting in the existing primary clarifiers at the MRWWTP in addition to or in place of constructing the CSO 102 Chlorination/ Dechlorination facility; and (2) assist in developing design assumptions for a future retention treatment basin (RTB), which will treat flows pumped out of a future deep CSO tunnel.

SAMPLING PROCEDURES

To understand the sampling procedure, a brief discussion on the intent of the Project and MRWWTP hydraulics is needed. A project is being constructed that will separate the South Omaha Industrial Area (SOIA) flow from the municipal flow. The SOIA flow contains a large portion of meat packing wastewater and contains very high concentrations of pathogens, biochemical oxygen demand (BOD), ammonia (NH_3-N), and total suspended solids (TSS). The MRWWTP CSO Improvements Project includes constructing a SOIA headworks and clarifiers to provide preliminary and primary treatment to the SOIA flow. A flow control structure will be constructed to ensure that SOIA flow is always sent to the secondary treatment facilities. An independent municipal headworks will also be constructed and the existing primary clarifiers will be used to treat the municipal flow. The MRWWTP will be capable of pumping 242 ML/d (64 mgd) to secondary treatment through the Transfer Lift Station. This will include up to 64 ML/d (17 mgd) of SOIA flow with the balance being municipal flow. Any additional flow brought into the plant during wet weather conditions will be chlorinated, dechlorinated, and discharged out CSO 102 to the Missouri River. Flows will be chlorinated to meet an effluent concentration of 126 MPN per 100 mL *E. coli* geometric mean for each storm event.

The SOIA flow, which will be separated from the municipal wastewater, currently enters the MRWWTP through the Monroe Street Lift Station, along with other municipal flows, and is pumped to the South Inlet headworks. The North Inlet headworks does not receive SOIA flow. Eight primary clarifiers are in operation at the MRWWTP; however, only three of the four southernmost clarifiers (South Bank) are operated during dry weather. Under wet weather conditions (when samples were collected), all four of the South Bank primary clarifiers were in service. Wastewater is conveyed from both the North Inlet and South Inlet to the Primary Splitter Structure, where it is possible for mixing of the flow streams to occur. The Primary Splitter Structure operation is complex and the hydraulics during a storm event have not been modeled to confirm flow patterns. Therefore, it was assumed that if the flow was higher from the North Inlet, as would be the case during wet weather, the hydraulics would be such that the majority of flow into the northernmost South Bank primary clarifiers would be only from the North Inlet. Therefore, primary effluent samples were collected at the launder of one of the northernmost South Bank primary clarifiers (Primary Clarifier No. 5) as a representative sample of the flow (without SOIA flow) that would be disinfected in the future by the CSO 102 Chlorination/Dechlorination facility. Samples were collected downstream of the screens at the North Inlet to provide a representative sample of flows that would be disinfected in the primary clarifiers and that will be treated in the RTB.

Following the analysis of the initial results, it was determined that the industrial flow was impacting the Primary Clarifier effluent samples. As such, it was determined that for future events, all samples would be collected at the North Inlet, which does not receive SOIA flow. A portion of the North Inlet sample would then undergo bench settling to provide a representative primary clarifier effluent sample.

All samples were collected by MRWWTP staff. Following the sampling events, for both the first flush analysis and the disinfection study, samples were gathered by CH2M HILL, transferred to the appropriate sample containers provided by the CH2M HILL Applied Sciences Lab (ASL), packed in coolers with ice, and shipped to ASL via FedEx overnight to ensure that samples were tested within the required EPA holding times. Upon receipt at the laboratory, the temperature of the samples was checked to ensure that they were kept cool in transit.

Samples were collected on six occasions. The first two samples were collected during wet weather events to conduct first flush testing. The third, fourth, and fifth samples were collected during wet weather events and underwent chlorination testing. The sixth sample was collected during dry weather and underwent various bench settling and chlorination testing. Three different sampling and analytical procedures were used throughout the study and are discussed as follows.

TABLE 1
Sample Collection Summary

Event	Date	Sample Location and Number of Samples Collected	Testing Conducted and Purpose
1	April 21, 2010	12 samples collected at 20-minute intervals at the North Inlet at the MRWWTP	First flush analysis to determine the magnitude of the higher loadings experienced during a first flush of a storm event and the duration of the first flush
2	May 10, 2010	24 samples collected at 10-minute intervals at the North Inlet at the MRWWTP	
3	June 8, 2010	Two samples at the North Inlet (t=0 and 180 minutes) and two samples at the Primary Clarifier Effluent (t=90 and 270 minutes)	Evaluate the disinfection efficacy of chlorine during a storm event. Testing focused on standard chlorine doses and contact times. Subsequently, monochloramine was primary disinfectant.
4	Nov. 12, 2010		
5	Nov. 29, 2010		
6	April 12, 2011 (Dry Weather)	One sample at the North Inlet. This sample was divided into three separate samples — raw water, settled water to simulate primary clarification, and chemically treated water to simulate a chemically enhanced primary treatment (CEPT) process.	Evaluate the disinfection efficacy of chlorine during dry weather. Testing focused on monochloramine disinfection and free chlorine disinfection. Chlorine doses were varied based on ammonia-nitrogen (NH_3-N) concentration of the sample to be disinfected.

First Flush Analysis (Events 1 and 2)

To determine the timing of sample collection for chlorination testing, a first flush analysis was conducted during Events 1 and 2. In theory, the high flows seen during wet weather events will re-suspend solids that have settled to the bottom of the gravity sewers and force mains in the collection system and provide sufficient scour velocity to re-suspend solids that are attached to the interior surfaces of the collection system pipes. The first flush analysis was conducted to determine if a first flush existed, the magnitude of the higher loadings, and the timing and duration of the first flush peak.

The North Inlet at the MRWWTP receives flow from the South Interceptor Force Main (SIFM). Several large combined sewers contribute flow to lift stations connected to the SIFM. Diversion structures connected to these sewers allow flow to be bypassed during a storm event to ensure that the flow to the plant does not exceed the plant headworks screening capacity. Lift stations are bypassed based on their relative pollutant concentrations in order to minimize impacts on receiving waters. Based on storm intensity and the rate of increase at the treatment plant, lift stations are sequentially taken out of service and flow is sent to the Missouri River.

First flush samples were collected on April 21, 2010 (Event 1), and May 10, 2010 (Event 2). For the April 21, 2010 event, 12 samples were collected at 20-minute intervals starting at the time that a lift station was bypassed. As shown in Figure 1, the concentration peaks for TSS, chemical oxygen demand (COD), and NH_3-N occurred approximately 3 hours and 20 minutes after the CSO event began. It was calculated that the approximate time for flow to get from the north end of the SIFM to the MRWWTP during this storm, based on the pipe volume of 15.1 ML (4.0 MG) and average flow of 106 ML (28 MGD) during the event, was also 3 hours and 20 minutes. Therefore, a high portion of the first flush likely comes from the sewers feeding the northern end of the SIFM, where the lift stations are first taken out of service. As shown in Figure 2, the hydraulic peak occurred at the start of the event, at a maximum flow that triggers the City to begin taking lift stations out of service. The flow then decreases as lift stations are subsequently taken out of service. The flow increases and stabilizes as the wet weather event progresses.

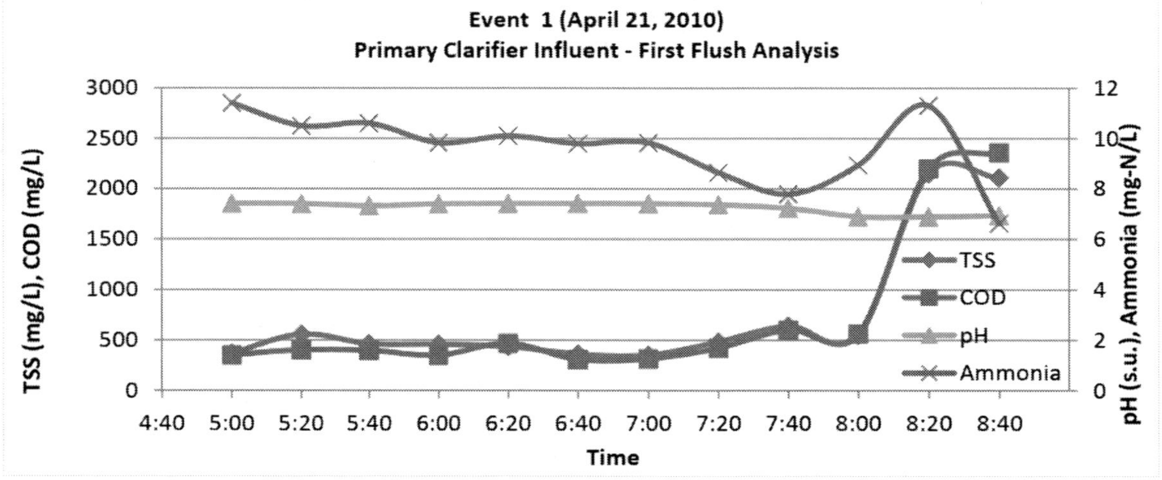

Figure 1: Event 1 First Flush Analysis – Constituent Concentrations

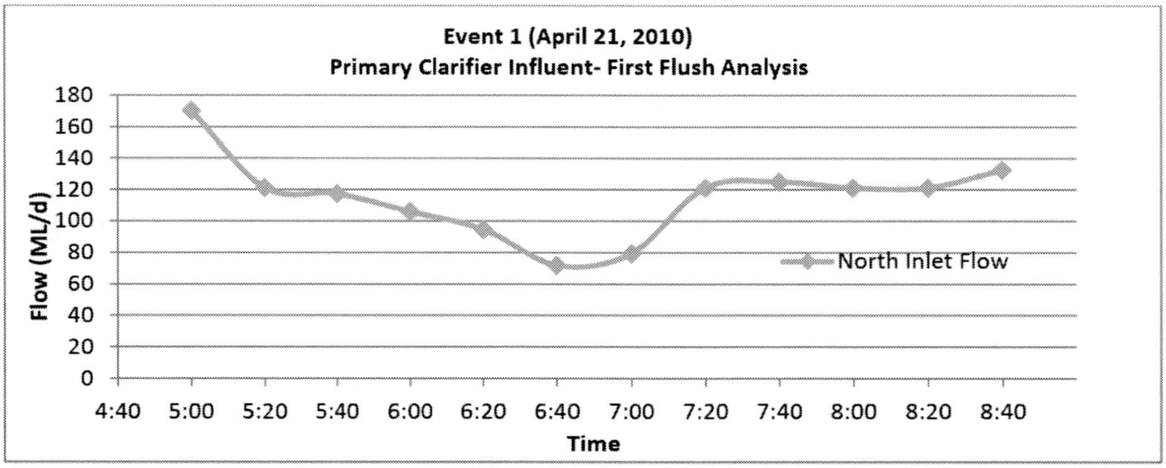

Figure 2: Event 1 First Flush Analysis – North Inlet Flow

For the May 10, 2010 event, the sampling procedure was modified. To capture the entire concentration peak, samples were not collected until 90 minutes after a lift station was taken out of service. For this event, the first lift station was taken out of service at 9:07 am, and the first sample was collected at 10:37 am. As shown in Figure 3, the concentration peak varied slightly by constituent. A TSS peak, NH_3 peak, and a COD peak occurred at the 11:47 am sample, or 2 hours and 40 minutes after the start of the event. The retention time in the SIFM during this event, based on an average flow rate of 129 ML/d (34 mgd), was 2 hours and 50 minutes, again close to the first flush peak. Figure 4 shows a relatively level influent flow, which would be expected since sampling started 90 minutes after the event started. The decrease and then spike shows a lift station being brought back into service. *E. coli* data was also obtained, as shown in Figure 5, to verify the assumption that the *E. coli* concentration profile matches that of TSS.

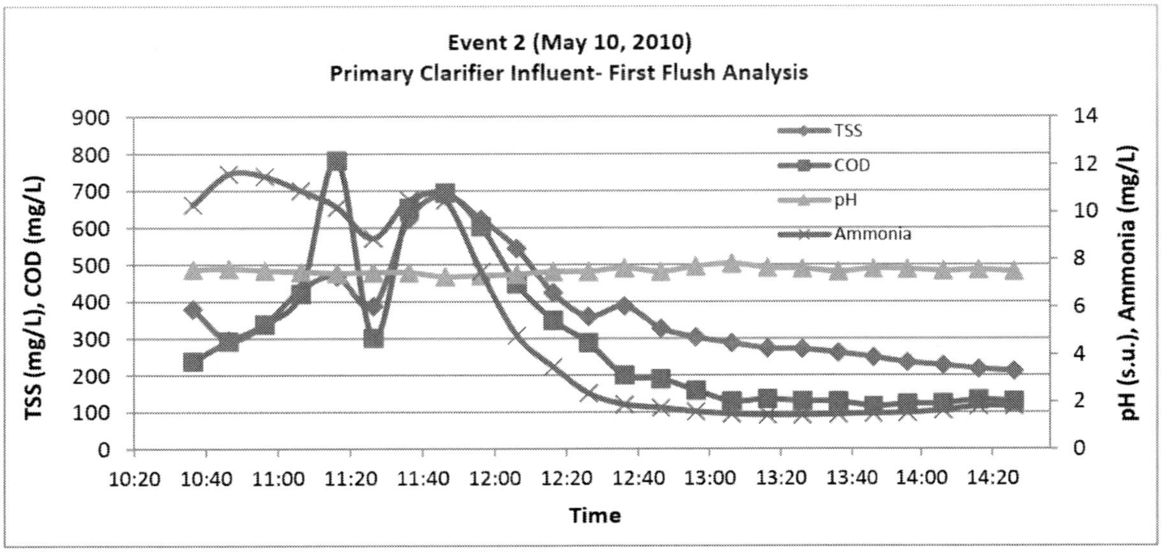

Figure 3: Event 2 First Flush Analysis – Constituent Concentrations

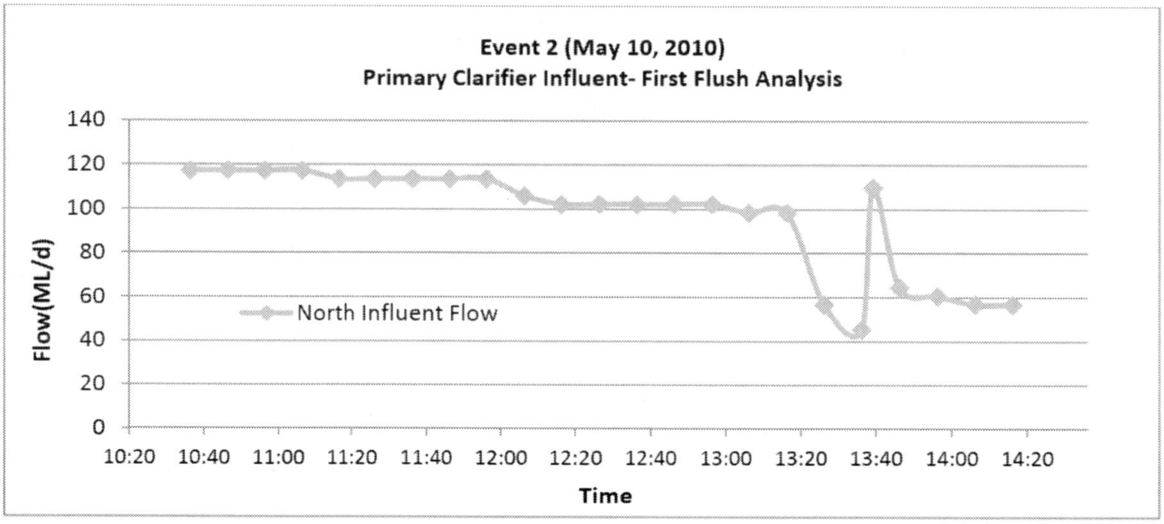

Figure 4: Event 2 First Flush Analysis – North Inlet Flow

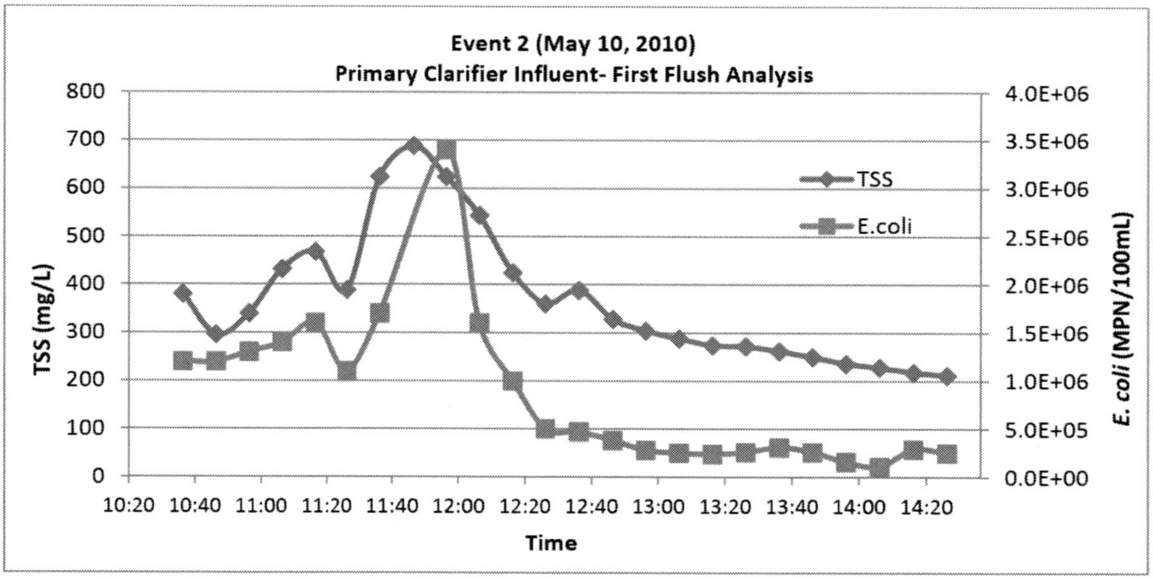

Figure 5: Event 2 First Flush Analysis – E. coli and TSS

The variance in loading magnitude seen between the two events should also be noted. For the first event, the maximum TSS concentration was about 2,200 mg/L. For the second event, the maximum TSS was only 700 mg/L. A similar amount of variance can be seen for COD between the two events. The NH_3 was relatively consistent for the two events. This is likely a result of the time between rain events. The longer the duration between rain events, the more solid material can accumulate on the interior surface of the sewers. The Burt Izard basin, specifically, has large sewers with velocities typically lower than 0.3 m/s (1 fps), meaning that solids in the basin are likely to settle to the bottom of the pipes. Further, the intensity of the event could impact the concentrations because the more-intense rainfall would be more likely to suspend and convey solids in the drainage area (from streets, yards, roofs, etc.) to the sewers. This confirms that COD is mostly solids associated, as it peaks similar to TSS. NH_3, being soluble, does not follow the same pattern or vary as much by storm event as TSS or COD.

Chlorine Disinfection Analysis – Original Sampling Procedure (Events 3, 4, and 5)

For the disinfection study samples taken during Events 3, 4, and 5, as soon as an upstream lift station began discharging CSO flow to the Missouri River (t=0), a 5-gallon container of sample was collected at the North Inlet to represent the hydraulic peak at the MRWWTP prior to wet weather. Ninety minutes later (t=90), a Primary Clarifier effluent sample was collected, which corresponds roughly to the hydraulic retention time of the primary clarifiers just prior to wet weather event. This sample represents the initial flow (prior to wet weather) that would be treated in the CSO 102 Chlorine Contact Basin (CCB). After another 90 minutes (t=180), another sample was collected at the North Inlet. This sample corresponds with the first flush raw influent loading seen in the first flush analysis described previously. Lastly, after another 90 minutes (t=270), a second Primary Clarifier effluent sample was collected to represent primary treated wet weather flow. Samples were collected by auto-samplers that required an operator to engage the sampling procedure once a CSO event occurred. The operators also had to switch out 5-gallon containers between the North Inlet and Primary Clarifier effluent samples. The bulk disinfection samples were not preserved with chemicals. Samples were labeled as shown in Table 3.

TABLE 3
Sample Types

Event		Sample Type			
		Raw Influent Prior to Wet Weather (t=0)	Primary Effluent Prior to Wet Weather (t=90)	Wet Weather (First Flush) Raw Influent (t=180)	Primary Effluent During Wet Weather (t=270)
3	June 8, 2010	3-1 INF	3-2 EFF	3-3 INF	3-4 EFF
4	Nov. 12, 2010	4-1 INF	4-2 EFF	4-3 INF	4-4 EFF
5	Nov. 29, 2010	5-1 INF	5-2 EFF	5-3 INF	5-4 EFF

Chlorine Disinfection Analysis – Modified Sampling Procedure (Event 6)

Following analysis of the initial chlorination results, as discussed later, it was determined to modify both the sampling and testing procedures. Rather than collecting Primary Clarifier effluent at the effluent launder of Primary Clarifier No. 5, as done for Events 3, 4, and 5, samples were collected at the North Inlet. This provided greater assurance that the industrial flows would not impact the collected samples.

It was further determined to collect samples during dry weather. It was assumed that dry weather concentrations were similar to the "worst-case" scenario flows to disinfect during wet weather events. Three 5-gallon cubitainers of sample were collected at the North Inlet sequentially over a short time period. The dry weather sample was collected on April 12, 2011.

TESTING PROCEDURES

Testing procedures for both the original chlorination testing procedure and the revised chlorination testing procedure used in the lab are provided as follows. A list of the standard methods used is provided in Table 4.

TABLE 4
Standard Testing Methods

Test Parameter	Methodology
pH	EPA 150.1, SM4500-H BElectrometric (pH probe)
Alkalinity	EPA 310.1, SM 2310 B Titration
TSS	EPA 160.2, SM2540 D
Volatile Suspended Solids (VSS)	EPA 160.4, SM2540 E
E. coli	SM 9223 B MPN
NH_3-N	EPA 350.1, SM 4500-NH_3 G Flow Injection Analysis
Total Organic Carbon (TOC)	EPA 415.1, SM 5310 B Combustion Method
BOD, 5-Day	SM 5210
COD	EPA 410.4, SM 5220 D
Chlorine Demand (CLD)	SM 2350 B, SM 4500-Cl G (Chlorine Measurement)
Sodium Bisulfite Quenching of Chlorine	SM 2350 B Modified

Original Chlorine Disinfection Testing Procedure (Events 3, 4, and 5)

Samples were received by the CH2M Hill Applied Sciences Laboratory and characterization tests were conducted on all four of the bulk samples for each event. Characterization tests included pH, oxidation reduction potential (ORP), COD, BOD, NH_3, TSS, VSS, alkalinity, and TOC. Samples from the first disinfection study event, collected on June 8, 2010 (Event 3), underwent chlorine breakpoint testing to assist with choosing doses for disinfection testing to be conducted in future events. The breakpoint curves are shown in Figures 6 through 9 of this TM. Based on the results of the chlorine breakpoint curves, chlorine doses of 5, 15, 30, and 60 mg/L were chosen to provide disinfection with a monochloramine residual. At this stage of the testing, the focus was on evaluating the disinfection efficacy using monochloramine rather than free chlorine because of the high chlorine doses necessary (>100 mg/L) to produce a free chlorine residual through breakpoint chlorination.

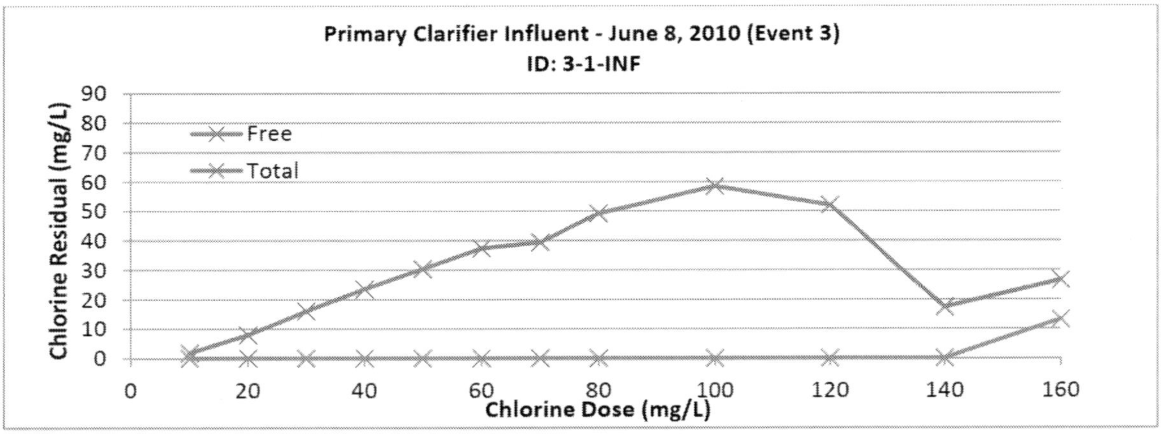

Figure 6: Chlorine Breakpoint Curve – 3-1-INF (NH3-N=10.4 mg/L, TSS = 338 mg/L, pH=7.41)

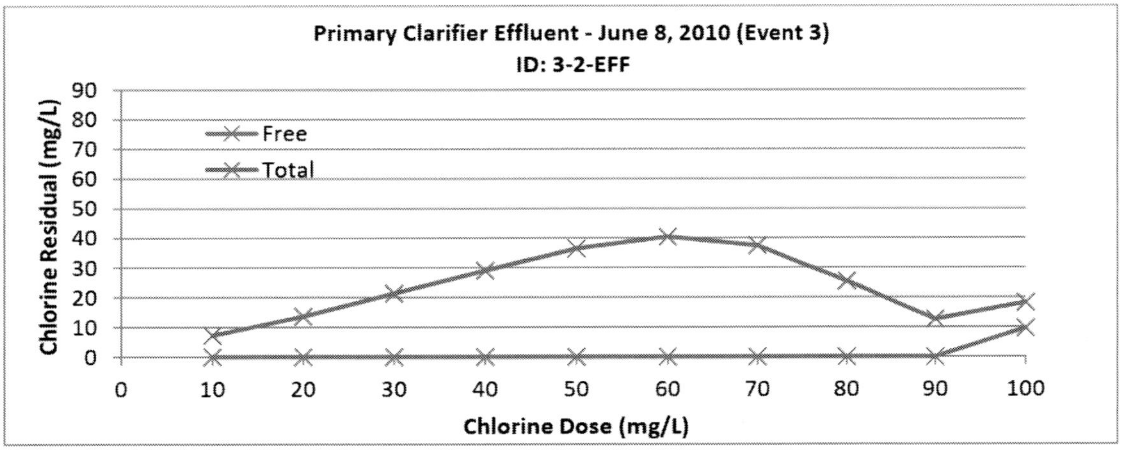

Figure 7: Chlorine Breakpoint Curve – 3-2-EFF (NH3-N = 7.52 mg/L, TSS = 124 mg/L, pH= 7.52)

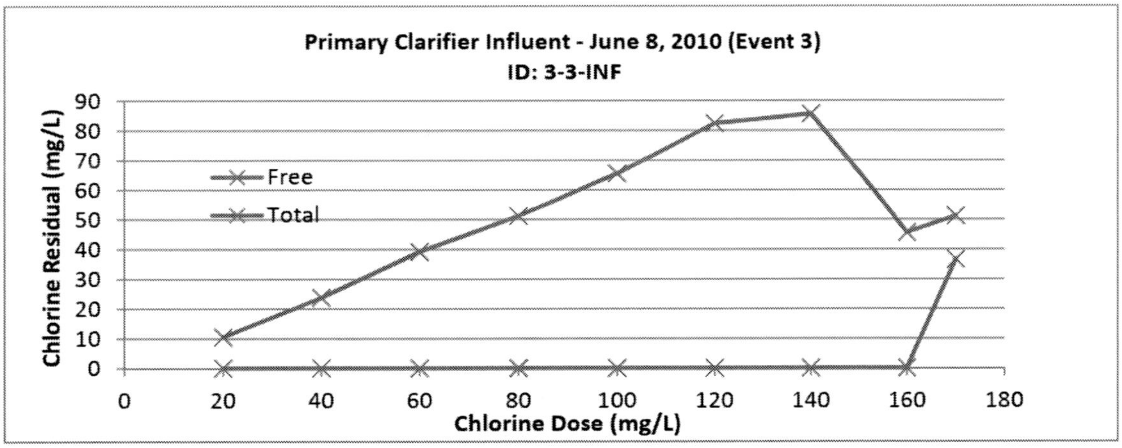

Figure 8: Chlorine Breakpoint Curve – 3-3-INF (NH3-N = 0.81 mg/L, TSS = 882 mg/L, pH = 7.75)

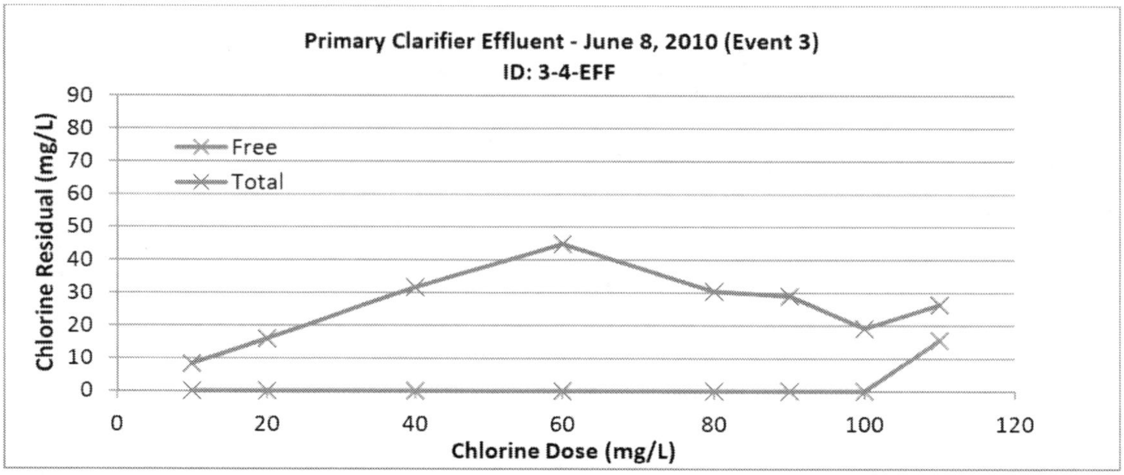

Figure 9: Chlorine Breakpoint Curve – 3-4-INF (NH3-N = 7.98 mg/L, TSS = 122 mg/L, pH = 7.22 mg/L)

Each of the doses was tested at various contact times. Two contact times were used for the Primary Clarifier influent samples—30 minutes (assumed for the RTB) and the average hydraulic retention time of the clarifier during the event based on flow data (assumed for disinfecting in the primary clarifiers). For the Primary Clarifier effluent samples, a batch reaction time of 15 minutes was used (based on the design assumption of 15 minutes for the CCB for Event 3. For Events 4 and 5, a batch reaction time of 30 minutes was also tested. The batch reaction times used for the original chlorine disinfection testing are shown in Table 5.

TABLE 5
Tested Batch Reaction Times for Original Testing Procedure – Events 3, 4, and 5

Sample Type	Batch Reaction Time	Notes
Primary Influent	30 minutes	Assumed contact time of future RTB
	Hydraulic Residence Time of Primary Clarifiers	Test disinfection in primary clarifiers
Primary Effluent	15 minutes	Test contact time for CSO 102 Chlorination facility, Events 3, 4 and 5
	30 minutes	Test impacts of increased contact time for CSO 102 Chlorination facility for Events 4 and 5 only.

Each of the samples was divided into four 2-L beakers. Samples were chlorinated by adding the predetermined volume of stock sodium hypochlorite concentration to each stirred sample to achieve the test dose. Immediately following sodium hypochlorite addition, the samples were poured into two amber bottles. At each test batch reaction time, the total residual chlorine (TRC) and ORP were measured. The samples were quenched with sodium thiosulfate to consume any remaining germicidal chlorine.

Modified Chlorine Disinfection Testing Procedure (Event 6)

Upon reviewing the results from chlorination testing for Events 3, 4, and 5, the testing procedure was modified. First, as discussed previously, all samples were collected at the North Inlet, rather than collecting samples at the Primary Clarifier effluent launder, to provide greater assurance that industrial flows were not impacting the samples. Of the three 5-gallon cubitainers collected at the North Inlet, one (labeled 6-INF-RAW) underwent chlorination testing directly. A second cubitainer was mixed in a 5-gallon bucket and settled for 40 minutes, corresponding to an average hydraulic retention time of the primary clarifiers during wet weather events, and the supernatant (to simulate clarifier effluent) was collected for chlorination testing. This sample was labeled 6-DW-SETT. The third cubitainer was dosed with 1 mg/L of anionic polymer and 9 mg/L of ferric chloride. The ferric chloride dose was a function of the measured orthophosphate concentration of the sample, multiplying the orthophosphate concentration by 9.0 to get the ferric chloride dose. This sample was labeled 6-DW-CEPT and represented effluent from a chemically enhanced clarification process.

For the chlorination testing procedure, the chlorine doses and batch reaction times were modified. Rather than using predetermined doses of chlorine, the chlorine dose was a function of the measured NH_3 concentration in the wastewater with the focus of testing the disinfection efficacy of both monochloramine and free chlorine. Ratios of chlorine-to-NH_3 of 3 and 5 were used to generate a monochloramine residual and 12 and 16 were used to generate a free chlorine residual. The reason for using these ratios is discussed further in the Discussion, Primary Disinfectant section. Furthermore, batch reaction times of 5, 15, 30, and 45 minutes were tested.

Sodium hypochlorite was added to the samples as described in the previous section. For the revised protocol, the monochloramine and free chlorine were measured at each batch reaction time, in addition to the TRC.

For *E. coli* testing for the revised protocol, the IDEXX Quanti-Tray/2000 process was used, which complies with Standard Method 9223 B. This method was preferable to that used in the original protocol due to the high quantity of *E. coli* tests conducted. Three dilutions were tested in triplicate for each of the disinfection samples. It should be noted that for samples for which monochloramine was the primary disinfectant, different dilutions resulted in different *E. coli* values for the same sample. For instance, for the 6-DW-RAW sample dosed with 38.8 mg/L of chlorine and a batch reaction time of 15 minutes, the undiluted sample resulted in MPN values of 533.5, 424.5, and 372.4 MPN per 100 mL; the 10:1 dilution resulted in MPN values of <10.0, 20.1, and <10.0 MPN per 100mL and the 100:1 dilution resulted in 99, 99, and <100. In this case, and for similar cases, the dilution series resulting in the maximum MPN was used in developing the results.

RESULTS

Original Chlorine Disinfection Testing Procedure (Events 3, 4, and 5)

The characterization results for each of the three wet weather events are provided in Table 6. As can be seen in this table, the storm events varied significantly in terms of loadings to the MRWWTP.

The results of Events 3, 4 and 5 have been combined and presented as *E. coli* log removal versus CT, where CT is the batch reaction time multiplied by the residual chlorine concentration. Results are presented by sample type (X-1-INF, X-2-EFF, X-3-INF, and X-4-EFF) and primary disinfectant. The primary disinfectant, either combined chlorine residual (ie, monochloramine) or free chlorine, was determined based on a calculated ratio. The Cl_2:NH_3-N ratio at which free chlorine is formed was calculated using the breakpoint curves created for Samples 3-1-INF, 3-2-EFF, and 3-4-EFF (3-3-INF was not used as the influent NH_3 did not correspond to the curve). The breakpoint was found to occur at a weight ratio of 12.7 (the average ratio of the three samples tested) versus a theoretical weight ratio of 7.6. Thus, to determine the ratio at which dichloramine likely begins to form, the theoretical weight ratio of 5.1 was multiplied by the factor of 12.7 over 7.6; this gives an actual weight ratio of 8.5. The ratios obtained for Event 3 were used to identify the primary disinfectant for the subsequent events.

		Event 3 – June 8, 2010				Event 4 - November 12, 2010				Event 5 – November 29, 2010			
		3-1-INF (t=0)	3-2-EFF (+90)	3-3-INF (+180)	3-4-EFF (+270)	4-1-INF (t=0)	4-2-EFF (+90)	4-3-INF (+180)	4-4-EFF (+270)	5-1-INF (t=0)	5-2-EFF (+90)	5-3-INF (+180)	5-4-EFF (+270)
pH	Units	7.41	7.52	7.75	7.22	7.13	6.99	7.21	7.07	7.15	7.13	6.93	7.06
ORP	mV	-106	+251	+268	+94	-276	-212	-142	63	6	-40	Not Reported	-8
COD	mg/L	411	215	133	169	782	353	224	154	258	542	436	592
BOD	mg/L	96.4	49.9	11.7	37.9	254	120	37.4	40.0	120	311	223	362
TOC	mg/L	30.9	22.3	6.22	29.9	43.4	46.2	14.4	14.5	29.4	84.3	45.6	86.7
TSS	mg/L	338	124	882	122	1,250	80	710	170	124	132	236	184
VSS	mg/L	176	48.0	80.0	36.0	300	<50	70	<50	132	136	196	172
Ammonia	mg/L NH_3-N	10.4	7.52	0.81	7.98	15.6	11.8	0.61	2.06	12.8	30.1	14.4	29.3
Alkalinity	mg/L $CaCO_3$	325	228	100	125	335	215	72.9	74.1	273	315	292	342
E. coli	MPN/100mL X 10^6	1.73	2.13	0.220	1.47	1.37	2.03	0.183	Not Reported	5.60	8.63	5.60	14.3

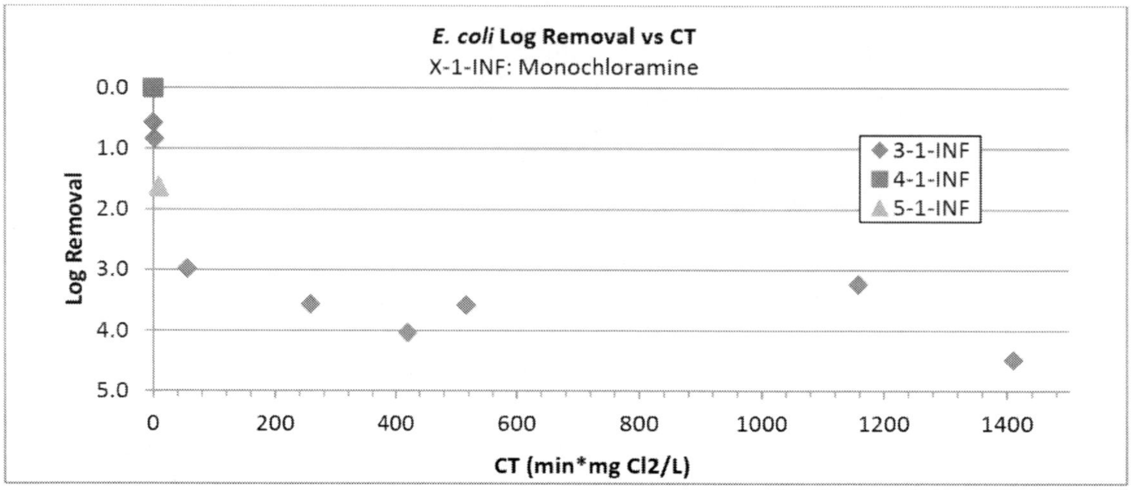

Figure 10: E.coli Log Removal vs CT, Primary Clarifier Influent Samples, Start of Storm Event samples, Monochloramine

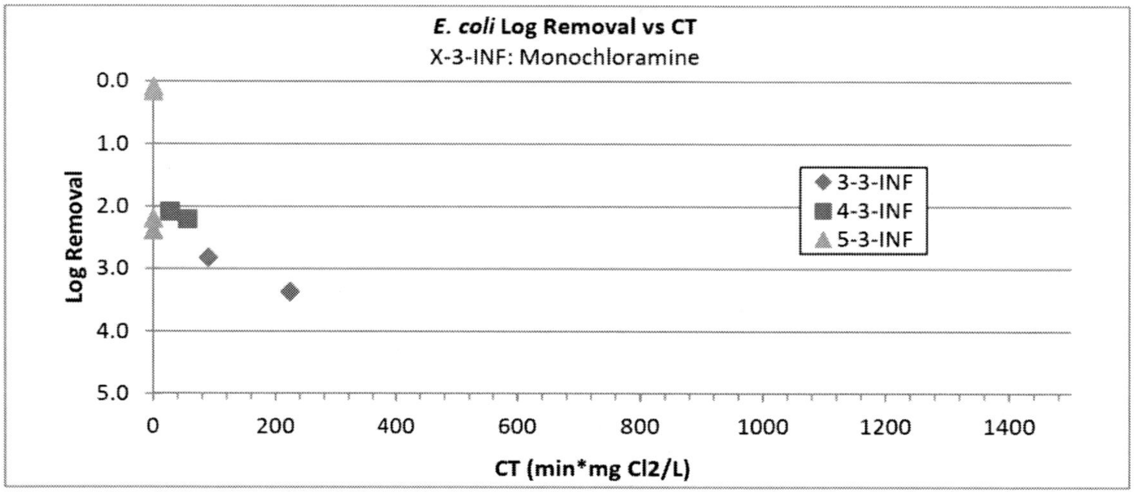

Figure 11: E. coli Log Removal vs CT, Primary Clarifier Influent, First Flush Samples, Monochloramine

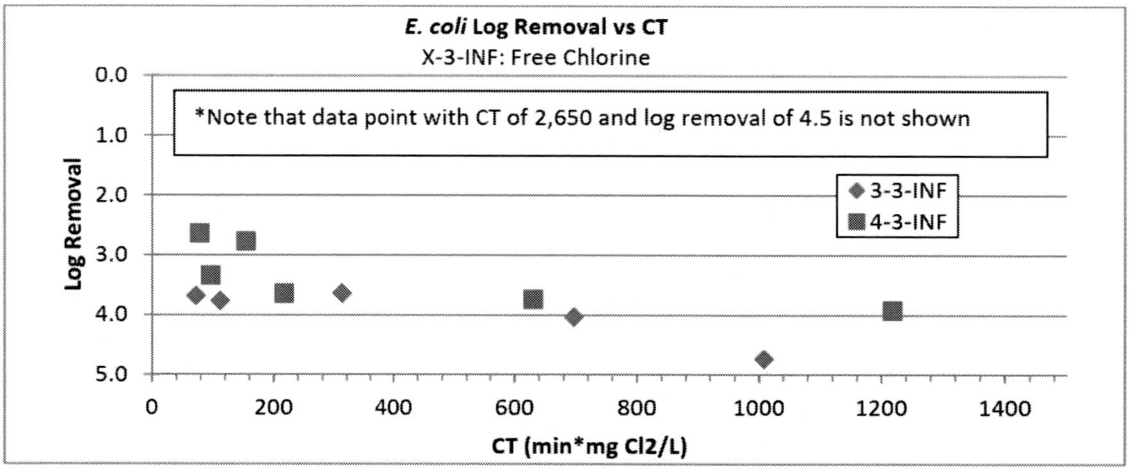

Figure 12: E. coli Log Removal vs CT, Primary Clarifier Influent, First Flush Samples, Free Chlorine

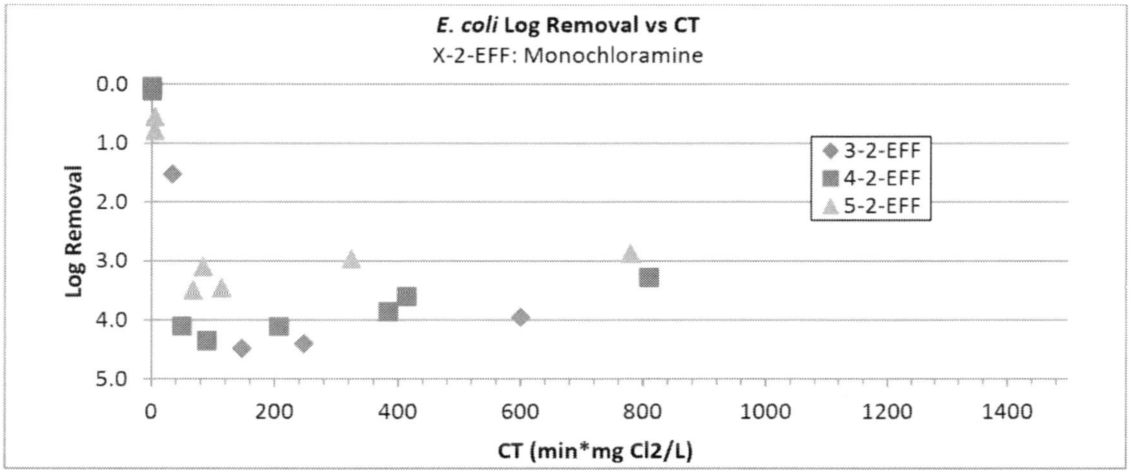

Figure 13: E. coli Log Removal vs CT, Primary Clarifier Effluent, Start of Storm Event Samples, Free Chlorine

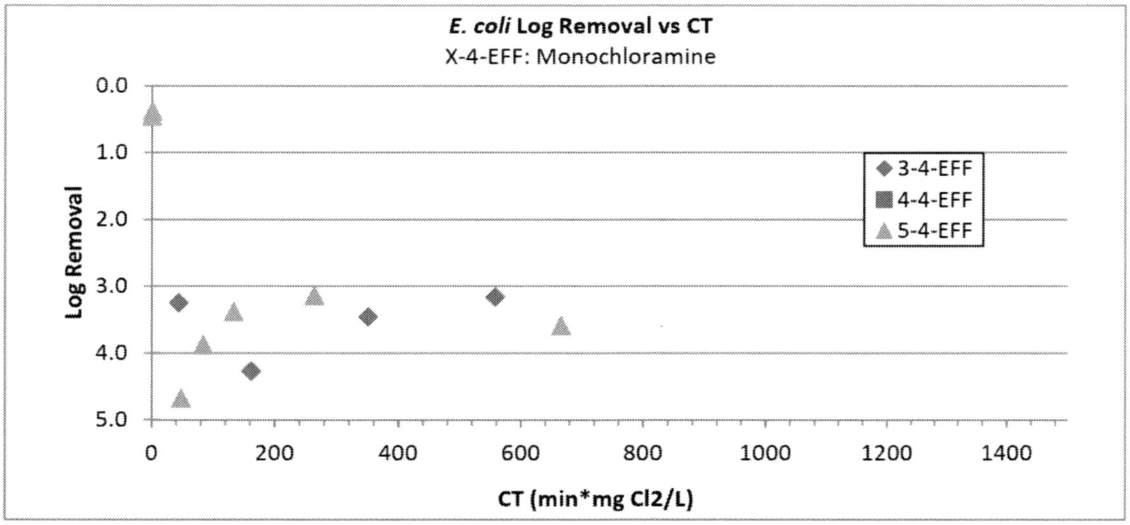

Figure 14: E. Coli Log Removal vs CT, Primary Clarifier Effluent, First Flush Samples, Monochloramine

Modified Chlorine Disinfection Testing Procedure (Event 6)

The characterization results for each of the samples are provided in Table 7. Results are provided for each of the three samples that underwent chlorination testing—6-DW-RAW, 6-DW-SETT, and 6-DW-CEPT. For each of these samples the following results are provided: *E. coli* concentration versus batch reaction time for each dose; chlorine residual versus chlorine dose for each batch reaction time; the chlorine residual versus time for each dose; and the *E. coli* log removal versus CT. Furthermore, plots of *E. coli* log removal versus monochloramine CT and *E. coli* log removal versus free chlorine CT are provided.

TABLE 7

Wastewater Characteristics – Event 6

		6-DW-RAW	6-DW-SETT	6-DW-CEPT
pH	units	7.03	7.10	7.05
ORP	mV	-29	152	218
COD	mg/L	273	-	-
BOD	mg/L	107	48.6	64.1
TOC	mg/L	39.9	-	-
TSS	mg/L	142	81	81
VSS	mg/L	114	-	-
Ammonia	mg/L NH_3-N	12.9	12.9	12.4
Orthophosphate	mg/L P	1.01	-	-
Sulfide	mg/L	0.183	-	-
Alkalinity	mg/L $CaCO_3$	271	-	-
E. coli	MPN/100mL X 10^6	1.24	1.32	8.42

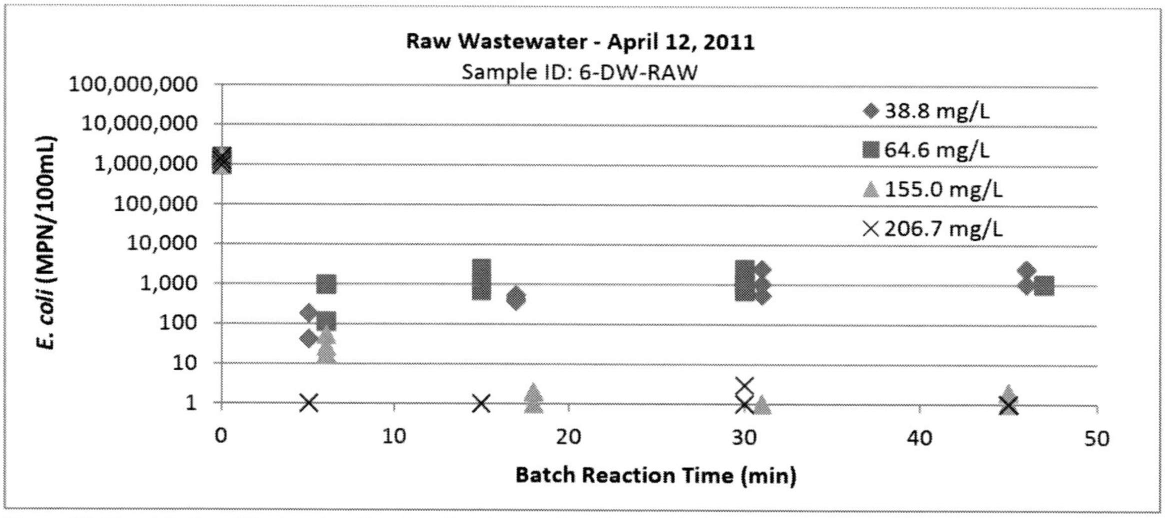

Figure 15: Event 6, Raw Wastewater, E. coli vs Batch Reaction Time at Various Chlorine Doses

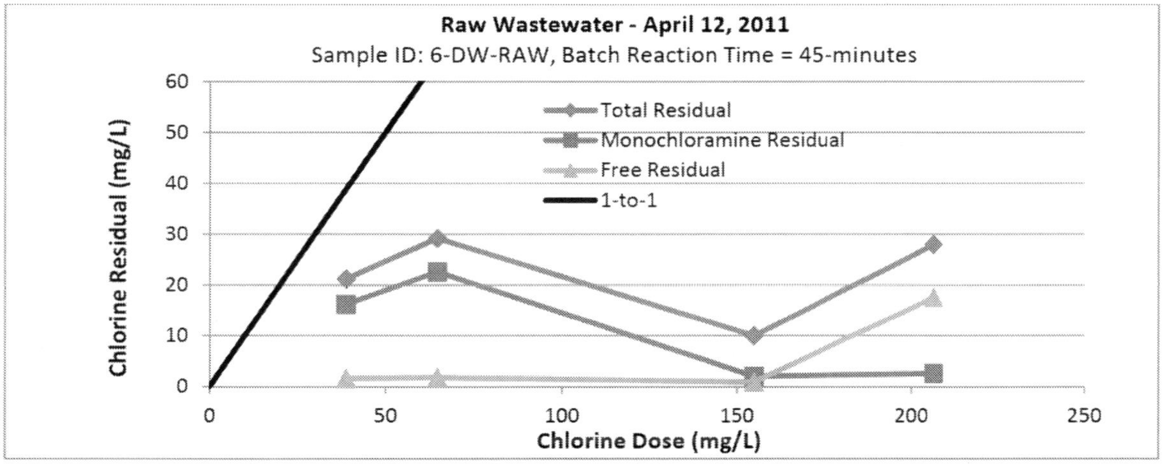

Figure 16: Event 6, Raw Wastewater, Chlorine Residuals after 45 minutes

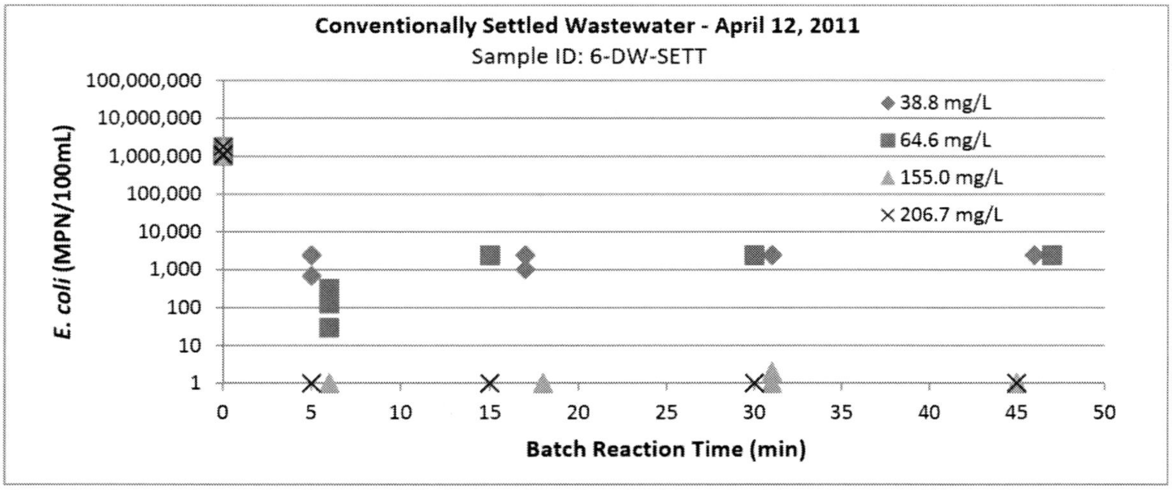

Figure 17: Event 6, Conventionally Settled Wastewater, E. coli vs Batch Reaction Time at Various Chlorine Doses

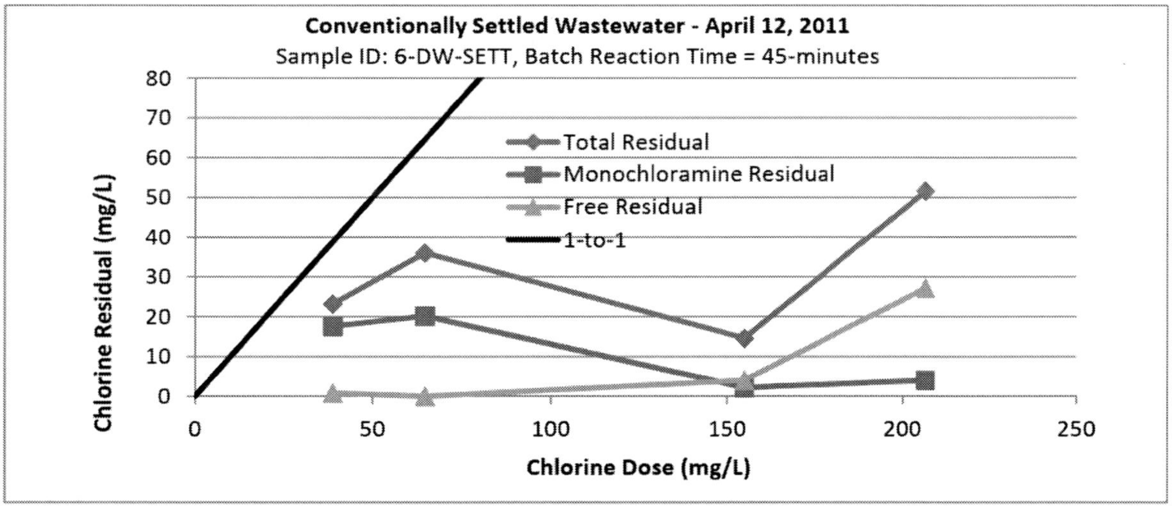

Figure 18: Event 6, Conventionally Settled Wastewater, Chlorine Residuals after 45-minutes

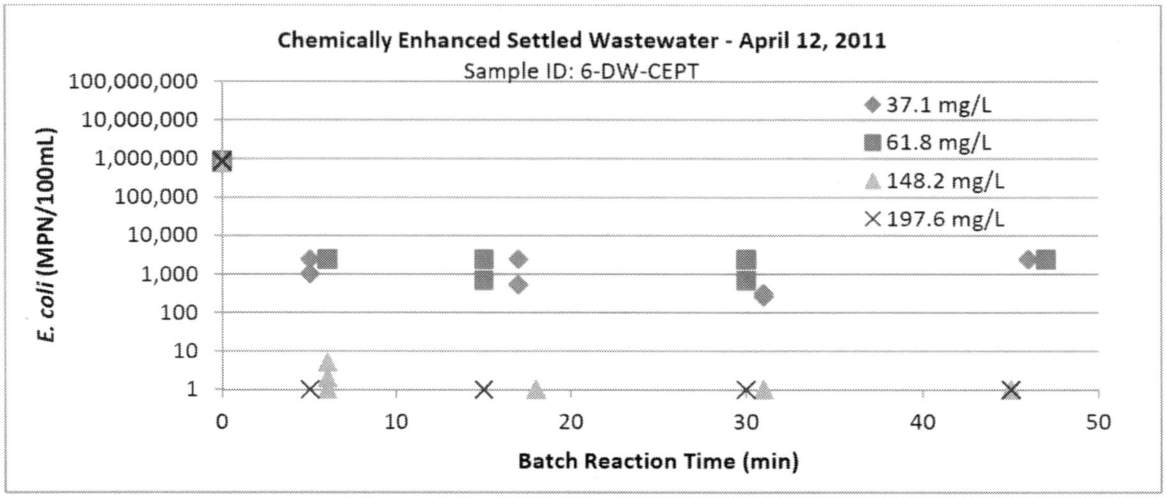

Figure 19: Event 6, CEPT Settled Wastewater, E. coli vs Batch Reaction Time at Various Chlorine Doses

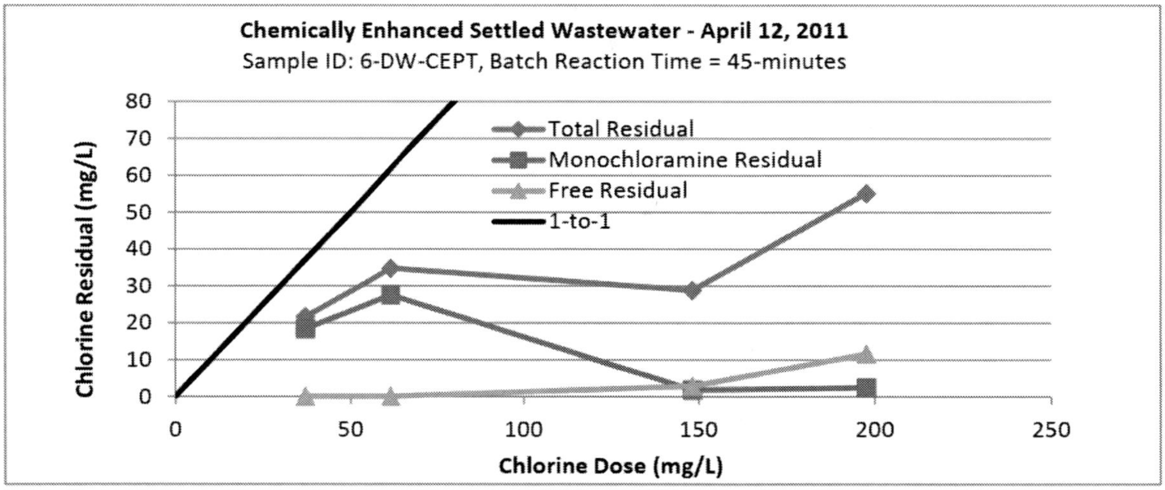

Figure 20: Event 6, Conventionally Settled Wastewater, Chlorine Residuals after 45-minutes

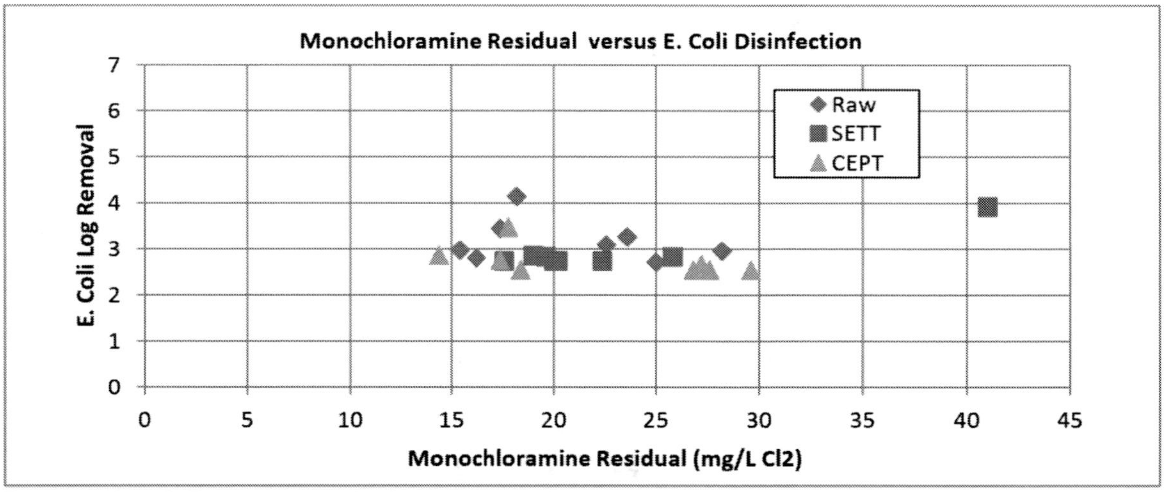

Figure 21: Event 6, E. coli log removal vs Monochloramine Residual

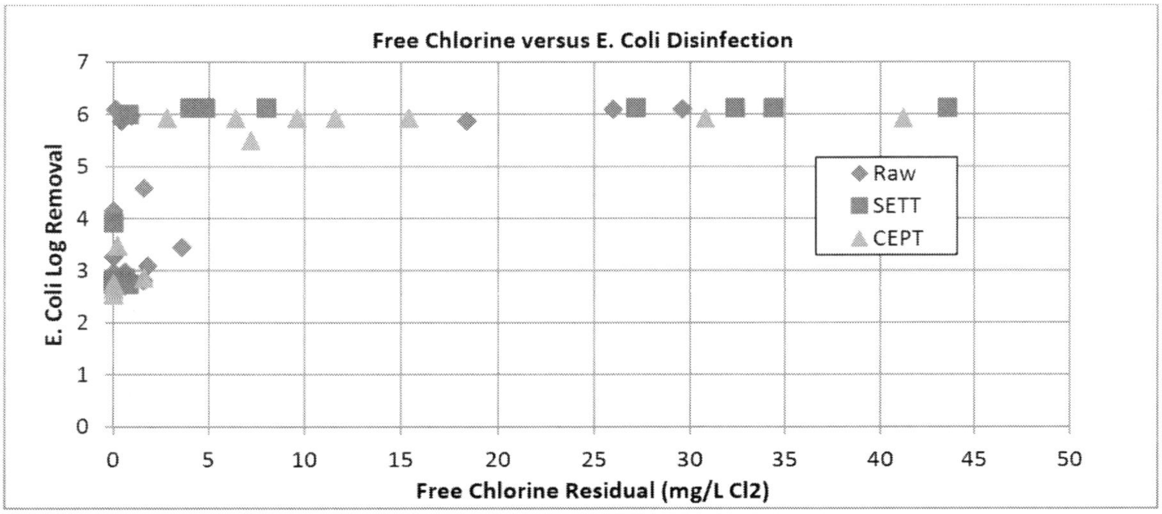

Figure 22: Event 6, E. coli Log Removal vs Free Chlorine Residual

DISCUSSION

Original Chlorine Disinfection Testing Procedure (Events 3, 4, and 5)

Based on a review of the results obtained for Events 3, 4, and 5, four items of particular interest are noted for further discussion—the varying characteristics by storm event, the impacts of SOIA and DAFT subnatant flow, the primary disinfectant for the samples, and removal of *E. coli*.

Varying Storm Characteristics

As is shown in Tables 4 and 5, the parameters measured to characterize each storm event (Events 3, 4, and 5) varied significantly. This was also the case for both of the first flush events, Events 1 and 2, discussed previously. With regards to solids loading, Event 4 had the highest TSS concentration, which occurred during the first sample of the event (t=0). Events 3 and 4 had similar first-flush influent (Sample X-3-INF) TSS/VSS ratios (at t=180) of approximately 10. This high TSS/VSS ratio is indicative of a first flush event; the majority of the solids are inorganic material, most likely resulting from storm water runoff and re-suspension of solids that accumulate in the collection system. For a typical non-wet weather wastewater sample, a majority of the solids would be organic material The influent for Event 5 had a much lower TSS/VSS ratios for all samples, including the first flush sample (t=180). For t=0 and t=90, the TSS/VSS ratio was less than 1, which implies sampling or analytical error.

In terms of organic loading, Event 4 had the highest COD and BOD concentrations. For Events 3 and 4, the COD, BOD, and TOC concentrations were higher for the first (t=0) sample (X-1-INF) than for the sample representing the first flush (X-3-INF). For Event 5, the first flush sample (5-3-INF) had much higher COD, BOD, and TOC than the 5-1-INF sample. Also, Primary Clarifier effluent COD, BOD, and TOC were much higher than the influent samples for Event 5.

Event 4 had the highest influent NH_3 concentration of 15.6 mg/L. For both Event 3 and Event 4, the first flush samples (3-3-INF and 4-3-INF) had very low NH_3 concentrations. Similar to the organic characteristics for Event 5, the NH_3 concentration in the Primary Clarifier effluent was higher than the Primary Clarifier influent samples. This was also seen for the first flush samples for Events 3 and 4.

For *E. coli*, the highest influent concentration was 5.60 x 10^6 MPN per 100 mL during Event 5 for both the 5-1-INF and 5-3-INF samples. For Events 3 and 5, the Primary Clarifier influent (at t=0) and effluent (at t=90) *E. coli* values were of the same order of magnitude, indicating no appreciable Primary Clarifier removals. For these two events, subsequent sampling showed that the primary effluent (t=270) *E. coli* level was an order of magnitude higher than the corresponding influent sample (t=180). For Events 3 and 4, the first flush samples (3-3-INF and 4-3-INF) were almost an order of magnitude lower than the initial influent samples (3-1-INF and 4-1-INF).

Impacts of SOIA and Dissolved Air Flotation Thickener Subnatant Flow

Comparing the results of the Primary Clarifier influent characteristics with the Primary Clarifier effluent characteristics, summarized in Tables 2 and 3, it appears that the effluent samples were impacted by the SOIA flow, the dissolved air floatation thickener (DAFT) subnatant flow, or both. For all three of the events' "first flush" samples, the effluent NH_3-N concentration was higher than the influent indicating that high NH_3-N flow from the South Inlet or DAFT subnatant mixes with the North Inlet flow prior to entering Primary Clarifier No. 5 (i.e., the assumption that samples from the Primary Clarifier No. 5 effluent launder would not be impacted by the industrial flow or DAFT subnatant flow is likely not valid).

As described previously, high-strength wastewater from local industries currently enters the South Inlet of the MRWWTP. The SOIA flow will be separated and treated at the new SOIA headworks and clarifiers and an air gap will be provided to ensure that the SOIA flow does not enter the CSO 102 Chlorination/Dechlorination facility. It was assumed that if the North Inlet flow was higher than the South Inlet flow, the industrial flow would be kept in the southernmost primary clarifiers, Primary Clarifier No. 6 and No. 7. Therefore, the Primary Clarifier effluent sample was taken from the effluent launder of Primary Clarifier No. 5.

During wet weather events, it is assumed that the North Inlet flow will be higher than the South Inlet flow because there is a much larger drainage area and higher pumping capacities for the SIFM (which feeds the North Inlet) than the Monroe Street Lift Station (which feeds the South Inlet); however, if the south area experiences a higher volume or more-intense rain event, the South Inlet flow may be higher than the North Inlet flow. Flow data provided by the MRWWTP, shows that the North Inlet flow was higher for each of the sampling events. During dry weather conditions, the South Inlet and North Inlet see similar flow rates, and at times, the South Inlet flow rate is higher than the North Inlet. In addition, only three clarifiers are in operation. As such, industrial flow enters Clarifier No. 5, and some of this "lingering" industrial flow may be included in the wet weather clarifier effluent samples. Additionally, the North Inlet and South Inlet flows may mix in the Primary Splitter Structure more than originally expected.

The characteristics of the influent samples could be due to the DAFT operation. Two DAFTs at the MRWWTP thicken secondary sludge prior to digestion. Sludge belt filter press filtrate, which has a very high concentration of NH_3-N, is also sent to the DAFTs. The subnatant from the DAFTs is conveyed to the Primary Splitter Structure with the South Inlet flow. The DAFT flow has a very high NH_3-N concentration and may mix with the North Inlet flow in the Primary Splitter Structure and enter Primary Clarifier No. 5, potentially increasing the NH_3-N concentration in the clarifier effluent samples.

Inactivation of E. coli

Graphs of log removal versus CT (where CT is TRC final x batch reaction time) were plotted for each sample type (initial primary influent, first flush primary influent, initial primary effluent, and first flush primary effluent) and labeled as X-1-INF, X-3-INF, X-2-EFF, and X-4-EFF. Each of the curves contains results from different events, and separate curves are presented for each of the primary disinfectants, monochloramine or free chlorine, based on the Cl_2:NH_3-N ratio. It should be noted that a free chlorine curve is only provided for X-3-INF. The NH_3-N concentrations of all the X-1-INF and X-2-EFF samples were high enough that monochloramine was the primary disinfectant. For X-4-EFF, the NH_3-N concentration of Event 4 was low enough that free chlorine was formed for some doses; however, the initial *E. coli* concentration was not reported by the lab, so log removals could not be calculated. Looking at the log removal curves provided it can be noticed that for monochloramine disinfection, a 3-log removal is achievable at a CT of 200 milligrams per minute per liter (mg-min/L). Furthermore, the data indicates that using a CT greater than 200 does not ensure that greater log removal will be achieved. This is the case for both Primary Clarifier influent and effluent samples.

At a batch reaction time of 15 minutes and a dose of 30 mg/L (values consistent with initial design assumptions of the CSO 102 Chlorination/Dechlorination facility), a final *E. coli* concentration of 126 MPN per 100 mL was not achieved reliably. Table 8 shows the results for Primary Clarifier effluent samples for Events 3 and 4, which are believed to have similar constituent concentrations as to what will be treated at the CSO 102 Chlorination/Dechlorination facility, at a batch reaction time of 15 minutes. Results from Event 5 are not included in this table because the high NH_3-N concentrations measured in the primary effluent samples of Event 5 indicate that they contain a large portion of SOIA and/or DAFT subnatant flow. *E. coli* values are an average of the three geometric mean samples taken for each contact time. As shown in Table 8, only two tests met the *E. coli* limit of 126 MPN per 100 mL; these tests were done on Sample 4-4-EFF, which had a significantly lower NH_3-N concentration than the other samples (2.06 mg/L versus >7.5 mg/L), which could have resulted in free chlorine being generated (rather than monochloramine), which would likely provide better disinfection. Because of this result, testing for Event 6 (see next section for discussion) was designed to generate a free chlorine residual to test its disinfection efficacy.

TABLE 8
Final *E. coli* as a Function of Dose for 3-2-EFF, 3-4-EFF, 4-2-EFF, and 4-4-EFF

Sample ID	Final *E. Coli* Concentration (MPN/100 mL)			
Dose ->	5 mg/L	15 mg/L	30 mg/L	60 mg/L
3-2-EFF	64,000	71	85	324
3-4-EFF	833	80	770	1,000
4-2-EFF	1,900,000	167	165	555
4-4-EFF	370,000	<100	1	1

Note:
Monochloramine was predominantly formed in Samples 3-2-EFF, 3-4-EFF, and 4-2-EFF at all chlorine doses. Monochloramine was predominantly formed in Sample 4-4-EFF at chlorine doses of 5 and 15 mg/L and free chlorine was likely formed in Sample 4-4-EFF at chlorine doses of 30 and 60 mg/L.

For the Primary Clarifier effluent samples, a longer contact time at a similar dose did not provide significantly improved *E. coli* destruction. Table 9 shows the final *E. coli* concentration for similar samples at contact times of 15 and 30 minutes. The TRC is provided for comparison and shows that the TRC is similar between the contact times. The table shows that the longer contact times did not substantially improve disinfection for Samples 4-2-EFF, 5-2-EFF and 5-3-EFF. The only instance where the longer contact time showed improved disinfection is for the 5 mg/L dose of the 4-4-EFF dose. It should be noted again that for the 4-4-EFF sample, free chlorine is the primary disinfectant compared to monochloramine for the other three samples.

TABLE 9
Comparison of contact times for Primary Effluent Samples

	Batch Reaction Time = 15-min		Batch Reaction Time = 30-min	
Chlorine Dose	TRC (mg/L Cl2)	*E. coli* (MPN/100 mL)	TRC (mg/L Cl2)	*E. coli* (MPN/100 mL)
4-2-EFF				
5	0.0	1,866,667	0.0	1,566,667
15	3.3	167	3.0	96
30	13.8	165	12.8	307
60	27.6	555	27.0	1,000
4-4-EFF				
5	2.70	370,000	2.55	133
15	9.30	100	8.90	100
30	11.20	2	10.60	3
60	15.20	1	12.60	10
5-2-EFF				
5	0.30	2,533,333	0.16	1,433,333
15	4.50	2,867	3.80	4,497

TABLE 9
Comparison of contact times for Primary Effluent Samples

	Batch Reaction Time = 15-min		Batch Reaction Time = 30-min	
Chlorine Dose	TRC (mg/L Cl2)	*E. coli* (MPN/100 mL)	TRC (mg/L Cl2)	*E. coli* (MPN/100 mL)
30	5.60	6,833	10.80	10,000
60	21.60	>24,000	26.00	15,450
5-4-EFF				
5	0.06	5,833,333	0.00	5,100,000
15	3.20	293	2.80	2,033
30	8.90	5,867	8.80	10,000
60	22.40	>2,400	22.20	4,600

Note: Monochloramine was predominantly formed in Samples 4-2-EFF, 5-2-EFF, and 5-4-EFF at all chlorine doses. Monochloramine was predominantly formed in Sample 4-4-EFF at chlorine doses of 5 and 15 mg/L and free chlorine was likely formed in Sample 4-4-EFF at chlorine doses of 30 and 60 mg/L.

Modified Testing Procedure (Event 6)

The results obtained for Event 6 add strong evidence that free chlorine residual is needed to meet the *E. coli* concentration requirement. Similar to the results of Events 3, 4, and 5, monochloramine disinfection resulted in log removals of 2.5 to 3.5. With free chlorine residuals of 5 mg/L or greater, a log removal of 6 or greater was readily achieved, resulting in final *E. coli* concentrations less than 10 MPN per 100 mL.

Characterization Relative to Wet Weather Samples

To investigate the assumption that a dry weather sample would be consistent with a "worst case" wet weather scenario, Table 10 provides a summary of constituent concentrations measured for each of the wet weather samples as well as the dry weather raw wastewater sample (6-DW-RAW).

TABLE 10
Summary of Constituent Concentrations Measured for Each of the Wet Weather Samples

	pH	ORP	COD	BOD	Ammonia	TSS	VSS	Alkalinity	TOC
Sample	units	mV	mg/L	mg/L	mg/L as NH_3-N	mg/L	mg/L	mg/L as $CaCO_3$	mg/L
3-1-INF	7.41	-106	411	96.4	10.4	338	176	335	30.9
3-3-INF	7.75	+268	133	11.7	0.81	882	80.0	72.9	6.22
4-1-INF	7.13	-276	782	254	15.6	1,250	300	335	43.4
4-3-INF	7.21	-142	224	37.4	0.61	710	70	72.9	14.4
5-1-INF	7.15	6	258	120	12.8	124	132	273	29.4
5-3-INF	6.93	NR	436	223	14.4	236	196	292	45.6
6-DW-RAW	**7.03**	**-29**	**273**	**107**	**12.9**	**142**	**114**	**271**	**39.9**

As can be seen in Table 10, the COD, BOD, TSS, VSS, NH_3-N, and TOC values measured for the dry weather sample are lower than all of the values measured for the wet weather Sample 4-1-INF. The concentrations for wet weather Sample 5-1-INF are similar to the dry weather sample, and the other wet weather samples show some constituents at higher concentrations than dry weather and some at lower concentrations. Therefore, while the dry weather sample is consistent with one of the wet weather events, it is not a "worst-case" scenario.

With regards to NH_3-N, which has the greatest impact on chlorination as discussed previously, the highest NH_3-N concentration measured in the wet weather events is 15.6 milligrams of nitrogen per liter. Dosing at a 12:1 Cl_2:NH_3-N ratio for this sample would require a chlorine dose of 187 mg/L Cl_2; a 16:1 C12:NH_3-N ratio would require a 250 mg/L Cl_2 dose.

Chlorine Speciation

As predicted, C12:NH_3-N ratios of 3:1 and 5:1 resulted in the production of monochloramine residual while ratios of 12:1 and 16:1 resulted in the formation of free chlorine residual. The chlorine residual curves provided in the results section show that at the ratio of 12:1, small amounts of free chlorine residual form. At ratios of 16:1, the free chlorine residual increases linearly at close to a 1-to-1 ratio to the chlorine added, meaning that nearly all of the chlorine added greater than a ratio of 12:1 results in a free chlorine residual.

Chlorine Dose

The chlorine dosages tested ranged from 40 to 200 mg/L. These doses are much higher than doses presented in literature for chlorine disinfection of CSOs. The results for the dry weather event, as presented in Figures 21 and 22 and summarized in Table 11 below, further suggest that a free chlorine residual is needed to meet the permitted *E. coli* discharge limit.

TABLE 11
Comparison of *E. coli* removal for RAW, SETT, and CEPT samples

Average *E.coli* in Raw Wastewater	1,241,200	MPN/100mL
Disinfection Target	126	MPN/100mL
Log Removal Required	3.99	

C12:NH_3-N Ratio	**Average *E. coli* Log Removal Achieved During Testing**		
	Raw	**SETT**	**CEPT**
3:1	3.07	2.77	2.96
5:1	2.92	2.77	2.58
12:1	5.98	6.08	5.93
16:1	6.02	6.12	5.93

Note:
Average log removal is for 15-, 30-, and 45-minute times since disinfection improves little after 15 minutes.

While dry weather wastewater was being disinfected rather than wet weather flows, it was previously determined that, at the start of a CSO event, the influent wastewater at the MRWWTP is very similar in strength to dry weather wastewater. During Events 1 and 2, the influent NH_3-N concentration for the first 3 hours of the storm ranged from 8 to 11.5 mg-N/L. The NH_3-N concentrations for Event 6 samples ranged from 12.4 to 12.9 mg-N/L.

The first flush analysis in Events 1 and 2 shows that the wastewater strength starts to decrease after approximately 3 hours. Figure 3 shows that the NH_3-N concentration drops to 2 mg/L later in the event. As such, dosing at a ratio of 12:1 would require a dose of only 24 mg/L. Therefore, a dose of 30 mg/L should be capable of meeting the *E. coli* requirement at later stages of a storm event. Chlorine doses in excess of 150 mg/L would be required at the beginning of a storm event to breakpoint all of the initial NH_3-N. Sophisticated instrumentation and control would be required to properly respond to the changing NH_3-N concentration by regularly changing the chlorine dose.

Batch Reaction Time

Similar to Events 3, 4, and 5, it was shown for Event 6, contact times greater than 15 minutes do not improve disinfection. The results of Event 6 indicate that there may be some improvement in disinfection between batch reaction times of 5 and 15 minutes. At the dose ratio of 12:1, 5 minutes of contact time resulted in *E. coli* values between 10 and 100 MPN per 100 mL, whereas at 15 minutes, the *E. coli* concentration was reliably less than 10 MPN per 100 mL.

Impacts of Settling

Compared to the conventionally settled (SETT) and chemically enhanced settled (CEPT) samples, the results of disinfecting the raw, unsettled sample were similar (see Table 11). As discussed previously, the ability to disinfect depends greatly on the NH_3-N concentration of the wastewater. Because NH_3-N is soluble, it is not removed by settling.

It should be noted that the CEPT sample was very similar in wastewater characteristics to the SETT sample. This was due to under-dosing ferric chloride for this sample. The CEPT results should not be considered indicative of the disinfectability of a truly CEPT settled sample; rather, the CEPT sample in this case is consistent with a conventionally settled sample. Additional CEPT testing was performed on unsettled samples to refine the ferric dose for future tests. Future CEPT testing should use significantly higher ferric chloride doses, as suggested by the significant TSS and *E. Coli* removed during a jar test conducted that investigated the performance of higher ferric chloride doses (Figure 23). Chlorine disinfection was not conducted on these jar tests, but the good performance of the ferric chloride suggests that chlorination may be very successful following high doses of ferric chloride. This assumption could be validated with further testing.

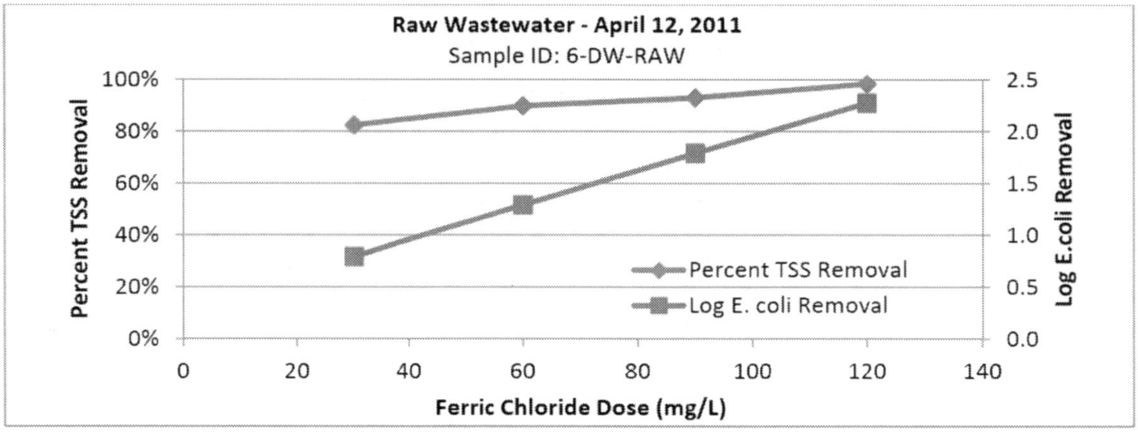

Figure 23: TSS and E. coli removal versus ferric dose, 45-min settling time, polymer dose = 1 mg/L

It was expected that settling should improve disinfectability and allow for lower chlorine doses for meeting the *E. coli* requirements. By removing portions of the organic-nitrogen, organic compounds and solids, it was expected that the initial chlorine demand could be significantly reduced through settling. In addition, removing solids can improve disinfection by reducing the amount of particle shielding that can occur. The point at which the free chlorine residual forms is, as discussed, tied directly to concentration of NH_3-N in the wastewater. Therefore, the benefits of settling prior to disinfection are limited with regards to the required chlorine dose. As such, disinfecting in the primary clarifiers may be a viable option relative to *E. coli* destruction.

Oxidation Reduction Potential

Test results of the dry weather event shows that at ORP values greater than 600 millivolts (mV), free chlorine residual is present. As such, ORP measurement may be beneficial in controlling the chlorine dose to the CSO 102 Chlorination/Dechlorination basin. A graph of free chlorine residual versus ORP is shown in Figure 24.

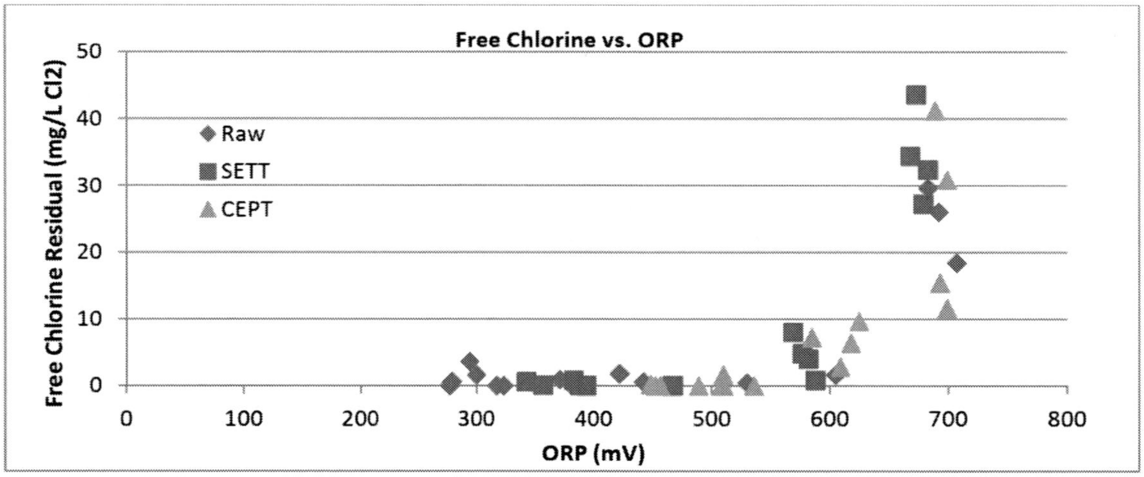

Figure 24: Event 6, Free Chlorine Residual versus ORP

DESIGN CONSIDERATIONS

Study results show that the ability to disinfect to the *targeted E. coli* concentration is dependent upon the ammonia concentration of the wastewater. The study results further show that the concentration of ammonia in flows received at the CCB will vary widely over the course of a storm event. Dynamic hydraulic analysis using CH2M Hill's Replica Dynamic Simulation software was performed and determined that using primary clarifiers for flow equalization shaves the peak ammonia concentration to approximately 10 mg/L. Further, the E. coli concentration could be shaved by one order of magnitude, from approximately 10^6 to 10^5 MPN/ 100 mL.

While the study considered chlorine doses as high as 200 mg/L, the maximum practical chlorine dose for the CCB was determined to be 60 mg/L, based on building footprint constraints as well as maximum doses at chlorination facilities throughout North America.

At the beginning of an overflow event, the initial chlorine dose will be set at 40 milligrams per liter (mg/L), which will result in the formation of a monochloramine residual in almost all cases (because of the resultant chlorine-to-ammonia ratio). This dose will be maintained as long as the ammonia concentration is greater than 6 mg/L, regardless of the influent ammonia concentration. The ammonia concentration is expected to decrease with time, due to dilution, as the wet weather reaches the plant. When the ammonia concentration decreases to 6 mg/L, the chlorine dose will be increased to 60 mg/L to achieve a free chlorine residual, which provides better disinfection. Ammonia concentration will be measured in the CSO 102 Primary Effluent Flow Diversion Structure located immediately upstream of the CCB. The following steps specifically illustrate how the chlorine dose will be controlled throughout a wet-weather event:

Step 1: The initial chlorine dose will be 40 mg/L. This assumes an initial ammonia concentration between 6 and 10 mg/L while trying to stay in the monochloramine range of the curve. This dose will be maintained regardless of the ammonia concentration (and subsequent chlorine-to-ammonia ratio) as long as the CCB influent ammonia concentration remains greater than 6 mg/L. Based on bench testing, a chlorine dose of 40 mg/L under this condition is expected to result in a monochloramine residual of about 20 mg/L after 15 minutes of contact time, which would result in a monochloramine CT (monochloramine residual X contact time) of 300 milligram-minutes per liter (mg-min/L). The initial 40 mg/L chlorine dose will be an Operator adjustable input; after additional data has been collected through operation of the system, this initial dose may be modified to achieve a different chlorine-to-ammonia ratio based on typical initial ammonia concentrations observed in the system or a different breakpoint ratio (e.g., an initial chlorine dose of 50 mg/L might be selected if the breakpoint ratio is more in the range of 12:1, which would result in a 72 mg/L dose in Step 2, and it may be decided that the large dose increase from 40 mg/L to 72 mg/L should be avoided).

Step 1 Delay: The total chlorine analyzer at the end of Pass 1 must record a concentration in excess of 5 mg/L before the chlorine dose of 40 mg/L is adjusted. This delay and measurement prevents the Step 1 process from moving to Step 2 (which requires chlorine residual measurement for control) until the CCB has filled up to a level where the sample pumps can deliver disinfected primary effluent from the CCB to the chlorine analyzers for measurement of the chlorine residual.

Step 2: When the CCB influent ammonia residual decreases to 6.0 mg/L (as determined by two consecutive readings by the CCB influent ammonia analyzer), the chlorine dose (flow paced) will be increased to 60 mg/L (Operator adjustable), which assumes that a 10:1 chlorine-to-ammonia ratio will achieve breakpoint. The 60 mg/L dose should be held constant as the ammonia concentrations continue to fall (with more wet-weather dilution) until a free chlorine residual of 10 mg/L is achieved. The 10 mg/L free chlorine residual will be an Operator adjustable set-point.

The initial target for free chlorine residual of 10 mg/L has been determined based on the bench-top testing results presented in the Chlorination Study, which suggests that a free chlorine CT of approximately 150 mg-min/L would typically be required to provide between 5 and 6 log inactivation of *E. Coli*. Designing the disinfection system to be able to maintain the free chlorine residual at 10 mg/L also provides a robustness in the system sizing that can address potential ammonia spikes (even 0.5 to 1 mg/L) in the wet-weather overflow that could significantly reduce the total chlorine concentration to extremely low levels (approaching breakpoint), thereby greatly limiting the disinfection achieved through the CCB. Full-scale operational experience may allow the Operator to adjust the 10 mg/L set-point to a lower value.

Step 3: The cascade proportional–integral–derivative (PID) control loops will be engaged to adjust the chlorine dose to maintain free chlorine residual of 10 mg/L, as measured at the end of Pass 1 in the CCB.

CONCLUSIONS

The results of the study show that chlorinating primary clarifier effluent flows at the MRWWTP presents several unique challenges, including:

1. Varying wastewater characteristics from storm event to storm event;
2. Varying wastewater characteristics during storm events and the presence of a first flush phenomenon; and
3. Raw wastewater strength ammonia concentrations at the start of a storm event, followed by diluted ammonia concentrations as the event progresses.

Due to the varying ammonia concentrations expected at the CCB during a storm event, both monochloramine and free chlorine disinfection will be utilized. This requires a robust dosing control process that will provide flexibility for plant operators to optimize dosing strategies as the system is operated over time.

REFERENCES

Black and Veatch Corporation, 2010. *White's Handbook of Chlorination and Alternative Disinfectants*. 4th Edition. New Jersey: John Wiley and Sons.

United Stated Environmental Protection Agency (EPA). 1979. *Disinfection of Combined Sewer Overflows, Syracuse, NY*. National Environmental Research Center, Office of Research and Development, U.S. Environmental Protection Agency, Cincinnati, OH 45268. EPA-600/2-79-134.

United Stated Environmental Protection Agency (EPA). 1999. *Combined Sewer Overflow Technology Fact Sheet Chlorine Disinfection.* EPA-832-F-99-034.

Water Environment Research Foundation (WERF), 2005. *Identifying Technologies and Communicating the Benefits and Risks of Disinfecting Wet Weather Flows.* Alexandria, VA, Water Environment Research Foundation Federation.

The Use of Ozone to Treat Wet Weather Flows in Springfield, MO

G. L. Hunter[1*], J. J. Neemann[1], N. Burns[2], J. L. Pruss[1], Kelly Green[3]

[1]Black & Veatch, 8400 Ward Parkway, Kansas City, Missouri, 64114, USA
[2] Black & Veatch, 8400 Ward Parkway, Kansas City, Missouri, 64114, USA
[3]Springfield Clean Water Services, 840 Boonville Avenue, Springfield, Missouri 65802 USA

[*]To whom correspondence should be addressed. Email: huntergl@bv.com.

ABSTRACT

Ozone's use in wastewater treatment has experienced new interest with additional disinfection requirements and the potential to remove compounds of emerging concern. One of the most critical aspects of its application is determining the ozone demand, which is important in sizing the system. The most significant factors that influence the ozone demand of wastewater effluent are chemical oxygen demand (soluble COD), inorganic oxidant demand (nitrite), influent bacteria density, and target effluent bacteria density. As regulations have been tightened and nutrient limits have been lowered, ozone demand has decreased and dosages as low as 5 mg/L can be used to meet a high-level of disinfection. The lower dosages, coupled with the advances in ozone generating and dissolution technology, have made ozone as a disinfectant for wastewater cost-competitive with UV.

The City of Springfield, Missouri currently uses ozone for disinfection of its secondary treated effluent. Recent permit requirements require the plant to treat both dry and wet weather flows. Testing indicated that ozone was capable of disinfecting the wet weather flows. Recent construction added an ozone application point for wet weather flows.

KEYWORDS: ozone, disinfection, compounds of emerging concern.

INTRODUCTION

While the first plant-scale use of ozone was in 1893 in Oudshoorn, Netherlands, most people consider Nice, France, which was installed in 1906, as the birthplace of ozone for drinking water treatment. Ozone has been used more extensively throughout Europe than in any other part of the world because of its benefit as a disinfectant and oxidant. Thousands of communities have installed ozone in Europe during its 100 year history. The first installation in North America was in 1908, however widespread use of ozone did not happen until the 1970's and 1980's when chlorination byproducts were beginning to be regulated and people were looking for an alternative to chlorine. Also, the transition from compressed air to high purity oxygen has allowed ozone generation efficiency to improve and its use to be more widespread.

Wastewater

In the early 1970s, utilities in the US began to use ozone as a disinfectant for treated wastewater effluent. From 1975 through 1983, it was estimated that ozone was used at 41 wastewater plants in the United States. However, only nine of these facilities (Table 1) were still using ozone in

2008 (Black & Veatch 2010). Many facilities had installed systems that were undersized and others found disinfection difficult to achieve as the quality of the treated effluent was inconsistent. Early research indicated that ozone doses greater than 10 mg/L were required to achieve specific levels of disinfection. The US EPA Wastewater Disinfection Manual included the dosage ranges to meet disinfection performance and those values are listed in Table 2.

Table 1: Wastewater Utilities Using Ozone for Disinfection (Black & Veatch 2010)

Startup(year)	Facility	Capacity (mgd)	Installed Ozone Capacity (ppd)	Firm Dosage (mg/L)
1978	Meander Creek WWTP Mahoning County, OH	8	500	7.5
1978	Southwest WWTP Springfield, MO	64	3,200	4.5
1980	Frankfort WWTP, Frankfort, KY	40	1,000	3.0
1982	Belmont WWTP, Indianapolis, IN	150	12,000	3.2
1982	Hagerstown WPCP Hagerstown, MD	12	1,400	4.7
1982	Southport WWTP, Indianapolis, IN	150	12,000	3.2
1985	Fred Hervey WWTP, El Paso, TX	12.5	900	8.6
1997	WWTP, Trion, GA	8	1,800	13.5
2003	F Wayne Hill Water Reclamation Plant, Gwinnett County, GA	60	2,000	4.0

Table 2: Ozone Dose Required for Various Disinfection Targets (USEPA 1986)

Effluent Quality	Absorbed Ozone Dosage for 2.2 cfu/100mL total coliform	Absorbed Ozone Dosage for 70 cfu/100 mL total coliform	Absorbed Ozone Dosage for 200 cfu/100 mL fecal coliform
Filtered Secondary	35 to 40	15 to 20	12 to 15
Filtered and Nitrified	15 to 20	5 to 10	3 to 5

Historically, utilities that considered or installed ozone faced many challenges with its implementation and use. These challenges where generally caused by the following factors:

- Ozone was usually not the least costly disinfection alternative.
- Upsets in treatment and variable water quality made disinfection challenging.

- Fouling of diffuser stones resulted in additional maintenance and other transfer technologies such as surface mixing resulted in poor disinfection and a lack of controls.
- Lack of automation and controls to gauge disinfection performance.

Over the past several years, wastewater facilities have been required to meet a number of goals ranging from high level disinfection (4 of 7 non detect for fecal coliforms) to N-nitrosodimethylamine (NDMA) treatment. Now utilities are facing the need to remove micro-pollutants and personal and pharmaceutical care products. As the focus of using ozone has moved from single goal: disinfection, to multiple goals: disinfection and micro-pollutant removal, ozone is gaining in popularity. As a disinfectant, ozone is becoming more cost-competitive with UV, but when considering the capability to remove micro-pollutants, ozone is clearly a technology that will be considered extensively in the coming years.

OZONE DEMAND

The most significant factors that influence the ozone demand and required ozone dosage of wastewater effluent are chemical oxygen demand (soluble COD), influent bacteria density, and target effluent bacteria density. As regulations have been tightened and nutrient limits have been lowered, ozone demand has decreased and dosages as low as 5 mg/L can be used to meet a high-level of disinfection. The lower dosages, coupled with the advances in ozone generating and dissolution technology, have made ozone as a disinfectant for wastewater cost-competitive with UV.

Because ozone is very reactive, ozone demand is usually expressed as the dose that is required to develop a measurable dissolved residual. This is determined by adding various dosages of ozone to water and measuring the resultant dissolved ozone residual. The ozone demand is then determined based on a linear regression of the transferred ozone dosages and residuals (Figure 1). The x-intercept value corresponds to the ozone dosage required to satisfy the ozone demand. Bench-scale testing of a wastewater effluent showed the ozone demand varied from approximately 2.0 to 3.5 mg/L. Testing was done at different flow rates and different seasons to evaluate the variability. One thing to note is that once a residual is developed, adding more ozone does not result in a 1:1 increase of ozone residual (slope of the regression line). Instead what is observed in practice is that only 50 to 75 percent of the ozone added after demand is met results in measurable ozone residual. The slope of the line will depend on many factors such as the water temperature, pH, and concentration of organics. Where high water temperature, high pH values, and more organics will shift the line to the right and decrease the slope.

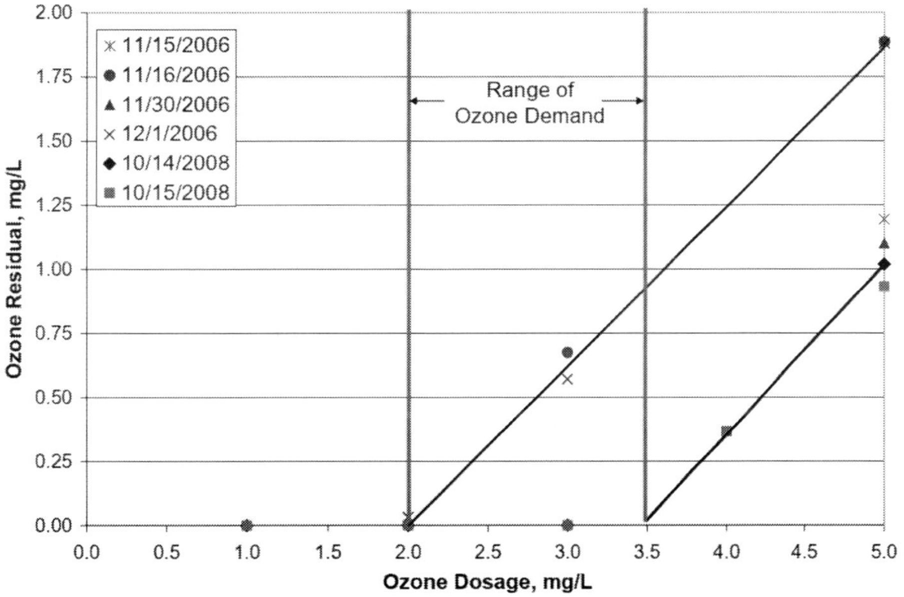

Figure 1. Ozone Demand During Bench-scale Testing.

Ozone Demand of Methanol and Nitrite

If a facility is practicing nitrification or denitrification, the resultant concentrations of nitrite will impact ozone demand. Also if the facility is using a carbon source for the denitrification, over feeding that carbon source will impact the ozone demand. A summary of the impact on treatment and ozone demand is shown in Table 3. The stoichiometric methanol dosage (in mg/L) for complete removal of nitrate can be calculated using the coefficients below and the influent concentrations of nitrate (mg/L as N), nitrite (mg/L as N), and dissolved oxygen (mg/L):

Methanol Dosage = (2.55) (Nitrate) + (1.53) (Nitrite) + (0.87) (DO)

Bench-scale testing has been conducted to determine the empirical ozone demand of methanol and nitrite. Methanol was experimentally found to have an ozone demand of 0.1 mg ozone per mg methanol. Nitrite was experimentally found to have an ozone demand of 2.1 mg ozone per mg nitrite as N. The stoichiometric ozone demand (mg ozone per mg nitrite) of 3.5 is higher than the 2.1 determined experimentally.

Table 3: Impact of Methanol Feed on Treatment and Ozone Demand

Problem	Indication	Impacts on Treatment
Underfeeding Methanol	• Nitrate concentration is above 1.5 mg/L as N	• Incomplete reduction of nitrate • Nitrite carryover • Large increase in ozone demand as a result of nitrate carryover
Slightly Overfeeding Methanol	• None	• Small increase in ozone demand as a result of methanol carryover
Grossly Overfeeding Methanol	• Operator can smell hydrogen sulfide	• Large increase in ozone demand as a result of hydrogen sulfide carryover

Impact of Denitrification Filter Operation on Ozone Demand

In addition to the water quality challenges, the operation of the denitrification filters can also make disinfection challenging at times. The filters must be backwashed, which involves taking them out of service, draining them to 1 foot above the media, scouring the media with air, and then washing a small portion of the scoured material from the media using process water. The backwashing procedure takes approximately 50 minutes and is done when the filter can no longer achieve the target flow. Filters can also undergo a "bump," which involves taking each filter out of service for about 10 minutes and reversing the flow of water through it to release trapped nitrogen gas.

Effluent from full-scale denitrfication filters was collected during normal operation, 5 minutes following a bump, and 10 minutes after backwashing. The water was dosed with 5 mg/L of ozone and the residual was measured. The ozone residual data, presented on Figure 2, indicate that bumping a filter may actually improve the water quality with respect to ozone demand. The ozone demand increased (ozone residual decreased) following backwashing.

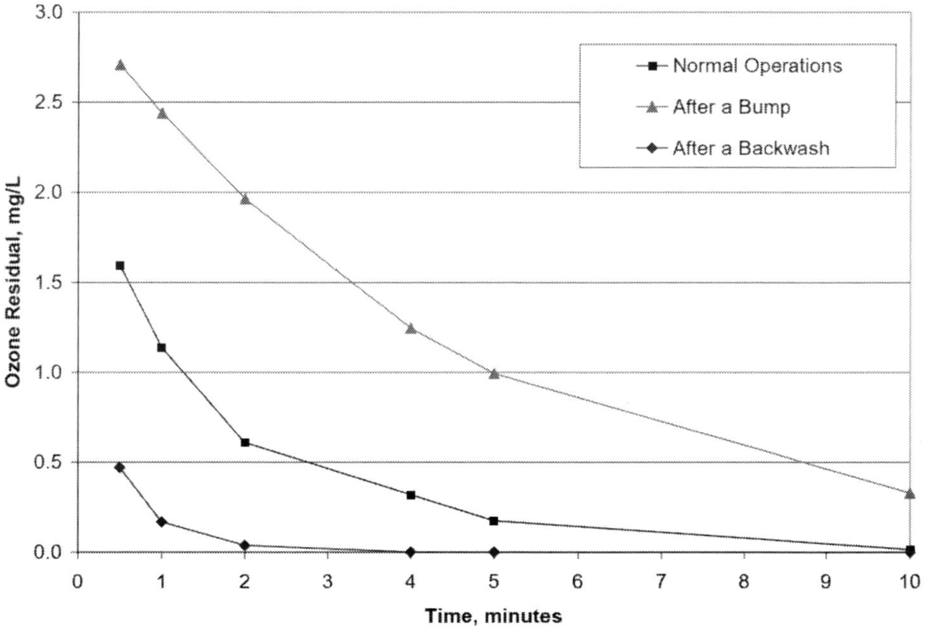

Figure 2. Ozone Residual Data for Denitrification Filter Effluent.

OZONE DISINFECTION

Ozone is a strong disinfectant that can be used to meet primary disinfection requirements. Bench-scale testing of a full-scale secondary effluent was performed to determine the ozone dosage for disinfection, samples were dosed with 1 to 5 mg/L of ozone. A dosage of 3 mg/L was found to reduce the fecal coliform concentration to less than 200 cfu/100 mL and E-coli to less than 126 cfu/100 mL, the monthly average NPDES permit limits (Figures 3 and 5).

The fecal coliform concentrations are plotted on Figure 4 as a function of the ozone demand. The same data for E-coli are plotted on Figure 6. The dosage is equivalent to the ozone demand at a value of zero. Ozone dosages that exceed the demand result in positive values, those that failed to meet ozone demand in the water are negative numbers. The relationship between obtaining a measurable residual and meeting permit limits was well pronounced. At no time during testing were monthly average permit limits exceeded when ozone demand was met.

The concentration of fecal coliforms and E-coli was lower than the permit limits when an ozone residual was detected; therefore, detection of an ozone residual can be used as an indicator that disinfection requirements are being met. If the dissolved ozone residual begins to decline, the dosage can be increased to maintain current levels of disinfection. This residual trim operating strategy can be used as a simple method to control disinfection with an ozone system.

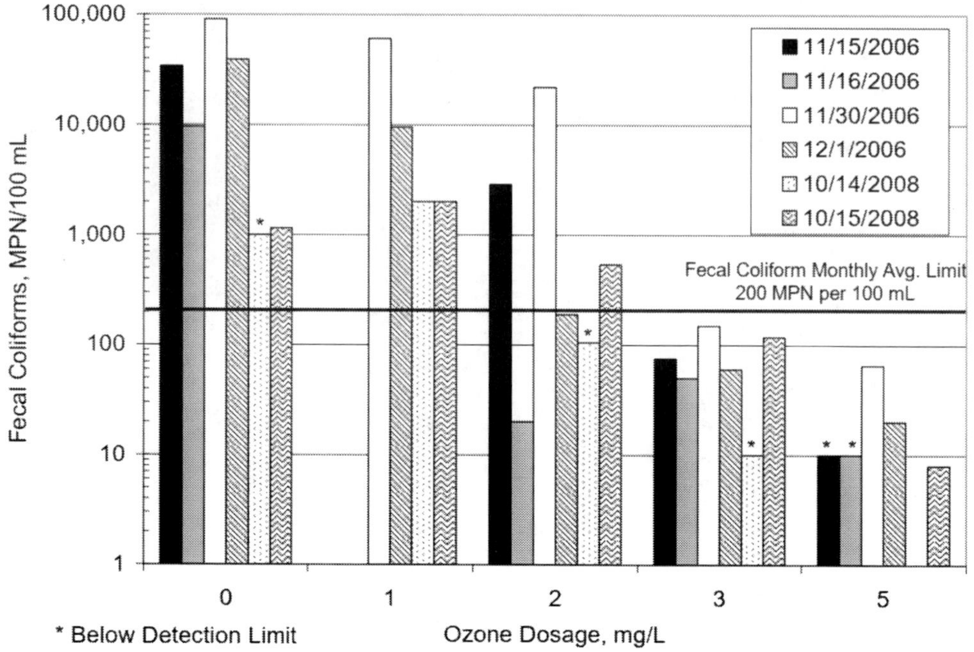

Figure 3. Bench-scale Fecal Coliform Testing Results – Ozone Contactor Influent

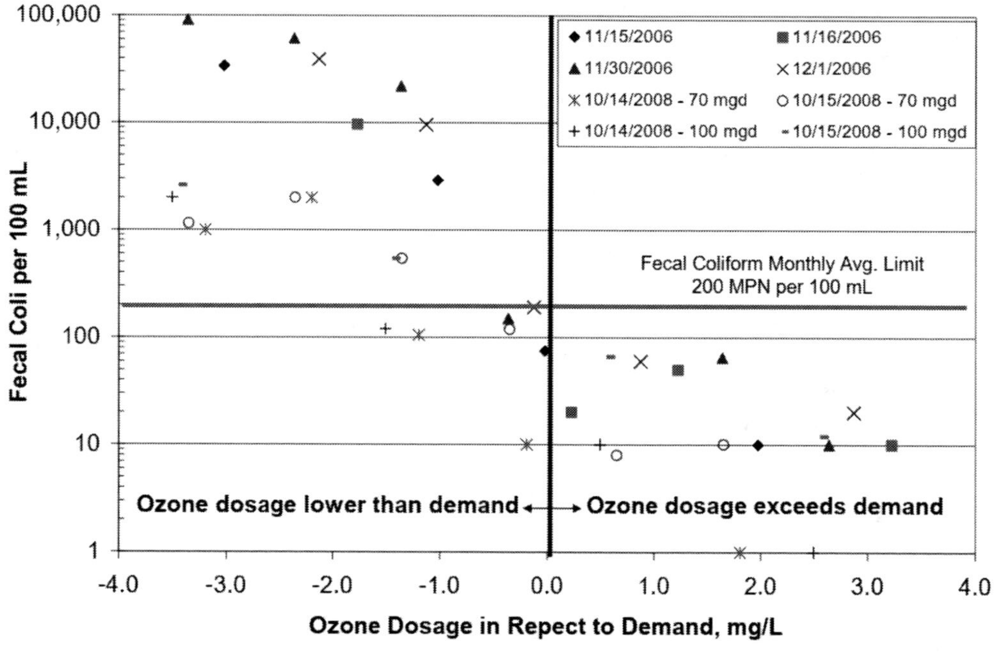

Figure 4. Fecal Coliform Concentration as a Function of the Ozone Demand

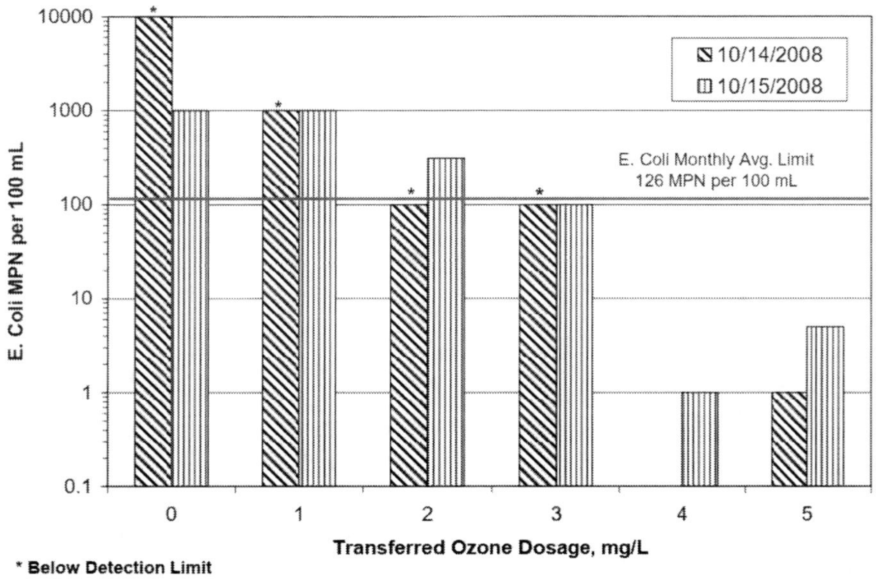

Figure 5. E-Coli Data – Simulated Blend of Ozone Contactor Influent

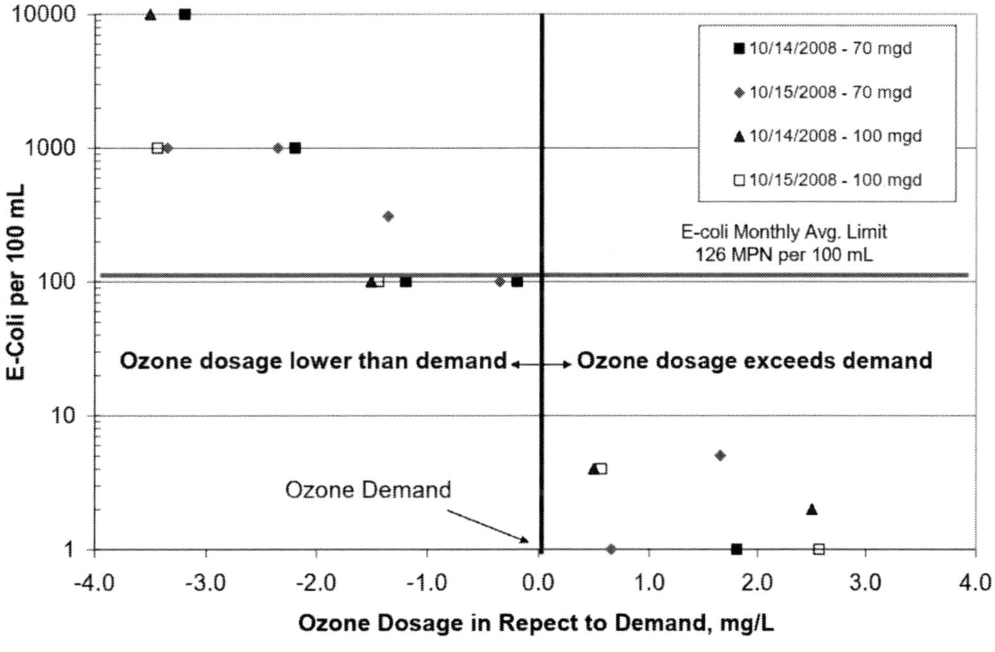

Figure 6. E-coli Concentration as a Function of Ozone Demand

Disinfection of Wet Weather Flows

To simulate the use of ozone for disinfection of wet weather flows, primary effluent samples were dosed with 5 to 10 mg/L of ozone. Based on these results, an ozone dosage greater than 10 mg/L would be necessary to reduce the fecal coliform concentration to less than 400 cfu/100 mL (Figure 7).

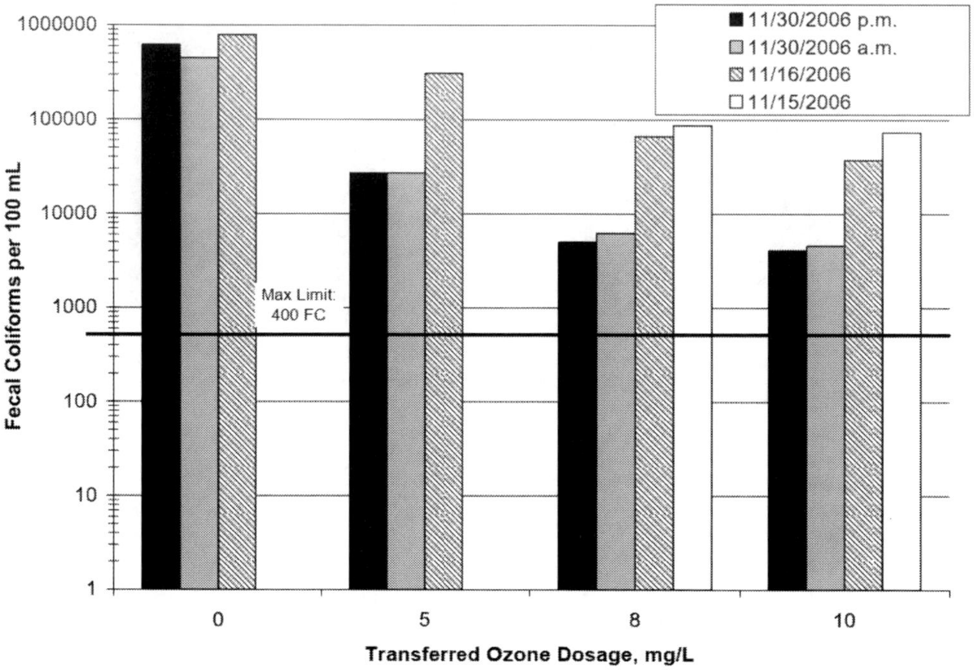

Figure 7. Primary Effluent Fecal Coliform Concentrations

Based on these results, during a recent upgrade of the ozone system, a feed point was established for the wet weather flow conditions. Due to the downturn in the economy, the work on the ozone systems has just recently been completed and is in the process of becoming operational.

CONCLUSIONS

Ozonation of drinking water continues to be strongly established as the number of installations exceeds 400. Testing result indicated that ozone could be effectively used to disinfect wet weather flows.

REFERENCES

Black & Veatch Corporation, "White's Handbook of Chlorination and Alternative Disinfectants". Fifth Edition. John Wiley and Sons, Inc. 2010.

US Environmental Protection Agency, Wastewater Disinfection Manual, 1986.

Ried, A.; Mielcke, J.; Kampmann, M.; Ternes T.A.; Teiser, B. (2004) Ozone and UV processes for additional wastewater treatment to remove pharmaceuticals and EDCs. *Proc. IWA Leading Edge Technologies Conf.*

Snyder, S.A., Wert, E.C., Lei, H., Westerhoff, P. and Yoon, Y. (2007) Removal of EDCs and Pharmaceuticals in Drinking and Reuse Treatment Processes. Awwa Research Foundation, Denver, Colorado.

Wintgens, T.; Lyko, S.; Melin, T.; Schäfer, A.I.; Khan, S.; Sherman, P.; Ried, A. (2004) Removal of estrogenic trace contaminants from wastewater and landfill leachate with advanced treatment processes. *Proc. IWA Leading Edge Technologies Conf.*

In-Line Disinfection of a CSO First Flush Basin in Detroit

Curtis D. Courter, Khamis Al-Omari and Nichole M. Sajdak (Hazen and Sawyer)

Mirza Rabbaig (Detroit Water and Sewerage Dept)

KEYWORDS: CSO, In-Line Disinfection, First Flush Basin, Demonstration Project, DWSD

ABSTRACT

The Detroit Water and Sewerage Department (DWSD) developed a Long Term CSO Control Plan for the City of Detroit in 1996. That plan recommended several uncontrolled outfalls be connected to a CSO storage tunnel. However, the cost of the project escalated coincident with the global recession of 2009, a time that the City of Detroit and DWSD were experiencing unprecedented economic distress. In June 2009, DWSD submitted a request to the state regulating agency to pursue an analysis of alternative, less costly CSO control options for the Upper Rouge River outfalls. This request was subsequently granted to DWSD and the result of the analysis was a recommendation to construct seven "First Flush" Capture Basins for CSOs with In-Pipe Disinfection and Disposable Nets to treat the remaining overflow volumes associated with peak flow rates up to 10-year, 1-hour recurrence interval.

First flush capture with in-pipe disinfection and netting of overflows is an unproven approach to controlling CSO's and achieving State of Michigan outfall standards. Disinfection of the CSO discharge will be achieved by dosing at an upstream in-pipe location(s) to allow the disinfectant contact time with the combined sewer flow to achieve an adequate kill of harmful organisms before overflow occurs to the river. The permitting agency approved this approach as a demonstration pilot facility to determine the effectiveness of these methods. This paper focuses on the challenges involved in assessing flow rates and correctly dosing and monitoring disinfection of highly variable sewer flows.

Due to concerns with locating chemical storage facilities within the dense residential neighborhood surrounding the in-pipe application points, the chemical storage and pumping equipment will be centrally located at the first flush capture tank site. From this site, NaOCl will be pumped to the application points in the sewer system.

CSO facilities operation is much different than wastewater treatment plant operations as the facility is operated intermittently and at higher flow rates. Due to the dilute nature of CSO flow, disinfection of CSO can be achieved with less contact time than wastewater treatment with various studies throughout the country suggesting that 5 minutes or less is adequate. A target contact time of five minutes was chosen for this facility.

Due to the layout of the existing collection system in the area, a minimum of two dosing points were required to provide adequate detention time. The dosing points are located along two large sewers: one dosing point is 2100 LF upstream of the outfall on a 4'-0" sewer and the other is 2700 LF upstream along a 6'-6" cylindrical sewer. Each dosing location was determined by computing 5 minutes of travel time to the outfall through each leg of the sewer using the time of travel for the 10 year, 1-hour storm peak flow rate.

INTRODUCTION

The Detroit Water and Sewerage Department (DWSD) developed a Long Term CSO Control Plan for the City of Detroit in 1996. That plan recommended that the uncontrolled outfalls between Warren Avenue and Pembroke Avenue be connected to a CSO storage tunnel. However, the cost of the project escalated coincident with the global recession of 2009, a time that the City of Detroit and DWSD were experiencing unprecedented economic distress. In June 2009, DWSD submitted a request to the Michigan Department of Natural Resources and Environment (MDNRE) to pursue an analysis of alternative, less costly CSO control options for the Upper Rouge River outfalls (note that permitting and compliance responsibility has since been shifted to the Department of Environmental Quality - MDEQ). This request was subsequently granted and DWSD undertook an analysis of alternative CSO control measures that resulted in the recommendation to construct Seven "First Flush" Capture Basins for CSOs with In-Pipe Disinfection and Disposable Nets to treat the remaining overflow volumes associated with peak flow rates up to 10-year, 1-hour recurrence interval.

Because first flush capture basins with disposable nets and in-pipe disinfection are an unproven approach to controlling CSO's and achieving State of Michigan outfall standards the DWSD and MDEQ agreed to developing a full scale demonstration pilot facility to determine the effectiveness of treatment. This paper specifically examines the disinfection system requirements and design considerations for the demonstration facility to be constructed at the Puritan CSO to provide adequate disinfection for varying sewer flow rates and contact times.

In order to meet discharge permit requirements, monthly average effluent *Fecal Coliform* bacteria counts must be below 200-cts/100-ml, and daily samples below 400-cts/100-ml. The low effluent bacteria counts and limited available land in these previously developed residential areas made in-pipe disinfection with chlorine an appealing solution. Disinfection of the CSO flow will occur through dosing of sodium hypochlorite solution at upstream locations in the conveyance system. This approach uses the travel time in the sewer system upstream of the DSO diversion and outfall as the contact time necessary to achieve an adequate kill of harmful organisms before the combined sewer overflows to the river.

DESIGN OBJECTIVES

The existing CSO outfall discharges into the Rouge River which is protected for total body contact recreation and is required to meet effluent bacteria concentrations for the 10-year 1-hour storm event. Disinfection requirements were determined using DWSD's Greater Detroit Regional Sewer System (GDRSS) model with a river level of 128.0, and surcharge criteria in the Northwest Interceptor (NWI) of not greater than 10-feet. The design flow rate to the CSO is 340 CFS.

In order to meet permit, the monthly average effluent *Fecal Coliform* bacteria counts must be below 200-cts/100-ml, and daily samples below 400-cts/100-ml. While wet weather flow characteristics, including influent bacterial concentrations, were not available for Puritan East outfall, similarities were assumed between the Puritan East flow and other DWSD operated CSO facilities.

DWSD uses NaOCl solution at all of its currently operating CSO facilities and has extensive operational experience using to disinfect CSOs. Therefore, the Puritan disinfection system is designed to accept delivery of 12 to 15% NaOCl solution and dilute it to 8% for long-term storage. In general, due to the dilute nature of CSO flow, disinfection can be achieved with less contact time than conventional wastewater treatment. Experience has shown that long disinfection contact times necessary at wastewater treatment plants are neither necessary nor appropriate for the treatment of CSO. In addition, due to the very high peak flow rates and low contact times, high chemical feed rates would be required to meet the same CT values required for disinfection of wastewater. Previous research[1] in Detroit shows that by using high-rate disinfection with appropriate mixing, contact times can be reduced from 15-30 minutes to 5 minutes or less.

To provide in-pipe disinfection within the existing collection system upstream of the outfall, multiple dosing points located in residential neighborhoods are required. For the purposes of this demonstration project, it is assumed that disinfection of inflows from trunk sewers downstream of the selected feed points will be achieved through diffusion of free chlorine residuals and secondary mixing (dispersion) within the sewers. This is consistent with the pilot nature of this facility and will be confirmed through full scale operation and monitoring. The system was designed to apply an appropriate mixing energy at the dosing points. Mixing was evaluated using computational fluid dynamics modeling to assure appropriate diffuser configuration.

DETERMINING CT REQUIREMENTS

While wet weather influent flow characteristics were not available during design at the Puritan East outfall, similarities were assumed between the water quality and sample analyses performed at nearby Baby Creek CSO. Understanding that the Puritan CSO facility will be constructed to fully capture the first flush of CSO that contains elevated pollutant concentrations, wet weather flow sampling at Baby Creek and its tributaries was evaluated to determine *fecal coliform* levels during mid to late stages of storms and densities varied from 25,600-825,600 cfu/100mL. Based on influent concentrations, the Baby Creek samples require a 3.5-log reduction to meet the monthly average effluent bacteria concentration of 200-cfu/100-ml.

Dose-response studies had been previously conducted on the Baby Creek wet weather flow to determine the chlorine dose necessary to meet the daily and monthly fecal coliform limits (400 and 200 fecal coliform per 100 mL, respectively) at various contact times. The data showed that with 5 minutes of contact time at Baby Creek, a chlorine dose of 15 to 20 mg/L was needed to achieve a 3.5 log reduction and meet the *fecal coliform* effluent limits. Therefore a CT of 100 mg/L*min has been used as the basis of design for the Puritan outfall. Further confirmation of the CT goal of 100 mg/L*min was gained from reviewing previous field scale pilot testing at the Conner Creek CSO Control Facility in Detroit where greater than 4-log reductions in fecal coliform bacteria were achieved using similar NaOCl doses with 5 minutes of contact time.

The Selleck – Collins model was utilized to calculate an empirical or predicted CT requirement for varying log reductions of E Coli concentration. It is widely used to model the declining rate of disinfection such as that observed in the disinfection of coliform bacteria at wastewater treatment plants using free and combined chlorine. The mathematical function predicts future concentrations of indicator organisms based on chlorine residual concentrations and system

---

1 Fujita, Rabbaig, Moffa, Davis, Courter, Hocking; The Challenges of Abating a Very Large Combined Sewer Overflow (CSO)"; WEF Collection Systems Conference, 2006

contact time. The equation can be rearranged and used to identify a system's required CT for a given target bacterial log reduction. To validate the use of the model in a wet weather or CSO situation, field data collected for DWSD's Conner Creek CSO Facility are compared with the model's predicted values in Figure 1.

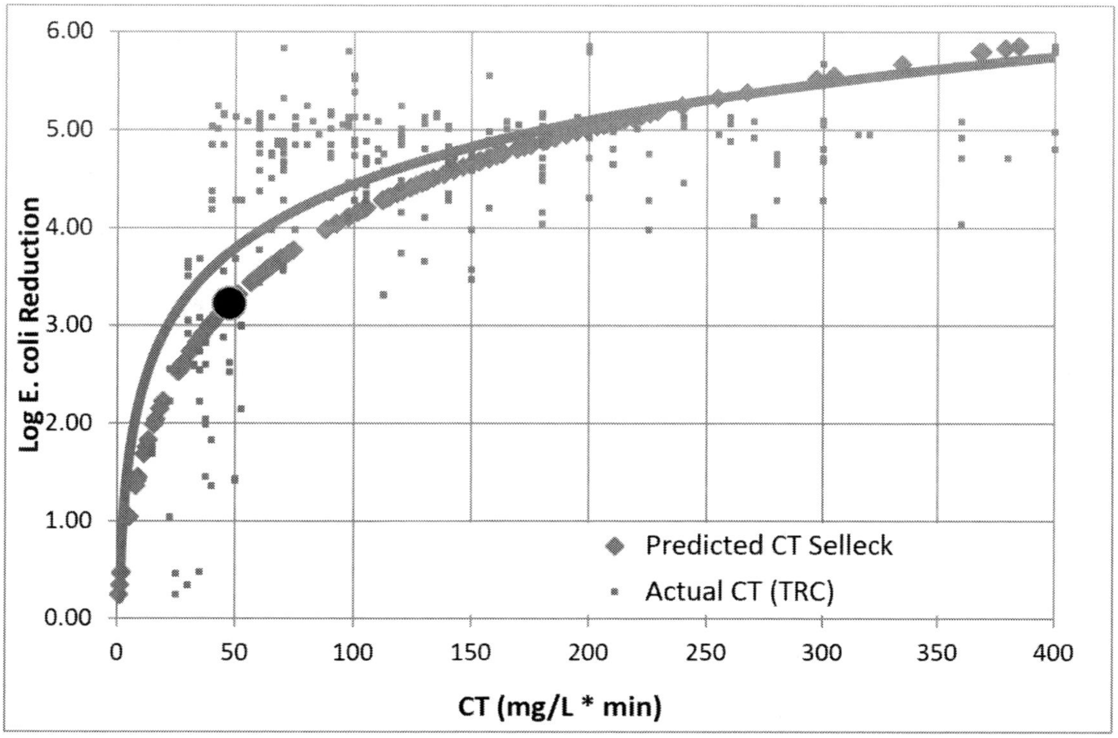

Figure 1: Conner Creek CSO, Detroit, Michigan
Pilot Test Results vs. CT Log E. Coli Reduction

The model predicts that the required 3.5-log reduction for the Puritan CSO can be achieved by using a CT of 50 mg/L*min. For this demonstration facility, a conservative CT of 100 mg/L*min was used as the basis of design to make sure that adequate disinfection equipment is available to adequately pilot this technology. This is not only due to the fact that this is a first of its kind demonstration facility, but also to a number of uncertainties, including:

- Sampling and bench scale testing information used in developing the basis of design is from a different location;
- The highly variable nature of CSO flow provides for substantial uncertainty even if sampling at the facility location was available; and
- Applying disinfectant in the collection system upstream of the overflow weir and numerous trunk sewer connections is untested at any scale.

In order to achieve a CT of 100 mg/L*min with a design minimum contact time of 5 minutes in the basin, a maximum dose of 20 mg/L of NaOCl will be required.

DOSING LOCATION AND SUBSEQUENT CONTACT TIMES

The Northwest Interceptor (NWI) conveys flow to the DWSD Wastewater Treatment plant and is regulated and excess flows discharged to the Rouge River at a number of locations including Puritan Street. Figure 2 shows the NWI upstream of the Puritan CSO. The 4'-0" diameter NWI runs south on Beaverland from McNichols Road to Puritan Street. At Puritan Street, the NWI joins with a 6'-6" sewer running east on Puritan Street. Additionally, 15" to 20" trunk sewers connect to the 6'-6" sewer at almost every intersection along Puritan Street starting at Beaverland and moving west upstream

As noted above, the disinfection system was designed to provide a minimum of 5 minutes of contact time at the peak 10-year, 1-hour flow rate. This required determining the points upstream in the conveyance stream providing 5 minutes of travel time at the design flow rate. During dry weather flow, the NWI conveys the majority of flow. However, during rain events the majority of the combined flow is delivered to the NWI through a 6'-6" cylindrical sewer under Puritan. The point of confluence of the two sewers occurs too close to the outfall to provide 5 minutes of contact time. Therefore, due to the layout of the existing collection system in the area of Puritan East outfall, a minimum of two dosing points are required. The dosing points are located along each of the two largest sewers: one on the 4'-0" cylindrical NWI on Beaverton at Florence Avenue and one along the 6'-6" cylindrical sewer under Puritan between Dacosta and Dolphin Streets. Each dosing location was determined by computing 5 minutes of travel time to the outfall through each leg of the sewer using the time of travel for the 10 year, 1-hour storm peak flowrate.

Due to concerns with locating chemical storage facilities within the dense residential neighborhood surrounding these application points, the chemical storage and pumping equipment is going to be centrally located at the first flush capture tank site adjacent to the Puritan outfall. From this site, NaOCl will be pumped approximately 2,100 feet to the dosing point at Beaverton and Florence (Figure 2) and approximately 2,700 feet to the dosing point at Puritan between Dacosta and Dolphin (Figure 3).

For flow rates less than the design flow rate, the contact time will be greater than 5 minutes prior to discharging at the outfall. The dosing algorithm for the disinfection system was designed to maintain a CT of 100 mg/L*min in order to minimize dosing and associated total residual chlorine (TRC) levels at the outfall for a given flow rate. To maintain the design CT, the chemical dose will be flow paced and self-adjusting to reflect changes in both flow rate and the resulting contact time. Chemical feed pumping will be initiated prior to filling the first flush capture tank to account for travel time of the chemical from the central storage and feed point to the dosing locations and of the chlorinated flow from the dosing locations to the outfall. The feed rate will be flow paced using a permanent flow meter installed within the NWI upstream of the new diversion dam (see Figure 2 for location).

Figure 2: Proposed NWI Dosing Point at Beaverland Street and Florence Street

Figure 3: Proposed Dosing Point at Puritan Street between Dacosta and Dolphin Streets

Feed rates have been determined based on the design flow rate of 340 CFS. The total feed rate necessary for disinfecting the 10-year, 1-hour storm with 8% sodium hypochlorite is 38 GPM. Using the design storm hydrographs, the maximum feed rates to each dosing point were determined as shown in Table 1 below.

Table 1: Maximum Feed Rates to Dosing Locations For Puritan CSO Disinfection System

	Peak Flow (CFS)	Peak Flow (MGD)	Peak Dosage (mg/L)	8% Sol. Feed Rate (GPM)
U/S NWI	70	45	7.7	3.0
U/S Puritan	324	209	18.7	34.9

Because the system is designed using CT setpoint control, the design feed rate is only necessary to disinfect the peak flow rate to the facility. The majority of the time, during lower flows and larger available contact times, the influent flow rate to the facility will be much smaller. At lower flower rates, the velocity will be somewhat lower and thus the contact time available in the conveyance system will be larger and a lower feed rate will be necessary. The chemical feed system therefore requires a large turndown capability.

As noted above, there will be two feed points. The GDRSS model was used to evaluate the proposed conditions at the outfall and at each upstream dosing point. Figure 4 includes hydrographs at the outfall and two dosing points, which were used to evaluate feed rate requirements to each point. The hydrograph for Beaverland and Florence (Upstream NWI) shows that the NWI flow peaks out and remains nearly constant throughout the event. This is likely due to upstream regulation that occurs near McNichols. For the point upstream of Dacosta along Puritan (Upstream Puritan), the hydrograph shows significant flow variation throughout the duration of the event. Also note that the combined flow rate in the collection system upstream of the diversion is higher than the facilities design flow of 350 cfs. The disinfection system is sized to treat the full peak flow with part of the disinfected flow staying in the collection system and continuing downstream toward the WWTP. With only one flow monitor, downstream of the confluence of the two sewers but upstream of the diversion weir, special care had to be taken to develop the dosing algorithm for each dosing location.

The dosing algorithm for the dosing point on Beaverland at Florence was the most challenging since the model currently shows a flat response during extreme wet weather. This means that there is some flow threshold at which larger total flows to the facility, where the only flow monitor is located, do not result in higher flows at the dosing point. Below that threshold it is anticipated that flows at the dosing point will be variable in nature, similar to the variation seen at the dosing point on Puritan upstream of Dacosta.

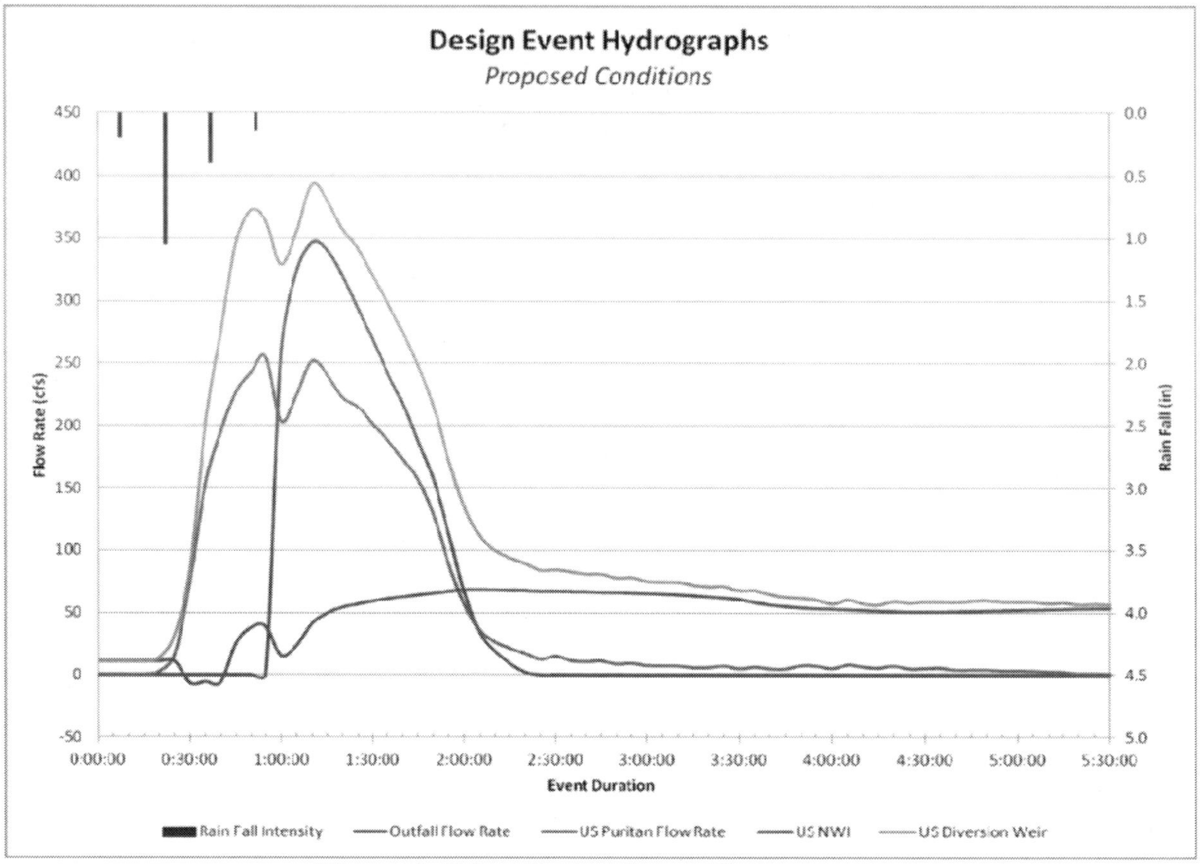

Figure 4: Select Design Event Hydrographs

With a minimum of 5 minutes of contact time, achieving good mixing in close proximity of the nozzles at both points of application is important. Initially, a diffuser was planned using several nozzles of 1"-2" diameter at various elevations along one side of the sewer. Through CFD analysis of the pipe flow and application points the hydraulics of injecting the chemical stream into the sewer flow were studied. The analysis showed that the chemical jet stream was unable to penetrate the combined sewage flow and therefore disinfectant did not mix well with the sewer flow. Rather, the disinfectant plume got trapped along the pipe walls and the flow traveled downstream a significant distance before adequate mixing occurred.

Based on the results of the CFD, the configuration of the dosing point was changed to use a single duckbill style valve angled upstream into the sewer flow path. By using a single nozzle, the disinfectant had increased momentum from increased flow and velocity, and the jet penetrated further into the sewer flow and provided more energy which lead to better mixing. Figure 5 below shows the chlorine concentration through a cross section of the pipe as flow continues laterally down the pipe with mixing occurring within the first 200 LF of pipe after addition. Note that this model run does not include the laterals entering the pipe, which introduce a significant amount of turbulence that further aids mixing.

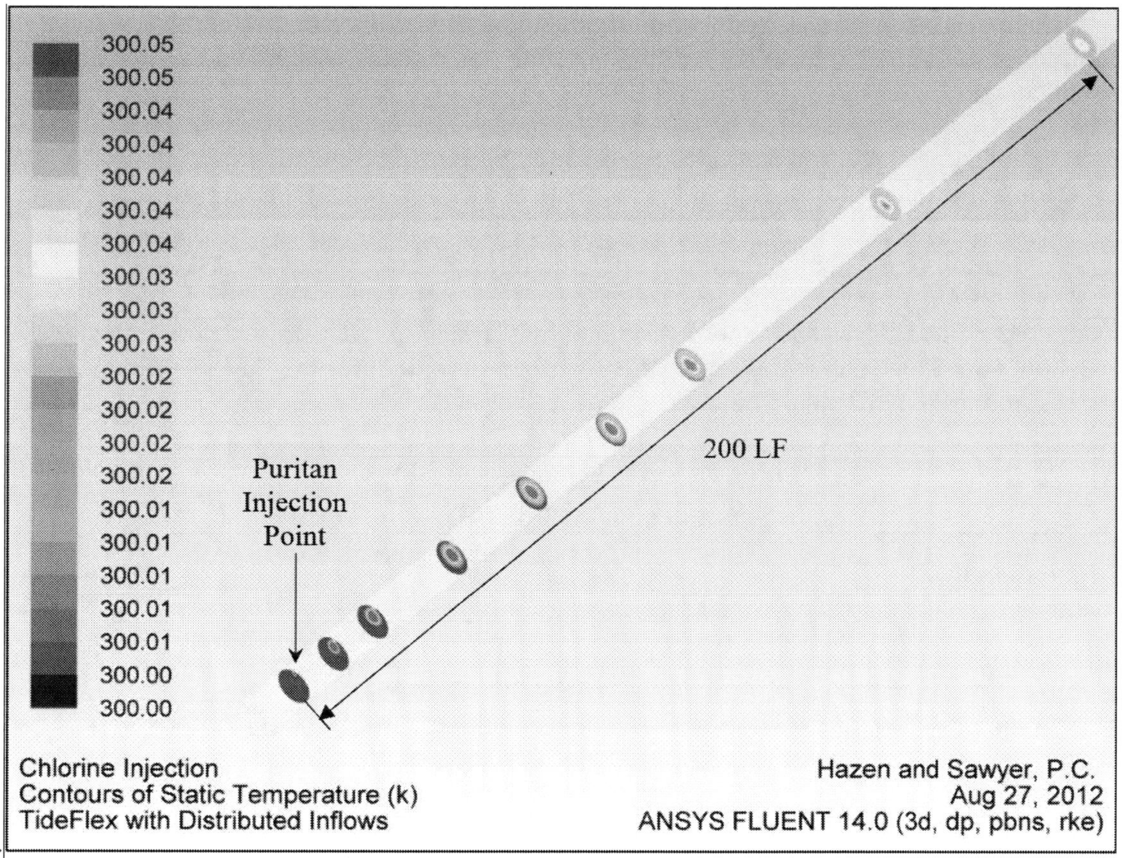

Figure 5: Mixing at Puritan Dosing Location

CONCLUSIONS

In conclusion, the disinfection system for the Puritan East CSO is based on the 10-year, 1-hour storm design flow rate of 340 CFS. The system is designed to disinfect combined sewage using sodium hypochlorite (NaOCl) stored at 8% strength.

Table 2: NaOCl Disinfection System Design Criteria

Delivered NaOCl Strength	Minimum 12.5% by weight
Stored NaOCl Strength	Assume 8% (to be confirmed from long term model simulation)
Design Contact Time	Greater than 5 minutes at 340 CFS
Design Dose	20 mg/L at 340 CFS

Because the first flush will be captured prior to disinfection, less than a 4-log reduction in fecal coliform is required. The disinfection system will be designed to meet water quality bacteria standards for *Fecal Coliform* Bacteria: 200-cts/100-ml (monthly), 400-cts/100-ml (Daily). Due to uncertainties related a lack of site specific data and the varying nature of wet weather flow, a

CT of 100 mg/L*min has been used for the preliminary basis of design. Coupled with the selection of dosing points that provide a minimum contact time of 5 minutes in the sewers, a maximum dose of 20 mg/L of NaOCl will be required. Pump sizing is based on this maximum dose, coupled with the design NaOCl strength.

Based on the existing sewer orientations, two dosing points will be required, one along the NWI on Beaverton at Florence and one along Puritan between Dacosta and Dolphin Streets. Dosing points were selected to provide greater than 5 minutes of contact time prior to discharge from Puritan East Outfall No. 72 at the maximum potential design flow rate of 340 CFS.

Chemical dose will be flow paced and determined by CT setpoint and flow metering upstream of the diversion dam. The basis of design for the disinfection system for the Puritan CSO Control Facility is summarized in Table 3 below.

Table 3: NaOCl Storage and Feed

Design Summary for 340 CFS Design Storm

Feed Line Length	NWI: 2100 LF Puritan: 2700 LF
Dilution/Carrier Water	Non-Potable Water; Local city pressure with back up booster pump
Type of NaOCl Feed Pumps	Gear pump
Feed Control	CT Setpoint, Flow Paced
Number/Volume of Storage Tanks	2 @ 2,500 gallons
Number/Capacity of Feed Pumps	5 : 2 large (+/- 60 GPM) 3 small (<3 GPM turndown capability)

Pilot Study Results for Peracetic Acid and Sodium Hypochlorite Disinfection of Combined Sewer Overflows

Jay Brady, P.E., Stanley Consultants

ABSTRACT

The City of Fort Madison (City), Iowa is required by a United States Environmental Protection Agency (USEPA) consent order to mitigate its combined sewer overflows (CSOs). The selected CSO mitigation method is chemical disinfection of the overflows to reduce *Escherichia Coli* (*E. Coli*) bacteria to 1073 most probable number per 100 milliliters (MPN/100 mL). The consent order required a full-scale pilot CSO disinfection system be constructed and operated to demonstrate the feasibility of disinfecting highly variable CSO discharges. As part of the design process, a bench study of peracetic acid (PAA) and sodium hypochlorite (SH) disinfectants was performed by the University of Iowa on behalf of Stanley Consultants, the City's wastewater engineer, to establish potential dose rates and contact times. The bench study utilized jar testing to examine dose concentration-time relationships and impact of suspended solids on disinfection effectiveness. A pilot chemical disinfection system consisting of chemical feed systems and disinfection contact pipe was designed and constructed.

Bench study work under ideal and well mixed conditions suggest that a SH dose of 20 mg/L with 5 minutes of contact time or a dose of 10 mg/L with 10 minutes of contact time should reduce E. Coli below 126 mpn/100 mL.

Bench study work under ideal and well mixed conditions suggest a PAA dose of 15 mg/L or higher with 3 minutes of contact time or a dose of 10 mg/L with slightly greater than 5 minutes of contact time should reduce *E. Coli* below a target of 126 mpn/100 mL.

Preliminary pilot study results indicate that sodium hypochlorite appears to be effective at meeting the CSO disinfection goal of reducing *E. coli* in CSO water to below 1,073 mpn/100 mL. Unfortunately dose data is not available.

Preliminary pilot study results suggest PAA can be an effective disinfectant for meeting CSO *E. Coli* reductions. The data provides evidence that doses that provide residuals of 0.5 mg/L or greater reduces *E. Coli* to down to approximately 100 mpn/100 mL.

KEYWORDS

Peracetic Acid Disinfection, Sodium Hypochlorite Disinfection, Combined Sewer Overflows

INTRODUCTION

The City of Fort Madison, Iowa, a community of 11,000, resides on the Mississippi River in southeastern Iowa. The City entered into a consent agreement with the USEPA to mitigate the impact of its CSOs on the Mississippi River, namely bacteria as evidenced by *E. Coli*. The mitigation approach is to disinfect the overflows to reduce the levels of *E. Coli* bacteria entering the river. The consent agreement requires a full scale pilot demonstration of the CSO disinfection system to determine the impacts of CSO flow, duration, and pollutant variability, and the overall feasibility of concept.

The City's combined sewers are confined to the older eastern portion of the community. The combined sewer system shown in Figure 1 consists of a series of trunk sewers that run from the toe of the upland bluffs at the northern fringe of the city to its Mississippi riverfront. Historically, these trunk sewers discharged directly into the river. In 1966, a riverfront interceptor was constructed to collect the dry weather flows from the trunk sewers and convey them via the 10th Street Pump Station to the wastewater treatment plant at 20th Street. Leaping weir control structures were installed at each of the trunk–interceptor sewer interface locations to control the amount of flow entering the interceptor during higher wet weather flows. The flow that exceeds the volume allowed by the weirs is discharged to the river via the original trunk sewer outfalls.

Figure 1 Fort Madison Combined Sewer System

Flow monitoring, automated sampling, and floatable control structures were added to the each of the CSO locations to meet the City's nine minimum CSO control obligations. Monitoring and study work including hydraulic modeling performed in conjunction with preparation of the City's CSO long term control plan (LTCP) indicated that the environmental impacts of the CSOs on the Mississippi River are negligible due to low pollutant concentrations and the further dilution of the already dilute CSO water by the much larger Mississippi River. However, *E. Coli* levels in the CSO discharges generally exceeded regulatory limits.

The City decided to approach CSO mitigation by disinfection due to lower initial capital cost. Disinfecting CSOs is challenging due to intermittent flows, highly variable flow rates, and the overall variability of the CSO water quality. The historic CSO flowrates at 6th Street vary between less than 0.0283 cubic meters per second (cm/s) (1 cubic feet per second (cfs)) to a peak of 0.354 cm/s (12.5 cfs).

The CSO water quality also varies. Table 1 data illustrates the variability inherent to Fort Madison's CSOs.

Table 1 CSO Water Quality

Parameter	Minimum mg/L	Median mg/L	Maximum mg/L
BOD_5	2	20	145
TSS	1	135	1480
NH3N	<0.1		1.4
pH	6.9 S.U	7.7 S.U.	8.4 S.U.

BOD5 = 5-Day Biochemical Oxygen Demand
TSS = Total Suspended Solids
NH3N = Ammonia Nitrogen

USEPA, noting these types of challenges for CSO disinfection, required that the City perform a full scale pilot system to demonstrate that the CSOs could be consistently disinfected to meet permit limits for *E. Coli*.

An engineering feasibility study was performed by Stanley Consultants to evaluate potential disinfection technologies appropriate for CSO disinfection with minimal pretreatment upstream of the disinfection process. Chemical disinfection using sodium hypochlorite (SH) or a newly emerging wastewater disinfectant, peracetic acid (PAA), showed the most promise for success. PAA, a very strong oxidant, is a solution of hydrogen peroxide and acetic acid in equilibrium in water. The chemical formula of PAA is CH_3C_3H

Typical solution strength is 15 percent PAA and 22 percent hydrogen peroxide. PAA has been used in previously in the hospital and food/beverage industries as a biocide. PAA offers several advantages over SH including being a faster acting oxidant requiring lower doses and producing no residual byproduct. PAA also has a longer shelf life of six months to a year. However, PAA is expensive relative to SH on a unit cost basis, has limited availability, and is new to regulators.

Additional investigation was necessary to determine the probable values of critical design parameters including chemical dose and contact time needed with anticipated suspended solids concentrations. The University of Iowa Department of Civil and Environmental Engineering was retained to perform bench testing to gain more insight into these design parameters.

METHODOLOGY

Bench Study Methodology

The study objective was to determine appropriate doses and reaction rates for independent use of sodium hypochlorite (SH) and peracetic acid (PAA) and the potential impact of suspended solids levels on disinfection results through bench study work. The results of the bench study were utilized to size and design a full-scale pilot chemical disinfection system at one of the city's CSOs. The pilot system is capable of feeding either SH or PAA allowing either chemical to be studied.

Jar testing was performed on simulated CSO effluent to develop chemical disinfection dose rates at varying suspended solids concentrations representative of typical Fort Madison CSO effluent. The simulated CSO effluent consisted of primary clarifier effluent with fine sand addition to attain target total suspended solids (TSS) concentrations of 50, 125, and 200 mg/L. Sodium hypochlorite dose was varied from 10 to 40 mg/l as available chlorine and the PAA dose was varied from 10 to 20 mg/L (Parkin, 2009). The testing was performed to determine doses needed to reduce *E. Coli* to a target of 126 mpn/100 mL in about 5 minutes to minimize contact volume requirements. The target *E. Coli* number was the preliminary discharge limit identified by regulators.

After preparation of the simulated CSO effluent, an initial sample was taken for *E. Coli* and TSS testing. The chemical dose was then applied. The test jars were mixed at 175 rpm to maintain all solids in suspension during the testing. For a given trial, samples were taken from the jar at 10 seconds, 2, 5, 10, 20, and 30 minutes for *E. Coli* testing to develop a dose – time response curve. The samples were immediately dosed with an excess of sodium thiosulfate to quench the oxidation reactions. *E. Coli* was measured using Standard Method 9223B (Parkin, 2009).

Pilot System

The full scale pilot system, consisting of a chemical feed building, SH, PAA, and sodium bisulfite chemical feed systems, debris pretreatment, and disinfection contact chamber, was designed and constructed in 2009 and 2010. Field study work commenced in the fall of 2010 and continued through December 2012. Field study work was hampered by several construction-related issues including flow short circuiting, and drought conditions requiring the extended pilot time period.

The pilot system is located in the City's Riverfront Park on the 6th Street CSO. A contact chamber was provided by constructing a 305 meter (1000 foot) long, 900 mm (36 inch) and 1500 mm (60 inch) diameter polypropylene contact pipe to provide 18 minutes of contact time at 0.354 cm/s (17.4 cfs) as shown in Figure 2.

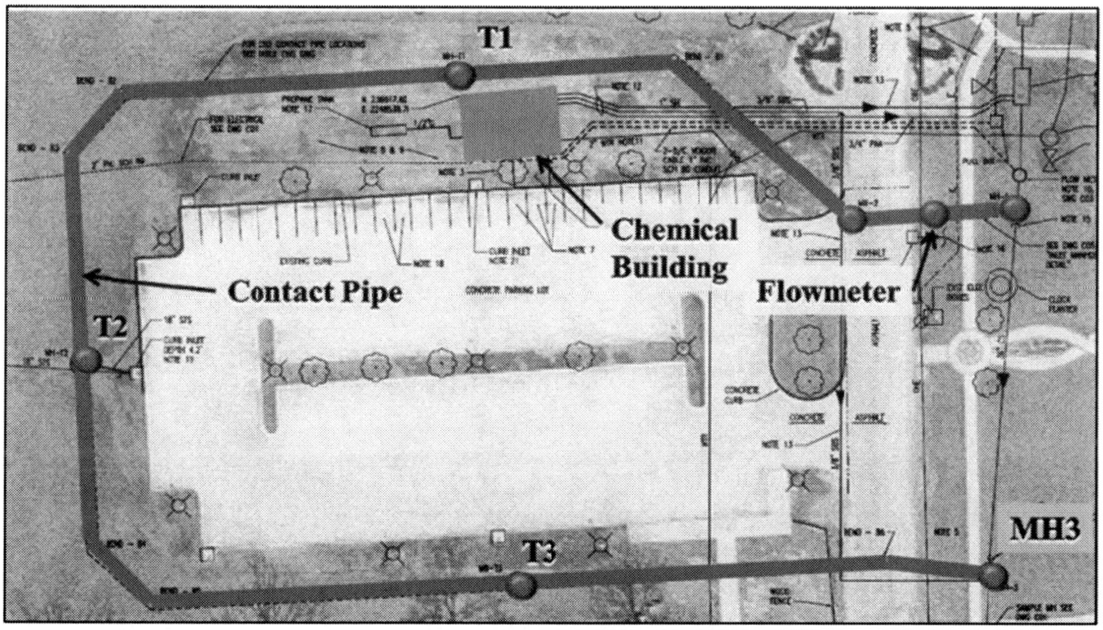

Figure 2 – Pilot Disinfection System Contact Pipe Layout

The contact pipe is equipped with an outlet weir to separate river backwater from the pipe contents and to maintain contact volume and time when the river is low. The pipe is equipped with five manholes for access for cleaning and sampling. Flow is pretreated by a baffle chamber equipped with coarse static screen system for debris and floatables reduction and the flow rate measured with an area velocity flow meter.

The chemical feed building consists of a flood-proofed masonry structure housing SH storage tanks and feed pumps. The PAA feed system consisting of totes and feed pumps are located within the building in a bay separated from the SH system. The building features chemical containment in case of spills or leaks. A sodium bisulfite feed system consisting of totes and feed pumps is also provided to quench the action of the SH or PAA. Separate chemical feed pipes are routed from the building to the diffuser feed point to allow either SH or PAA to be fed. No supplemental mechanical mixing is provided for dispersing the chemicals. Sodium bisulfite piping is routed to the discharge weir. The chemical feed rate is flow-paced based on the upstream flow meter. A data logger records flows and the chemical feed systems are equipped with chemical totalizers. Figure 3 presents the system schematic. Figures 4 – 6 show photographs of various system components.

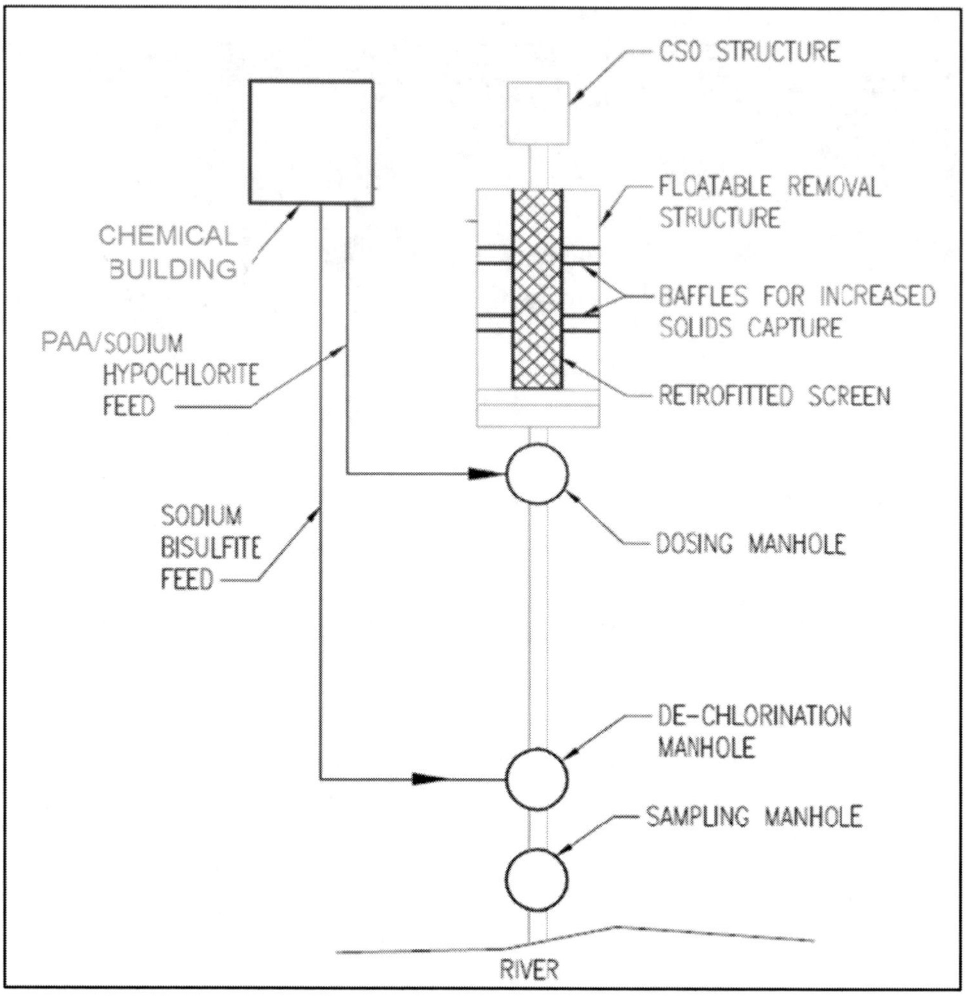

Figure 3- System Schematic

Figure 4 – Contact Pipe Installation

Figure 5 – Chemical Feed Building

Figure 6 – PAA Feed System

The chemical feed system is triggered upon the flow meter sensing CSO flow. The chemical feed pump initially runs at 100 percent capacity for a short period of time to charge the chemical feed piping. The pump then switches to preset flow-paced dosing which continues until CSO flow stops.

Grab samples are obtained upstream of the contact system at the overflow structure (river gate location), the flow meter manhole, T1, T2, T3, and MH-3 and are tested for *E. Coli*. The City strives to start sampling within 30 minutes of the CSO event start. Laboratory analysis of the samples is performed by the City of Fort Madison.

RESULTS AND DISCUSSION

Bench Study Data

Figure 7 presents the results obtained with SH doses as available chlorine at 10, 20, 30, and 40 mg/L with a high TSS concentration of approximately 200 mg/L. The data indicates that a SH dose of 20 mg/L or higher should reduce *E. Coli* below a target of 126 mpn/100 mL within 5 minutes in a well-mixed environment. A SH dose of 10 mg/L will require approximately 10 minutes of contact time to achieve the same target *E. Coli* concentration (Parkin, 2009).

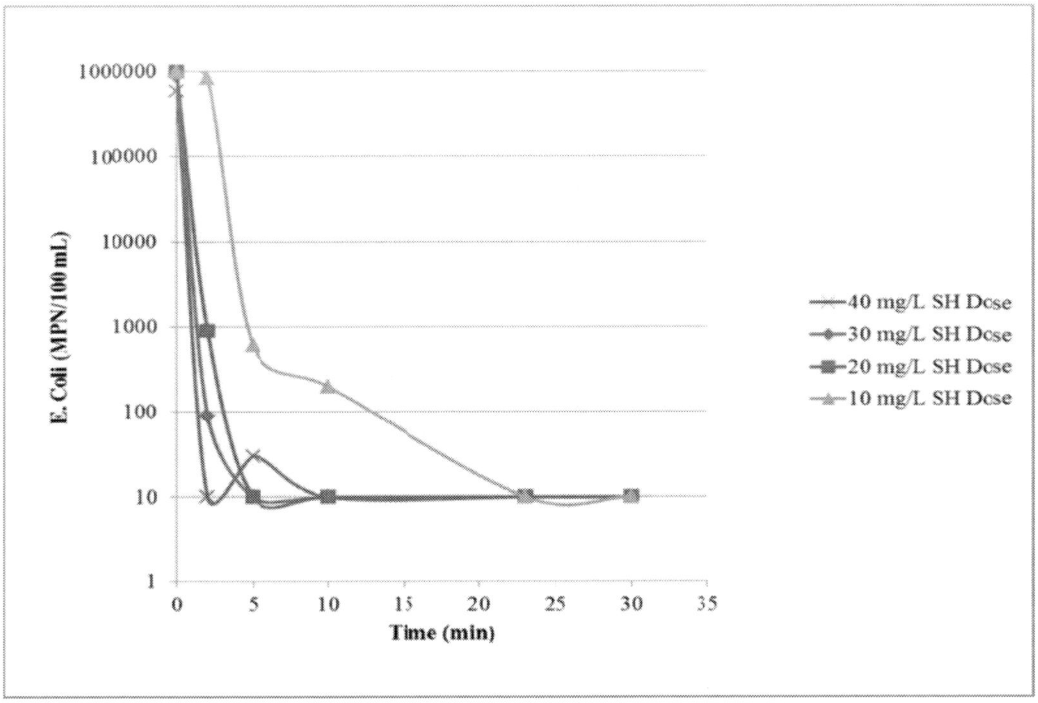

Figure 7 – Results of Variable SH Doses at High TSS Concentration

Figure 8 presents the results obtained with PAA doses at 10, 15, and 20 mg/L with a high TSS concentration of approximately 200 mg/L. The data indicates that a PAA dose of 15 mg/L or higher should reduce *E. Coli* below a target of 126 mpn/100 mL within 3 minutes in a well-mixed environment. A PAA dose of 10 mg/L will require approximately 5 minutes of contact time to achieve the same target *E. Coli* concentration. It should be noted that the 10 mg/L PAA dose consistently achieved near the target bacteria level in 5 minutes, but for some experiments, bacteria levels were somewhat above this target level (Parkin, 2009).

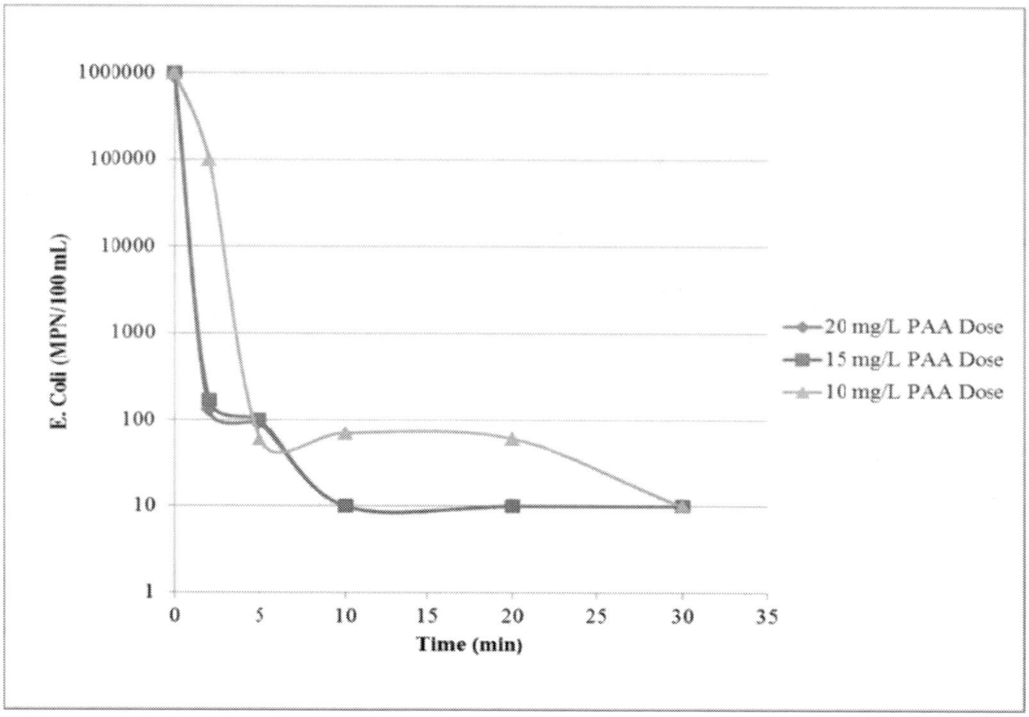

Figure 8 – Results of Variable PAA Doses at High TSS Concentration

Pilot Data

Field study work was hampered by several construction related issues, initial operational challenges, and drought conditions. However, these unintended situations have provided additional insight into the performance of the system and provide real world implications for other systems configured for disinfecting marginally clean wet weather flows.

Thirty-two CSO events occurred at the 6th Street CSO for the period of the pilot study from June 2010 to December 2012. The quality of the overflow water during the pilot study consisted of 5 to 529 mg/L TSS, 3 to 36 mg/L of BOD_5, and 0.1 to 1.5 mg/L of NH3N. The pH varied from 6.2 to 8.4.

The pilot system was operated using SH as the disinfectant agent from June 2010 to September 2010. The disinfectant agent was switched to PAA in September 2010 and used until the end of the study.

Data from SH disinfected events is sparse due to initial sampling procedural issues and overall feed system control issues. In addition, the overall dose level is uncertain due to chemical feed piping leakage stemming from faulty installation. Apparent sodium hypochlorite doses as available chlorine range from 32 to 255 mg/L for three events with resulting *E. Coli* concentrations ranging from 17 to 167 mpn/100 mL.

Figures 9 and 10 present SH results and flows from a CSO event that occurred on July 19, 2010. The sodium hypochlorite was effective at reducing the E. Coli from about 10,000 mpn/100 mL to about 167 mpn/100 mL. Unfortunately, dose amount is not able to be ascertained due to equipment issues. The grab samples used for analysis were taken about 10:30 a.m. during the lower flow 1st peak of this two peak event shown in Figure 10. The flow was not of sufficient volume for the event's water to have reached these sampling locations with perhaps the exception of T-1. Therefore, while Figure 9 presents promising results, they are not conclusive proof of successful disinfection.

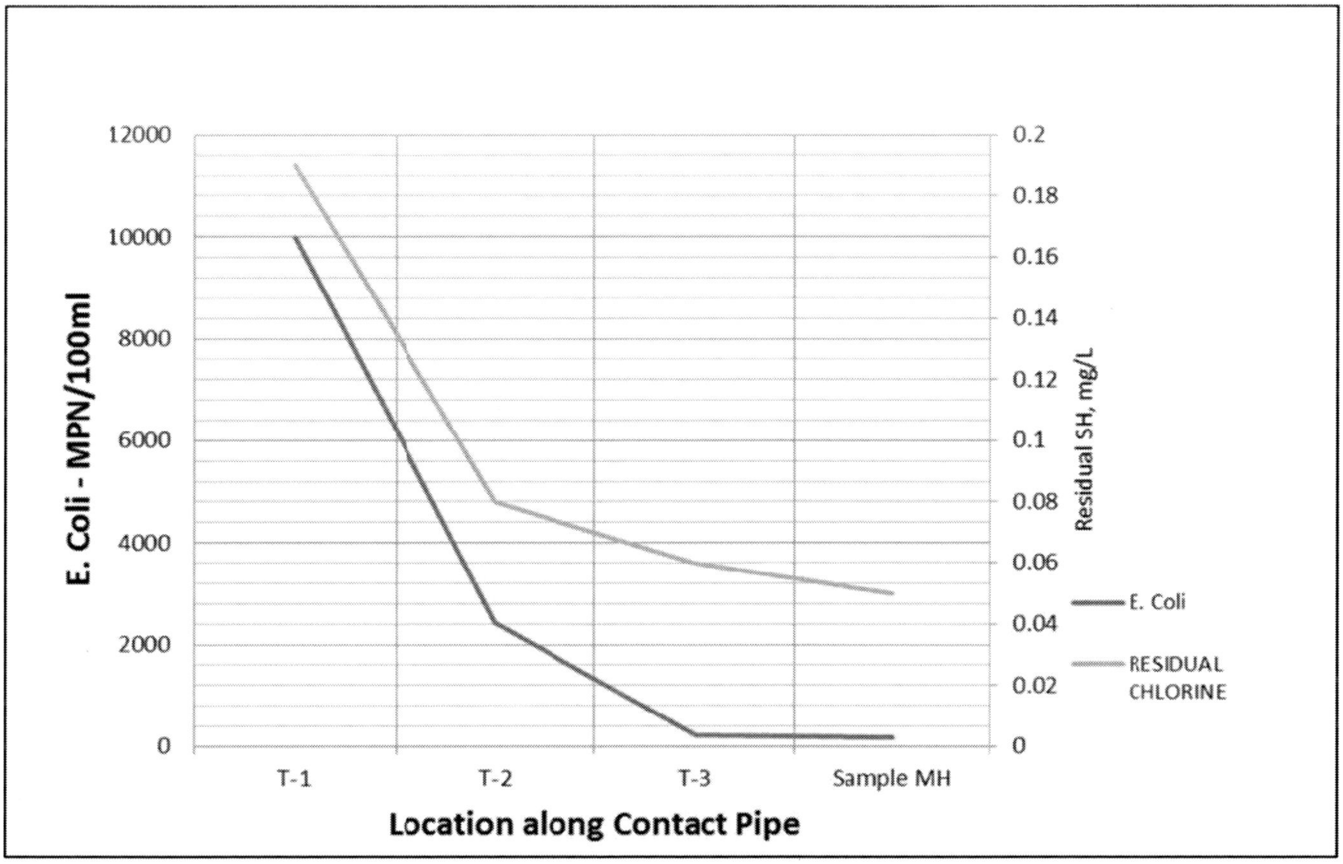

Figure 9 – SH Results for July 19, 2010 CSO

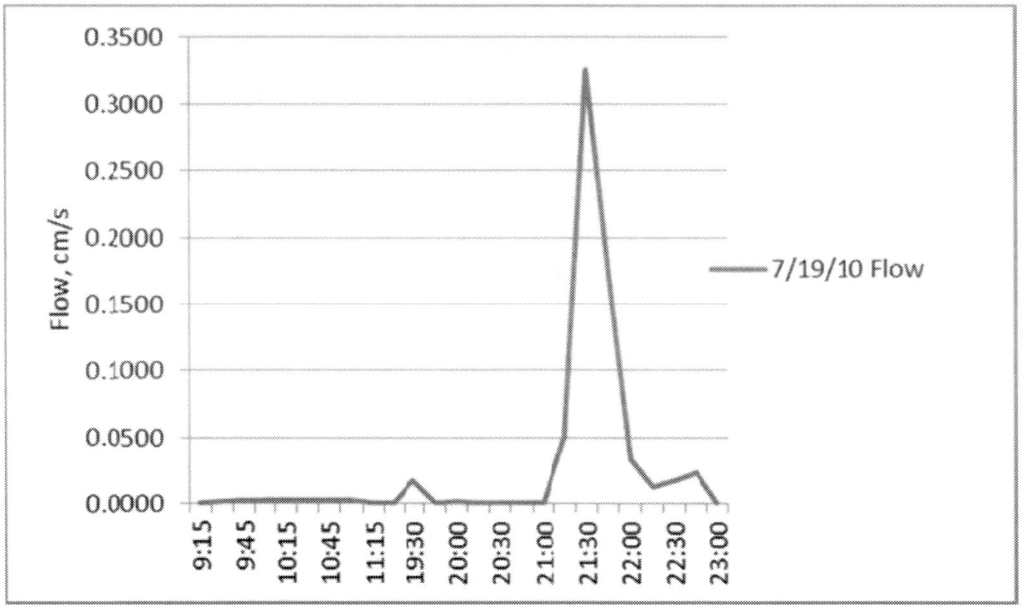

Figure 10 – July 19, 2010 CSO Event Flow

As shown in Figure 11, PAA results are variable with some events being effectively disinfected and others having variation in *E. Coli* concentrations along the length of the contact pipe. Some of the variability is due to the overall performance of the chemical feed pumping system. The chemical feed pumps for the PAA are diaphragm pumps which, unfortunately, were not equipped with degassing heads. At times, the pumps became vapor-locked causing the pumps to not start chemical feed or not deliver the full dose expected. The apparent PAA dose delivered was highly variable ranging from 0.9 to 37 mg/L and when combined with variable CSO water qualities produced a range of residual PAA concentrations.

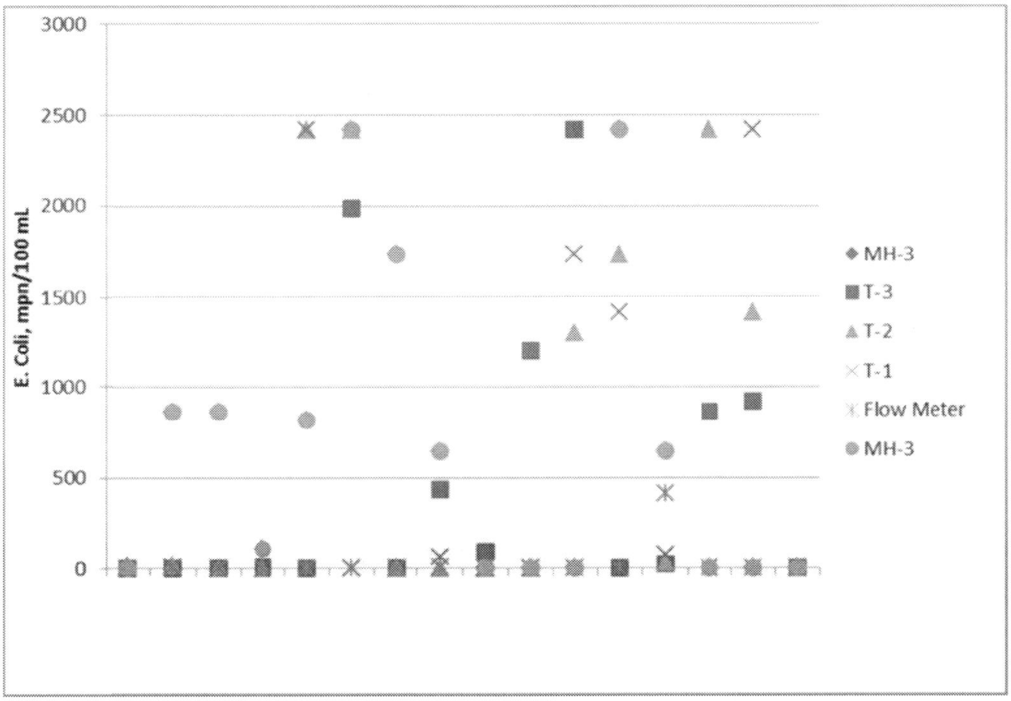

Figure 11 – July 19, 2010 CSO Event Flow

However, the data presented in Figures 12 and 13 does suggest PAA is effective when a residual of 0.5 mg/L or greater can be achieved. Figure 12 shows *E. Coli* concentrations compared to PAA residual. This figure generally indicates that a PAA residual of 0.5 mg/L or more reduces *E. Coli* to 100 mpn/100 L or less. Figure 13 shows *E. Coli* concentrations compared to PAA residual for E. Coli samples below 1000 mpn/100 mL. This figure shows increasing *E. Coli* as PAA residual drops below about 0.5 mg/L.

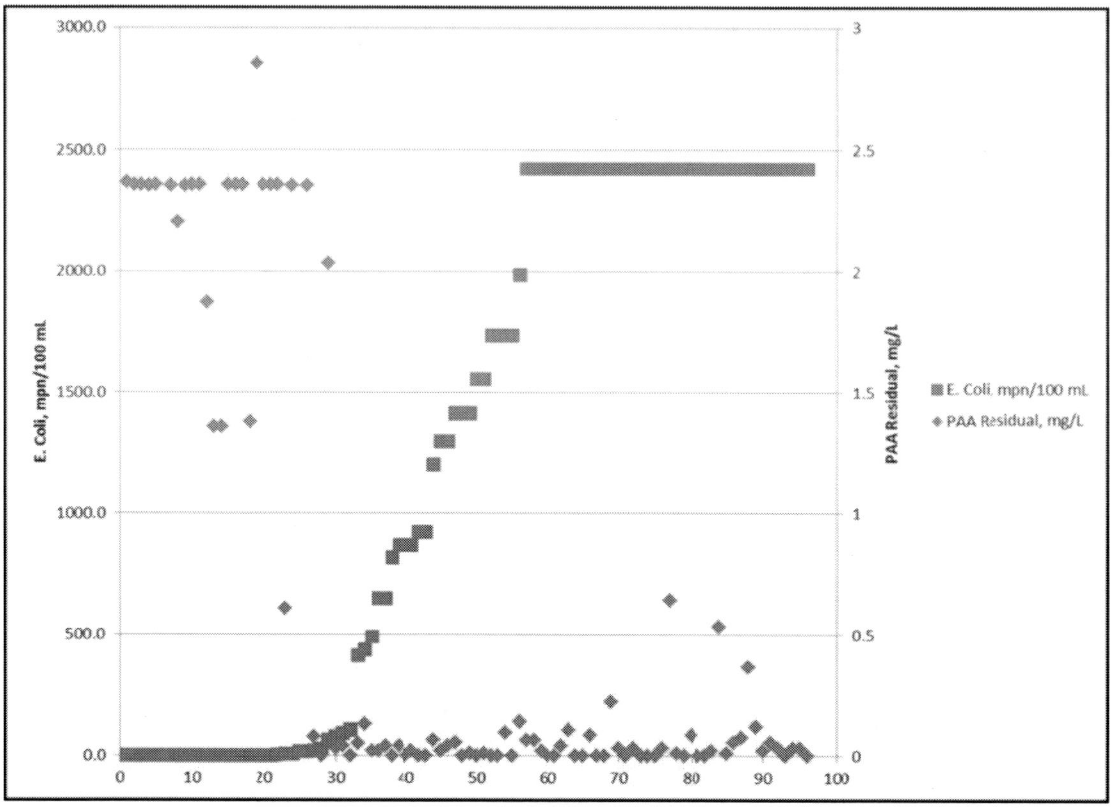

Figure 12 – *E. Coli* Compared to PAA Residual

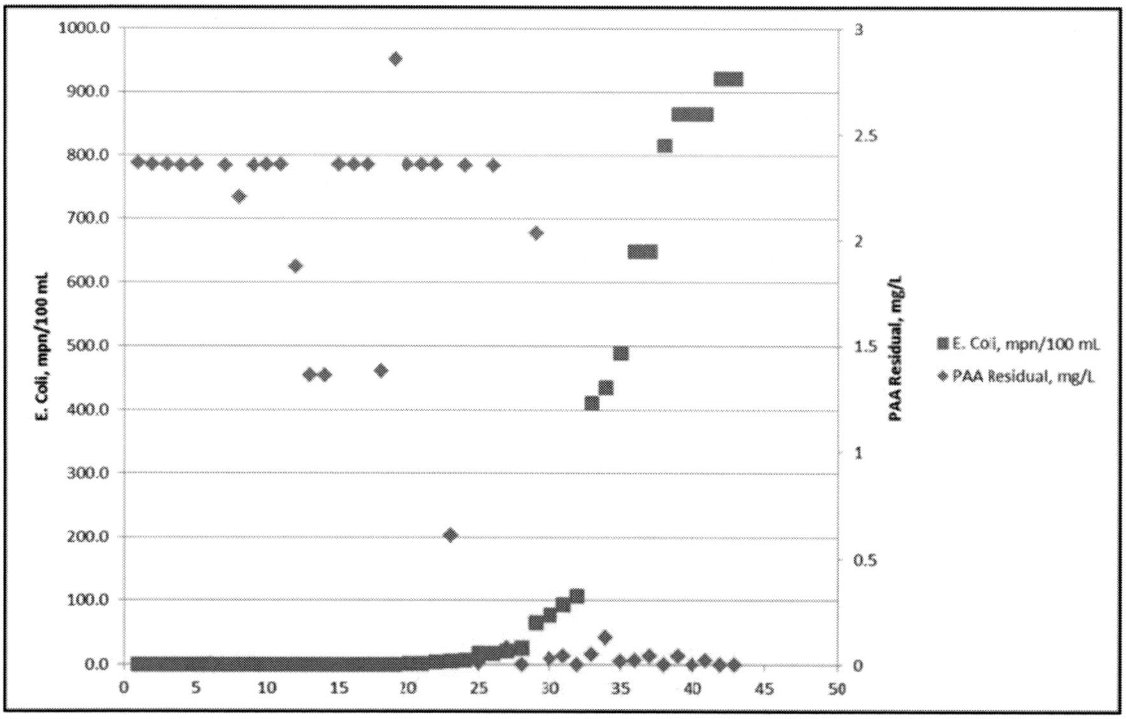

Figure 13 – *E. Coli* Concentrations Below 1000 mpn/100 mL Versus PAA Residual

Figure 14 shows a regression for 2012 and late fall 2011 CSO events indicating that a PAA residual of about 0.5 mg/L is necessary for effective reduction of *E. Coli*.

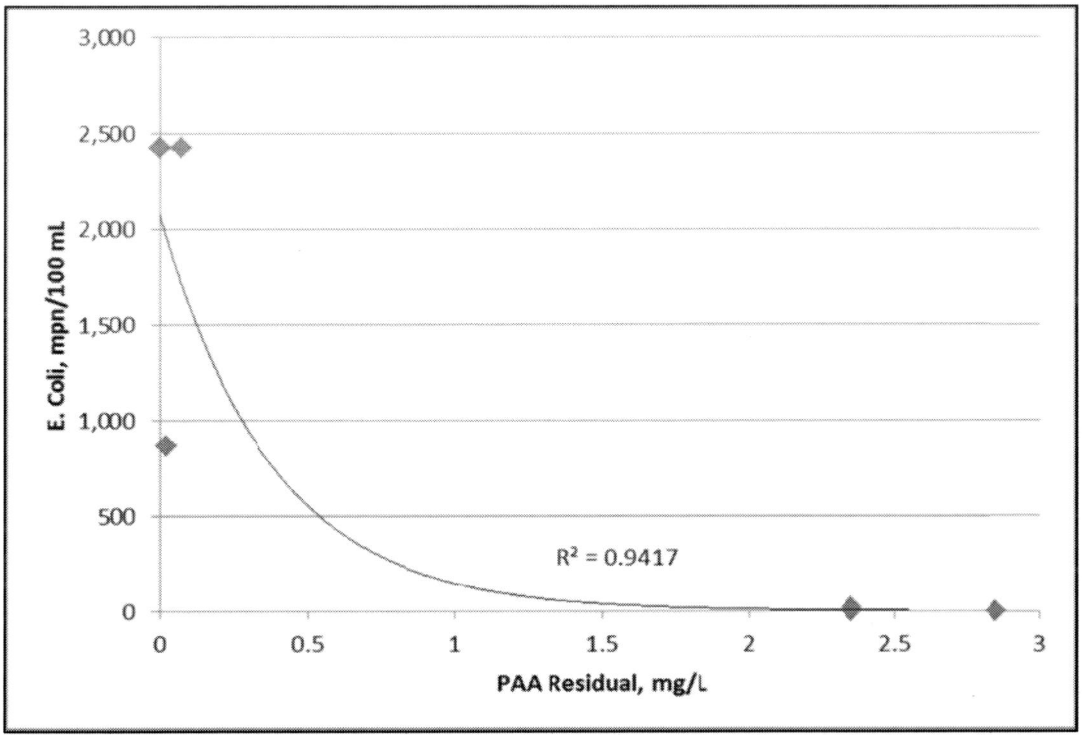

Figure 14 – *E. Coli* vs PAA Residual for 2012 CSO Events

Other factors aside from the chemical pumping issue contribute to the overall variability observed in the PAA results.

The large contact pipe and the feed configuration contribute to the variability. The large contact pipe sized for an extreme flow event holds approximately 379 cubic meters (100,000 gallons). Often the CSO events are a fraction of that volume resulting in the freshly disinfected water traveling only partially down the length of the pipe. Over time the disinfectant is consumed by organics in the water potentially allowing regrowth of bacteria. Some solids settling occurs producing a thin film of bottom solids that can also act as a source for bacteria.

Another issue is the location of the flow meter downstream of the chemical feed point and associated chemical pump control. The chemical pumping is initiated when the flow meter senses flow. A relatively small volume (11.4 cubic meters) of water enters into the contact pipe before the chemical feed starts resulting in non-disinfected water entering the pipe with each CSO event.

Another contributing factor until spring of 2012 was the parking lot storm drain pipe was connected into the contact pipe at T-2 location. The extraneous flows from the parking lot storm water runoff disrupted the hydraulics of the contact pipe and provided another potential source of bacteria.

Figures 15 and 16 present results from a CSO event that occurred on September 1, 2012. This CSO event had relatively low flows and was a two peak storm event with a lull between peaks. The total CSO volume was 322 cubic meters (85,140 gallons). Grab sampling occurred at approximately 10 a.m. The water from this event did not travel much beyond T-2 location. Figure 15 shows good *E. Coli* levels until the T-3 location. The upstream (river gate) location also shows very low *E. Coli* levels which may be attributable to backflow from the chemical feed system since actual CSO flow at this time was Zero. Figure 15 shows higher *E. Coli* concentrations at T-3 and MH-3 which could be due to regrowth in the pipe, the intermittent nature of the disinfectant application, or prior chemical feed problems.

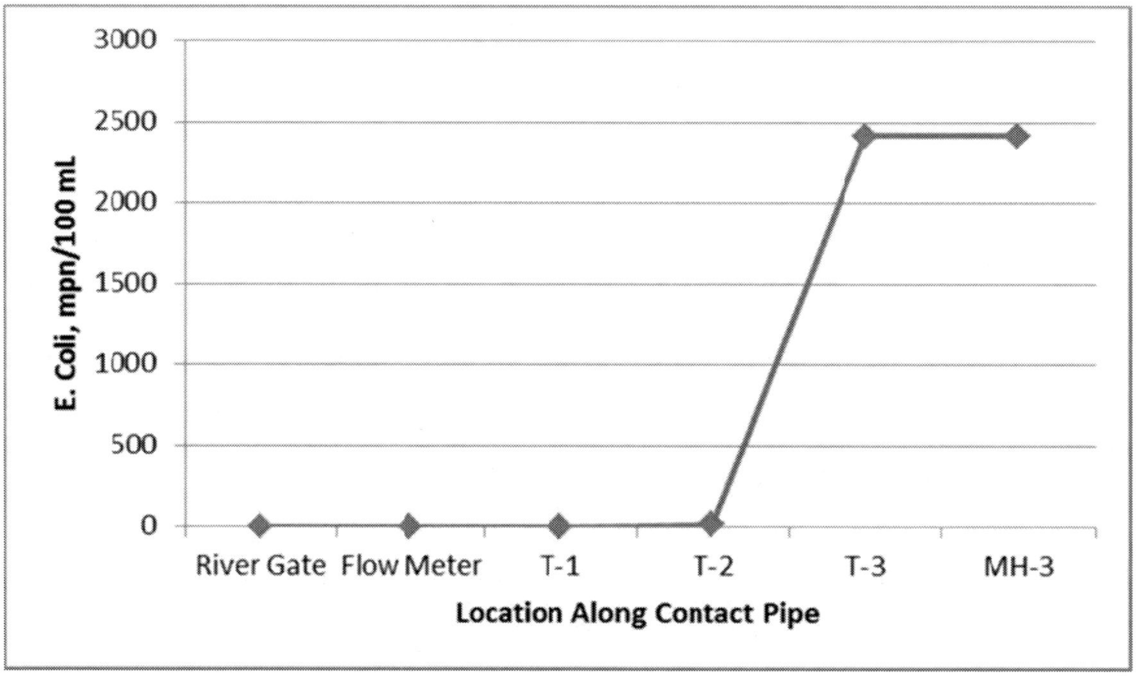

Figure 15 – September 1, 2012 CSO Event

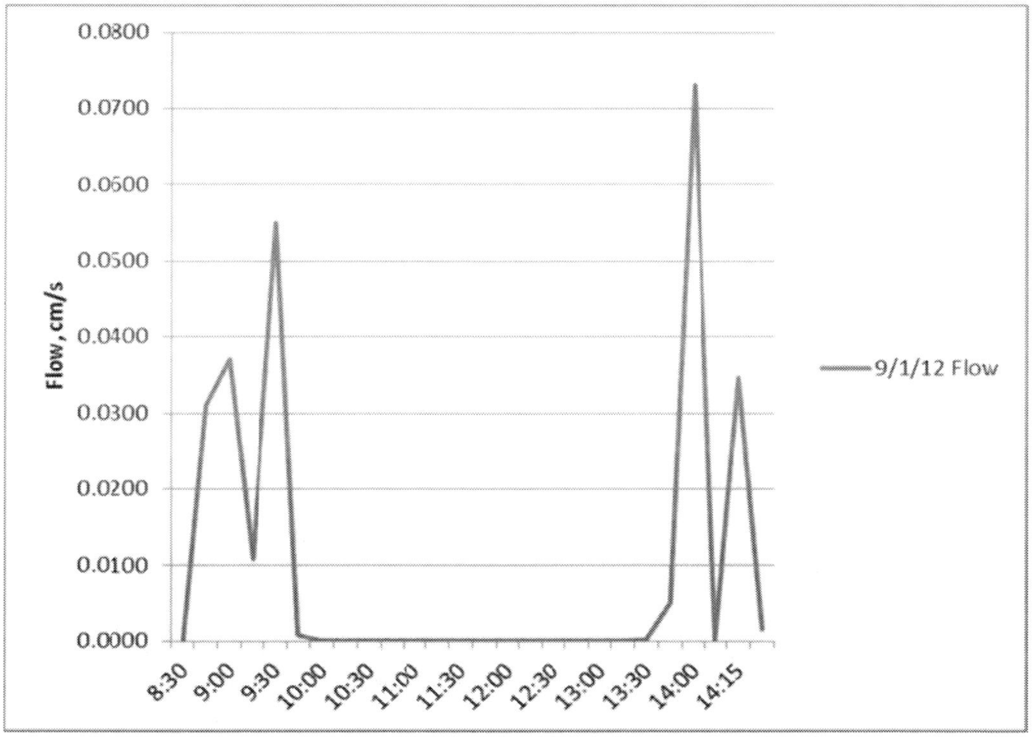

Figure 16 – September 1, 2012 CSO Event Flow

Figures 17 and 18 present results from a CSO event occurring on May 4, 2012. This CSO event had higher flows and was a single peak storm event. The total CSO volume was 964 cubic meters (254,700 gallons). Grab sampling occurred at approximately 3:10 and 3:50 a.m. Some of the water from this event travelled the length of the contact pipe and was discharged into the river. Figure 17 shows good *E. Coli* levels until the T-2 location. Figure 17 shows higher *E. Coli* concentrations at T-2 which could be due to storm water entering from the parking lot storm drain.

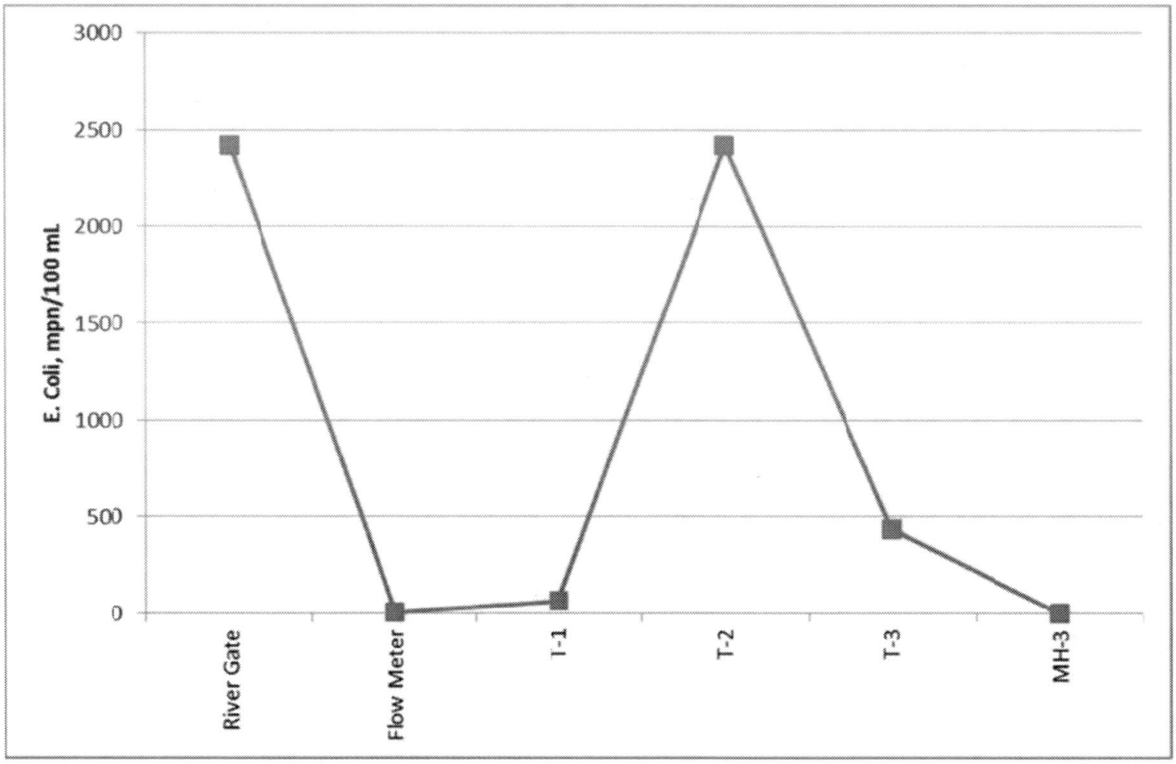

Figure 17 – May 4, 2012 CSO Event

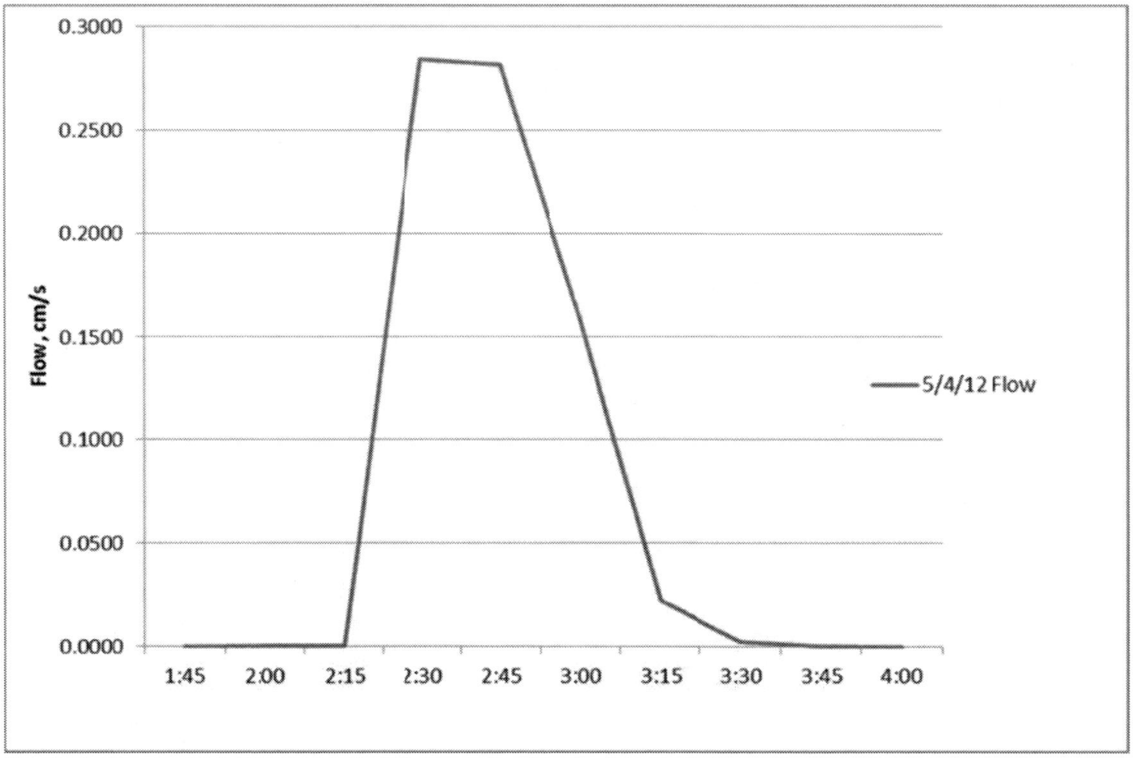

Figure 18 – May 4, 2012 CSO Event Flow

DISCUSSION AND CONCLUSIONS

Bench study work under ideal and well mixed conditions suggest that a SH dose of 20 mg/L with 5 minutes of contact time or a dose of 10 mg/L with 10 minutes of contact time should reduce E. Coli below 126 mpn/100 mL.

Bench study work under ideal and well mixed conditions suggest a PAA dose of 15 mg/L or higher with 3 minutes of contact time or a dose of 10 mg/L with slightly greater than 5 minutes should reduce *E. Coli* below a target of 126 mpn/100 mL.

Preliminary pilot study results suggest that sodium hypochlorite appears to be effective at meeting the CSO disinfection goal of reducing *E. coli* in CSO water to below 1,073 mpn/100 mL. However, the amount of data is small and additional testing is warranted. Unfortunately dose data is not available.

Preliminary pilot study results suggest PAA can be an effective disinfectant for meeting CSO *E. Coli* reductions. The data provides evidence that doses that provide residuals of 0.5 mg/L or greater reduces E. Coli to down to approximately 100 mpn/100 mL.

Inadequate chemical dispersion especially at low, less turbulent flows may be contributing to the observed variable performance. Consistent chemical disinfectant feed with complete dispersion into the water so the disinfectant contacts all of the water is very important in the overall effectiveness of the disinfection process. Addition of mechanical mixing at the point of application should be considered to improve chemical dispersion.

Bacteria regrowth in the contact pipe and associated sediment may be a factor in the inconsistent results. Draining the pipe between CSO events may reduce this potential source and also allow storage of some CSO water reducing total CSO volumes discharged.

Continuous chemical dose feedback via oxidation reduction potential sensor or similar sensor technology would enhance overall dose control and system performance.

PAA shows promise as a disinfectant agent in wastewater especially in non-ideal wet weather systems. The chemical has a long shelf life of greater than six months. The PAA being used in Fort Madison is from the initial purchase over two years ago at the beginning of the pilot demonstration. Periodic testing of the solution strength indicates that it has only declined by about a percent. This type of shelf life is very advantageous for intermittent wet weather flow applications where a system may be go for a month or more without being used.

ACKNOWLEGEMENTS

1. City of Fort Madison, Iowa especially Rick Kuhljuergen, Wastewater Superintendent; Jeff Helling, Assistant Wastewater Superintendent; the dedicated laboratory staff; and the operations staff who have to go out and sample.
2. University of Iowa, Professor Gene Parkin, Graduate Students, Jessica Beswick, Matt Reusswig.

REFERENCES

Enviro Tech Chemicals, Inc, 2009, *Peracetic Acid*

Kitis, M., 2003, *Disinfection of Wastewater with Peracetic Acid: A Review*

Meakim, J.; Rolak Jr., J.; Howarth, J., 2008, Point Source Disinfection of New Jersey Wet Weather Events: Stormwater and Combined Sewer Overflows (CSO's).

Parkin, G.F.; Beswick, Jessica; Reusswig, Matt, 2009, *Peracetic Acid Disinfection of Simulated Combined Sewer Overflow* Prepared by the University of Iowa for Stanley Consultants (Unpublished).

Stanley Consultants, 2009, *Disinfection Engineering Study, Combined Sewer Overflows* Prepared for the City of Fort Madison, Iowa.

Stanley Consultants, 2006, *Final Combined Sewer Overflow Long Term Control Plan* Prepared for the City of Fort Madison, Iowa.

United States Environmental Protection Agency, 1999, *Combined Sewer Overflow Technology Fact Sheet Alternative Disinfection Methods,* EPA 832-F-99-033, Washington D.C.

Future Directions Biosolids Management of The Second Decade of the 21st Century

By

Robert S. Reimers[1], Yue Xu[1], Suresh D. Pillai[2], Kari B. Fitzmorris-Brisolara[3], Jan A. Oleszkiewicz[4] and Gerald Seidl[5]

[1]Tulane University School of Public Health and Tropical Medicine, Department of Environmental Health Sciences, New Orleans, Louisiana
[2] Texas A&M University, Institute of Food Science and Engineering, College Station, Texas
[3]Lousiana State University School of Public Health, Department of Environmental Health Sciences, New Orleans, Louisiana
[4]University of Manitoba School of Engineering, Department of Civil Engineering, Winnipeg, Canada
[5]HeadworksBIO, Houston, Texas

Abstract

Over the next ten years, the direction of biosolids management is becoming more holistic. The assessment of biosolids disinfection, stabilization and microconstituents impact is utilizing microcellular and biological assays, which will help in public outreach and education. These concerns relate to potential environmental impact in their application as value-added products or renewable energy sources. The criteria of assessment are listed below:

- Pertinent indicator organism monitoring for disinfection,
- More restrictive metals and microconstituents levels for land application and air emissions,
- Usage of molecular and biological assays,

As a result of these concerns, the different value-added product options are under closer public scrutiny. They are assessed with respect to carbon foot printing, potential emerging pathogens and emerging chemicals of concern.

Key Words

Sustainability, Disinfection, Stability, Emerging Chemicals of Concern, Indicator Organisms, Value-Added Products, Biofuels

Introduction

The biosolids management required a holistic waste approach related to sustainable resources management. At that time (ten years ago), the direction of the biosolids management was leaning to agricultural applications in the United States. At this time, the energy options were being addressed in Germany, Japan, Belgium and Western Poland (Englande and Reimers, 2001).

The main element of biosolids has and is still being directed to sustainability, and in 1987, the United Nations defined sustainable development as meeting the present needs without compromising the ability of future generations to meet their needs. There are a range of other definitions related to the quality of life in a community related to economic, social and environmental elements affecting the sustainability. They are elucidated below:

1. The **environmental element** covers environmentally robust, and supported by consistent, applicable and legal requirements.
2. The **economic element** elucidates economically affordable, technically feasible and operationally viable.
3. The **social element** covers socially desirable, culturally acceptable and psychologically nurturing.

Currently, the United States has been operating unsustainable situation. This is when the total sum of natural resources is used up faster than it can be naturally replenished. In this situation, this long-term environmental degradation could impact the sustainability of human life. This is noted in Figure 1 (Adams, 2006).

-

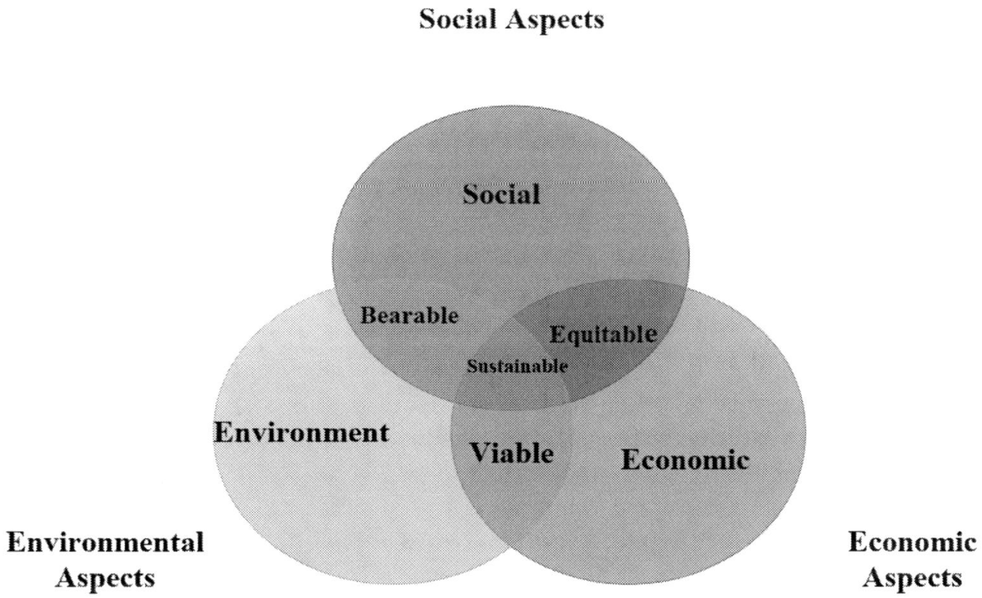

Figure 1: Interconnection of Elements Affecting Sustainability (Adams, 2006)

As noted in Figure 1, the development of a sustainable sludge management program requires the integration environmental, economic and social aspects. A movement toward more sustainable approaches must be promoted through an integrated direction and includes the following:

1. Better utilization of current technologies
2. Innovative and creative planning protocols
3. Development of self-sustaining energy systems
4. Optimization of the production reused products with multi-applications
5. Reuse products from biosolids with local niche markets

The goal of optimized sludge management programs should have the aim to reducing the amount of sludge disposed and potential environmental impacts with a given reuse option. An important factor in the United States is the control of toxins input into the collection system by the implementation of pretreatment guidelines. Over the last six years, there has been a major issue that has arisen which has changed the direction of biosolids management (Spinosa, 2010). They fall into the below categories:

1. The movement to Class A disinfected products and leading to value-added products relating disinfection and stability
2. Better management of carbon footprinting noting the reduction of GHG production by production of biogenesis fuels from biosolids and optimization of the agricultural applications
3. Improved assess and monitoring of microconstituents such as EDCs and PCPs
4. Rationale management of utilizing human waste as a sustainable resource in agricultural settings for developing countries.

Development of Sustainability in the Application of Municipal Sludge

Disinfection

Advances in disinfection and stabilization processes have lowered costs for the treatment of municipal sludge. With the increased understanding of chemical/biological processing and the optimization of their production, the costs of these processes have dropped an order of magnitude. Table 1 illustrates the cost for Class A biosolids treatment processes (Reimers et.al. 2005).

Table 1. Economics of Existing and Innovative Disinfection/Stabilization Municipal Waste Residual Processing

Process	Cost per Dry Ton ($2005 US)	Cost per Dry Ton as VA Product ($2005 US)
Incineration/Co-generation	600 to 1100	3 to 30 as Ash Reuse
Thermal Treatment	500 to 1200	50 to 150 Usage as a Fuel or Fertilizer
Modified Commercial Fertilizer	450 to 700	500 to 600 as a Commercial Fertilizer
Biological Anaerobic Digestion	350 to 650	30 to 200 as Natural Gas /Residual Soil Amender
Biological Aerobic Digestion	350 to 700	30 to 70 as a Residual Soil Amender
Advanced Alkaline Stabilization	350 to 550	80 to 120 as a Ag Lime Agent
Acid Stabilization/Disinfection	350 to 550	30 to 70 as a Residual Soil Amender
E-Beam Disinfection/Stabilization	100 to 250	30 to 70 as a Residual Soil Amender

Source: Reimers et.al. 2005.

To understand residuals stabilization and disinfection, one must address physical, chemical, and biological factors which assist stabilization and benefit the disinfection process (Smith et.al., 2008, Acquisto et.al., 2006, Smith and Milliner, 2001). There are various papers that elucidate the above points. These disinfection processes utilize matrices of stressors, which are elucidated in this section. Almost all processes developed to inactivate pathogens utilize a combination of physical, chemical, and biological stressors. For example, Table 2 illustrates the combination of these agents (Acquisto et.al., 2006).

Table 2. Disinfection Processes and Some Stressors that Influence their Effectiveness

Process	# of Stressors	Exposure Time	pH	Irradiation	Temperature	Solids Concentration	NH_3	Organic By-products	Drying	Cavitation /Ultrasound
Composting	6	+	-	-	+	+	+	+	+	-
Anaerobic Digestion	6	+	-	-	+	+	+	+	-	+
Aerobic Digestion	4	+	-	-	+	+	-	-	-	+
Lagoon Storage	5	+	-	-	+	+	+	+	-	-
Air & Heat Drying	5	+	-	+	+	+	-	-	+	-
Alkaline Stabilization	7	+	+	-	+	+	+	+	+	-
High Energy Irradiation	6	+	-	+	+	-	+	+	-	+

Source: Authors.

Investigations related to integrating these stressors in the process design of disinfections indicate that most of these processes involve at least 3 to 7 stressors that affect pathogens (Smith et.al.).

The Class A disinfection processes for biosolids have been related to a technology basis and are not derived from a risk-based framework. This has been due to lack of data and the low precision and accuracy of pathogen assays to quantify and determine the viability of helminth eggs (eggs other than *Ascaris* spp.) and protozoan oocysts, along with the respective assays for viruses. The Class A processes are fairly effective, as noted by fairly rigorous operational testing required by the US EPA pathogen equivalency committee (PEC), but there are concerns with bioaerosols and odors under specific operating conditions. Even so, viable treatment schemes are being developed in both economic and resource management modes utilizing as many stressors as possible (Smith and Milliner, 2001). One should note that the more stressors involved in a process the harder to control.

Stabilization (Vector Attraction Control)

Although there has been some dispute over the exact definition of stabilization, the process has three factors, which have to be covered to verify that a residual is stabilized:

1. Control pathogen regrowth after disinfection
2. Eliminate offensive odors and vectors
3. Minimize potential for putrefaction

Stabilization or vector attraction control is required for a process to be certified as Class B or A disinfected. Vectors and pathogen regrowth are related to the potential food source of the putrescible organics in the residuals. Vector attraction is related to the putrefaction process, which can cause odors, attracting flies, mosquitoes, rodents, and other pests to the waste. To achieve stabilization, both biological and/or physical treatment processes can be utilized. Biological stabilization is determined by volatile solids reduction, specific oxygen uptake rate or carbon dioxide release, which is elucidated in the 503 regulations (Acquisto et.al. 2006). These regulations are currently under upgrading and review. Biological activity can be lowered or inhibited by reducing the moisture content in the residual or raising the pH. To achieve this, moisture must be extracted from the product to less than 10% for raw municipal sludge, or less than 25% for digested biosolids. Adjustment of pH can also inhibit biological activity by raising the pH above 12.

Emerging Chemicals of Concern

Unlike metals, which have been regarded as a major concern since 1993 by EPA, PCBs and dioxins were eliminated from the 503 regulation in the late 1990s because their levels were non-detectable or lower than the criteria for endangering human health. In 2002, the National Research Council of the National Academies completed a report concerning the potential risks of application of biosolids to the land (National Research Council, 2002). The report stated that there was a need for further studies concerning a list of chemicals used in several daily personal care products; the use of such chemicals has increased since the late 1980s. Since then, EPA has been looking at the following groups of compounds of emerging concern:

1. Pesticides and agrochemicals: 73 pesticide chemicals are currently out for public comments and review.
2. Industrial and household chemicals, especially cleaning agents because they have been observed to biomagnify in mother's milk and human tissues (Goodman, 2007, Hunt, 2007, Guardia et.al., 2007).
3. Nanoparticles: particles in the range of 1 to 100 nm. They are composed of substances such as carbon fullerenes, titanium dioxide, and zinc oxide (Ganesh and Leong, 2007).

4. Constituents of pharmaceuticals: those found in municipal wastewater come from sources such as contraceptives, antibiotics, heart medicines, and fragrances (Goodman, 2007, Hunt, 2007).
5. Plastics: phthalates, and bisphenol-A have been detected in sewage (Goodman, 2007).

Some organic compounds from domestic use can cause environmental impacts in relatively low concentrations; are persistent in the environment due to low degradability; and can contaminate a wastewater treatment plant's effluents or residuals. Therefore, many compounds as discussed below have been increasingly studied (Boyd et.al., 2004,Boyd et.al., 2003).

After the National Academes of Science Report, EPA conducted its biennial assessment of new pollutants for potential inclusion in the 503 regulation. In their initial review of the original 803 chemicals, they picked 40 chemicals for risk assessment evaluation and 9 chemicals were recommended for further review. The chemicals recommended for further assessment are noted in the below table 3.

Table 3: Potential New 503 Pollutants and Their Receptors (CDM, 2011)

Chemical	**Land Application**	**Surface Disposal**
Barium	Aquatic Community	Adult, Child
Beryllium	Aquatic Community	
4-Chloroaniline	Aquatic Invertebrates	Adult, Child
Fluoranthene	Aquatic Community, Sediment Biota	
Manganese	Aquatic Community	Adult, Child
Nitrate		Adult, Child
Nitrite		Adult, Child
Pyrene	Aquatic Community, Sediment Biota, Soil Biota	
Silver	Aquatic Community, aquatic Invertebrates, Fish, Adult, Child	

Table 3 lists some of the organic compounds found in municipal biosolids. . In January 2009, the EPA releases the results of the Targeted National Sewage Sludge Survey (TNSSS), which sampled the biosolids from 79 different WWTPs throughout the United States. In the survey, 15 metals, benzo(a)pyrene, 2- methylnaphthalene, bis (2-ethylhexyl)phthalate, fluoride, water-extractable phosphorus (WEP), 11 polybrominated diphenyl ethers (PBDEs), 97 pharmaceuticals, steroids, and hormones analyzed since these pollutants are of emerging interest.

The Targeted National Sewage Sludge Survey conducted by U.S. EPA in 2009 released the Statistical Analysis Report (EPA, 2009) which provides technical background, statistical methods, and resulting estimates of pollutant concentrations in treated sewage sludge ("biosolids") that represent Publicly Owned Treatment Works (POTWs) in the contiguous United States with flow rates of at least 1 million gallons per day (MGD). Estimates were produced using data from a national probability sample of 74 POTWs that statistically represent 3,337 POTWs that met the study criteria.

The endocrine disrupting chemicals (EDCs) comprise pharmaceuticals, personal care products, surfactants, various industrial additives and numerous chemicals purported to be endocrine disrupter. Since the current wastewater treatment system is not effective in elimination of these different classes of emerging contaminants, these undesirable compounds can be released into the environment conveyed by effluent and sludge/biosolids.

Conventional methods do not have much efficiency on eliminating endocrine disrupting compounds, while sorption plays a great role when dealing with hydrophobic EDCs and PPCPs since they could be removed from the aquatic system by attaching to the solids. Current wastewater treatment plants are normally designed for carbon, nitrogen, and phosphorus removal but a partial EDCs removal is often achieved simultaneously. Although transformation and/or degradation processes may remove some EDCs from wastewater at variable levels, large uncertainty persists on the actual EDCs removal processes mechanism. For instance, removal pathways of EDCs during secondary biological treatment include adsorption onto microbial flocs and removal through the waste sludge, biological or chemical degradation, transformation, and volatilization during aeration. Campbell et al. (2006) reviewed the estrogeic endocrine disrupting compounds in water of their fate and transport. Many EDCs have moderate to high log Koc values, so the mass that does not remain soluble often ends up in organic complexes in, or sorbed to, sediments or suspended organic material. In the sediments there is the potential for biological uptake, degradation and transformation to less mobile or more mobile forms. If mobilized, the EDC complexes may move back into the water column or downward toward groundwater.

There has recently been an investigation of the occurrence and elimination of micro-organic pollutants including endocrine disrupting compounds (EDCs) and pharmaceuticals and personal care products (PPCPs) in a full-scale anaerobic/anoxic/aerobic-membrane bioreactor (A^2/O MBR) process. It was found that the fast sludge adsorption, which indicated by mass-transfer rates greater than 10 for most compounds, is responsible for the rapid decline of the aqueous concentration of the targets in the anaerobic tank as noted in figure 2 (Xue et al. 2010). As has been suggested by many researchers (Stasinakis 2008), the adsorption of these compounds by activated sludge could be an important removal pathway in many cases.

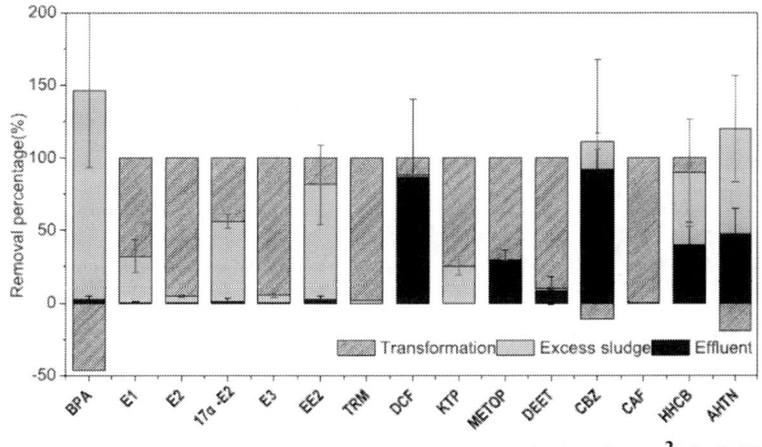

Figure 2: Mass balance of target compounds in the A^2/O-MBR process.

A comparative study on coagulation and sorption (activated carbon) in eliminating estrogens has proven that sorption by powdered activated carbon and granular activated carbon was more efficient than coagulation, even in a hybrid system with nano-filtration membranes (Bolong et al. 2009).

Regarding the nature of estrogens, it is addressed that unlike excreted conjugated forms (which are less toxic and more polar), free estrogens are more likely to accumulate in sewage sludge, due to their moderate to high hydrophobic properties (log K_{ow} = > 3-5) (Díaz-Cruz et al. 2009).

In this regard, the role of adsorption was investigated via analyses of sludge-water distribution coefficient (K_d) and organic carbon normalized distribution coefficient (K_{oc}). Digested sludge was examined for the solid-water distribution of pharmaceuticals and estrogens it contains (Carballa et al. 2008). The work determined the solid-water distribution coefficient (K_d) and the organic carbon normalized distribution coefficient (K_{oc}) of several pharmaceuticals and three estrogens (estrone, 17b-estradiol and 17a-ethinylestradiol) in digested sludge. Sorption can be one of the key factors controlling the removal of organic pollutants in sewage treatment plants (STPs). Several approaches have been used to determine the affinity of a given substance to solids, which are based on the octanol-water partitioning coefficient (K_{ow}) and on the organic carbon-based coefficient (K_{oc}) of the respective compound. Both are reasonable assumptions for non-specific lipophilic interactions. However, concerning PPCPs, recent studies revealed significant deviations from available K_{ow} and K_{oc} coefficients, which could be explained by the fact that pharmaceuticals often contain polar functional groups (i.e. carboxylic moieties, aldehydes and amines), which might interact only with special parts of organic matter (OM) or with minerals, and hence the K_{ow} and K_{oc} values are inappropriate to simulate their sorption properties in sludge samples. Sorption on digested sludge is a relevant process for estrogens (log: 2.1-2.9). Due to the elevated total suspended solids concentrations (TSS) of digested sludge (around 35 g/L), estrogens are sorbed to 80 to 99% on particles and less than 20% are dissolved. However, for pharmaceuticals, sorption on digested sludge is of minor importance (log K_d: 0.7-1.9). Contrary to the expected, different anaerobic digestion operation conditions do not change the resulting sludge in such a way that the sorption properties are affected (Carballa et al. 2008).

Study of EDCs in municipal and industrial wastewater water treatment plants was conducted in Northern Greece, the occurrence and fate of EDCs from a sewage treantment and two industrial wastewater treatments plants from textile and tannery factories were investigated. Results showed that NP and NP1EO exhibited high concentrations in sludge (1089 ng/g and 1080 ng/g respectively), followed by TCS (461 ng/g), tOP (179 ng/g) and BPA (30 ng/g). BPA with high concentrations up to 1750 ng/g was determined in Greek WWTPs and much lower 3.1-3.8 ng/g in Australian WWTPs (Stasinakis et al. 2008). The concentrations of TCS in sludge ranged from 0.09 to 16.79 ug/g in various WWTPs (Stasinakis et al. 2008). Another study conducted in Greek was to investigate the occurrence of five endocrine-disrupting compounds (EDCs), namely 4-n-nonylphenol (4-n-NP), nonylphenol monoethoxylate (NP1EO), nonylphenol diethoxylate (NP2EO), triclosan(TCS) and bispenol A (BPA) in the raw, treated wastewater and sewage sludge of eight sewage treatment plants (STPs) in Greece. . With average concentrations in the raw and treated wastewater ranged from 0.23 (4-n-NP) to 5.76 ug/L (NP1EO) and from 0.15 (BPA) to 1.84 ug/L (NP2EO), respectively, a great part of the detected EDCs was sorbed on suspended solids. In sewage sludge, the average concentrations ranged between 0.17 (4-n-NP) and 12.3 ug/g d.w. (NP1EO). A mass balance (Figure 2) was performed for the largest sewage treatment plant (STP) examined in the study to investigate the fate of EDCs during wastewater treatment and to estimate the role of degradation and sorption on their removal.

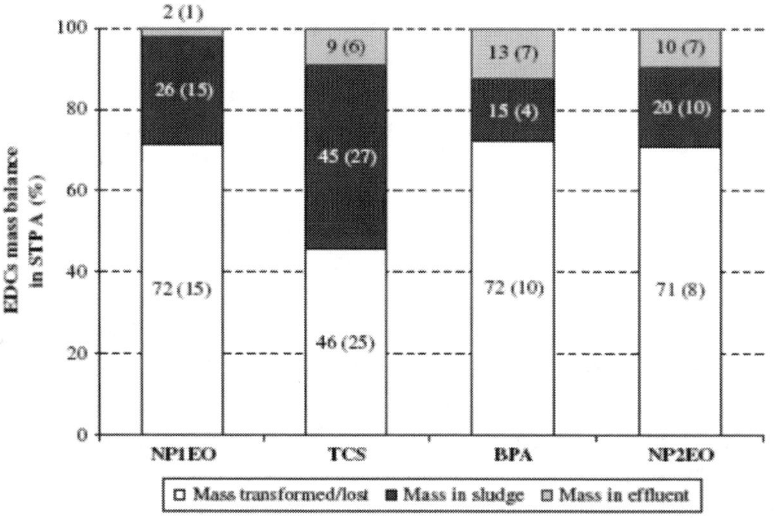

Figure 3: EDC's mass balance in sewage treatment plant (standard deviations are given in parentheses).

The review of microbial transformation of synthetic estrogen 17a-ethinylestradiol discussed the different efficiencies of bacterial degradation of 17a- ethynylestradiol under aerobic and anaerobic conditions, the role of sulfate-, nitrate-, and iron-reducing conditions in anaerobic degradation, and the role of sorption. In experiments testing EE2 sorption to nitrifying activated sludge, biosorption of EE2 by the biomass was shown to be affected by the initial ammonia concentration. At higher initial ammonia concentrations the role of biodegradation became more important in EE2 removal than its biosorption. Correlation of EE2 sorption to activated sludge biomass and temperature was described (Xu, 2008).

Amphoteric pharmaceuticals were found to be released to the environment through land disposal of livestock wastes and municipal biosolids. These wastes are rich in dissolved organic carbon (DOC), which may bin and potentially enhance pharmaceutical transport in the environment (Carmosini and Lee 2009). These results indicated that pharmaceuticals could survive wastewater treatment processes, and accumulate in sewage sludge and biosolids. Subsequent land application of the contaminated biosolids could lead to the dissemination of pharmaceuticals in soil and water environment, which poses potential threats to at-risk populations in the receiving ecosystems (Carmosini and Lee 2009). Le-Minh et al. (2010) reviewed existing knowledge of a wide range of treatment processes to the ultimate fate of antibiotics during conventional and advanced wastewater treatment. Observed behaviors of groups of antibiotics are much more difficult to characterize due to varying removal efficiencies reported from studies undertaken in different parts of the world. The most reasonable interpretation appears to be that specific design and operating conditions of individual WWTPs is an important determinative factor for antibiotic chemical removal. Unfortunately, insufficient information is currently available to thoroughly assess these differences. (Le-Minh et al. 2010)

While studies have revealed the tendency of many endocrine-active compounds to partition to the sludge flocs over the course of conventional wastewater treatment, relatively little work has been conducted to systematically track the fate of EDCs during various common sludge stabilization processes, such as anaerobic or aerobic digestion, alkaline treatment and composting, via chemical analysis. Far less research has focused on using in vitro bioassays to determine the total endocrine-active potency of biosolids, as opposed to pinpointing the concentrations of specific known and suspected EDCs. (Citulski and Farahbakhsh 2010)

Emerging Pathogens and Validity of Indicator Organisms

As a result of studies conducted at Tulane University, Cornell University, University of Manitoba and the literature, it is recommended that the surrogate indicators vary depending upon the disinfection process as shown in Table 4. The surrogate indicators can be utilized to assess the performance of these processes to inactivate helminth eggs, viruses and bacteria. The process categories are heat drying, anaerobic digestion, aerobic digestion, advanced alkaline stabilization and acid treatment. This discussion will be broken down into the EPA-approved target organisms, Ascaris suum and Enteroviruses.

Table 4: Classification of Matrix and Proposed Surrogates

Target	Heat Drying	Anaerobic Digestion	Aerobic Digestion	Advanced Alkaline Stabilization*	Acid Treatment*
Helminth	Plant and fungal surrogates **	Somatic bacteriophage	Plant and fungal surrogates **	Aerobic endospores	*C. perfringens*
Viruses	Reoviruses Somatic bacteriophages	Reoviruses Somatic bacteriophages	Reoviruses Somatic bacteriophages	Reoviruses Somatic bacteriophages	Reoviruses Somatic bacteriophages
Bacteria	Fecal coliform Enterococci *E. coli*	Fecal coliform Enterococci *E. coli*	Fecal coliform Enterococci *E. coli*	Fecal coliform Enterococci *E. coli*	Fecal coliform Enterococci *E. coli*

*High concentration of non-charged biocidal agent
**Vascular and nonvascular plant seeds, and fungal spores and propagules are also included in Advanced Alkaline Stabilization and Acid treatment

The helminth eggs are generally the most resistant organism to be inactivated. These eggs can survive in concentrated acids at a pH of 1 or in caustic at a pH of 13 or greater (Bowman, et .al. 2003). They can be inactivated by either heating to greater than 60°C or by a non-charged biocidal agent, such as ammonia, nitrous acid and fermentation by-products. When non-charged biocides are present, the usage of bacterial spores has been observed to be a potential surrogate indicator. In alkaline systems, the aerobic endospores appear to be a good indicator with a 3 order of magnitude reduction or a 99.9 percent reduction (Fitzmorris, et. al., 2007). Under acid conditions, the spore forming anaerobic spores, such as *C. perfringens* may be used. Yet, they are not effective for alkaline conditions since they are sensitive to the hydroxide ion (Reimers, et. al., 2004 and Bujozek, et. al., 2002). In heat drying and anaerobic and aerobic digestion, the bacterial spores were not effective but somatic bacteriophages, vascular /non-vascular plant seeds and fungal may be effective surrogate indicators, which are sensitive to heat and integrated fermentation by-products. It should be noted that the seed and fungal spores could be an excellent surrogate indicator, yet more research needs to be carried out to determine its efficiency as a surrogate organism (Bowman, 2004).

The viruses currently assessed are polio viruses. There are concerns with potential infection form the polio spikes. Recent work has observed that reoviruses and somatic bacteriophages could be excellent surrogate indicators (Fitzmorris, et. al., 2007).

Fecal coliform bacteria were utilized as an indicator for bacterial inactivation as a result of Yanko's study in 1987 (Yanko, 1987). In the 1990s, *E. coli* has replaced fecal coliforms in wastewater disinfection, yet there have been observations that Enterococci are a more resistant bacterial indicator in biosolids. Therefore, Enterococci may become a better surrogate indicator for bacterial inactivation. Data has noted Enterococci to be more resistant in lagoons, alkaline stabilization and digestion processes (Reimers, et. al. 2001).

In summary, for each type of disinfection process, the performance of its efficacy can be assessed by a different surrogate indicator. For the advanced alkaline disinfection, aerobic endospores and somatic bacteriophages mimic disinfection of helminths and viruses inactivation, respectively. In the neutral pH range, the somatic bacteriophages appears to be an excellent indicator for indicator for both viruses and helminths inactivation. In the acidic disinfection processes, *C, perfringens* and somatic bacteriophages appear to be potential surrogate indicator organisms for helminths and viruses, respectively. As a result of these observations, it can be concluded that there is not one individual surrogate indicator that is appropriate to assess the inactivation of pathogens in municipal sludge or manure.

Future Assessment Techniques

This movement has resulted from existing methods of incineration, advanced alkaline stabilization, heat drying, innovative chemical or thermophillic processes insuring disinfection, stabilization and the destruction/biodegradation of refractory organics along with overcoming existing or perceived liability. These treated manures and municipal waste residuals are further treated to produce a environmental safe product.

The disinfection criterion is found in the 40 CFR 503 (USEPA, 1999). Table 5 shows the broad requirements for demonstrating equivalency of an innovative or alternative technology.

Table 5. USEPA Criteria for Disinfected or Partially Disinfected Biosolids

PSRP Equivalency or Class B	PFRP Equivalency or Class A
> 1 log reduction of *Salmonella* sp. or > 2 log reduction of fecal coliforms	≥ 3 log reduction of enteroviruses
> 1 log reduction of enteroviruses	≥ 2 log reduction of viable *Ascaris* spp. ova
Final product contains < 2,000,000 fecal coliforms/g	Final product contains < 1000 fecal coliforms or < 3 *Salmonella* sp./4 g; < 1 pfu/4g of enteroviruses and < 1 helminth ova/ 4g

In the next ten years, the determination of disinfection will utilize indicator organisms such as *Clostridium perfringens,* aerobic endospores, and somatic bacteriophages. The usage of ferrate, ferric salts or ferrous salts may be short-term constituents, which will assist in stability at dosages of 50 to 10 mg/kg of dry solids (Pillai and Reimers, 2010).

In the 503 regulations (40 CFR Parts 257 and 503), the stabilization of municipal sludge is ascertained for the long term by either the Saturated Oxygen Uptake Rate (SOUR) test or the volatile suspended solids (VSS) reduction. The short-term stabilization of alkaline processes is based on high alkaline dosages or solids content. In the future, the content of ferrate or CO_2 release will be incorporated. When looking at carbon dioxide respiration and oxygen uptake rates, we will most likely base our procedure on the current carbon dioxide respiration with regard to compost stability (Epstein, 1957; Bartha, 1965). Our results may be evaluated using the Compost Stability Index, which is based on the level of microbial activity in a sample by monitoring respiration. This index will be similar for the SOUR test and will require a similar acclimatization step. See Table 6 below:

Table 6: Compost Stability Index (Epstein, 1957; Bartha, 1965)

RESPIRATION RATE (mg CO2-C/g compost carbon day)	**RATING**	**CHARACTERISTICS**
<2	Very stable	▪ Well cured compost ▪ No odors ▪ No continued decomposition
2 – 5	Stable	▪ Cured compost ▪ Limited odor potential ▪ Minimal impact on soil carbon and

		nitrogen dynamics
5 – 10	Moderately Unstable	▪ Uncured compost ▪ Minimal odor production ▪ Addition to soil may result in nitrogen immobilization ▪ High phytotoxicity potential ▪ Not recommended for growing plants from seed
10 – 20	Unstable compost	▪ Very immature compost ▪ High odor and phytotoxicity potential ▪ Not recommended for growing plants from seed
>20	Unstabilized material	▪ Extremely unstable material ▪ Very high odor and phytotoxicity potential ▪ Not recommended for use

The levels of metals allowed by the US EPA are well above the criteria for Germany and the United Kingdom. Two countries (Denmark and Spain) have more restrictive criteria. In the United States, the metals concentrations at equivalent (EQ) levels or lower are not restrictive at any loading rate, whereas in other countries or in the food industry the allowable levels are much more restrictive, as noted in Table 7 (Reimers et.al., 2008). In fact, when assessing contaminated soils, the metals limits are very restrictive for residential soils but not as limiting for industrial soils. These criteria are used to determine whether a soil or an industrial residual is a solid waste or a non-contaminated soil. These data are shown in Table 7.

Table 7. Metal Comparison (mg/kg dry weight)

Category	As	Cd	Cr	Cu	Pb	Hg	Mo	Ni	Se	Zn
EPA Part 503 (15) Upper Limit	75	85	3,000	4,300	840	57	75	420	100	7,500
AAPFCO (13)	13	10			61	1	42	250	26	420
AAPFCO	78	60			366	6	252	1,500	156	2,520
SST (USA) (14)	12	3.4	23	310	100	2.3		16	20	2,300
EPA Part 503,3 Excellent Quality (EQ) (15)	**41**	**39**	**1,200**	**1,500**	**300**	**17**		**420**	**36**	**2,800**
Germany		1.5	100	60	100	1		50		200
Spain		1	100	50	50	1		30		150
UK		3	400	135	300	1		75		200
Denmark		0.5	30	40	40	0.5		15		100

Source: Reimers et.al., 2008

For ashes and sludges, the levels of metals will have to meet the soil screening testing protocol plus pass the synthetic acid precipitation leaching test (USEPA SW-846 Method 1312), which assimilates acid rain level in the United States. This test will determine whether a ash and sludge could be contaminated; if the test shows that the soil and sludge is not contaminated, these residuals can be taken out of the solid waste category.

With respect to EDCs and PPCPs, there are possibly 200 to 300 of these compounds in the ng/kg range, and the measurement of the concentration of a few EDCs would be meaningless so it recommended utilizing molecular assays. The organic compounds described in the section on emerging chemicals of concern could perform as carcinogens or bind with certain gene receptors. MVLN/MCF-7 cell proliferation assays and yeast estrogen screen (YES) cell assays have been used for endocrine screening.

MVLN/MCF-7 cell assay

The MCF-7 cell line, which was developed at the Michigan Cancer Foundation in the early 1970s, derives from the pleural effusion of a 69-year-old human female in the late stages of metastatic mammary carcinoma. Early investigations found it to express the estrogen receptor, androgen receptor, progesterone receptor, glucocorticoid receptor, vitamin D receptor, and retinoic acid receptor. The MCF-7 human cell line has been used widely throughout the last 23 years in the study of cancer biology, steroid hormone biochemistry, and, more recently, toxicology. One of the most common applications of MCF-7 is for the characterization of estrogenic compounds (Gray et.al., 1997).

Yeast estrogen screen (YES) assay

Recombinant yeast strains expressing nuclear receptors (NRs) and ligand-dependent reporter genes have been useful for testing environmental components (air, soil, waste and water), drugs, herbal supplements, plant extracts, foods and other materials for the presence of NR agonists or antagonists (Collins et.al., 1997, Bovee et.al., 2006, Escher et.al., 2005).
The yeast *Saccharomyces cerevisiae* is a useful tool for the study of NR signal transduction pathways. It provides a relatively simple and well-defined eukaryotic system for the expression of genes from other organisms (Strausberg et.al., 2001). Numerous studies, using yeast strains designed to express metazoan NRs and restriction endonuclease (RE)-driven reporter genes have demonstrated ligand dose dependent activation of NR-responsive reporter genes (3). Such ligand-dependent NR reporter assays have been extensively used to analyze signaling from a wide variety of NRs, including, but not limited to, estrogen and glucocorticoid receptors, aryl hydrocarbon receptor (AhR) and ecdysone receptor (Collins et.al., 1997, Miller, 1999, Routledge et.al., 2000).

Emerging Processes of the Future

Conservation of man-made waste by transformation into a consumable product has evolved from existing processes. Again, there has been a movement from incineration and heat drying to advanced alkaline stabilization, and innovative chemical or thermophillic processes. This progression must ensure disinfection, stabilization, and the destruction/biodegradation of refractory organics, while overcoming existing or perceived liability as noted in the previous sections. These treated municipal waste residuals are further refined to fabricate a value-added product. The specific factors in developing a value-added product include:

1. Optimizing heat/chemical stabilization/disinfection processes for municipal waste, manures and other waste residuals;
2. Developing cost effective processes to convert municipal waste, manures and other waste residuals and other residuals to value-added products, such as engineered soils, ornamental horticultural fertilizer, turf grass, ion exchange resins, synthetic coal, and activated carbon;
3 Evaluating value-added products against comparable commercial products on an economic and risk assessment basis;
4. Developing engineering, economic, and outreach plans to utilize the value-added products in the marketplace.

An area that has not been assessed with fertilizer/soil amender utilization of municipal waste residuals, manures and other waste residuals is the nature of crop yields, crop qualities, and nutrient value. Preliminary data indicates that usage of biosolids could have as much input as biotechnology. Therefore, this area has to be addressed in the near future. Additional variables are elucidated below:

1. Inactivate emerging pathogens such as prions, helminth eggs, and bacterial spores,
2. Stabilize these treated residuals chemically by use of oxidants such as chlorine dioxide and ferrate,
3. Destroy or alter endocrine disrupting chemicals (EDCs) so that they are mineralized or in a form that is degradable.

Currently, the renewable energy resources field is a fast growing research area across the World and in some parts of the United States. Renewable energy resources are related to natural gas production, ethanol production, biodiesel fuels, and pyrolysis production of fuels and other energy forms from heat drying processes. Processes can capture energy from the degradation of waste matter, production of processed crops, pyrolysis of municipal sludge, digester microbes and, algae from waste sludge lagoons: all of which are emerging biofuels technology. The integration of these technologies (Yurtsever et.al., 2009) is shown in the below figure4 and table 8.

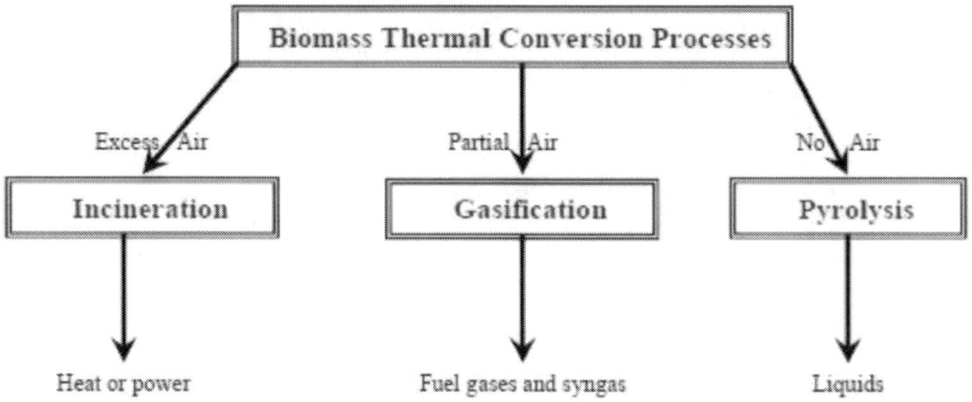

Figure 4: Biomass Thermal Conversion Processes (Yurtsever et.al., 2009)

Table 8: Summary of Advantages and Disadvantages of Thermal Processes

Process	Advantages	Disadvantages & Risks	Conclusions
Incineration	• Only ash to dispose of • Proven technology available with extensive records • Potential for power and heat recovery • Low greenhouse gas emissions	• Strict regulations • Planning process and public perception of incineration • High capital and O&M costs	Proven technology can provide a cost-effective solution for medium to large scale installations. May not be cost-effective for small-scale applications.
Gasification	• Only ash to dispose of • Potential for power and heat recovery • Low greenhouse gas emissions	• Lack of information on capital and O&M costs • Preparation of biosolids prior to the process	Not as many commercially proven installations as those for incineration technology
Pyrolysis	• Promoted as cost-effective at a much smaller scale than incineration	• Lack of commercially proven technologies for biosolids handling • Preparation of biosolids prior to the process	Too early in development to be considered proven technology

In addition, there has been a movement of enhancing the methane production in anaerobic digestion process by the either breaking up the suspended solids before entering the digestion process by hydrolysis processes (thermal, biological, mechanical, chemical (caustic or chlorine dioxide) and E-beam treatment with or without chlorine dioxide), the addition of fats, oils and greases (FOGs) and the enhancement of the anaerobic digestion process.

As the liability begins to rise, then the movement to commodity products or value-added products has moved out of arranger liability arena. The following commodity products produced from biosolids are listed below:

1. Commercial Fertilizers
2. Activated Carbon
3. Ion Exchange Resins
4. Light Weight Aggregate
5. Aglime Agents

Summary

Over the next ten years, the direction of biosolids management is becoming more holistic. The assessment of biosolids treatment and usage requires assessing disinfection, stabilization and microconstituent content. These concerns relate to their application as value-added products or renewable energy sources. The criteria of assessment are listed below:

- Indicator organism monitoring for disinfection
- More restrictive metals and microconstituent levels for land application and air emissions.
- Usage of molecular assays
- Optimization of biosolids treatment processes utilizing ultrasound, electron beam, chemical oxidants and other innovative approaches.

Currently, the topic of renewable energy resources is a fast-growing area across the world and some parts of the United States. Renewable energy resources are related to natural gas production, ethanol production, biodiesel fuels, pyrolysis production of fuels, and other energy forms from heat-drying processes. The energy can come from degradation of waste matter, production of processed crops, pyrolysis of municipal sludge, and digester microbes and algae (in essence, generators of biofuels) from waste sludge lagoons. The following commodity products are produced from biosolids, manures and plant residuals: 1) commercial fertilizers, 2) activated carbon, 3) ion exchange resins, 4) light weight aggregate, 5) aglime agents. As the potential liability from hazards present in waste matter rises, the transition to commodity products or value-added products can minimize this liability.

References

Adams, W.M., (2006). "The Future of Sustainability: Re-thinking Environment and Development in the Twenty-first Century," Report of the IUCN Renowned Thinkers Meeting, 29-31 January, 2006.

Acquisto, B.A., R.S. Reimers, J.E. Smith, and S.D. Pillai. (2006). "Factors Affecting Disinfectionand Stabilization of Sewage Sludge." in WEF Annual Proceedings. 2006. Water Environment Federation, Alexandria, Virginia CD-ROM.

Barnabé, S., S. K. Brar, R. D. Tyagi, I. Beauchesne & R. Y. Surampalli (2009) Pre-treatment and bioconversion of wastewater sludge to value-added products--Fate of endocrine disrupting compounds. Science of The Total Environment, 407, 1471-1488.

Bartha, R. and D. Pramer. 1965. "Features of a flask method for measuring the persistence and biological effects of pesticides in the soil." Soil Sci. 100:68-70.

Bolong, N., A. F. Ismail, M. R. Salim & T. Matsuura (2009) A review of the effects of emerging contaminants in wastewater and options for their removal. Desalination, 239, 229-246.

Bovee TF, Bor G, Heskamp HH, et al. Validation and application of a robust yeast estrogen bioassay for the screening of estrogenic activity in animal feed. Food Addit Contam; 2006:23(6):556-568.

Bowman, D.D. (2004) Personal Communication of Invention.

Bowman, D. D., Little, M. D. and Reimers, R. S., (2003) "Precision and Accuracy of an Assay for Detecting Ascaris eggs in Various Biosolids Matrices". Water Resources, Vol. 37, Elsevior Sciences, Ltd., London, pp. 2063-2072.

Boyd GR, Palmeri JM, and Zhang S, (2004), "Pharmaceuticals and personal care products (PPCPs) and endocrine disrupting chemicals (EDCs) in stormwater canals and Bayou St. John in New Orleans, Louisiana," USA. Sci Total Environ; 333(1-3): 137-148.

Boyd GR, Reemtsma H, and Grimm DA, (2003). Pharmaceuticals and personal care products (PPCPs) in Surface and Treated Waters of Louisiana, USA and Ontario, Canada. Sci Total Environ;:311(1-3):135-149.

Cajthaml, T., Z. Kresinová, K. Svobodová, K. Sigler & T. Rezanka (2009) Microbial transformation of synthetic estrogen 17[alpha]-ethinylestradiol. Environmental Pollution, 157, 3325-3335.

Campbell, C. G., S. E. Borglin, F. B. Green, A. Grayson, E. Wozei & W. T. Stringfellow (2006) Biologically directed environmental monitoring, fate, and transport of estrogenic endocrine disrupting compounds in water: A review. Chemosphere, 65, 1265-1280.

Carballa, M., G. Fink, F. Omil, J. M. Lema & T. Ternes (2008) Determination of the solid-water distribution coefficient (Kd) for pharmaceuticals, estrogens and musk fragrances in digested sludge. Water Research, 42, 287-295.

CDM Report, (2010). "Charting the Future of Biosolids Management: Background Materials for Discussion," Report to National Biosolids Partnership and Water Environmental Federation. Alexandria, VA.

Carmosini, N. & L. S. Lee (2009) Ciprofloxacin sorption by dissolved organic carbon from reference and bio-waste materials. Chemosphere, 77, 813.

Citulski, J. A. & K. Farahbakhsh (2010) Fate of Endocrine-Active Compounds during Municipal Biosolids Treatment: A Review. Environmental Science & Technology, 44, 8367-8376.

Collins BM, Malachlan JA, Arnold SF. The estrogenic and antiestrogenic activities of phytochemicals with the human estrogen receptor expressed in yeast. Steroids; 1997:62:365-372.

Díaz-Cruz, M. S., M. J. García-Galán, P. Guerra, A. Jelic, C. Postigo, E. Eljarrat, M. Farré, M. J. López de Alda, M. Petrovic & D. Barceló (2009) Analysis of selected emerging contaminants in sewage sludge. TrAC Trends in Analytical Chemistry, 28, 1263-1275.

Englande, A.J. and Reimers, R.S., (2001). "Biosolids Management- sustainable Development Status and Future Direction," Water Science and Technology, Vol. 44, No. 10, IWA Publishing, London, England, pp. 41-46.

Epstein, E. and H. Kohnke. 1957. "Soil aeration as affected by organic matter application." Soil Sci. Soc. Amer. Proc. 21(6):585-588.

Escher BI, Bramaz N, Maurer M, et al. Screening test battery for pharmaceuticals in urine and wastewater. Envion Toxicol Chem; 2005:24(3):750-758.

Fitzmorris, K.B., Reimers, R.S., Oleszkiewicz, J.A., and M.D. Little, 2007. "Decrease of time for pathogen inactivation in alkaline disinfection systems using pressure." Vol. 79, Water Environment Research, Alexandria, Virginia, pp. 388-395.

Ganesh R, and Leong LYC. (2007) "Manufactured Nanoparticles in the Environment: Evaluation of Sources, Fate, Analyses and Treatability." WEF Compounds of Emerging Concern. Water Environment Federation Alexandria, VA; CD-ROM..

Goodman A. (2007). "Sources and Origins of Compounds of Emerging Concern." WEF Compounds of Emerging Concern. Water Environment Federation (CD-ROM): Alexandria, VA; CD-ROM.

Gray LE, Kelce WR, Wiese T, et al. Endocrine screening methods workshop report: detection of estrogenic and androgenic hormonal and antihormonal activity for chemicals that act via receptor or steroidogenic enzyme mechanisms. Reprod Toxicol; 1997:11(5):719-750.

Guardia MJL, Hale RC, and Mainor TM, (2007). "Brominated Flame Retardants (BFRs) - Residual Trends in Sewage Sludge." WEF Compounds of Emerging Concern. Water Environment Federation: Alexandria, VA; CD-ROM.

Hall S, and Olthof M., (2007). Toxicity of Nanoparticles to Aquatic Life and Potential Treatment Methods. WEF Compounds of Emerging Concern. Water Environment Federation : Alexandria, VA; CD-ROM.

Hunt G.T., (2007). "Persistent Organic Pollutants as Compounds of Emerging Concern." WEF Compounds of Emerging Concern. Water Environment Federation: Alexandria, VA, (CD-ROM).

Langdon, K. A., M. S. J. Warne, R. J. Smernik, A. Shareef & R. S. Kookana (2011) Selected personal care products and endocrine disruptors in biosolids: An Australia-wide survey. Science of The Total Environment, 409, 1075-1081.

Le-Minh, N., S. J. Khan, J. E. Drewes & R. M. Stuetz (2010) Fate of antibiotics during municipal water recycling treatment processes. Water Research, 44, 4295-4323.

Miller CA III. A human aryl hydrocarbon receptor signaling pathway constructed in yeast displays additive responses to ligand mixtures. Toxicol Appl Pharmacol; 1999:160(3):297-303.

Mohapatra, D. P., S. K. Brar, R. D. Tyagi & R. Y. Surampalli (2010) Physico-chemical pre-treatment and biotransformation of wastewater and wastewater Sludge - Fate of bisphenol A. Chemosphere, 78, 923-941.

Nie, Y., Z. Qiang, H. Zhang & C. Adams (2009) Determination of endocrine-disrupting chemicals in the liquid and solid phases of activated sludge by solid phase extraction and gas chromatography-mass spectrometry. Journal of Chromatography A, 1216, 7071-7080.

Nieto, A., F. Borrull, E. Pocurull & R. M. Marcé (2008) Determination of natural and synthetic estrogens and their conjugates in sewage sludge by pressurized liquid extraction and liquid chromatography-tandem mass spectrometry. Journal of Chromatography A, 1213, 224-230.

National Research Council. (2002) "Biosolids Applied to the Land-Advancing Standards and Practices." The National Academies Press. Washington D.C., pp. 345.

Pillai, S.D. and Reimers, R.S., 2010. "Disinfecting and Stabilizing Biosolids using E-Beam and Chemical Oxidants," Water Environment Research Foundation Report Number U4R06, WERF Publishing, Alexandria, Virginia.

Pothitou, P. & D. Voutsa (2008) Endocrine disrupting compounds in municipal and industrial wastewater treatment plants in Northern Greece. Chemosphere, 73, 1716-1723.

Reimers RS, Weber DA, and Oleszkiewicz JA, (2008). Residuals management drying technology produces noncombustible ammonium sulfate inorganic/organic fertilizer. European Conference on Sludge Management (ECSM'08). Liege, Belgium; 2008, September 1-2.

Reimers, R.S., S.D. Pillai, D.D. Bowman, K.B. Fitzmorris, and L.S. Pratt. (2005). "Stressors Influencing Disinfection and Residuals." in Disinfection 2005 - Sharing Disinfection Technologies: Water, Wastewater, and Biosolids. 2005. Alexandria, VA: Water Environment Federation.

Reimers, R.S., Fitzmorris, K.B., Smith, J.E., Boyd, G.R., and Bowman, D. D., (2004) "State of Art in Treatment and Survival of Pathogens in Biosolids," Journal of Residuals Science and Technology, Volume 1, Number 2 , DEStech Publications Inc., Lancaster, Pennsylvania, pp. 93-104.

Reimers, R.S. Bowman, D.B., Schafer, P.L. Tata, P. Leftwich, B. and Atique, M.M. "Factors Affecting Lagoon Storage Disinfection of Biosolids," (2001) WEF Residuals and Biosolids Specialty Conference, water Environment Federation, Alexandria, VA CD-ROM.

Routledge EJ, White R, Parker MG, et al. Differential effects of xenoestrogens on coactivator recruitment by estrogen receptor (ER) alpha and ERbeta. J Biol Chem; 2000: 275(46):35986-35993.

Smith, J.E. , Surampalli, R.Y., Reimers, R.S., Tyagi, R.D. and Lohani, B,N., (2008)."Disinfection Processes and Stability Refinements to Biosolids Treatment Technologies," American Society of Civil Engineering, Practice Periodical of Hazardous, Toxic and Radioactive Waste Management. Washington D.C., Volume 12, Issue 1, pp. 10-17.

Smith, J.E. and P.D. Millner. (2001). "Workshop on Emerging Infectious Disease Agents and Issues Associated with Animal Manures, Biosolids, and Other Similar By-

Products." in USEPA Contract No. C7-C8-0011, Work Assignment 3-98. 2001. Cincinnati, OH: USEPA National Risk Management Research Laboratory.

Snyder, E. H., G. A. O'Connor & D. C. McAvoy (2010) Fate of 14C triclocarban in biosolids-amended soils. Science of The Total Environment, 408, 2726.

Stasinakis, A. S., G. Gatidou, D. Mamais, N. S. Thomaidis & T. D. Lekkas (2008) Occurrence and fate of endocrine disrupters in Greek sewage treatment plants. Water Research, 42, 1796-1804.

Strausberg RL, Strausber SL. Overview of protein expression in Saccharomyces cerevisiae. Curr Protoc Protein Sci; 2001:Chapter 5,Unit 5.6.

Spinosa, L., (2010). "Sustainability in Sludge Management: a Combined Approach to meet Future Needs," Water 21, December 2010, IWA Publishing, London, England, pp. 31-33.

Tan, B. L. L., D. W. Hawker, J. F. Müller, L. A. Tremblay & H. F. Chapman (2008) Stir bar sorptive extraction and trace analysis of selected endocrine disruptors in water, biosolids and sludge samples by thermal desorption with gas chromatography-mass spectrometry. Water Research, 42, 404-412.

USEPA, *Biosolids Generation, Use and Disposal in the United States.*, Municipal and Solid Waste Division, Office of Solid Waste; 1999:81.

Xue, W., C. Wu, K. Xiao, X. Huang, H. Zhou, H. Tsuno & H. Tanaka (2010) Elimination and fate of selected micro-organic pollutants in a full-scale anaerobic/anoxic/aerobic process combined with membrane bioreactor for municipal wastewater reclamation. Water Research, 44, 5999-6010.

Yanko, W.A., (1987) "Occurrence of Pathogens in Distribution and Marketing Municipal." EPA 600-1-87-014. NTIS PB88-15273/AS. Health Effects Research Laboratory, U.S. Environmental Protection Agency., Research Triangle.

Yurtsever, D., Rowan, J.M., and Santha, H., (2009). 'comparison of Gasification, Pyrolysis and Incineration Technologies for Residuals Management: Future of Advanced Biosolids Processes," Residuals and Biosolids 2009, Water Environment Federations, Alexandria, Virginia ppl80 - 91.

Application of High Energy Electron Beam and Chemical Oxidants to Destroy Endocrine Disrupting Compounds and Inactivate (Pathogenic) Microorganisms in Wastewater Effluent and Biosolids

Charlotte Rambo, Chandni Praveen, Palmy Jesudhasan, Robert Reimers, and Suresh D. Pillai

Introduction

Wastewater effluent and biosolids from municipal wastewater treatment facilities can contain a variety of infectious microorganisms as well as man-made and natural chemicals and their metabolites. Wastewater effluent is often released into receiving streams and rivers, and can eventually end up in groundwater and agricultural irrigation water. Biosolids are often land-applied for land reclamation projects and agriculture. Pathogens in biosolids can create health risks for humans if the biosolids are not appropriately managed. Endocrine disrupting chemicals (EDCs), such as estrogenic compounds, and pathogens such as *Salmonella* and enteric viruses in wastewater effluent and biosolids are cause for concern around the world. Additionally, the U.S. Environmental Protection Agency (EPA) and the World Health Organization (WHO) have identified endocrine disrupting chemicals, such as estrogen and its metabolites, to be a critical emerging environmental concern.

The objective of this study was to demonstrate the efficacy of high energy (10MeV) electron beam (e-beam) to break down EDCs and reduce pathogens in wastewater effluent and biosolids. Developing new technologies to address pathogens and emerging contaminants are extremely important in order to sustain the wastewater industry and also find environmentally sustainable solutions for biosolids and wastewater effluent. Our underlying hypothesis is that e-beam technology is an appropriate technology for disinfecting biosolids and for addressing the emerging issues associated with endocrine disrupting chemicals.

E-beam Irradiation

Wastewater effluent and biosolids samples were collected from both aerobic and anaerobic wastewater treatment plants. The samples were collected from the digester outlets in order to determine whether e-beam technology could be used to treat Class B biosolids to Class A biosolids. The samples were spiked with a known concentration of the model estrogen, 17-β-estradiol (E2) (Sigma, St. Louis, MO). Preliminary studies had shown that the digester samples did contain certain indicators organisms (*Escherichia coli*, and aerobic and anaerobic spore formers) in elevated levels and therefore did not require inoculation. However, certain pathogens and test organisms were below detectable levels and there required their inoculation. The microbial pathogens and indicator organisms that were included in these studies were

Salmonella Typhimurium, phi X 174 and MS2 phage, and poliovirus). The samples were evenly mixed and aliquots were placed in sterile Whirl-Pak® bags (Nasco, New York, NY) and heat sealed. Each bag was triple bagged and then transported to the National Center for Electron Beam Research (NCEBR) at Texas A&M University. Samples were irradiated using a 10MeV e-beam linear accelerator at doses ranging from 2 to 12 kGy. Absorbed doses were measured using alanine dosimeters and the Bruker e-scan electron paramagnetic resonance spectrometer (Bruker, Billerica, MA).

Estrogenic Compounds

After e-beam treatment, estrogenic compounds were extracted from wastewater effluent using the dichloromethane solvent extraction method. Samples were allowed to air-dry and resuspended in DMSO. For the biosolid samples, the liquid and solid portions had to be extracted and analyzed separately. Estrogenic compounds for the biosolids were extracted using the solid-phase extraction method. The estrogenic activity was measured using two different bioassays, the breast cancer cell line MCF-7 and the YES (Yeast Estrogen Screening) assay. The reduction of estrogenic activity in wastewater effluent after e-beam irradiation ranged from 71-79% depending on the absorbed dose. There was no reduction of estrogenic activity in biosolids after e-beam irradiation, even at 12 kGy. These results suggest that high energy e-beam irradiation is capable of destroying estrogenic compounds in wastewater effluent but not in biosolids. A possible reason for this is the fact that there is the EDC gets adsorbed to the organic matter in the biosolids sample thereby being protected from the e-beam irradiation. Having organic matter does attenuate the effects of ionizing irradiation since there are fewer electrons to interact with, resulting in fewer interactions with the estrogenic compound. Nonetheless, the results indicate that e-beam irradiation shows strong potential for the elimination of estrogenic activity in wastewater effluent.

Microorganisms

Following e-beam irradiation, the samples were analyzed for the various target organisms. Briefly, for *Salmonella* and *E. coli*, the samples were serially diluted and plated in tryptic soy agar (TSA) and EC-MUG media respectively. For aerobic and anaerobic spore formers, the samples were heated to 64°C for 15 min (to kill all the vegetative cells), serially diluted and plated onto TSA and anaerobic spore former media respectively. The viruses were extracted using 3% beef extract, filtered using 0.22μm filters, and enumerated. Enumeration of coliphages was carried out using the Single Agar Layer method (Method 1602, US EPA, 2001) with the host bacteria *E. coli* CN- 13 for somatic coliphages and *E. coli* Famp+ for male specific coliphages. Poliovirus was enumerated using tissue culture methods with BGMK (Buffalo Green Monkey Kidney) cell line as the host.

An e-beam dose of 1 kGy resulted in a 3 to 4.5 log reduction of *E. coli* and *Salmonella* in both aerobic and anaerobic biosolids. An e-beam dose between 6-7 kGy resulted in a 1 to 3 log reduction of aerobic and anaerobic spore formers in both types of biosolids. An e-beam dose of 9 kGy resulted in a 2.5 to 4 log reduction of somatic and male-specific coliphages in both aerobic and anaerobic biosolids. An e-beam dose of 8 kGy resulted in a 3.3 log reduction in anaerobic biosolids. Considerable variation in susceptibility was observed for the pathogens and indicator organism used in this study. In the presence of E-beam irradiation, significant microbial inactivation was achieved even at relatively low doses. Bacteria were the most susceptible organisms, followed by aerobic and anaerobic spore formers. Viruses were more resistant than bacteria to e-beam irradiation. Among the viruses, poliovirus was more susceptible than coliphages. It appears that the larger the organism, the more susceptible it is to e-beam irradiation. The results also indicate that organic matter imparts protection to the microorganisms, and it reduces the effectiveness of e-beam to break down estrogenic compounds. Overall, the results suggest that at 10 kGy (the ionizing irradiation dose that the US EPA has mandated as a PFRP process) e-beam irradiation is capable of significant reductions of the regulated and the emerging contaminants. When economic analyses of the e-beam process for wastewater treatment were performed, the results suggest that the process is significantly cheaper than currently employed PFRP processes.

This research shows that e-beam irradiation greatly reduces estrogenic activity in wastewater effluent. E-beam irradiation also reduces microorganisms in biosolids, hence reducing the risk of infectious diseases. To reduce the potential for adverse environmental and human impacts, it is critical that novel approaches be investigated so that municipal sewage sludges or biosolids can be disinfected and stabilized to reduce the pathogen loads, reduce the levels of estrogenic compounds and make the material suitable for land-application. However, it should not be overlooked that sewage sludges are also significant pools of energy substrates and nutrients as well. Wastewater treatment plants of today should be viewed as "*Sustainable Resource Recovery Facilities*". Overall, this research study supports the immense potential of industrial use of high energy e-beam irradiation for municipal waste water and biosolids treatment.

Keywords:

Electron beam irradiation, wastewater, biosolids, estrogens, bacteria, virus, coliphages

Novel BioElectro™ Process for Sterilization of Biosolids

Eham Safaei Takhtehfouladi[1], Maria Elektorowicz[1], Robert S. Reimers[2] and Jan A. Oleszkiewicz[3]

[1]Department of Building, Civil and Environmental Engineering, Concordia University
[2]Department of Environmental Health Sciences, Tulane University
[3]Department of Civil Engineering, University of Manitoba

ABSTRACT

Sterilization, pasteurization and disinfection of biosolids by novel BioElectro™ Process is shown. The process is based on applying, enhanced by conditioners, DC field to biosolids. The reduction of *Clostridium perfringens* spores by 8 log was demonstrated within less than two hours while 3 log reduction was reached within less an hour. It was concluded that presence of unionized ammonia molecule along with the elevated temperature generated due to exothermic reaction within the system may sterilize biosolids. Also, an empirical model was developed by non-linear programming approach to quantify disinfection kinetics of process. The proposed model was able to predict the spore inactivation very accurately. The new process is ready to be applied into full scale facilities.

KEYWORDS

BioElectro™, Biosolids, *Clostridium perfringens* spore, electrokinetics, sterilization, disinfection

INTRODUCTION

Current regulations restrict the amount of pathogens in biosolids subjected to reuse or dispose. Generally, sludge contains high levels of pathogens which may threaten human health and environment if biosolids is not being used safely in accordance with good practice (LeBlanc *et al.*, 2006). Appropriate fecal indictor organisms such as *Clostridium perfringens* spore can be used to evaluate the hygienic characteristics of biosolids indirectly. Sufficient reduction of potentially pathogenic bacteria can be accomplished by means of physical, chemical or biological treatments. In the last several years, the environmental engineering community has shown interest in an innovative application of Electrokinetic (EK) technology to treat biosolids (Elektorowicz and Oleszkiewicz, 2009). In this study, the enhanced EK process (BioElectro™) was assessed through which C. *perfringens* spore inactivation in anaerobically digested biosolids was achieved. It should be noted that under anaerobic conditions *C. perfringens* spores is the most resistant microbe and with a three log reduction indicates Class A disinfection with respect to viruses, bacteria and helminth eggs (Blanker et.al., 1992).

METHODOLOGY

Anaerobically digested (AnD) biosolids was sampled from the Robert O. Pickard Environmental Center, Ottawa. Sludge total solids (TS) averaged 2.4% ($\pm$0.2%), pH was 8.17 units; *C. perfringens* spore ranged from 6.4 to 7.3 log CFUs/g TS. The BioElectro™ disinfection testes were carried out in a laboratory size prototype BioElectro™ equipment consisted of a regulated power supply (Genesys 1500W, TDK Lambda Americas Inc) and a 3100 mL Plexiglas rectangular EK reactor with internal dimensions of 214 mm length × 214 mm width × 74 mm height. The power supply transforms alternating current (AC) from a utility line (120-V, 60 Hz) into high voltage AC, then rectifies to DC signals. The EK reactor was equipped with four perforated stainless steel (316SS) electrodes (10 mm diameter, 102 mm long) located at a distance of 172 mm from each other acting as anode and cathode. Underneath each electrode a 200 mL Nalegene bottle was installed to collect electroosmotic dewatering flow during the experiment. Four glass red-dye thermometers (0-100 °C) were installed at the immediate vicinity (10 mm) of electrodes to monitor heating rate. The heat rate at the middle of the reactor was measured using a HACH probe (HQ40d Digital Multi-Parameter Meter).The disinfection tests were run under the DC gradient of less than 5V/cm. An ammonium salt (AS) and Bioxy S™ (BS) (Atomes company, St Laurent, QC) used as enhancement agents. Previous studies have been shown the effectives of AS (Elektorowicz et al., 2008) and BS (Elektorowicz et al., 2007; Safaei et al., 2009) in EK disinfection systems used for treatment of different sewage sludge matrices. BS is a solid source of peracetic acid (PAA) (Dagher and Dagher, 2006) with enhancing effects on conductivity of biosolids. Test samples and a control were homogenized with agitation at 200 rpm for 5 minutes, then AS and BS were applied, mixed at 300 rpm for two minutes and poured into the EK reactors. No mixing was done during the course of experiments. At the end of each treatment (2 hours), samples were taken for microbial analyses. Spores were counted based on method developed by (Reimers et al., 2002). All tests were performed in quadruplicates, and no standard errors are higher than ±11%.

RESULTS AND DISCUSSION

The results showed that BioElectro™ process was effective in significantly eradicating *C. perfringens* spores (almost by 8 log and 3 log when high, but less than 5V/cm, and low voltage gradients were applied respectively) over 100 min period. The control cell without DC field did not show spore inactivation in anaerobically digested biosolids matrix (Figure 1). This also can be observed evidently by transmission electron microscopy examination of spores' samples after exposure to BioElectro™ process (Figure 2).

Furthermore, the temperatures have risen to 90°C within 1 to almost 2 hours, and just from the temperature data, this process meets the requirements for pasteurization alone. The inactivation temperature is a result of the exothermic reaction caused by the BS and AS due the catalysis by the electrolytic current. The results of this exothermic catalysis are shown in tables 1 and 2.

Based on the temperature profiles it is concluded that the primary stressor for disinfection is temperature caused by ohmic heating. In fact, the process may actually sterilize the biosolids. The rate of heating is directly proportional to the square of the electric field intensity and the electrical conductivity of biosolids. As mentioned earlier, application of BS and AS played an important role in enhancing electrical conductivity of the system. The electrical conductivity rises with increasing temperature, vindicating that ohmic heating becomes more efficient as temperature rises, which could theoretically contribute to runaway heating.

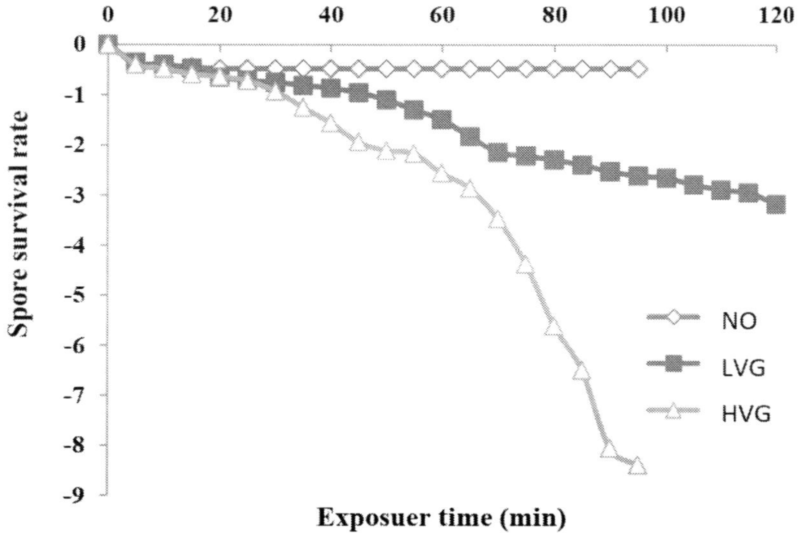

Figure 1- Effect of DC high voltage gradient (HVG), low voltage gradient (LVG) and no DC field (NO) as well as duration of exposure to DC field (min) on spore survival rate (log reduction) in the presence of AS and BS

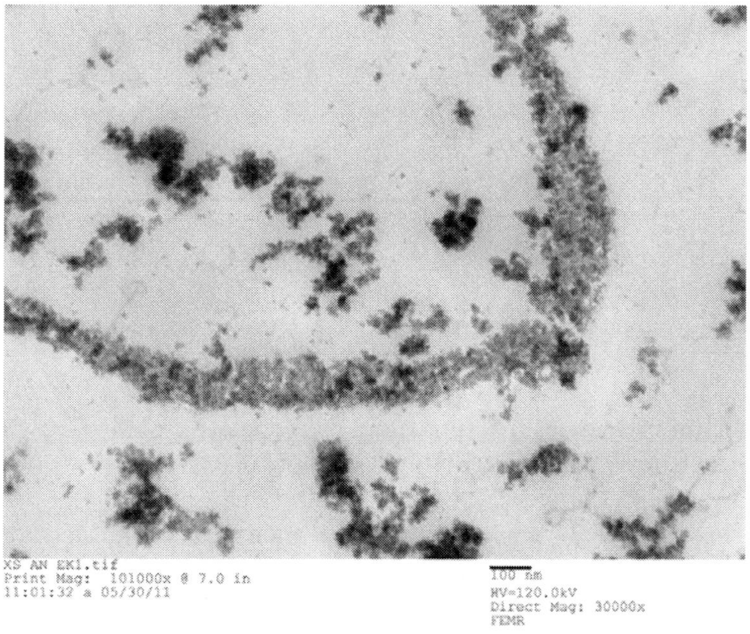

Figure 2 - Micrograph shows creation of nicks in spore membranes after treatment by Electro-disinfection process

Table 1: Temperature rise which can be achieved by BioElectro^{TM} process in the presence of additives (BS, AS) and low voltage gradient

	Lower voltage gradient		
Time (min)	**Log reduction (*C. perfringens* spores)**	**Temp. (°C)**	**Comments**
20	<1	35-40	
40	1	45-55	Near **pasteurization**
60	2	65-75	At **pasteurization**
80	3	85	Indicates that **disinfection** and **pasteurization** has occurred.
100	5	95	Near **sterilization** with the assistance of ammonia.

Table 2: Temperature rise which can be achieved by BioElectro^{TM} process in the presence of additives (BS, AS) and high voltage gradient (but less than 5 V/cm)

	Higher voltage gradient		
Time (min)	**Log reduction (*C. perfringens* spores)**	**Temp. (°C)**	**Comments**
20	1	45-55	
40	2	60-70	Near **pasteurization**
60	3	80-90	Indicates that **disinfection** and **pasteurization** has occurred.
80	6	95	Near **sterilization** with the assistance of ammonia.
100	8 (at DTL)	95	Near **sterilization** with the assistance of ammonia.

This process should work for aerobically digested biosolids and raw municipal sludge. The levels of pathogen inactivation can be categorized as disinfection, pasteurization and sterilization and are defined below:

1. **Disinfection processes** inactivate just the vegetative pathogens.
2. **Pasteurization processes** inactivate all the vegetative microbes except the spore forming bacteria (greater than 70°C for 30 minutes).
3. **Sterilization processes** inactivate all the microbes (including spores).

The AS and BS has enhanced the ionic strength, causes an exothermic reaction and increases the biocidal stressors/constituents ammonia and PAA. Disinfection for the biosolids pathogens occurs at 70°C or at the range defined for pasteurization, but at temperatures around 95°C, the combination of AS and BS in the BioElectro^{TM} Process indicates near sterilization of the biosolids inside 2 hours. At a temperature of 95°C, the pKa for NH_3/NH_4^+ should be in the range of 7 (Table 3), which means that the ammonia concentrations should be a viable disinfectant for the spores and assist in the sterilization process.

Table 3: Computed pK_a and NH_3 in each experimental conditions of BioElectro™ process

EFI (V/cm)	Temp. (°C)	pKa	pH	f*	$(NH_3)t$(mg/L)	NH_3(mg/L)
2.8	95	7.46	7.80	0.69	489	337
2.5	95	7.50	8.15	0.82	194	159

* f = Fraction of total ammonia that is unionized

Another related factor in the close system of this study was vapor pressure. As elucidated by Henry's Gas Law, a rise in temperature in concurrence with the presence of ammonia increases the vapor pressure of the ammonia gas within the biosolids. This raise in vapor pressure serves to raise the diffusion of the ammonia into the biosolids, consequently permitting greater contact with pathogens and higher inactivation effects.

To evaluate direct effects of PAA produced by BS its demand and decay rate was examined (Figure 3). The kinetic analysis of disinfectant residuals showed that there was a substantial decay of PAA, decay rate of 0.293 (geometric mean) min^{-1}. Based on these results PAA decomposed within 15 minutes. Therefore, it cannot be considered as a major contributing factor in the disinfection process.

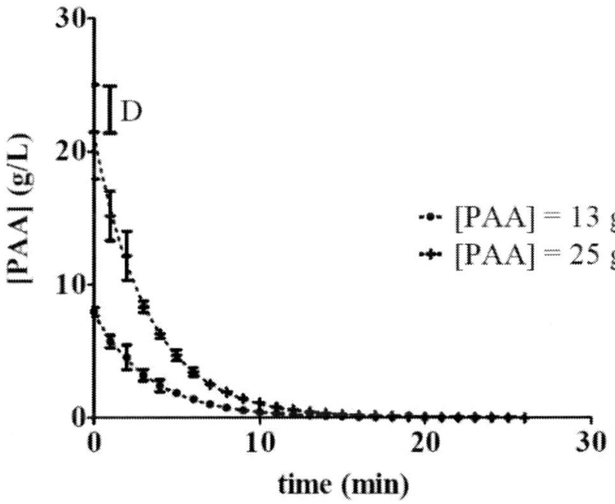

Figure 3- PAA consumption curve. Dotted lines refer to the 1st order model with instantaneous demands (D) ≠ 0 (bars indicate minimum and maximum PAA concentration values)

To explain the overall mechanism of *C. perfringens* spore inactivation in BioElectro™ process kinetic model was defined (eq. 1). The model defines the relationship between predictors, to wit:

E (applied voltage gradient), BS, AS and time and the response variable (spore survival). Furthermore, it explains the effect of initial microbial density on disinfection.

$$\log\left(\frac{N}{N_0}\right) = (b_1 + a_1 . BS^{n_1} . t^{k_1})(b_2 + a_2 . AS^{n_2} . t^{k_2})(b_3 + a_3 . E^{n_3} . t^{k_3}) \quad \text{(eq. 1)}$$

where N: spore densities at time t (CFU/ g TS), N_0: initial spore density (CFU/ g TS), AS: applied ammonium salt dose (g/L), BS: applied BIOXY S^{TM} dose (g/L), E: applied voltage gradient (V/cm), t: exposure time (min) and a_i, b_i, d_i, k_i, n_i are the reaction rate parameters.

CONCLUSIONS

BioElectro^{TM} is a breakthrough technology which presents a high potential of practical applications for the advanced biosolids treatment and management. The research showed that process can achieve a near sterilization level of biosolids in less than two hours. The *Clostridium perfringens* spores were inactivated after treatment in all samples of anaerobically digested biosolids in EK reactors. Biosolids disinfection would occur at temperatures 70°C inside 60 minutes, but the *Clostridium perfringens* inactivation results at 90 °C indicates that sterilization is occurring related to the double stressors of temperature and ammonia. From this study, PAA was not a major factor. The results and review of the data affirmed process reliability and product quality that exceeds Class A regulatory compliance. The BioElectro^{TM} process was patented (Elektorowicz *et al.*, 2012) and is ready for full scale applications in both batch and continuous flow systems. It can also be applied in combination with EKDIM technology (Elektorowicz and Oleszkiewicz, 2009), which can additionally assure dewatering and removal of metals.

ACKNOWLEDGEMENTS

The authors thank the financial support from Natural Science and Engineering Research Council of Canada (NSERC) Discovery Project granted to Dr. M. Elektorowicz. The authors also appreciate the cooperation of the employees of the Robert O. Pickard Environmental Center, Ottawa, ON, Canada.

REFERENCES

Blanker, E.M.; Little, M.D.; Reimers, R.S.; Akers, T.G (1992) Evaluating the Use of *Clostridium perfringens* spores as indicator of the presence of viable *Ascaris* eggs in chemically treated municipal sludges; Proceedings of Municipal Sludge (Biosolids) Management - Where We Are and Where We're Going (Volume I), Water Environmental Specialty Conference Series, Alexandria, VA, pp. 187-201

Dagher, F. and Dagher, D. (2006) Powdered composition for the generation of peracetic acid and use thereof to sanitize surface, Patent CA2569025.

Elektorowicz, M.; Oleszkiewicz J (2009) A method of treating sludge material using electrokinetics, applied US Patent 8329042.

Elektorowicz, M.; Oleszkiewicz, J.; Huang, J.; Safaei, E.; Abdoli, H. and Habel, H. (2008) EKDIM: an electrokinetic process treats municipal wastewater sludge for safe land application. Volume 7, pp 276-287, Seoul, South Korea.

Elektorowicz, M.; Safaei, E.; Reimers, R. Oleszkiewicz, J, and Dagher F, 2012, Process for treatment of residuals, Provisional US Patent. 6171838
Elektorowicz, M.; Safaei, E.; Oleszkiewicz, J. and Reimers, R (2007) Electrokinetic remediation of biosolids through inactivation of *Clostridium perfringens* spores, Vigo, Spain.
LeBlanc, R.J , Matthews, P. and Richard. R.P. (2006) Global atlas of excreta, wastewater sludge, and biosolids management: Moving forward the sustainable and welcome uses of a global resource. Technical report, Greater Moncton Sewage Commission.
Safaei, E.; Elektorowicz, M. and Reimers, R (2009) Bench-scale trial of electrokinetic treatment on anaerobically digested biosolids; microbial inactivation. 8th World Congress of Chemical Engineering, Montréal, Canada
Reimers, R.S.; Oleszkiewicz, J.A.; Bowman, D.D and Faulmann, E.L (2002) PEC application for the national classification of the J--Vap process to the category of PFRP, Report for Paradigm International, Inc. Louisiana, USA.

Developments in Disinfection Using Advanced Alkaline Stabilization

Kari Fitzmorris Brisolara, ScD[1] and Robert S. Reimers, PhD[2]

[1]Louisiana State University Health Sciences Center, School of Public Health, New Orleans, Louisiana
[2]Tulane University, School of Public Health and Tropical Medicine, New Orleans, Louisiana

Innovation in biosolids treatment can be a challenging process. The development of a new or innovative biosolids disinfection process not only has to function accordingly, but also must adhere to ever-increasing restrictions imposed by the federal government in addition to local and state governments. The purpose of this project was to conduct experiments to define an advanced alkaline process as PFRP by demonstrating the ability of the system to handle different sludge types and verifying the suggested optimum operating conditions indicated by previous tests. Prior to PEC approval, Class A was being achieved via the pasteurization option and Vector Attraction Reduction is still met via Option 6, pH 12 for 2 hours, pH 11.5 for an additional 22 hours. In order to achieve a Class A product, increased temperatures and/or ammonia addition is required to inactivate the highly resistant viruses, protozoan spores, and helminth eggs. The primary physical parameters of interest are pH, temperature and ammonia concentration with regard to disinfection. Studies at Tulane have observed disinfection under closed systems at 55°C instead of 70°C due to the containment of ammonia in a closed system (Reimers et al., 2004). In the disinfection of biosolids along with temperature caused by the exothermic reactions of quicklime and acid trimming, the non-charged disinfectants are a major player. For alkaline disinfection, ammonia is the primary constituent. The key parameters of interest to determine the efficacy of the process are the pathogens: *Ascaris,* fecal coliform and poliovirus. Aerobic endospores were also monitored as a potential indicator organism for alkaline processes. The data demonstrates the effectiveness of the closed alkaline system at inactivating helminth ova within 30 – 40 minutes resulting in a greater than 2 log reduction. Most importantly, a greater than 3 log reduction was noted for all samples with regard to poliovirus and fecal coliform. The resistance of the aerobic endospores again proved to be higher than that of the other pathogens of concern. The data for the final runs conducted in the summer of 2010 substantiated the results of the previous studies conducted by Tulane. This data indicates meeting the criteria for Class A disinfected biosolids. The data related to pathogen testing, the required reductions, and indicator organism usage, will be presented along with the major hurdles to be overcome in pathogen recovery.

Key Words: alkaline, biosolids, disinfection, pathogen inactivation

INTRODUCTION

The process that new sludge disinfection technologies must undergo to gain acceptance by state, local and federal regulatory officials can be a long and arduous one. The purpose of the PEC within US EPA is to ensure all new technologies meet the requirements set forth in 40 CFR 503 (USEPA 1992) sewage sludge standards and in turn protects human health. The purpose of this project was to conduct experiments to define an advanced alkaline process as PFRP by demonstrating the ability of the system to handle different sludge types and verifying the suggested optimum operating conditions indicated by previous tests thereby demonstrating equivalency to a Class A disinfection process under Alternative 6: Use of a Process Equivalent to PFRP [503.32(a)(8)]. This project sought approval from EPA's Pathogen Equivalency Committee for the operational conditions to be changed from 70°C to 55°C on a nationwide basis with a time of 40 minutes. Tests were conducted at the Lakeland, Florida operation to provide the basis for a matrix that will lead to nationwide EPA-PEC approval. The conditions examined included: pH, time, temperature, ammonia exposure, sludge type, percent solids, and pathogen inactivation rate. Initial testing began in 2002 with national equivalency awarded in 2011. There have been many challenges over the years ranging from analytical methods for pathogen analysis to the introduction of test organisms at the full-scale level. The data related to pathogen testing, the required reductions, and indicator organism usage will be presented along with the major hurdles to be overcome in pathogen recovery.

MATERIALS AND METHODS

The advanced alkaline process referenced in this project is the Schwing Bioset Process, which mixes dewatered biosolids with calcium oxide (Quicklime) and sulfamic acid and continuously pumps it into an enclosed plug flow reactor maintaining temperature and pressure for greater than 30 minutes. Because of the closed system, (1) the sludge cake becomes thixotropic; (2) the calcium oxide is thoroughly mixed; and (3) long-term stability is enhanced with lower lime doses.

All testing was conducted at the Lakeland, Florida Glendale Street plant utilizing the Schwing-Bioset system that was modified to include sampling ports. The final description of the runs is indicated below and this series was conducted for the baseline data and repeated for the final spike confirmation.

- Waste Activated Sludge – Glendale Plant (Lakeland, FL); One solids content level (as generated by the belt press); Five time options (T0 – Final)

- Anaerobically Digested Sludge – Glendale Plant (Lakeland, FL); One solids content level (as generated by the belt press); Five time options (T0 – Final)

- Aerobically Digested Sludge – Auburndale Regional (Auburndale, FL); **Two** solids content levels (as generated by the centrifuge run at differing speeds); Five time options (T0 – Final)

Initial baseline measurements were taken for each sludge type using aerobic endospores as microbiological indicators to confirm timing between ports and serve as trial runs prior to spiking with *Ascaris* and poliovirus. Once the optimum conditions were determined, four runs were conducted utilizing a spike with *Ascaris* and Poliovirus to verify the aerobic endospore data.

The following parameters were measured only, with no changes or adjustments studied ammonia (TNT831 - EPA 350.1), pH (APHA Method 4500-H+), calcium (EPA 200.7 / APHA Method 3120 B), and temperature (APHA Method 2550). Standard methods as defined by EPA were used for the microbiological parameters including aerobic endospores (Modification of APHA Method 9218), fecal coliform (EPA Method 1681), *Ascaris* (Modification of method described in

Appendix I of EPA/625/R-92/013), and poliovirus (Method described in Appendix H of EPA/625/R-92/013). It is important to note that separate runs were not necessary for the different times due to the installation of the sampling ports along the length of the unit. By utilizing sampling ports, it was possible to conduct sampling as the material passed through the unit and not just at the final output. The pressure at the reactor inlet was also monitored. The pressure is significant in that it establishes the reactor is a closed system to maintain the contact of the dissolved ammonia with the biosolids within the unit. Sampling locations are noted on Figure 1 below.

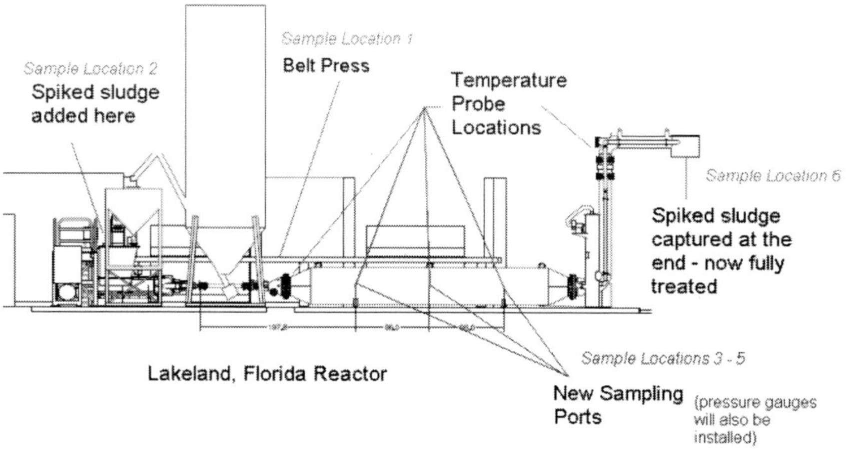

Figure 1: Sample location map

RESULTS AND DISCUSSION

Pilot testing was conducted at Tulane University in 2004 and indicated that the Bioset process met the criteria for a PEC process to disinfect to Class A (Fitzmorris et al. 2007). This data is capsulated in Table 1.

Table 1. Capsulation of the Preliminary Pilot Studies

Plant	Log Reduction of Viable *Ascaris*	Log Reduction of Aerobic Endospores	Log Reduction of Polio Viruses	Log Reduction of Somatic Coliphage
Kingwood, TX (8.2 % Solids)	2.5	3.0	4.8	4.8
Hollywood, Fl (17.9 % Solids)	2.9	2.5	3.6	4.8
Lakeland, FL (10.5 % Solids)	2.1	6.0	3.0	4.7
Kissimmee, FL (14.7 % Solids)	2.7	3.6	2.8*	5.4
Morgan City, LA (21 % Solids)	1.8*	3.0	N/A	3.1
Sulphur, LA (2.9 % Solids)	2.4	N/A	N/A	N/A

* The log reductions noted for Kissimmee and Morgan City were the only ones below the required three log reduction for viruses and two log reduction for helminth ova. This was because of low recoveries of virus and the ova in the influent, not because of the lack of kill.

The primary physical parameters of interest are pH, temperature and ammonia concentration with regard to disinfection. Calcium was measured to confirm mixing. The pH of the material was consistently above 12 with the minimum reading of 12.38 in one sample of the anaerobic run. The goal for temperature for the runs was 55°C. This level was determined by previous testing to be optimum and is in accordance with the site-specific approval granted for the Kingwood, Texas. See results above in Table 1 for Kingwood data. Significant variation was noted in the ammonium levels between sludge types with the highest levels found in the anaerobic material (7.02 mg NH_4^+/g dry wt). Previous studies have indicated the importance of ammonia as a disinfectant (Pecson and Nelson 2005) particularly as related to alkaline systems. The variation of ammonia noted in this study is similar to the variability found in previous testing which ranged from 0.6 mg NH_4^+/g dry wt at Kissimmee to 9.7 mg NH_4^+/g dry wt at Hollywood. The physical parameters for the latest studies are in Table 2 below.

Table 2. Physical Parameters of Interest

Sample ID	Reactor Pressure (PSIG)	Temperature (°C)	pH	Ammonia (mg NH_4^+/ g dry wt)	Calcium (mg/L)
Waste Activated Sludge (Raw)					
Influent		33.9		0.1	44
Effluent	20	55.1	12.7	0.7	709
Anaerobic					
Influent		31.7		3.8	123
Effluent	17	52.6	12.5	7.0	686
Aerobic – Run 1					
Influent		31.2		2.8	65
Effluent	17	54.6	12.6	2.3	803
Aerobic – Run 2					
Influent		33.1		0.0	109
Effluent	25	56.1	12.5	1.9	766

Calcium was chosen as a parameter of interest to confirm the consistency of mixing. With each of the four types: AER1, AER2, ANA, and WAS, five random samples were selected from a hypothetical infinite population of samples and replicated within each sample. In this type of experiment, the hypothesis of interest is H0: σ2sample = 0 verses Ha: σ2sample ≠0. Failure to reject H0 supports that the samples are homogeneous and this was the case for all four sludge groups.

The data presented below in Table 3 demonstrates the effectiveness of the advanced alkaline process at inactivating helminth ova within 30 – 40 minutes resulting in a greater than two log reduction. The percent solids presented above for the influent were directly measured. The values for the effluent samples are calculated based upon the direct measurement of the percent solids of the neutralized biosolids and the amount of 6N H_2SO_4 and deionized water added to a known sample weight. Most importantly, a greater than three log reduction was noted for all samples with regard to poliovirus and fecal coliform.

Table 3. Microbiological Data with Log Reduction

Sample ID	Time (h:mm:ss)	Percent Solids (%)	*Ascaris* Larvated (% Viable)	Fecal Coliform (MPN/g dry wt)	Poliovirus (PFU/4g dry wt)
Waste Activated Sludge (Raw)					
Influent	0:00:00	11.0	519 (91.1%)	>1.5E+06	1.2E+07
Effluent (A)	0:17:22	20.2	0.67 (23.0%)	<1.6	0.45
Effluent (B)	0:40:15	20.0	0 (0%)		
Log Reduction			2.69	5.98	7.42
Anaerobic					
Influent	0:00:00	11.0	281 (90.8%)	6.1E+03	2.8E+06
Effluent (A)	0:13:38	18.1	0.33 (100%)	<1.8	<1.0
Effluent (B)	0:30:02	18.0	0 (0%)		
Log Reduction			2.40	3.52	6.44
Aerobic – Run 1					
Influent	0:00:00	12.3	2492 (90.9%)	>1.3E+05	1.1E+07
Effluent (A)			*No Sample*	*No Sample*	*No Sample*
Effluent (B)	0:34:47	29.3	0 (0%)	<1.0	<1.0
Log Reduction			3.37	5.11	6.99
Aerobic – Run 2					
Influent	0:00:00	15.3	1353 (91.4%)	7.3E+03	2.8E+07
Effluent (A)	0:14:20	22.4	0 (0%)	<1.3	<1.0
Effluent (B)	0:31:30	22.5	0 (0%)		
Log Reduction			3.13	3.58	7.28

CONCLUSIONS

The data for the current runs substantiated the results of the previous studies conducted by Tulane. This data indicates meeting the criteria for Class A disinfected biosolids. This data is elucidated in Table 4.

Table 4. Capsulation of the Microbiological Data

Sludge Type	Log Reduction of Viable *Ascaris*	Log Reduction of Polio Viruses	Log Reduction of Fecal Coliform
Raw (WAS)	2.7 (<1*)	7.4 (0.45**)	7.4 (<1.6***)
Anaerobic	2.4 (<1*)	6.4 (<1.0**)	3.5 (<1.8***)
Aerobic I	3.4 (<1*)	7.0 (<1.0**)	5.1 (<1.0***)
Aerobic II	3.1 (<1*)	7.3 (<1.0**)	3.6 (<1.3***)
Class A Standards (USEPA 1992)	*2 log reduction and <1 viable ova per 4 gram dry weight*	*3 log reduction and <1 PFU per 4 gram dry weight*	*3 log reduction and <1000 MPN per gram dry weight*

*Ova/4 grams dry weight
**PFU/4 grams dry weight
*** MPN/gram dry weight

The resulting treated municipal sludge meets not only the log reduction requirements but is below the criteria for Class A disinfected biosolids. The biosolids produced with this process yielded a long-term stabilized product with a 30 percent reduction in anhydrous lime as a result of the excellent mixing. The projected operational parameters to obtain these levels are related to pH, contact time, temperature with a closed system resulting in positive pressure. Based upon the data, all criteria for PFRP equivalency were met. In addition, previous testing further confirms the capabilities of this process in addition to demonstrating the performance of the process in multiple (6) locations across the U.S.

In addition, the development of more accurate and reproducible analytical methods for viruses along with better pathogen indicators would help reduce costs and increase the speed of the application process.

The original timeline set out for development and regulatory approval of the process was exceedingly optimistic. The two major reasons cited for the delay by the company were:

1. Difficulties in demonstrating enteric virus viability due to poor recoveries even with high-titer spikes during the pilot-scale/full-scale testing.
2. Clarity in expectations from the PEC, which has now been addressed through the establishment of a website and the QAPP process.

REFERENCES

Fitzmorris, K. B., Reimers, R. S., Oleszkiewicz, J. A., Little, M. D. (2007) Decrease of time for pathogen inactivation in alkaline disinfection systems using pressure. *Water Environ. Res.*, **79**(4), 388-395.

Pecson, B. and Nelson, K., (2005) Inactivation of *Ascaris suum* Eggs by Ammonia. *Environmental Science Technology*. **39**: 7909-7914.

USEPA (1992) "Standards for the Used and Disposal of Sewage Sludge," 40 CFR Part 503, November 25.

Development of Heat-dried Biosolids Product for Application in the Wetlands

Yue Xu, Robert S. Reimers, Andrew J. Englande, Jr.

Department of Global Environmental Health Sciences
School of Public Health and Tropical Medicine
Tulane University
1440 Canal Street, Suite 2133
New Orleans, Louisiana 70112
yxu1@tulane.edu
Tel: 504-810-6957
Fax: 504-988-1726

Abstract

In this paper, ferrate and ferric chloride were added to New Orleans municipal sludge in the dewatering step before heat drying. The study noted that ferrate was instable in the sludge cake, which has been substantiated by the deactivation EDCs and disinfection of New Orleans wastewater effluent going to the wetlands. For this study, the optimum dosage for ferric chloride was observed to be 2.3% iron/total solids (TS) of sludge by weight.

The iron treated heat-dried biosolids inhibits putrefaction for over 5 to 6 days under a worst-case scenario (moisture content of 98 to 99% and seeded with microbes). The heat-dried biosolids inactivated vegetative microbes since it simulates pasteurization. Therefore, this ferric chloride could resist the putrefication process for over 2 to 3 weeks. The rationale for inhibition is that the iron addition will buffer the ORP with an electron pk_e similar to an acid/base reaction, and the ferric hydroxide has a high affinity to react with the sulfide and thiocompounds to reduce odor emission. The usage of ferrous and ferric salts has been used in for over the last 20 years in controlling odors in collection systems.

In addition, ferrate and ferric salts have a high affinity for sulfur compound or functional groups as noted in the literature. The EDC activity of the heat-dried biosolids and E-beam treated biosolids appears to be lower than that in the sludge cake and puts its activity similar to those in wetland sediment as a result of ferrate and ferric hydroxide addition.

Key words

Iron-based salts; Heat drying process; Respirometry; SOUR; EDCs.

Introduction

The treatment of wastewater treatment plant residuals to create value-added product is fast becoming a reality as a result of public opinion and regulatory concerns. This mandate has given impetus to the need for developing innovative approaches for municipal waste residuals' treatment based on the possible end uses and recycling modes such as commercial fertilizer, sediment amender for dredge sediment for wetlands and remediation of lead contaminated soils.

This movement is facilitated by advances in solids processing, including: incineration, co-generation of biosolids treatment, heat dried biosolids, advanced alkaline stabilization, innovative chemical or thermophilic processes (which insure disinfection), stabilization and the destruction/biodegradation of refractory organics. These treated municipal waste residuals may be further treated to produce a value-added product that reduce or eliminate pre-existing or perceived liability.

Results of this study will move these tested value-added products beyond the criteria developed in the 503 regulations. The criteria, which led to the development of the USEPA 503 regulations originated from research conducted in the 1970'sand 1980's.

Details of applicable process related to this study are shown in Table 1.

Table 1. Heat Drying Process for Municipal Sludge Processes with Ferric Chloride Addition

Process	Co-generation Incineration plus Heat Drying
Pathogen Inactivation	Heat Drying
Stabilization Short-term and Long-term	Short –term with solids and long-term with ferrate
Destruction/ Biodegradation of Refractory Organics	Incinerated or altered with ferrate to stabilize and deactivate EDCs
Economics Cost per Dry Ton in US Dollars	400 to 500
End Use	Soil and Sediment Amender; potential Lead Remediating

In this study, the Cogeneration/Heat Drying processed biosolids blended with ferrate and/or ferrous chloride were selected for assessment. The treated biosolids would be blended with the dredged sediment. This Class A biosolids would add nutrients and remediate any potentially toxic heavy metals.

Vector Attraction and Stabilization

According to 40 CFR Part 503, biosolids must be stabilized prior to land application in order to reduce pathogens, odors, potential putrefaction processes, and vector attraction concerns (USEPA, 2003). Over the past ten years, the development of several short-term and long-term stabilization techniques has been achieved. Long-term stabilization is obtained when the readily degradable organics are decomposed. Processes currently recognized as capable of obtaining long-term stabilization is limited to the biological processes, such as aerobic digestion, anaerobic digestion, composting, lagoon storage and pile storage. Physical and chemical treatment processes, such as acid treatment, alkaline stabilization, and heat drying, are currently delineated as short-term stabilization processes that only inhibit the putrefaction process. With acid and alkaline-stabilized biosolids, long-term storage may yield odors and attract pests, and pathogen regrowth can occur. Windrow alkaline composting of these biosolids types can result in long-term stabilization and reduced vector attraction (Burnham et. al., 1992).

The indicators used to determine whether biosolids have been treated adequately to achieve stabilization depend on the method of stabilization used and the grade of the biosolids desired (Switzenbaum et al, 1997). Treated biosolids are divided into Class A and Class B biosolids on the basis of pathogen reduction, determined by microbiological testing. Other assessment methods are used to demonstrate adequate reductions in vector attraction and odor control, such as the specific oxygen uptake rate (SOUR), percent volatile suspended solids (VSS) reduction, and the carbon dioxide release rates. Limited bench-scale experiments on sludges treated with both acid and alkaline systems have noted biosolids stability, as measured by decreased SOUR and reduced VSS. VSS reductions in the acid system ranged between 56% and 74%, which is comparable to the reduction of degradable organics; yet a total solids reduction was not appreciable due to some precipitation of calcium sulfate.

Methodology

This section includes details of experimental designs and corresponding analytical procedures to determine the stability and EDC deactivation of heat dried biosolids. The major objective is to assess ferric chloride's ability to stabilize sewage sludge; with the aid of ferrate addition affect EDC deactivations during heat dry process. Phase I is to optimize the dosage of ferric chloride/ferrate and decide the most plausible parameters for monitoring. Phase II was carried out with the intent to experiment on the chosen parameters with various dosage of ferric chloride/ferrate. Phase II also analyzes the effects of treatment on endocrine disrupting compounds. Phase III focuses on statistically analysis and correlations between the amount of chemicals and the readings of said parameters.

Sample collection

Thickened and return activated sludge samples were collected in 0.5-L sanitized brown bottles with Teflon-lined lids. Samples were kept at 4°C during transport and were brought back to ambient temperature before experiments would be conducted. All samples were collected on the same day of experiments. The required minimum sample volume was 0.4 liter for solids analysis.

Characterization of New Orleans municipal sewage sludge and return activated sludge

A number of parameters had been measured in order to characterize the thickened and return activated sludge. They include total solids (TS)/total suspended solids (TSS), volatile solids (VS)/volatile suspended solids (VSS), total organic carbon (TOC), pH and ORP value.

Optimization for sewage sludge stabilization

Tulane conducted tests on sludge with $FeCl_3$ and FeO_4^{-2} at doses of 10, 50, 100 and 200 mg/Kg dry weight of biosolids.

Thickened sewage sludge was poured into sterile 2-L Phipps&Bird B-Ker polycarbonate jars with a sampling port for flash mixing and flocculation in amounts of 1 liter to 2 liters. The ferric chloride/ferrate dose was be delivered by syringe to the jar-test beakers. Ferric chloride dose was be calculated by obtaining the total solids content of that day's sludge sample first and taking 2% of the result. Ferrate dose was calculated after computing the final synthesized concentration of ferrate in buffer solution. Both the doses were based on mass instead of volume. Ferrate solution strength is calculated based on mass instead of volume to minimize error in dose; this error is associated with off-gases from the ferrate solution changing the volume of ferrate within the syringe. Of the six jars, one served as control, namely raw sludge; and five testing dosage. First the ferric chloride was be added while the sludge samples are mixed at the speed of 300 rpm; after the through mixing which takes around 10 minutes, the ferrate was be added and the mixture was stirred at the same speed for 10 minutes. Well-mixed sludge was then transferred into centrifuge tubes and centrifuged, the thickened solids was collected and spread on backing pan for heat dry process for 30 minutes.

The N.O sludge was thickened before belt press to achieve a 5% solids content level. After ferric chloride or/and ferrate addition and dewatering by centrifugation, the resulting biosolids was placed in an oven at 80°C for 30 minutes that intended to mimic the heat drying process.

Respirometry Assessments

Once the optimum dose and ratio of ferric chloride and ferrate are ascertained, respirometry analysis was applied on the resulting biosolids so as to detect the stability. It is to determine the feasibility and scale-up dosages for stable biosolids. Untreated sewage sludge and treated ones were assured by respirometry assessments.

The heat-dried biosolids after the treatments mentioned above was tested for stability with 98-99% of moisture contents. There were three exogenous variables: **buffer, feed and seed**. **Buffer** was 1.5M KH_2PO_4 at pH 7.0. **Feed** was acetate acid and **microbial seed** was RAS being aerated from the moment samples reached lab till it was applied to the respirometric system. The cake was buffered to pH=7.0 and other cases it was seeded with RAS. The RAS from the New Orleans' POTW was used as seed since the microorganisms were already acclimated to it. Following the same logic, the effluent was selected as the dilution water to increased moisture content.

The aerobic conditions of respirometry assessments were designed to simulate the scenario that the treated biosolids were fully seeded with unlimited oxygen supply. The aeration maintained the microbial populations in the sample and also reduced them to endogenous state so the process of seeding did not provide any additional feed in the respirometry system.

The results were expressed as specific oxygen uptake rate (SOUR) that quantifies the activity of aerobic microorganisms and is an excellent indicator for aerobic heterotrophic/autotrophic reactions. pH and ORP values were assessed prior and post the tests.

EDC assessment

1. Sample Preparation and Extraction

Sample preparation and extraction will be conducted in accordance with procedures for chemical analysis of steroid hormones and alkylphenols and hormone assays. Iron treated heat-dried biosolids samples were subjected to a high-pressure, high-temperature solvent extraction utilizing Dionex® ASE 200 extraction system. The extracts were dried by Nitrogen gas flow and reconstituted by suitable solvent for the following biological assays.

QA/QC - QA/QC will include monthly laboratory blanks (extraction of sand or sodium sulfate and distilled laboratory water), matrix-spiked samples for every twenty field samples and a duplicate extraction/analysis for every twenty samples. Samples will be re-analyzed if the percent recovery for surrogate compounds falls outside of 50-150%, if the percent recovery for the matrix spiked sample falls outside 60-140% or if the % relative percent deviation is >25% for the duplicate sample.

2. Endocrine Disrupting Activity

Two types of analytical protocols were be utilized: recombinant yeast estrogen screen (YES) assay and MVLN human cancer cell line assay.

YES assay

YES assay was carried out using recombinant strains of the yeast *Saccharomyces cerevisiae* (Fox J E, Burow M E, McLachlan J A, and Miller C A III, 2008) in pre-sterilized 96-well microtiter plates (Fisher, PA, USA). Each plate contains triplicate extracted samples that are serially diluted, E2 standard solution series (ligand for ER), negative control (deionized water used in extraction), and a solvent blank (DMSO). The concentrations of E2 standard in the serially diluted wells was cover the range of concentrations of tested samples so the linear portion of the E2 standard curve could be used to calculate E2-equivalence levels in samples.

The plates were sealed with Parafilm to prevent evaporation and be incubated at 30°C for 18 hours. Thoroughly re-suspend the cells in the 96-well plate, remove certain amount from each well and transfer it to a new 96-well plate (sterile plates are not necessary in this step) and add Z buffer. Start a timer and place the new plate in a 37°C water bath till yellow colour shows. Time each reaction and stop it with 1 M sodium carbonate solution. Read the 96-well assay plate at a wavelength of 405 nm using the spectrophotometer to quantitate the yellow-coloured (nitrophenol) product.

The calculation of lacZ units for YES assay is as follows:

*((OD$_{405}$ of lacZ assay sample well – background OD$_{405}$) * 1000) / ((amount of time in minutes for assay to produce yellow colour * volume of cells added to well in ml)*(OD)$_{600}$ of lacZ assay sample well – background OD$_{600}$)*

MVLN assay

The same extraction protocol is also suitable for MVLN assay; the only different is ethanol is used as the reconstituting solvent. The work was conducted by Xavier University.

Data Assessment and Conclusion for Process Development

The influence of the iron and ferrate will be assessed to observe its ability to stabilize and deactivate EDCs in heat-dried biosolids. This work will be conducted so that the significance of these ingredients is determined. The viable costs can be ascertained. The optimized treated biosolids will then be utilized in the New Orleans Demonstration study. The results of this study will be utilized in application for patent to yield a long-term stabilization and EDC control for heat dried biosolids with potential reduction of the combustible concerns that is a problem in existing heat drying process especially direct drying processes.

Results

Assess the Stability of Heat Dried Municipal Biosolids

Preliminary assessments of thicken sludge samples (n=12) and return activated sludge samples (n=3) are detailed in Table 2, 3.

Table 2. Characteristics of Municipal Return Activated Sludge (RAS) From New Orleans' POTW

Parameters	Mean	St. D.	Number of Samples
TSS (%)	1.19	0.09	3
VSS (%)	0.63	0.08	3
TOC_{total}			
%	0.33	0.02	3
mg/L	3314	242	3
TOC_{liquid}			
%	0.01	0.005	3
mg/L	101	47	3
pH	8.2	0.01	3
ORP (mV)	-87	0.01	3

Return activated sludge had comparatively lower solids content: TSS and VSS, 1.19% and 0.63% respectively. It also contained lower TOC value of 3314 mg/L in total and 101 mg/L in supernatant. The pH and ORP values were higher than that of thickened sludge, 8.2 and -87 mV respectively.

The thickened sludge contains around 5.30 ± 0.6% of total solids and 2.80 ± 0.3% volatile solids. Total organic carbon measurements of sludge and supernatant are 10040 mg/L and 79 mg/L respectively. The pH values remain neutral, 6.8 with 0.4 of variance. The ORP values fall in the range of -166 ± 14 mV, the negative numbers are expected since there was foul odor emitting from the sludge.

Table 3. Characteristics of Thickened Municipal Sewage Sludge From New Orleans' POTW

Parameters	Mean	St. D.	Number of Samples
TS (%)	5.3	0.64	12
VS (%)	2.8	0.25	12
TOC_{total}			
%	1.0	0.2	12
mg/L	10040	1818	12
TOC_{liquid}			
%	0.008	0.0001	12
mg/L	79.02	9.9	12
pH	6.8	0.4	12
ORP (mV)	-166.1	14	12

To assess ability of heat drying process to produce a stable Class A biosolids as determined by VSS and SOUR reduction, stability was determined by measuring VSS/VS and SOUR of raw after treatment with/without ferrate and ferric chloride. Additional measurements will include pH, ORP, electrical conductivity, and temperature. The SOUR results of ferrate and ferric chloride addition are displayed below in Tables 4 through 7.

Table 4. SOUR Results of Ferrate Additions

Ferrate as Iron/TS (mg/Kg) (%)		Moisture	Buffer	Feed	Seed	SOUR (mg O_2/h/g TS)	pH	ORP (mV)	Number of Samples
12	1.2	98.5%	N	N	N	<1.5 (10-70h)	10.6	-154	3
13	1.3	99%, 98%	Y	Y	Y	>1.5 (0-5h); ≈1.5 (5-25h)	7.8	-168	3
24	2.4	98.5%	N	N	N	<1.5 (10-70h)	10.6	-154	3
49	4.9	98.5%	N	Y	Y	<1.5 (0-67h)	10.6	-154	3
49	4.9	98.5%	N	N	N	<1.5 (10-70h)	10.6	-154	3

Y- variable covered; N – variable not covered

As noted in Table 4, the results of the initial ferrate runs noted no stabilization at 1.3% ferrate as iron dose. The other tests indicates that of pH higher than 10. When the respirometry tests were not neutralized (buffer), the various dosages ranging from 1.2% to 4.9% all generated stable biosolids, even when acetate and seed had been added. The sludge was stabilized not due to the iron but pH itself since ferrate has a very strong alkalinity.

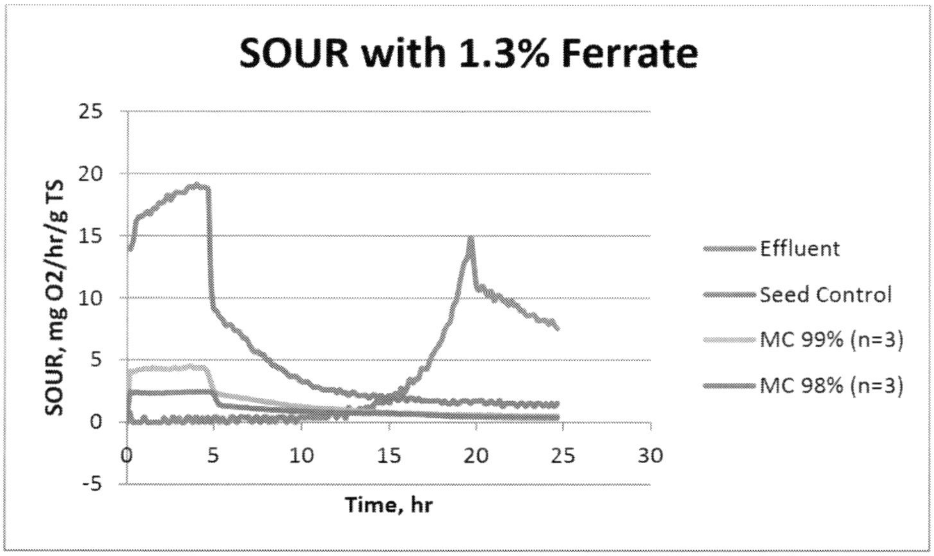

Figure 1. SOUR results of heat-dried biosolids with 1.3% ferrate dose (various moisture content (MC) with buffer, seed and feed in respirometer)

When dosed with 1.5% ferrate as iron by TS, two moisture contents (MC) were tested in the respirometer (Figure 1). Both showed inhibited oxygen demand and even lower when the substrate was consumed. The following figure indicated that SOUR was above the stable criteria requirement until hour 5 and 9, respectively for MC at 98% and 99%.

Table 5. SOUR Results of Ferric Chloride Additions (Seeded)

$FeCl_3$ as Iron/TS (mg/Kg) (%)		Moisture	Buffer	Feed	Seed	SOUR (mg O_2/h/g TS)	pH	ORP (mV)	Number of Samples
14	1.4	99%	N	N	Y	>1.5 (0-10h); <1.5 (10-48h)	5.3	-120.3	3
15	1.5	98%, 99%	Y	Y	Y	>1.5 (0-5h); <1.5 (5-18h)	7.7	23.4	3
18	1.8	98%	N	N	Y	<1.5 (0-24h)	7.3	74.2	3
20	2.0	99%	N	N	Y	>1.5 (0-10h); <1.5 (10-48h)	7.3	74.2	3
23	2.3	90%	N	N	Y	<1.5 (0-48h)	6.6	-201.3	3
31	3.1	90%	N	N	Y	<1.5 (0-48h)	6.6	-201.3	3

Y- variable covered; N – variable not covered

At doses at 1.4 to 1.5% ferric chloride, the heat-dried biosolids was not stable as results with SOUR values larger than 1.5 mg O_2/h/g TS for 5 to 10 hours when seeded. Ferric chloride addition at 1.5% inhibited the activity of bacteria after the exogenous nutrients had been consumed after five hours, which is noted in Table 5.

The pH values were naturally neutral without buffering; it could be the influence of heat drying process.

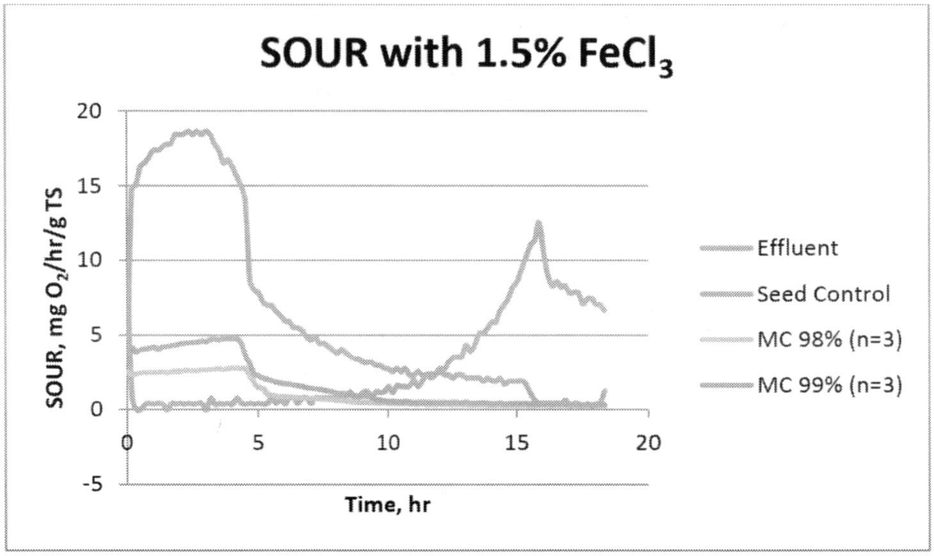

Figure 2. SOUR results of heat-dried biosolids with 1.5% ferric chloride dose (various MC with buffer, seed and feed in respirometer)

At dose larger than 2.3% ferric chloride, the heat-dried biosolids was observed to be stable for at least 2 days. This is elucidated in Figure 3.

Table 6. SOUR Results of Ferric Chloride Additions (Not Seeded)

$FeCl_3$ as Iron/TS (mg/Kg) (%)		Moisture	Buffer	Feed	Seed	SOUR (mg O_2/h/g TS)	pH	ORP (mV)	Number of Samples
9	09	98%	N	N	N	<1.5 (0-90h)	7.2	93.8	3
13	1.3	98%	N	N	N	<1.5 (0-24h)	7.2	99.9	3
15	1.5	98%	N	N	N	<1.5 (0-24h)	7.4	15.4	3
19	1.9	98%	N	N	N	<1.5 (0-24h)	7.4	15.4	3
20	2.0	98%	N	N	N	<1.5 (0-24h)	7.3	74.2	3
22	2.2	98%	N	N	N	<1.5 (0-90h)	7.2	99.9	3

Y- variable covered; N – variable not covered

Table 6 and Figure 4 indicated that the ferric chloride treated heat-dried biosolids was stable with no microbes or seed was added, but with the addition of seeds, the treated heat-dried biosolids were not stable when the ferric chloride doses were lower than 2.3% as iron.

Endocrine Disrupting Compounds (EDCs) assessments

Table 9 showed the EDCs concentrations obtained by MVLN assays on untreated sludge cake and incinerator ash form New Orleans POTW. The measurements of grab samples from local wetland are also included.

Table 7. EDCs assessment results of sludge cake, incinerator ash, and sediments from New Orleans' wetlands

Samples	EDCs concentration (as	ORP (mV)	Std.	pH Mean	Std.

	ng E2/g solid)	Mean			
Sludge Cake	82.10	-166.1	14.0	6.8	0.2
Incinerator Ash	-34.18	ND	ND	ND	ND
Wetland #1	5.31	108.9	37.5	7.9	0.7
Wetland #2	-0.10	28.9	70.3	7.8	0.3
Wetland #3	20.71	63.0	7.6	8.1	0.6
Wetland #4	9.98	29.4	82.0	7.7	0.6
Wetland #5	51.52	1.8	57.8	8.0	0.8
Wetland #6	-44.44	63.0	35.0	7.5	0.4
Wetland #7	33.39	63.0	10.3	8.1	0.4
Range of wetland	(-44.44, 51.52)	(1.8, 108.9)		(7.5, 8.1)	

It should be noted that the New Orleans raw sludge cake had EDCs activity close to that obtained from wetlands' sediments. But the ORP values were about 200 mV below the sediments due to the high organic contents of the cake.

Discussion

The sludge was assessed for stability and estrogen activity with the addition of iron-based salts and heat drying process.

Stability by iron based salts addition

The first observation is that ferrate decomposes within 60 seconds in biosolids to ferric hydroxide. This is substantiated by the work at Tulane on ferrate disinfection and EDCs deactivation (Skaggs, 2013; Sirsawat, 2013). They noted the decomposition ferrate inside one minute in wastewater effluents of TOC concentrations of 6 to 20 mg/L, with ferrate doses 6-12 mg/L (maximum TOC/ferrate=3.3). The TOC concentrations for New Orleans, municipal sludge is in the range of 1% (10,000mg/L) and the ferrate will not be stable at dosage at 1,250 mg/L. In the biosolids, the ferrate is very unstable and will decompose within one minute. Ferrate utilized in this study is at dose of TOC/ferrate equals 8.0.

Respirometry assessments

As studied earlier, the effect of this work is related to yielding short-term stabilization of heat-dried biosolids by the addition of iron salts (ferric chloride and ferrate) along with moisture content.

The respiration activity of incinerated ash had been studied prior to the study by respirometer with 98% moisture content, buffer and seed (Table 10 and Figure 5).

Table 8. Heat Dried Materials

Sample	Moisture	Buffer	Feed	Seed	SOUR (mg O_2/h/g TS)	pH	ORP (mV)
Municipal Sludge Ash	98%	Y/N	N	Y	<1.5 (0-24h)	7.3	131.3

Y- variable covered; N – variable not covered

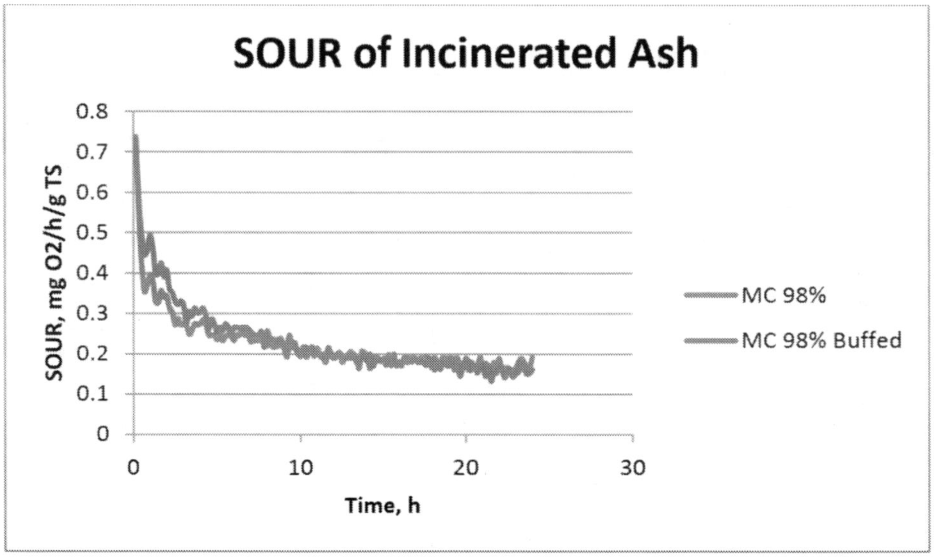

Figure 3. SOUR results of incinerated ash with 98% moisture content (seeded and buffered/non-buffered)

The ash is very stable and can be utilized as wetlands soil amenders. As the results in figure 5 indicated that high temperature during incineration exhausted the biologically available matters for putrefication to occur even when the samples were seeded with RAS. The temperature in heat drying process was 80°C that achieved pasteurization yet preserved the nutrients. Without microbes the biosolids maintained stable for the first 24 hours as shown in Figure 6. The stability stayed as long as 90 hours as mentioned in Table 7.

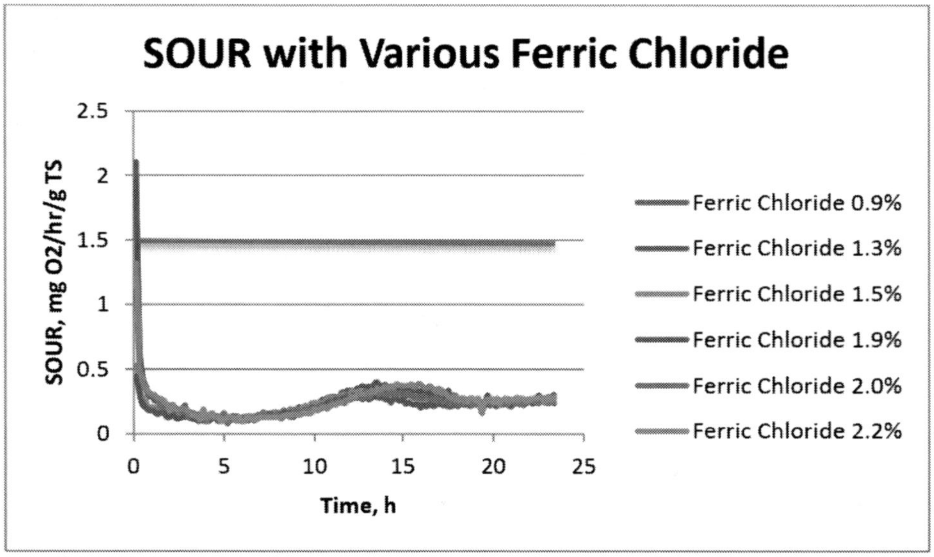

Figure 4. SOUR results of heat-dried biosolids with ferric chloride dose ranging from 0.9%--2.2% (MC at 98% with no buffer, seed or feed in the respirometer)

Once the seed was added, biosolids treated with iron/TS dosage lower than 2.3% was not stable for 5 to 10 hours depending on the moisture contents (Table 11, Figure 7&8).

Table 9. SOUR Results of Ferrate Additions (n=3) and Ferric Chloride Addition (n=3)

Ferrate as Iron/TS (mg/Kg)	(%)	Moisture	Buffer	Feed	Seed	SOUR (mg O_2/h/g TS)	pH	ORP (mV)
13	1.3	99%, 98%	Y	Y	Y	>1.5 (0-5h); ≈1.5 (5-25h)	7.8	-168
$FeCl_3$ as Iron/TS (mg/Kg)	(%)	Moisture	Buffer	Feed	Seed	SOUR (mg O_2/h/g TS)	pH	ORP (mV)
15	1.5	98%	Y	Y	Y	>1.5 (0-5h); <1.5 (5-18h)	7.7	23.4

Y- variable covered; N – variable not covered

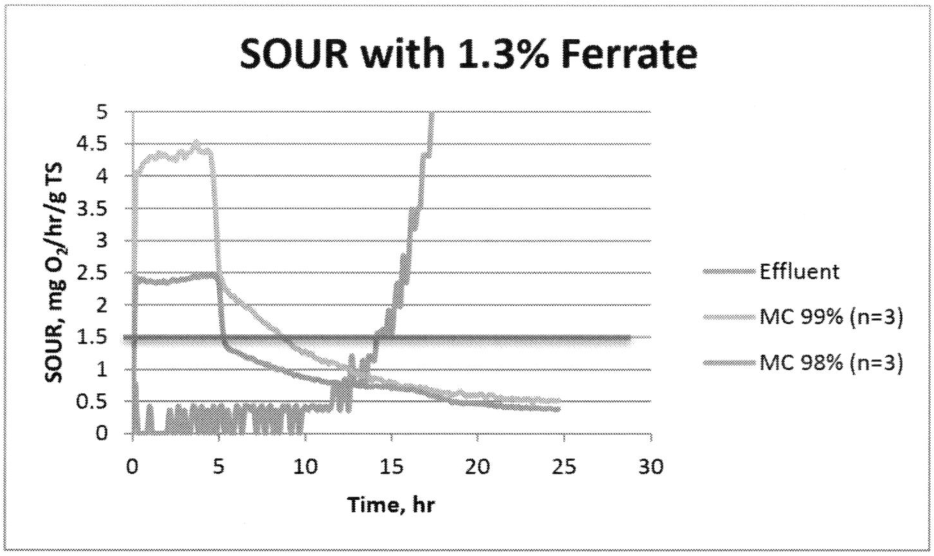

Figure 5. SOUR results of heat-dried biosolids with 1.3% ferrate dose expended (various MC with buffer, seed and feed in respirometer)

When dosed with 1.5% ferrate as iron by TS, two moisture contents (MC) were tested in the respirometer. Both showed inhibited oxygen demand and even lower when the substrate was consumed. The following figure indicated that SOUR did not reach below the limit requirement until hour 5 and 9, respectively for MC at 98% and 99%. The same trends when it is ferric chloride treatment.

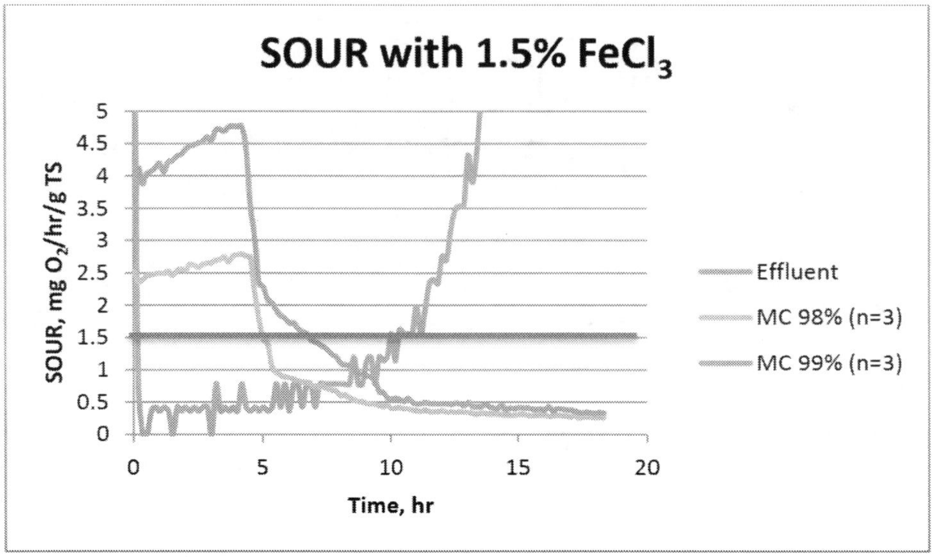

Figure 6. SOUR results of heat-dried biosolids with 1.5% ferric chloride dose expended (various MC with buffer, seed and feed in respirometer)

Comparing to the seed control (Figure 1&2), iron treated heat-dried biosolids inhibited the respiration activity: SOUR dropped below limit as soon as the acetate was consumed. When comparing to the effluent control, iron also held the SOUR down by inhibiting microbial proliferation as it happened in effluent.

It should be noted that the results in figure 5 and 6 mimicked each other for both ferrate and ferric chloride. Since the only similarity between these two test settings was the dosage expressed in iron/TS percentage, it could be conclude that stability agent was actually ferric hydroxide instead of ferrate.

When the iron/TS dosage was increased above 2.3%, stability was achieved as shown in Table 10 and Figure 7.

Table 10. SOUR Results of Ferric Chloride Additions (Seeded)

$FeCl_3$ as Iron/TS (mg/Kg)	(%)	Moisture	Buffer	Feed	Seed	SOUR (mg O_2/h/g TS)	pH	ORP (mV)	Number of Samples
23	2.3	90%	N	N	Y	<1.5 (0-48h)	6.6	-201.3	3
31	3.1	90%	N	N	Y	<1.5 (0-48h)	6.6	-201.3	3

Y- variable covered; N – variable not covered

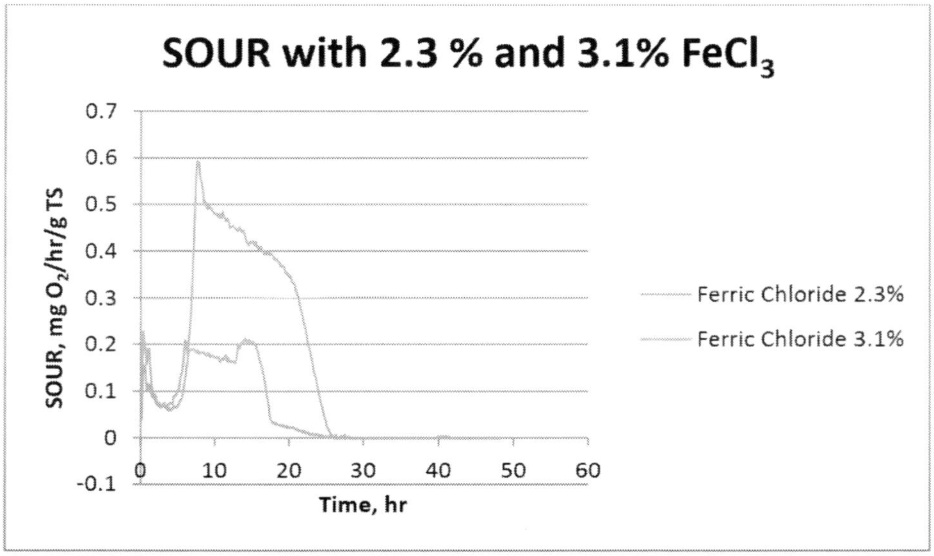

Figure 7. SOUR results of heat-dried biosolids with 2.3% and 3.1% ferric chloride dosage (MC at 90% with only seed in the respirometer)

Despite the testing conditions, the SOUR of each batch tended have a steep slop at the very beginning due to headspace or ferrous chemical influences, the system was steady afterwards within 10 minutes. Normally it does not create noise unless the intended SOUR measurements were small.

Seed control demanded higher oxygen uptake rate than biosolids with seed, it could be explained either as toxicity of biosolids towards seed which is unlikely, or simply that seed with biosolids provided higher solids content that rendered the SOUR smaller.

It should be noted that under the test conditions, Tulane dilute the dried biosolids 100 to 1 and seeded the cake with RAS. Even so, the cake appeared to be stable up to approximately 4 days. Obviously, the heat-dried biosolids would have to adsorb this moisture have a longer stabilization period.

ORP Buffer Effect of Iron

The addition of ferric chloride, which generates ferric hydroxide, acts as a electron buffer that hold the ORP of the heat-dried biosolids for 10 to 200 mV. This buffering capacity is shown by the ORP values of half reactions below (eq. 1,2,3):

$Fe^{3+} + e^- = Fe^{2+}$ $\qquad E_R^0 = 0.77V$ eq. 1

$Fe(OH)_3 + 3H^+ + 3e^- = Fe^{3+} + 3H_2O$ $\qquad E_R^0 = 1.06V$ eq. 2

$FeO_4^{2-} + 4H_2O + 3e^- = 3Fe(OH)_4 + 4OH^-$ $\qquad E_R^0 = 0.72V$ eq. 3

As shown in Table 11, both ferric chloride and ferrate will hold the ORP levels in the biosolids to around 0.1 V. This electron buffer enables the heat-dried biosolids to resist a drop of ORP to below -300mV. By this it means that cake will resist going to methonagenesis causing sulfide production.

Follow the reaction in eq. 1, the following relationship can be deduced:

$$p\varepsilon = p\varepsilon^o - log\frac{\{Fe^{2+}\}}{\{Fe^{3+}\}}$$

where

$$p\varepsilon = -\log\{e^-\}$$

$$p\varepsilon^o = logK$$

$$logK = \frac{FE^o}{2.3RT}$$

Here are the derived equations at temperature 25°C:

$$p\varepsilon = \frac{FE_H}{2.3RT} = 16.9E_H$$

$$p\varepsilon^o = 16.9E^o$$

the deduction is as follows:

$$pK_e = pE^o + f(pH)$$

$$pK_e = pE^o - \frac{1}{n}lg\frac{product(P)}{reactor(R)}[H^+]$$

$$pK_e = pE^o - \frac{1}{n}\lg[H^+]^n - \frac{1}{n}log\frac{P}{R}$$

$$pK_e = pE^o - \lg[H^+] - \frac{1}{n}log\frac{P}{R}$$

$$pK_e = pE^o + pH - \frac{1}{n}log\frac{P}{R}$$

$$pK_e = pE^o - f(pH)$$

since there is no hydrogen ion participating in eq. 1, pKe for iron is as follows:

$$pK_e = pE^o = 13$$

Figure 8a and 8b show the Fe(II), Fe(III) concentration as the function of pε/ORP if 0.001M of Fe(III) was dissolved in a liquid solution.

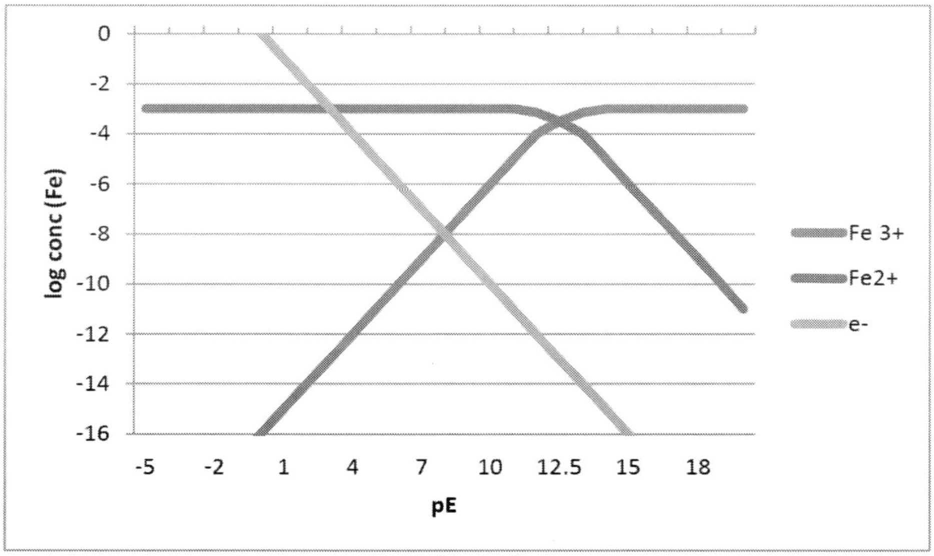

Figure 8a. Redox equilibrium Fe^{3+}, Fe^{2+} (Equilibrium distribution of a 10^{-3} M aqueous iron as a function of pE (acid solution))

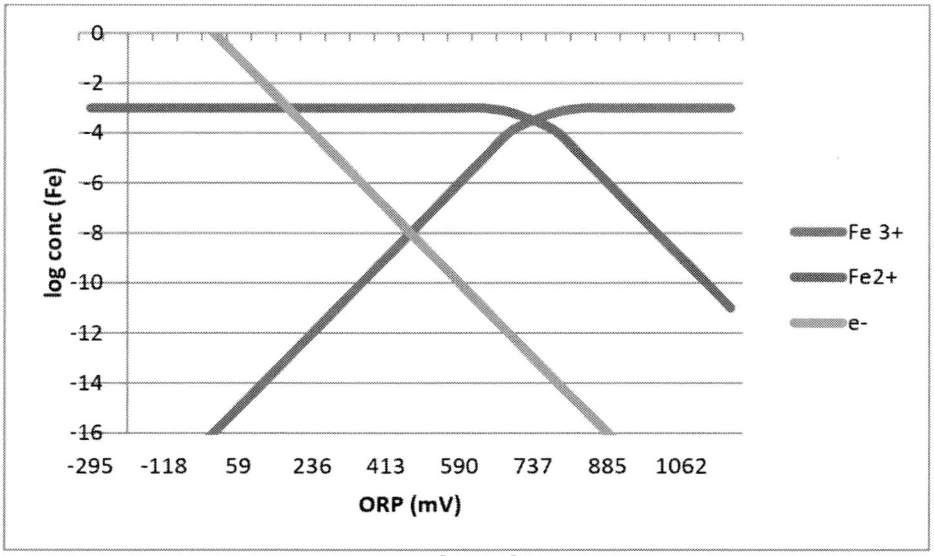

Figure 8b. Redox equilibrium Fe^{3+}, Fe^{2+} (Equilibrium distribution of a 10^{-3} M aqueous iron as a function of ORP (mV))

Table 11. ORP readings prior and post treatments

Iron/TS (%)	ORP_{prior} (mV)	ORP_{post} (mV)
1.3 (Ferrate)	-168.0	93.1
1.5 (Ferric Chloride)	-159.0	89.2

Odor control by iron reaction with sulfide

Ferric/ferrous salts have been applied for odor control in collection system since the 1970s since they react with sulfide. The Fe^{3+} will precipitate sulfide at a high level where the K_{sp} for Fe_3S_2 is $10^{-88.0}$. Fe(III) oxidizes sulfide to elemental sulfur in solution with pH values ranging from 0.88 to 1.96 at ambient temperature. The following reactions show the oxidation process (eq. 4, 5) (Zhai, et al. 2012) :

$$S^{2-} + 2H^{+} \rightarrow H_2S \qquad \text{eq. 4}$$

$$H_2S + 2Fe^{3+} \rightarrow S_0 + 2Fe^{2+} + 2H^{+} \qquad \text{eq. 5}$$

The reduced Fe(II) can subsequently produce FeS precipitant as the following reaction (eq. 6) shows (Zhang, et al. 2008) :

$$Fe^{2+} + HS^{-} \rightarrow FeS + H^{+} \qquad \text{eq. 6}$$

Under anoxic conditions that the treated heat-dried biosolids would be stored before application, sulfide could precipitate with heavy metals to form insoluble metal sulfides, i.e. iron sulfide. The form of sulfide in the wetland pore water can vary from S^{2-} to HS^{-} and H_2S depending on the pH, and a propensity to off gas of H_2S would occur in low pH systems with accompanying rotten-egg odor (Wu, et al. 2012) . The addition of iron (Van der Welle et al., 2007, 2008) and use of iron-rich soil matrix (Wu et al., 2012) were reported to control the dissolved sulfide concentration. In addition, the $Fe(OH)_3$ could react with sulfide in sediments among other pollutants, as noted in Table 14.

Table 12. Ferrate reactions time and products with selected pollutants

Pollutant	$T_{1/2}$	Product
Hydrogen Sulfide	2.7 msec	Sulfate
Thioacetamide	0.4 sec	Acetamide, Sulfate
Thiourea	0.6 sec	Urea, Sulfate
Cyanide	5.3 sec	Cyanate, Nitrite, Bicarbonate
Thiocyanate	1.7 min	Cyanate, Nitrite, Sulfate
Nitrite	37 min	Nitrate
Ammonia	196 min	Nitrate, Nitrite, Nitrogen

Conclusion

This study assessed the performance of iron-based salts on the stabilization of municipal sewage sludge with combination of heat drying process. Both traditional analytical assays and biological assays were conducted in order to assess the stability. The research also analyzed the evolution of EDCs in this treatment. Further analysis is currently being conduted. The following can be concluded:

- According to the SOUR results, the stabilizing agent is Fe(III) instead of Fe(VI) when the dosage is above 2.3% as iron/TS percentage by weight.

- Fe(III) increased the ORP value and created a ORP buffer to inhibit the hydrogen sulfur microbial production and release. In addition, Fe(III) reacts with sulfur.
- As noted with early study, the ferrate can assist in EDCs deactivation. Whether the ferric hydroxide might exhibit the same phenomenon is currently being tested.

References

Andersen, H., H. Siegrist, B. Halling-Sørensen, and T.A. Ternes. 2003. Fate of estrogens in a municipal sewage treatment plant, *Environ. Sci. Technol.*, 37(18):4021-4026.

Boyd, G.R., H. Reemtsma, D.A. Grimm and S. Mitra. 2003. Pharmaceuticals and personal care products (PPCPs) in surface and treated waters of Louisiana, USA and Ontario, Canada, *The Science of the Total Environment*, 311(1-2):135-149.

Boyd, G.R., J.M. Palmeri, S. Zhang and D.A. Grimm. 2004. Pharmaceuticals and personal care products (PPCPs) and endocrine disrupting chemicals (EDCs) in stormwater canals and Bayou St. John in New Orleans, Louisiana, USA, *Science of the Total Environment* 333(1-3): 137-147.

Burnham JC, Hatfield N, Bennett GF, Logan TJ. 1992. Use of Kiln Dust with Quicklime for Effective Municipal Sludge Pasteurization and Stabilization with N-Viro Soil Process. In *Innovations and Uses for Lime*, J. Walker et al., Editors. Philadelphia, PA: American Society for Testing and Materials, p. 128-41.

Çığgın, A.S., Insel, G., Majone, M., Orhon, D., 2012. Respirometric evaluation and modelling of acetate utilization in sequencing batch reactor under pulse and continuous feeding. Bioresour. Technol. 107, 61-69. doi: 10.1016/j.biortech.2011.12.029.

Daston GP, Cook JC, Kavlock RJ. 2003. Uncertainties for endocrine disrupters: our view on progress, *Toxicol. Sci.* 74: 245-252.

Daughton, C.G.; Ternes, T.A. 1999. Pharmaceuticals and personal care products in the environment: Agents of subtle change?, *Environ. Health Perspect.,* 107(Suppl.6):907-938.

Eikum, A.S., Paulsrud, B., 1977. Methods for measuring the degree of stability of aerobic stabilized sludges. Water Res. 11, 763-770. doi: 10.1016/0043-1354(77)90044-6.

Echenfelder WW, Ford DL, Englande AJ. (2009). *Industrial Water Quality*. 4th ed. New York, NY: McGraw Hill, pp 18-32.

Epstein E, Kohnke H. 1957. Soil aeration as affected by organic matter application. *Soil Sci Soc Amer Proc*. 21:585-588.

European Commission DG Environment [ECDGE]. 2001. Workshop report, European workshop on endocrine disrupters, sponsored by Swedish Ministry for Environment, Aronsborg (Bålsta), Sweden, June 18-20, http://europa.eu.int/comm/environment/chemicals/pdf/workshop_report.pdf

Favoino, E., 2006. The EU legislation and the requirements following for national organic waste management strategies and policies. In: First Baltic Biowaste Conference, European Compost Network/Organic Recovery and Biological Treatment association. <http://www.com- postnetwork.info/bbc/downloads/3%20Enzo%20Favoino.pdf>.

Gea, T., Barrena, R., Artola, A., Sánchez, A., 2004. Monitoring the biological activity of the composting process: Oxygen uptake rate (OUR), respirometric index (RI), and respiratory quotient (RQ). Biotechnol. Bioeng. 88, 520-527. doi: 10.1002/bit.20281.

Guillette, L.J. and E.A. Guillette. 1996. Environmental contaminants and reproductive abnormalities in wildlife: Implications for public health? *Toxicology and Industrial Health* 12(3-4):537-550.

Hanselman, T.A., D.A. Graetz, and A.C. Wilkie. 2003. Manure-borne estrogens as potential environmental contaminants: a review, *Environ. Sci. Technol.*, 37(24):5471-5478.

Holbrook, R.D., J.T. Novak, T.J. Grizzard, and N.G. Love. 2002. Estrogen receptor agonist fate during wastewater and biosolids treatment processes: A mass balance analysis, *Environ. Sci. Technol.*, 36:4533-4539.

Kolpin, D.W., E.J. Furloug, M.J. Meyer, E.M. Thurman, S.D. Zaugg, L.B. Barber, and H.T. Buxton. 2002. Pharmaceuticals, hormones, and other organic wastewater contaminants in U.S. Streams, 1999-2000 a national reconnaissance, *Environ. Sci. Technol.* 36(6):1202-1211.

Marsili-Libelli, S., Tabani, F., 2002. Accuracy analysis of a respirometer for activated sludge dynamic modelling. Water Res. 36, 1181-1192. doi: 10.1016/S0043-1354(01)00339-6.

McLachlan, J. A. (2001) Environmental signaling: What embryos and evolution teach us about endocrine disrupting chemicals. *Endocrine Reviews* 22 (3), 319-341.

National Research Council [NRC]. 2002. *Biosolids Applied to Land – Advancing Standards and Practices*, National Academy Press, Washington, D.C.

Okutman Tas, D., Pavlostathis, S.G., 2007. The influence of iron reduction on the reductive biotransformation of pentachloronitrobenzene. Eur. J. Soil Biol. 43, 264-275. doi: 10.1016/j.ejsobi.2007.03.003.

Ponsá, S., Gea, T., Alerm, L., Cerezo, J., Sánchez, A., 2008. Comparison of aerobic and anaerobic stability indices through a MSW biological treatment process. Waste Manage. 28, 2735-2742. doi: 10.1016/j.wasman.2007.12.002.

Pillai, S.D. and Reimers, R.S., 2010. "Disinfecting and Stabilizing Biosolids using E-Beam and Chemical Oxidants," Water Environment Research Foundation Report Number U4R06, WERF Publishing, Alexandria, Virginia.

Pratt-Ward LS. 2004. Development of an Oxidative-Acid Disinfection Process using Chlorine Dioxide and Nitrous Acid to produce Class A Biosolids, Doctoral Dissertation, Tulane SPHTM, New Orleans, Louisiana pp. 209.

Reimers RS, Pratt-Ward LS, Bradford HB, Mussari FP, Schmitz W. 2006. Development of the Neutralizer Process for Disinfection and Stabilization of Municipal Wastewater Residuals. WEF Residuals.

Scrimshaw, M.D. and J.N. Lester. 2003. Fate and behavior of endocrine disrupters in sludge treatment and disposal. In *Endocrine Disrupters in Wastewater and Sludge Treatment Processes*, IWA Publishing, London, UK, 145-176.

Snyder SA, Westerhoff P, Yoon Y, Sedlak DL. 2003. Pharmaceuticals, personal care products, and endocrine disruptors in water: implication for the water industry. *Environ Eng Sci* 20(5), 449-469.

Sánchez Arias, V., Fernández, F.J., Rodríguez, L., Villaseñor, J., 2012. Respiration indices and stability measurements of compost through electrolytic respirometry. J. Environ. Manage. 95, Supplement, S134-S138. doi: 10.1016/j.jenvman.2010.10.053.

Scaglia, B., Erriquens, F.G., Gigliotti, G., Taccari, M., Ciani, M., Genevini, P.L., Adani, F., 2007a. Precision determination for the specific oxygen uptake rate (SOUR) method used for biological stability evaluation of compost and biostabilized products. Bioresour. Technol. 98, 706-713. doi: 10.1016/j.biortech.2006.01.021.

Scaglia, B., Erriquens, F.G., Gigliotti, G., Taccari, M., Ciani, M., Genevini, P.L., Adani, F., 2007b. Precision determination for the specific oxygen uptake rate (SOUR) method used for biological stability evaluation of compost and biostabilized products. Bioresour. Technol. 98, 706-713. doi: 10.1016/j.biortech.2006.01.021.

Spinosa, L., 1985. Technological characterization of sewage sludge. Waste Manage. Res. 3, 389-398. doi: 10.1016/0734-242X(85)90132-6.

Switzenbaum MS, Moss LH, Epstein E, Pincince AB, Donovan JF (1997). Defining biosolids stability. *J Env Engineering* 123:1178-84.

Tas, D.O., 2010. Respirometric assessment of aerobic sludge stabilization. Bioresour. Technol. 101, 2592-2599. doi: 10.1016/j.biortech.2009.10.058.

U.S. Environmental Protection Agency (2003) Control of Pathogens and Vector Attraction in Sewage Sludge, EPA-625/R-92-013; U.S. Environmental Protection Agency, Office of Research and Development: Washington, D.C.

U.S. Environmental Protection Agency . 2003a. Multi-Year Plan (FY2000-2012) for Endocrine Disruptors, Office of Research and Development, December

U.S. Environmental Protection Agency [USEPA]. 2003b. Draft 2003 Strategic Plan, Washington, DC, Mar 5, http://www.epa.gov/ocfopage/plan/2003sp.pdf.

Van der Welle, M.E.W., Niggebrugge, K., Lamers, L.P.M., Roelofs, J.G.M., 2007a. Dif- ferential responses of the freshwater wetland species Juncus effusus L. and Caltha palustris L. to iron supply in sulphidic environments. Environ. Pollut. 147, 222–230.

Van der Welle, M.E.W., Roelofs, J.G.M., Lamers, L.P.M., 2008. Multi-level effects of sulphur–iron interactions in freshwater wetlands in The Netherlands. Sci. Total Environ. 406, 426–429.

Water Environment Federation & American Society of Civil Engineers, 1992. Sludge Stabilization, in Design of Municipal Wastewater Treatment Plants. WEF/ASCE, pp. 1255–1382.

World Health Organization [WHO]. 2002. *Global Assessment of the State-of-the-Science of Endocrine Disruptors*, International Programme on Chemical Safety, T. Damstra (ed.).

Wang, X., Waite, T.D., 2010. Iron speciation and iron species transformation in activated sludge membrane bioreactors. Water Res. 44, 3511-3521. doi: 10.1016/j.watres.2010.03.031.

Wu, S., Kuschk, P., Wiessner, A., Müller, J., Saad, R.A.B., Dong, R., Sulphur transformations in constructed wetlands for wastewater treatment: A review. Ecol. Eng. doi: 10.1016/j.ecoleng.2012.11.003.

Wu, S., Chen, Z., Braeckevelt, M., Seeger, E.M., Dong, R., Kästner, M., Paschke, H., Hahn, A., Kayser, G., Kuschk, P., 2012. Long-term dynam- ics of Fe(II), sulphide and phosphate in a pilot-scale constructed wetland treating sulphate-containing contaminated groundwater. Water Res. 46, 1923–1932.

Zhai, L., Song, W., Tong, Z., Sun, M., 2012. A fuel-cell-assisted iron redox process for simultaneous sulfur recovery and electricity production from synthetic sulfide wastewater. J. Hazard. Mater. 243, 350-356. doi: 10.1016/j.jhazmat.2012.10.046.

Zhang, L., De Schryver, P., De Gusseme, B., De Muynck, W., Boon, N., Verstraete, W., 2008. Chemical and biological technologies for hydrogen sulfide emission control in sewer systems: A review. Water Res. 42, 1-12. doi: 10.1016/j.watres.2007.07.013.

Zhu, J., Yan, X., Liu, Y., Zhang, B., 2006. Improving alachlor biodegradability by ferrate oxidation. J. Hazard. Mater. 135, 94-99. doi: 10.1016/j.jhazmat.2005.11.028.

Non-Deterministic CFD Modelling of Peracetic Acid Disinfection in Contact Tanks

Domenico Santoro*, Trojan Technologies, London, Canada
Giuseppe Raspa, La Sapienza University, Rome, Italy
Ferdinando Crapulli, Western University, London, Canada
Mehrdad Raisee, Teheran University, Teheran, Iran
Charles N. Haas, Drexel University, Philadelphia, USA
*Department of Research and Development,
Trojan Technologies
London, Ontario N5V 4T7
dsantoro@trojanuv.com, +1 519 630 6198

ABSTRACT

Currently, water treatment technologies are designed according to algorithms which are either empirical in nature or based on simulations performed for a unique set of input data. However, especially for wastewater applications, realistic operating conditions are dominated by water quality variability (and uncertainty in input parameters) which if not properly handled can increase the risk of noncompliance with existing regulations. In this paper, a non-deterministic modeling approach to peracetic acid (PAA) disinfection process analysis in secondary treated municipal wastewater is presented. The non-deterministic model consists of a stochastic sub-model describing the inactivation of *Escherichia coli* by PAA coupled with a flow transport sub-model obtained by means of Lagrangian particle tracking using computational fluid dynamics (CFD). The non-deterministic disinfection model was validated against pilot experiments allowing for the computation of probabilistic CT tables relating PAA dosage, contact tank average retention time, wastewater quality and *E.coli* inactivation. Our numerical analysis suggested that: (a) PAA can be considered a viable alternative to chlorination being bactericidal even concentration as low as 2 mg/L; (b) the CT dose needed for a desired PAA disinfection target is strongly dependent on wastewater quality and inactivation kinetic variability and (c) PAA quenching may be required if a max PAA effluent concentration of <1 mg/L has to be met at the point of effluent discharge.

KEYWORDS

Peracetic acid, non-deterministic CFD, *Escherichia coli*, stochastic inactivation kinetics, wastewater.

INTRODUCTION

While it is well known that chemical disinfection processes play an important role in wastewater reclamation schemes, the ascertained formation of chlorinated byproducts has raised new environmental concerns about the use of chlorine-based processes suggesting the investigation of alternative chemicals such as, for example, peracetic acid (PAA). PAA has been employed as a disinfectant for a variety of applications ranging from industrial wastewater treatment such as, for example, colour removal of textile effluents to surface sterilization of food processing plants and legionella control in hospitals. More recently, PAA has been tested for municipal wastewater disinfection, as reported by a number of investigations which have shown its microbial activity against a wide range of microorganisms including bacteria (Santoro et al., 2007), viruses (Koivunen, 2005) and protozoa (Lazarova et al., 1998). If applied to secondary settled biological effluent with low

COD and TSS and initial *E. coli* concentration, PAA doses as low as 4 mg/L and 10 minutes contact time were effective to reach the reuse target of 10 *E. coli*/100mL (Dell'Erba et al., 2004). Despite these promising results, variability in PAA disinfection efficacy is also reported in the literature. To our knowledge, our work if the first example of a non-deterministic CFD modeling approach applied to wastewater disinfection process analysis and specifically to *E. coli* inactivation by PAA in presence of uncertainties induced by wastewater quality and inactivation kinetic variability.

METHODOLOGY

The non-deterministic CFD model was obtained by coupling stochastic inactivation kinetics with steady-state particle trajectories obtained from the numerical solution of the mass conservation and Navier-Stokes equations under reactor-specified boundary conditions. The stochastic inactivation kinetic model was developed according to the following steps. First, the six parameters governing the *E.coli* disinfection kinetics were extracted from PAA dose-response curves obtained from batch studies (Santoro, 2007). Then, a stochastic model was adjusted on the set of kinetic parameters such that the correlation matrix and their marginal distributions were both honoured. Finally, a statistically significant number (e.g., 5000) of combinations of kinetic parameters was generated and coupled with flow distribution in a pilot contact tank using Fluent (Ansys, Canonsburg). The non-deterministic model was validated with pilot scale experiments using tracer tests (Fig.1), residual PAA and *E.coli* inactivation measurements.

RESULTS AND DISCUSSION

Some results of this work are shown in the following figures.

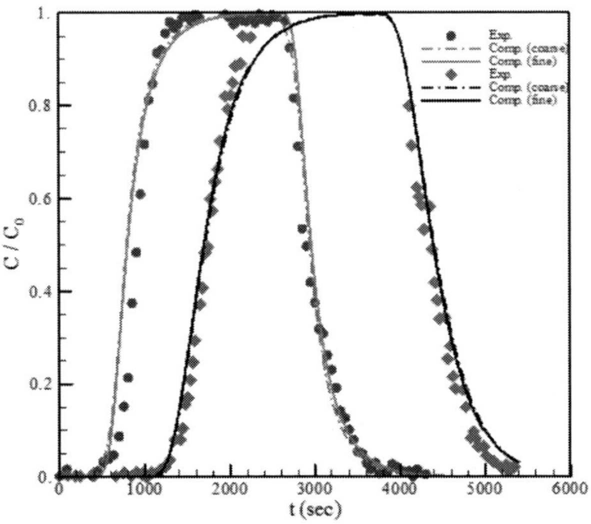

Fig. 1 – Comparison of simulated and experimental tracer tests in two locations of the contact tank (first and third pass)

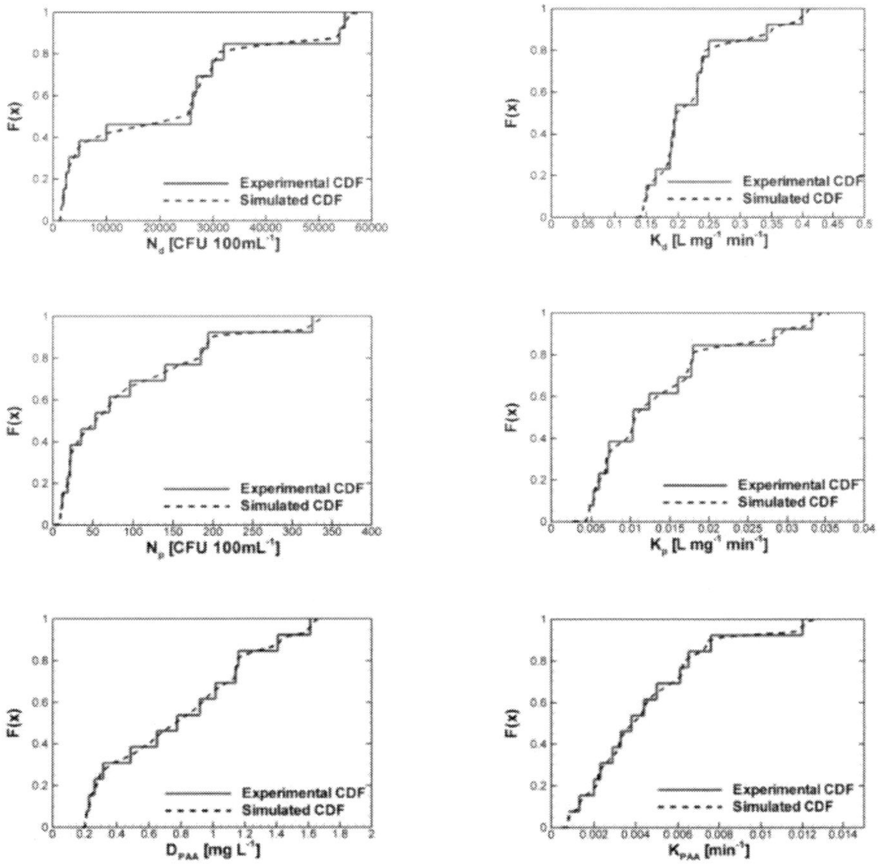

Fig. 2 – Comparison of simulated and experimental cumulative distribution function (CDF) of the six kinetic parameters involved in the *E.coli* inactivation model

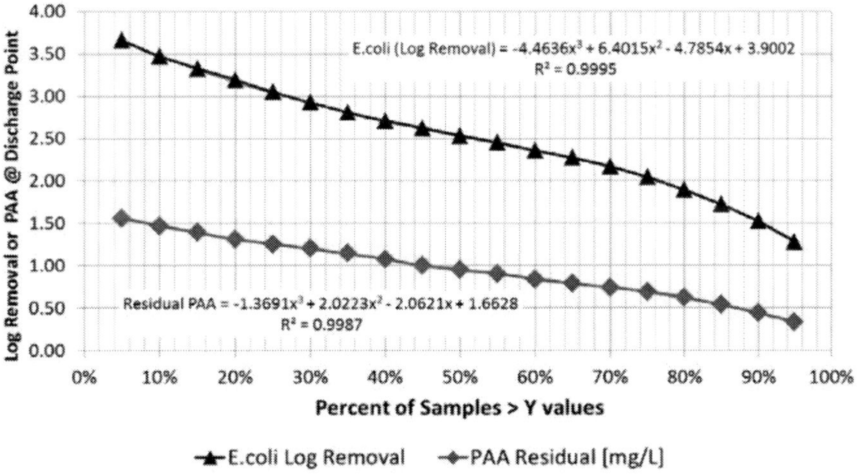

Fig. 3 – An example of probabilistic curve generated by the non-deterministic model (PAA dosage: 2 mg/L, Average Retention Time: 60 min)

CONCLUSIONS

Based on the study illustrated in this paper, the following conclusions can be made:

- A non-deterministic kinetic model (Fig.2) was successfully developed and integrated into CFD to describe PAA disinfection processes in secondary settled municipal wastewater.
- PAA is a viable alternative to chlorination and it is bactericidal even at concentration as low as 2 mg/L. However, the appropriate choice of CT dose (e.g., PAA dosage and average contact time) must take into account the important role played by wastewater quality variability (Fig.3).
- Given the low rate of PAA decay in secondary treated municipal wastewater, PAA quenching may be required for obtaining a maximum concentration of disinfectant at the point of discharge of 1 mg/L or below (Fig.3).

REFERENCES

Santoro, D.; Gehr, R.; Bartrand, T.A.; Liberti, L.; Notarnicola, M.; Dell'Erba, A.; Falsanisi, D.; Haas, C.N. (2007) *Wastewater disinfection by peracetic acid: Assessment of models for tracking residual measurements and inactivation.* Water Env. Research, 79(7), pp. 775-787.

Koivunen, J.; Heinonen-Tanski, H. (2005) *Peracetic acid (PAA) disinfection of primary, secondary and tertiary treated municipal wastewaters*. Water Research, 39(18), pp.4445-4453.

Lazarova, V.; Janex, M.L.; Fiksdal, L.; Oberg, C.; Barcina, I.; Pommepuy, M. (1998) *Advanced wastewater disinfection technologies: Short and long term efficiency*. Water Science and Technology, 38(12), pp. 109-117.

Dell'Erba, A.; Falsanisi, D.; Liberti, L.; Notarnicola, M.; Santoro, D. (2007) *Disinfection by-products formation during wastewater disinfection with peracetic acid*. Desalination, 215(1-3), pp. 177-186.

Design and Operation Considerations for PAA Disinfection Systems

Katherine Y. Bell, CDM Smith[1]
Todd Archer, CDM Smith
[1]CDM Smith
830 Crescent Centre Drive, Suite 400
Franklin, TN 37067
(615) 807-7800

ABSTRACT

Peracetic acid (PAA) is a wastewater disinfection alternative that is gaining interest due to its ability to provide bacterial inactivation performance at costs competitive to mature technologies. In addition to potential cost savings, the ability to provide treatment for challenging effluent while meeting stringent disinfection by-products (DBP) limits is driving this interest. To respond to this interest, it is necessary to develop engineering design criteria that will aid decision-makers understanding of the cost implications, as well as obtain local regulatory support for implementation of PAA disinfection. This paper provides an overview of the chemistry and current US applications of PAA. A discussion of engineering design and cost considerations for implementation is provided for this disinfection alternative.

KEYWORDS

Peracetic acid, PAA, wastewater disinfection, disinfection by-products

INTRODUCTION

With uncertainty regarding the requirements for future use of chlorine gas and potential future challenges associated with chlorinated disinfection by-products (DBPs), wastewater utilities are proactively looking for options to replace both gas and liquid chlorine with other disinfection alternatives. UV disinfection has gained a great deal of interest due to its ability to achieve bacterial discharge compliance; however, there are challenges with implementation of UV disinfection. Some of these challenges include high capital costs particularly for facilities with high peaking factors and its applicability to provide consistent disinfection performance for effluents that have low UV transmittance or total suspended solids that frequently exceed 20 mg/L. Another option for disinfecting wastewater effluent is peracetic acid (PAA), which is a chemical oxidant that has been applied to the food, beverage, medical and pharmaceutical industries as a disinfectant.

CHEMISTRY AND USE OF PAA FOR US MUNICIPAL WWTP APPLICATIONS

Peracetic acid has the chemical formula, CH_3CO_3H (**Figure 1**), and is produced as a solution in which PAA is in equilibrium with hydrogen peroxide, acetic acid and water. These solutions are generally prepared by reacting acetic acid with hydrogen peroxide in the presence of a catalyst. A range of PAA concentrations are formulated by adjusting the concentration and amounts of reagents during the manufacturing process. This is an important consideration because anecdotal evidence suggests disinfection efficacy is related to the PAA concentration applied.

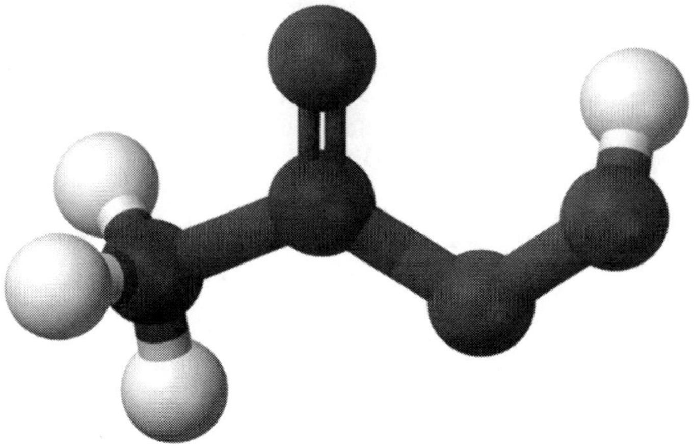

Figure 1. Chemical structure of PAA.

Peracetic acid is effective against microorganisms over a wide range of conditions and is most effective at pH values below its pKa (8.2), but has good biocidal activity under mildly alkaline conditions up to a pH of about 9. PAA also has good disinfection efficacy at low temperatures (5°C) and is relatively unaffected by organic matter compared to other chemical oxidants used for disinfection, e.g., chlorine or ozone (Block, 1991). Additionally, PAA has a long shelf life as a result of proprietary stabilizers that can be used in commercial solutions.

The USEPA has approved PAA for use as a disinfectant to treat wastewater; however, there is currently only one full-scale municipal wastewater treatment plant (WWTP) application in the US permitted to use this technology, in St. Augustine, FL. The full-scale PAA application was implemented as an alternative to UV or ozone to address compliance issues with disinfection by-product formation. The City of St. Augustine, with permission from Florida Department of Environmental Regulation conducted a full scale PAA pilot study to compare the disinfection performance and costs with chlorination-dechlorination. The 5.0 million gallons per day (mgd) WWTP has two identical chlorine contact chambers and this configuration allowed side-by-side comparison of the PAA and chlorination-dechlorination for disinfection performance, effluent toxicity and chemical costs (Keogh and Tran, 2010). The full-scale demonstration also tested the use of an automated dose control technology to determine if this process control approach could effectively reduce pathogen levels to achieve regulatory compliance while preventing formation of trihalomethanes (THMS) in the treated effluent. Results of the pilot demonstrated that PAA was consistently able to reduce pathogen loads to achieve regulatory compliance, prevent harmful DBP residuals from forming in the treated effluent, and reduce the chemical consumption and costs (Graham et al., 2012). In addition to this full-scale installation in Florida, the City of Frankfort, KY uses PAA as a full-scale back-up to the ozone system that serves as the primary method of disinfection at the WWTP.

ENGINEERING DESIGN CONSIDERATIONS

While there is a great deal of published literature on the results of PAA disinfection from laboratory studies and applications in the food processing and medical sterilization applications, there are limited PAA applications at WWTPs in the US, which has been a limitation to the implementation of this technology.

Thus, it is important to define the critical engineering considerations that could lead to successful testing, piloting and permitting of the technology. Engineering design requirements for PAA are similar to those for bulk sodium hypochlorite and consist primarily of a bulk storage tank with secondary containment, suction and discharge piping, chemical metering pumps and controls, residual analyzers, and an injection point followed by a contact volume. PAA is typically delivered in totes, but at least one supplier can now ship bulk deliveries (4,500 to 4,700 gallons); typical lead time for drums and totes is two weeks, but because bulk shipment is new, it is recommended to provide a 30-day chemical storage time. Other considerations for PAA applications include materials of construction; for example piping and fittings for chemical feed systems require passivated stainless steel. The chemical is compatible with other materials and depending on the bulk solution strength that is supplied, other materials including high density polyethylene may be used for chemical storage; however, replacement of these parts may be required on a more frequent basis.

Water quality may affect performance of PAA and, similar to other mature disinfection technologies, a review of effluent quality should include suspended solids, temperature, and pH and site specific testing is recommended to establish site-specific dosing requirements. Generally, for secondary effluent, initial design dosing is 1 - 2 mg/L of PAA with a 10 to 15 minute contact time; however, there are ongoing case studies to demonstrate that peak flow contact times of 5 minutes are adequate for achieving bacterial inactivation. This has a minimal impact for retrofit into facilities with existing chlorine contact systems which are designed for 15 – 30 minutes of contact at WWTP peak hydraulic flows. However, for facilities evaluating PAA as new process, the cost of constructing a smaller contact basin could be significant.

Using test data from a recent investigation, alternatives cost estimates were developed for a 21 mgd WWTP application for seasonal disinfection of fecal coliform to meet requirements that included a monthly geometric mean fecal coliform concentration of 200 cfu/100mL and a daily maximum concentration of 400 cfu/100mL. Four disinfection methods were evaluated and a summary of these costs are provided in Table 1. Costs were developed such that disinfection contact or UV reactors would be retrofit into an existing chlorine contact basin. All systems included construction of a building to include either electrical equipment for the UV system and all indoor chemical storage.

This evaluation demonstrated a low capital cost of the PAA system, which is a key advantage. However, costs of PAA are still several times the cost of bulk sodium hypochlorite, when compared on an active ingredient basis, and as result has a correspondingly high O&M cost. Although the EPA has approved the use of PAA for wastewater disinfection, there is no guarantee that individual states would permit use of it for this application. Therefore, it is important that regulatory support for PAA disinfection be obtained through rigorous testing and approval in several states so that wider application of the technology is accepted.

Table 1. Lifecycle Cost Evaluation of Alternative Disinfection Methods, Including PAA

Alternative	Total Capital Cost	20-Year NPV of Annual O&M	20-Year NPV
Bulk Hypochlorite	$5,740,000	$2,060,000	$7,830,000
Ozone	$16,990,000	$4,700,000	$21,690,000
UV	$9,940,000	$1,310,000	$11,250,000
PAA	$1,220,000	$5,110,000	$6,330,000

SUMMARY

Peracetic acid has been demonstrated to be an effective disinfectant over a wide range of wastewater effluent qualities, requiring low doses of chemical to achieve bacterial inactivation. A number of advantages of PAA include its fast kinetics, requiring very short contact times, and it does not produce problematic halogenated DBPs such as THMs; additionally, capital costs for retrofit of existing chlorine facilities is very low. The major disadvantage of PAA is its limited use when compared to other more mature disinfection technologies such as chlorination, UV disinfection, and ozone.

In summary, PAA is a disinfection alternative that is gaining a great deal of recent interest due to its ability to provide bacterial inactivation performance at lifecycle costs competitive to mature technologies. In addition to the potential cost savings, the ability to provide treatment for challenging effluent while meeting stringent DBP limits is driving interest in the technology. To respond to this interest, it is necessary to further develop engineering design criteria and have more demonstration projects that will aid decision-makers understanding of cost implications, as well as obtain local regulatory support for implementation of PAA disinfection.

REFERENCES

Block, S.S. (1991). Peroxygen compounds. In Block, S.S., ed. Disinfection, sterilization, and preservation, 4th ed. Lea and Felbinger, Philadelphia, PA, USA, pp. 167–181.

Graham, M.S., Lomax, G., Skipp, G. (2012). Peracetic Acid (PAA) versus Chlorination/De-chlorination – A Disinfection Comparison. WEF Technology and Exhibition Conference, Los Angeles, CA.

Keogh and Tran. (2010). Old City,New Ideas: Peracetic Acid inWastewater Disinfection at St.Augustine. *Florida Water Resources Journal*, April 2011: 44–48.

Kitis, M. (2004). Disinfection of wastewater with peracetic acid: A review. *Environment International,* 30(1):47–55.

THE CITY OF STEUBENVILLE OHIO

WHERE YOU ALWAYS HAVE A HOME

CITY OF STEUBENVILLE, OHIO WASTEWATER TREATMENT FACILITY

PAA Pilot Demonstration to determine PAA Efficacy as an alternative disinfectant to Chlorine and Sodium Bisulfite technology for wastewater

PERAGreen SOLUTIONS, LLC

CITY OF STEUBENVILLE, OHIO
WASTEWATER TREATMENT FACILITY

PAA Pilot Demonstration to determine PAA Efficacy as an alternative disinfectant to Chlorine and Sodium Bisulfite technology for wastewater

TABLE OF CONTENTS:

THE CITY OF STEUBENVILLE OHIO

WHERE YOU ALWAYS HAVE A HOME

PERAGreen SOLUTIONS, LLC

Section 1

Executive Summary

The City of Steubenville Waster Facility was prepared for a PAA Pilot disinfection demonstration beginning, March 26, 2012. A conceptual proposal was prepared on March 7, 2011 to evaluate Proxitane WW-12 on both "Capture" and "Event" occurrences during wet weather occurrences in September 2011. A full report was issued based on this data and subsequent flow studies to correct hydraulic anomalies with the under and overflow weirs. The current demonstration was focused on the secondary discharge stream only and the findings presented in this report.

The Proxitane WW-12 demonstration began Monday, March 26, 2012 with results summarized in the following report.

In summary, the pilot demonstration was extremely successful with expected PAA consumption and pathogen destruction from the injection point, through the contact chamber and ultimately to the river discharge. Pathogen, *Fecal Coliform,* levels of <10 CFU's/100 ml were achieved at the final outfall at a feed rate of 1.5 ppm of active PAA. Dosages were lowered throughout the demonstration to determine the lowest feed rate and PAA residual necessary to meet both summer and winter Fecal Coliform permit levels.

The study continued until April 30, 2012 and then switched back to chlorine (15% liquid bleach) and dechlorination (sodium bisulfite) to provide head to head treatment evaluations for both technologies. In summary, chlorine bleach and sodium hypochlorite could attain the same pathogen destruction but at levels far in excess as compared to peracetic acid making Proxitane WW-12 more cost competitive. The Ohio EPA was asked to extend the approved use of Proxitane WW-12 through April 30 and then until the end of June 2012 to further evaluate improved feed and control to fine tune product dosage requirements. The results of the entire demonstration are outlined in this report.

In summary, Proxitane feed rates could be set at 0.6-0.7 ppm with PAA residuals at levels down to 0.32 ppm in the final discharge and meet all NPDES permit pathogen requirements. Further tests conduced in June 2012 were designed to improve Proxitane WW-12 feed using modified pump controls to narrow the residual needed for maximum pathogen kill.

PERAGreen SOLUTIONS, LLC

Section 2

Disinfection Demonstration Protocol

The City of Steubenville wastewater treatment plant is capable of processing up to 36 (MGD) million gallons of wastewater and storm water per day. Total plant capacity through the secondary treatment system is 13.5 MGD. Nominal daily flow is 6 MGD and dry weather flow 4 MGD. The focus of this demonstration is to treat both wet and dry conditions from March 26 through April 30 only on the secondary treatment system. Please refer to the plant diagram in ATTACHMENT A.

PERAGreen Solutions conducted background demand and efficacy tests on primary and secondary streams at the City of Steubenville wastewater facility prior to the onsite evaluation beginning March 26, 2012. The uptake tests helped determine the starting PAA dosage required to meet NPDES discharge limits for *Fecal Coliform* to meet winter and summer geometric means.

Two (2) 3010 pound totes of Proxitane WW-12 were set above the mixing chamber and connected to a Walchem IX Series metering pump. The Walchem pump skid was designed to feed based on "totalized" flow from both the North and South contact chambers. The PAA was fed into a manhole, directly into the discharge pipe from the three (330,000 gallon) secondary clarifiers, twenty five (25) feet upstream from the contact mixing chamber. The Walchem pump was adjusted to feed at 3.65 gallons per hour at a flow of 7 million gallons per day or at a rate of 1.5 ppm active PAA.

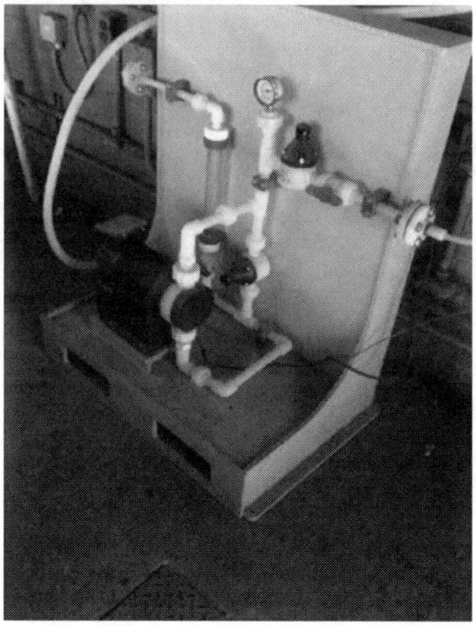

Proxitane WW-12 is an environmentally friendly stable equilibrium of peracetic acid (12%), acetic acid (15%), hydrogen peroxide (18.5%) and water. Peracetic acid has a proven record as a safe sanitizer in the food industry. It is used in the municipal market primarily as a cost effective replacement for chlorine gas and sodium hypochlorite (bleach) followed by dechlorination with sodium bisulfite of sulfur dioxide gas.

There are no special handling requirements for Proxitane WW-12 but proper safety precautions are always important with dealing with any acid.

Proxitane WW-12 Benefits

- Environmentally "Green" Product
- Cost effective compared to Chlorine/Dechlorination
- No known resistant organisms
- Long term efficacy
- Decomposes into non-harmful trace by-products (acetic acid, water, oxygen and trace amounts of organic acids)
- Final decomposition products are carbon dioxide and water
- Non-halogenated
- Forms no AOX
- Ready-to-use liquid
- Rapid Reactions
- In-pipe treatment
- Long term stability (12-18 month shelf life with no degradation)
- Equally effective in turbid and high solids waters
- No degradation by organics or ammonia
- Very stable in ambient temperatures
- Effective over the 5-8 pH range
- Low capital investment
- No dechlorination step

Sampling Locations, Monitored Parameters, Monitoring Frequency and Methods

Fecal coliform densities were measured at one recordable sampling location, the final effluent discharge from the combined North and South contact chambers. Two other sites (mixing chamber and ¾ point) were measured intermittently to provide additional data on PAA performance between the injection point and the final discharge into the Ohio River. Measurements included fecal coliform and PAA (indirectly through DPD total chlorine residual measurement). Effluent flow is based on totalized water meter readings from both the North and South chlorine contact chambers.

The three (3) test locations to determine residual PAA levels in the wastewater stream are outlined below in the attached chlorine contact chamber diagram found on the following page.

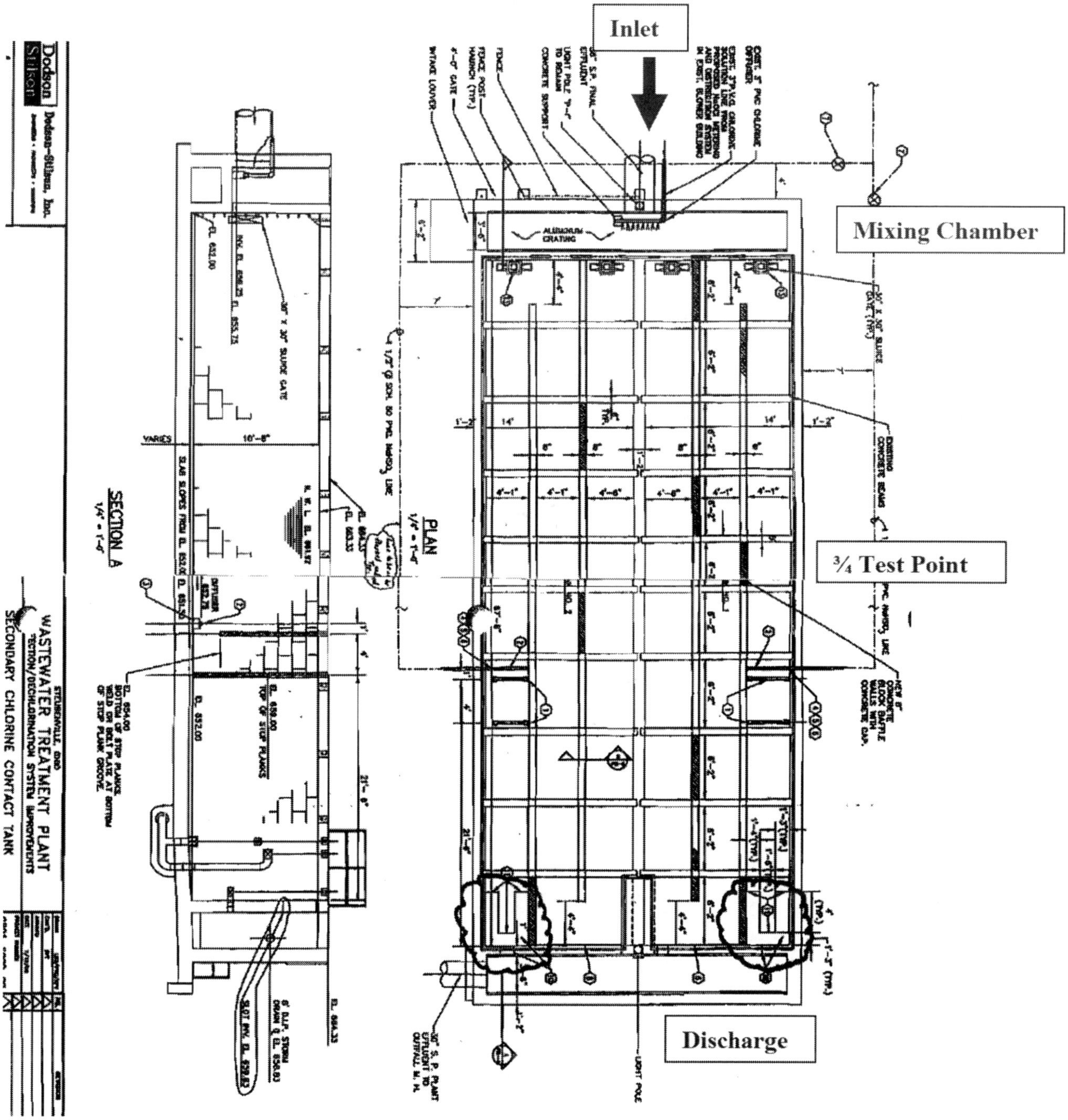
Inlet
Mixing Chamber
¾ Test Point
Discharge
Dodson-Stilson, Inc.
WASTEWATER TREATMENT PLANT
SECONDARY CHLORINE CONTACT TANK
PLAN
SECTION A

The demonstration criteria were as follows:

1. Inject Peracetic Acid (Proxitane WW-12) neat, directly into the secondary clarifier combined discharge. At an initial feed rate of 1.5 ppm active PAA based on total flow.
2. Allow for adequate mixing in the contact chamber mixing chamber.
3. Test ¾ point PAA residuals once per day. Test final effluent discharge every two hours and record PAA residual and total flow. Follow Low Level PAA test procedures below:

PERAGreen SOLUTIONS, LLC

Determination of **LOW LEVEL Peracetic Acid (PAA)** In Water

Equipment and Reagents

Chlorine, Total (0-2.00 mg/l) – Portable Colorimeter. Hach Model DR/890
DPD Total Chlorine Powder Pillows for 10 ml sample size. Hach Product number 21056-69. Sample cell, 10-ml.

Procedure for Total Chlorine (Method 8167)

USEPA approved for reporting water and wastewater analysis (USEPA Standard Method 330.5)

B. DR/890 setup

(1) Enter the stored program number for total chlorine (Cl2) powder pillows.
(2) Press: **PRGM**. The display will show: **PRGM ?**
(3) Press**: 9 ENTER**. The display will show **mg/l, Cl2** and the **ZERO** icon.

B. Sample Testing

(1) Fill a 10 ml sample cell with the PAA activated wastewater sample. This is the blank.

(2) Cap the blank cell and place it in the cell holder with the diamond mark facing you. Cover the cell compartment with the instrument cap to shield from stray light interferences and then press: **ZERO**.
(3) The instrument will turn on and the display will show: **0.00 mg/l CL2**.
(4) Remove the blank cell
(5) Immediately fill a second cell to the 10-ml mark with the PAA activated wastewater sample.
(6) Add the contents of one DPD Total Chlorine Reagent Powder Pillow to the cell (prepared sample). Cap and swirl the sample cell vigorously to dissolve the powder. (It is not necessary that all the powder dissolves)
(7) Insert the sample cell into the meter with the diamond mark facing you. Cover the cell compartment with the instrument cap to shield from stray light interferences.
(8) Immediately Press: **READ**. The cursor will move to the right, then the result in mg/l total chlorine (Cl2) will be displayed.
(9) This is the ppm total Cl_2 value. To calculate PAA, use the following formula:

<u>PAA Calculation</u>

$$\text{ppm PAA} = 1.07 \times (\text{ppm total } Cl_2 \text{ PAA value})$$

Multiply ppm total Cl_2 PAA value by 1.07 to express the result as ppm PAA based on the weight ratio of PAA to Cl_2 (76/71 = 1.07).

---

4. Test final effluent discharge for fecal coliform once per day on each of the North and South chambers and record.
5. Reduce PAA feed, record change in pump stroke and test residuals every two hours and fecal coliforms once per day.
6. Keep lowering PAA residuals until fecal coliform levels rise to winter discharge limits (1000 CFU's/100mil) and then increase pump stoke to maintain levels below summer limits of 200 CFU's/100ml continuously. Determine lowest PAA residual required to conform to discharge limits.
7. Maintain two hour fecal and flow monitoring until the Proxitane WW-12 is exhausted.
8. Once Proxitane WW-12 is empty, switch feed back to existing chlorine/dechlorination.

The criteria for success were as follows:

1. Determine the lowest PAA levels required to meet summer NPDES discharge pathogen levels.
2. Successfully pass the "WET" test on the PAA treated discharge water with 0.4 ppm residual PAA.
3. Determine that flow pacing is an adequate method of feeding Proxitane WW-12.
4. Calculate cost effectiveness versus chlorine bleach and dechlorination using sodium bisulfite.

PERAGreen SOLUTIONS, LLC

Section 3

Pilot Disinfection "Demonstration" Results and Conclusions

Demonstration Summary

Based on the pathogen and PAA residual tests conducted from March 26, 2012 through April 30, 2012 and comparing results to chlorine and dechlorination measurements immediately following the PAA demonstration, it was determined that Proxitane WW-12 out performed chlorine and dechlodrination on cost to achieve the same pathogen levels in the effluent discharge water from the secondary system.

It was also determined from "WET" testing that the PAA residual level of 0.4ppm active PAA had no acute effect on selected fresh water organisms. The "WET" testing was performed by Alloway Laboratories on April 5-9, 2012 and the Report of Acute Biomonitoring was issued on April 19, 2012. The report is found in ATTACHMENT B. The organisms tested were neonate water fleas (Ceriodaphnia dubia) and fathead minnow larvae (Pimphales promelas). *E.Coli* tests were also conducted by Alloway on April 18, 2012 for secondary discharge and final effluent. The results are also shown in ATTACHMENT B.

The demonstration results were as follows:

1. Proxitane WW-12 feed rates of 0.6-0.7 ppm yielded PAA discharge residuals of 0.4 ppm.
2. CBOD testing remained "constant" without any impact on the effluent discharge stream.
3. Fecal Geomean was below NPDES permit levels when PAA residuals were maintained above 0.4 ppm. (Please refer to the Permit to discharge issued June 21, 2010 found on the CD supplied with the report)
4. Flow pacing was verified to provide the proper Proxitane WW-12 feed rate to achieve PAA residuals for pathogen destruction.

General Discussion

Prior to the start of the Proxitane WW-2 demonstration, dye testing was performed on the chlorine contact chamber as well as the effluent into the Ohio River. The tests were performed to identify the contact time at various flows as well as visually determine the discharge plume into the Ohio River. Videos were taken at both locations and are included on the disk provided. The following picture demonstrates the minimal effect of flow near the river's edge as the effluent stream discharges into the Ohio River.

In addition to dye testing, PAA residual and background total halogen testing was conducted on April 12, 2012 to determine the PAA residual in the discharge plume. The results of the testing can be found in ATTACHMENT C. It was concluded that the PAA remained in the plume entering the Ohio River approximately 25 feet offshore before any substantial dilution occurred. Once the discharge plume contacted the main flow of the river, the PAA dissipated quickly.

Proxitane WW-12 "Demonstration"

The Proxitane WW-12 "Demonstration" was initiated Monday, March 26 with the bleach and sodium bisulfite feed being shut down and isolated to allow for PAA to be fed prior to the final chlorine contact chamber. Proxitane WW-12 residuals and flow were recorded every two hours during the demonstration and the results are shown in the following graph.

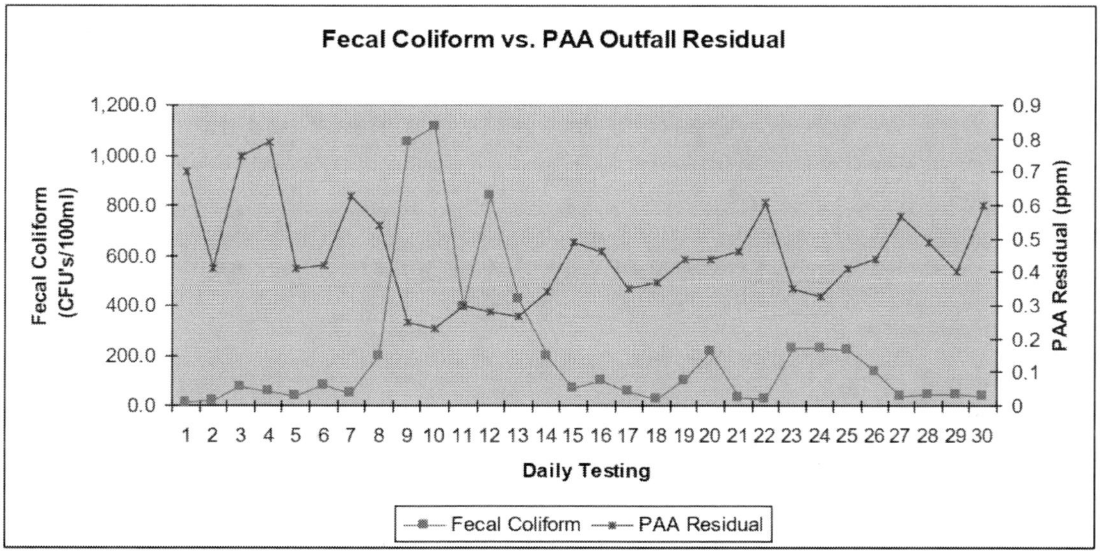

It is apparent from the graph that as PAA residuals were maintained above 0.4 ppm, fecal levels remained **BELOW** summer discharge limits. As PAA residual levels were reduced, correspondingly, fecal levels increased as seen during daily tests 8 through 14. Daily average effluent flow rates ranged from 3.59 to 5.99 MGD and secondary discharge fecal coliform levels ranged from 1700-42,000 CFU's/100 ml. The raw data can be found on the CD supplied with this report as well as the daily, two hour testing for PAA residuals and flows.

Geomean averages for fecal coliform versus average PAA residuals for the five weeks are also shown in the following graph.

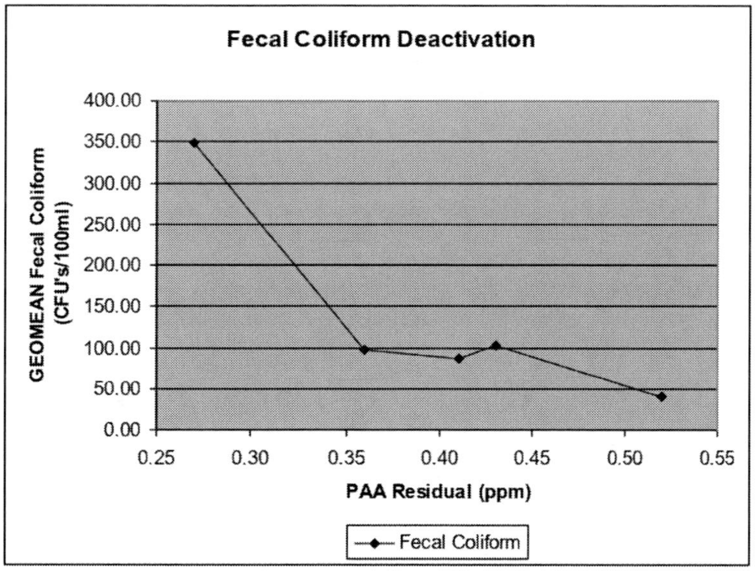

Results and Conclusions

1. Proxitane WW-12 effectively controlled effluent fecal coliform at a 0.4 ppm PAA residual to meet both summer and winter NPDES permit requirements.

2. "WET" testing concluded that at the 0.4 ppm PAA residual level, there was no effect on selected organism survivability in the discharge stream.

3. PAA could achieve fecal coliform discharge limits even al lower PAA residual dosages down to 0.32 ppm with improved chemical feed control.

4. Flow pacing is the desired method of feed and control of Proxitane WW-12 feed independent on incoming fecal coliform levels from the secondary system.

5. Proxitane WW-12 was cost competitive versus chlorine (bleach) and dechlorination (sodium bisulfite) by a minimum of 25%.

PERAGreen SOLUTIONS, LLC

Section 4

Attachment A
City of Steubenville Wastewater Treatment Facility

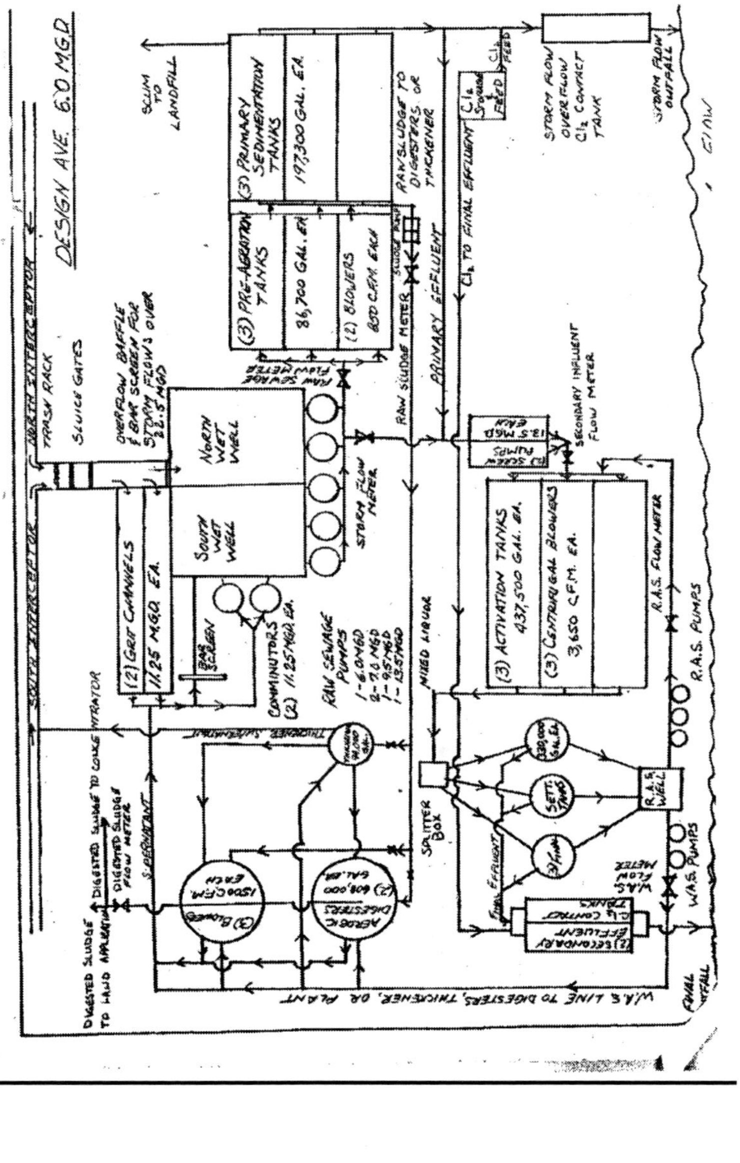

PERAGreen SOLUTIONS, LLC

Attachment B

Alloway

Report of Acute Biomonitoring

April 19, 2012

Alloway
Your Resource for Defensible Data

Report of Acute Biomonitoring

Samples Collected on:
4/4/12

Pimephales promelas
Ceriodaphnia dubia

Client: City of Steubenville WWTP
Address: 100 N. Water Street
Steubenville, OH 43952

Report Date:
4/19/12

Alloway Lab #: L12-13214-01

1101 N. Cole Street • Lima, Ohio 45805
419.223.1362 • Fax 419.227.3792
800.436.1243

508 Bissman Ct. • Mansfield, Ohio 44903
419.525.1644 • Fax 419.524.5575
800.635.3222

1776 Marion-Waldo Rd. • Marion, Ohio 43302
740.389.5991 • Fax 740.389.1481
800.873.2835

www.alloway.com

Alloway
Your Resource for Defensible Data

Date: 4/19/12

Alloway Lab #: L12-13214-01

Chuck Murphy
City of Steubenville WWTP
100 N. Water Street
Steubenville, OH 43952

Re: Acute Biomonitoring

Dear Mr. Murphy:

Enclosed is the report of the biomonitoring on the sample listed above. The objective of the acute biomonitoring was to determine the acute endpoint (mortality) from the test water. Fathead minnow larvae (*Pimephales promelas*) and neonate water fleas (*Ceriodaphnia dubia*) were the test organisms. Acute toxicity was determined by locating the median lethal concentration (LC50) and the median effective concentration (EC50).

Dissolved oxygen, pH, and temperature were recorded at the beginning of the test and at the conclusion of the test. Results may be found in Appendix B & C. Samples were analyzed for alkalinity, hardness and conductivity. These results may also be found in Appendix B & C.

Alloway Environmental Testing Services, Lima, Ohio performs biomonitoring testing in accordance with procedures established by the Environmental Protection Agency. The Standard Operating Procedures for the testing is on file with the Ohio EPA.

Please call if you have any questions or need additional information.

Paul J. Crerar
Biology Lab Manager

Richard F. Rupp
Senior Biologist

1101 N. Cole Street • Lima, Ohio 45805
419.223.1362 • Fax 419.227.3792
800.436.1243

508 Bissman Ct. • Mansfield, Ohio 44903
419.525.1644 • Fax 419.524.5575
800.635.3222

1776 Marion-Waldo Rd. • Marion, Ohio 43302
740.389.5991 • Fax 740.389.1481
800.873.2835

www.alloway.com

Alloway
Your Resource for Defensible Data

Reporting Date: 4/19/12

1. Facility Name	City of Steubenville WWTP
2. Facility Address	100 N. Water Street Steubenville, OH 43952
3. Ohio EPA Permit No.	0PD00014*JD
4. NPDES No.	OH0027511
5. Facility Contact	Chuck Murphy
6. Telephone No.	740-283-6150
7. Consulting Laboratory	Alloway
8. Lab Contact	Paul Crerar
9. Lab Telephone No.	419-223-1362
10. Receiving water(s) of discharge	Ohio River and Wells Run
11. Outfall(s) tested	001

12. Is your current Standard Operating Procedure (SOP) Manual on file with Ohio EPA? Yes
If yes, indicate date in which SOP was submitted to the EPA: 1/21/2005

I certify under penalty of law that I have personally examined and am familiar with the information submitted in this document and based on my inquiry of those individuals immediately responsible for obtaining information, I believe the submitted information is true, accurate and complete. I am aware that there are significant penalties for submitting false information, including the possibility of fine and imprisonment.

Signature ______________________ Date ______________________

Name (typed or printed) ______________________ Title ______________________

1101 N. Cole Street • Lima, Ohio 45805
419.223.1362 • Fax 419.227.3792
800.436.1243

508 Bissman Ct. • Mansfield, Ohio 44903
419.525.1644 • Fax 419.524.5575
800.635.3222

1776 Marion-Waldo Rd. • Marion, Ohio 43302
740.389.5991 • Fax 740.389.1481
800.873.2835

www.alloway.com

Alloway
Your Resource for Defensible Data

Customer:	City of Steubenville WWTP
Outfall:	001
Ohio EPA Permit No.:	0PD00014*JD
Date of Test:	4/5-9/12

Client Address: City of Steubenville WWTP
100 N. Water Street
Steubenville, OH 43952

Summary of Acute Toxicity Test Results
Pimephales promelas* and *Ceriodaphnia dubia

	4500 Form Value	MDL
Effluent/Outfall		
Acute Toxicity, *Ceriodaphnia dubia* TUa	AA	0.2
Acute Toxicity, *Pimephales promelas* TUa	AA	0.2
Upstream		
48-Hour Acute Toxicity *Ceriodaphnia dubia* - % Affected	AA	5
96-Hour Acute Toxicity *Pimephales promelas* - % Affected	AA	5

AA = below detectable

Conclusion/Comments:

1101 N. Cole Street • Lima, Ohio 45805
419.223.1362 • Fax 419.227.3792
800.436.1243

508 Bissman Ct. • Mansfield, Ohio 44903
419.525.1644 • Fax 419.524.5575
800.635.3222

1776 Marion-Waldo Rd. • Marion, Ohio 43302
740.389.5991 • Fax 740.389.1481
800.873.2835

www.alloway.com

OEPA Permit No.: 0PD00014*JD

Customer: City of Steubenville WWTP

Sample ID: 001

Date of Test: 4/5-9/12

Acute Toxicity Test Results

Results of a Pimephales promelas 96 hour Static Acute Toxicity Test

Test Solutions	Cumulative Percent Mortality (Cumulative* Percent Adversely Affected) 24-Hr	48-Hr	72-Hr	96-Hr
Primary Control Upstream	0 (0)	0 (0)	0 (0)	0 (0)
Secondary Control Lab	0 (0)	0 (0)	0 (0)	0 (0)
20%	0 (0)	0 (0)	0 (0)	0 (0)
40%	0 (0)	0 (0)	0 (0)	0 (0)
60%	0 (0)	0 (0)	0 (0)	0 (0)
80%	0 (0)	0 (0)	0 (0)	0 (0)
100%	0 (0)	0 (0)	0 (0)	0 (0)

LC50 Values (EC50 Values)	24-Hr	48-Hr	72-Hr	96-Hr
	>100 (>100)	>100 (>100)	>100 (>100)	>100 (>100)

LC50 95% Confidence Limits (EC50 95% Confidence Limits)	24-Hr	48-Hr	72-Hr	96-Hr
LL				
UL				
LL	( )	( )	( )	( )
UL	( )	( )	( )	( )

LL = Lower Limit
UL = Upper Limit

Calculated TUa Value (100/LC50 or 100/EC50):
TUa = AA (below detectable)

Method(s) used to Determine LC50, EC50, and Confidence Limit Values:

* - cumulative percent affected is the total percentage of test organisms observed dead, and exhibiting atypical behavior or appearance.

OEPA Permit No.: 0PD00014*JD

Customer: City of Steubenville WWTP

Sample ID: 001 **Date of Test:** 4/5-7/12

Acute Toxicity Test Results

Results of a Ceriodaphnia dubia 48 hour Static Acute Toxicity Test

* Test Solutions	Cumulative Percent Mortality (Cumulative Percent Adversely Affected) 24-Hr		48-Hr	LC50 Values (EC50 Values) 24-Hr	48-Hr
Primary Control Upstream	0 (0)		0 (0)	>100 (>100)	>100 (>100)
Secondary Control Lab	0 (0)		65 (65)	**LC50 95% Confidence Limits (EC50 95% Confidence Limits)** 24-Hr	48-Hr
20%	0 (0)		0 (0)	LL ___ UL ___	___ ___
40%	0 (0)		0 (0)	LL ( ) UL ( )	( ) ( )
60%	0 (0)		0 (0)		
80%	0 (0)		0 (0)	LL = Lower Limit UL = Upper Limit	
100%	0 (0)		0 (0)	Calculated TUa Value (100/LC50 or 100/EC50): **TUa = AA (below detectable)**	
				Method(s) used to Determine LC50, EC50, and Confidence Limit Values:	

* - cumulative percent affected is the total percentage of test organisms observed dead, and exhibiting atypical behavior or appearance.

OEPA Permit No.: 0PD00014*JD (City of Steubenville WWTP)

PIMEPHALES PROMELAS ACUTE TOXICITY TEST CONDITIONS

Summary of Toxicity Test Conditions (Ohio EPA 1998; EPA-821R-02-012, Oct. 2002)	
1. Test Species and Age:	*Pimephales promelas* 14 days old
2. Test Type and Duration:	Static non-renewal; 96 hours
3. Test Dates:	4/5-9/12
4. Test Temperature (°C):	25°C ± 1°C
5. Light Quality:	Ambient laboratory illumination
6. Photoperiod:	16 h light, 8 h dark
7. Feeding Regime:	Not fed during test
8. Size of Test Vessel:	500 mL clear plastic cup
9. Volume and Depth of Test Solutions:	400 mL, 10.0 cm deep
10. No. of Test Organisms per Test Vessel:	10 per test vessel
11. No. of Test Vessels per Test Solution:	2 per solution tested
12. Total No. of Test Organisms per Test Solution:	20 per solution tested
13. Test Concentrations (as percent by volume effluent):	100, 80, 60, 40 and 20% effluent (001)
14. Renewal of Test Solutions:	None
15. Dilution Control Water:	Upstream grab in receiving stream
16. Secondary Control Water:	Lab rearing unit water
17. Aeration During Test:	Single bubble aeration in concentrations where D.O. is below or may fall below 4.0 mg/L (see fish data sheet).
18. Endpoints Measured:	Mortality, and atypical behavior or appearance.
19. If secondary control water used as diluent due to toxicity in primary control water, indicate number of consecutive tests conducted with alternative diluent:	

OEPA Permit No.: 0PD00014*JD (City of Steubenville WWTP)

***CERIODAPHNIA DUBIA* ACUTE TOXICITY TEST CONDITIONS**

Summary of Toxicity Test Conditions (Ohio EPA 1998; EPA-821R-02-012, Oct. 2002)	
1. Test Species and Age:	*Ceriodaphnia dubia* <24 hours old
2. Test Type and Duration:	Static non-renewal; 48 hours
3. Test Dates:	4/5-7/12
4. Test Temperature (°C):	25°C ± 1°C
5. Light Quality:	Ambient laboratory illumination
6. Photoperiod:	16 h light, 8 h dark
7. Feeding Regime:	Not fed during test
8. Size of Test Vessel:	1 fl. oz. (30 mL) clear plastic cup
9. Volume and Depth of Test Solutions:	25 mL, 31 mm deep
10. No. of Test Organisms per Test Vessel:	5 per test vessel
11. No. of Test Vessels per Test Solution:	4 per solution tested
12. Total No. of Test Organisms per Test Solution:	20 per solution tested
13. Test Concentrations (as percent by volume effluent):	100, 80, 60, 40 and 20% effluent (001)
14. Renewal of Test Solutions:	None
15. Dilution Control Water:	Upstream grab in receiving stream
16. Secondary Control Water:	Lab rearing unit water
17. Aeration During Test:	None
18. Endpoints Measured:	Mortality, and atypical behavior or appearance.
19. If secondary control water used as diluent due to toxicity in primary control water, indicate number of consecutive tests conducted with alternative diluent:	

OEPA Permit No.: 0PD00014*JD City of Steubenville WWTP

Sampling Summary for Acute Toxicity Tests

Sample Location & Description		Sample	Sample Collection Beginning	Sample Collection Ending
Final Effluent:				
Outfall No.:	001	1st		4/4/12 @ 07:30
Type (Grab/Comp):	Comp.	2nd		
Volume Collected:	1 Gallon	3rd		
Upstream Station:				
Waterbody:	Ohio River - Marina Down			
Station No.:	801	1st		4/4/12 @ 08:00
Type (Grab/Comp):	Grab	2nd		
Volume Collected:	1 Gallon	3rd		

Samples collected by: City of Steubenville WWTP

Appendix A

Field Data

Alloway Your Resource for Defensible Data

Chain of Custody Record

This is a legal document that authorizes Alloway to perform testing on samples submitted under this agreement

1101 North Cole Street, Lima, OH 45805
(P) 419-223-1362 (F) 419-227-3792

Project: L12-13214

Report To:
Name: Steve Garay
Company: Steubenville W.W.T.P.
Address: 100 North Water Street
Steubenville, Ohio 43952

Invoice To (If Different):
Name: Same
Company:
Address:
PO#:

Phone #: (740) 283-6150 Fax #: (740) 283-6150
E-mail: wastewaterplant@cityofsteubenville.us

Project Name: VARIOUS - April

Sampler (Print): Steve Garay (Signature):

Matrix Codes: ww - wastewater, gw - groundwater, dw - drinking water, s - solid, w - water, oil - oil, sg - sludge, o - other

	Customer Sample ID / Sample Location	Sample Date	Sample Time	Comp.	Grab	Matrix	Number of Containers	Analysis Required	Alloway Lims # (For Lab Use Only)
1	Rvr 4412 / River - Marina Down	4/4/12	08:00	✓			1	Acute Toxicity	13214-01
2	Eff 4412 / Final Effluent	4/4/12	07:30	✓			1	Acute Toxicity	13214-02
3									
4									
5									
6									
7									
8									

	Relinquished by:	Received by:	Date	Time
1	[signature]	[signature]	4/4/12	855
2	[signature]	[signature]	4.4.12	5:20
3	[signature]			

Method of Sample Delivery: UPS/FedEx ☐ Client Delivery ☐ Other Alloway Pick-up ☐

Received for Laboratory By: (Signature) [signature] 4/05/12 07:03

Sample Receiving (For Lab Use Only)
Ice Present? Y ☑ N ☐
Proper Preservation? Y ☐ N ☐
Container Temperature:

Priority (for Client use) Note: Rush Charges May Apply
24 Hrs ☐
48 Hrs ☐
3 Working Days ☐
Routine (5-10 Working days) ☑

White - Lab Copy Yellow - Client Copy

Appendix B

Raw Data Sheets

96-Hour *Pimephales promelas* Static Acute Toxicity Test Data

Industry/Toxicant: Steubenville WWTP Contact: ______
Address: Steubenville, OH
Effluent Serial No: 001 Permit No.: 0PD00014*
Upstream Grab: collected 0800 (AM)/PM 4/4/12 (Date)
Effluent Comp.: collected from ____ AM/PM __/__/__ (Date)
to 0730 (AM)/PM 4/4/12 (Date)

Person(s) Conducting Test: PJC, RFR
Beginning: Date 4/5/12 Time 1110
Ending: Date 4/9/12 Time 1110
Test Organism: *Pimephales promelas*; Age: 14 days #14
Length: (X ± SD) 9.2 ± 0.7 mm
Weight: (X ± SD) .0046 ± .0004 g
Dilution Water Used: upstream

test vessel size = **500** mL; volume of test solution per cup = **400 mL**; test temperature = **25°C**.

Concentration or % Effluent	Test Rep. No.	Number of Live Organisms 0 h	24 h	48 h	72 h	96 h	Condition of Organisms 0 h	24 h	48 h	72 h	96 h	Temperature (°C) 0 h	24 h	48 h	72 h	96 h	Dissolved Oxygen (mg/L) 0 h	24 h	48 h	72 h	96 h	pH (s.u.) 0 h	96 h	Conductivity (µmho) 0 h	96 h	Hard. (mg/L) 0 h	Alk. (mg/L) 0 h
Lab Control	1	10	10	10	10	10	G	G	G	G	G	25.0	25.5	24.6	[illegible]	24.5	8.2	6.7	7.1	7.4	7.4	7.72	7.49	332	375	154	86
	2	10	10	10	10	10	G	G	G	G	G																
Upstream Control	1	10	10	10	10	10	G	G	G	G	G	25.1	25.4	24.7	24.7	24.6	8.5	6.9	6.9	7.5	7.4	7.47	7.52	345	388	137	86
	2	10	10	10	10	10	G	G	G	G	G																
20% Effluent	1	10	10	10	10	10	G	G	G	G	G	25.0	25.2	25.3	24.4	25.1	8.3	7.3	6.7	7.3	7.1	7.37	7.57	476	520		
	2	10	10	10	10	10	G	G	G	G	G																
40% Effluent	1	10	10	10	10	10	G	G	G	G	G	25.0	25.1	25.4	24.9	25.2	8.2	6.8	6.5	6.9	6.7	7.29	7.59	607	623		
	2	10	10	10	10	10	G	G	G	G	G																
60% Effluent	1	10	10	10	10	10	G	G	G	G	G	25.0	25.1	25.1	25.2	25.3	8.2	6.9	6.9	6.9	6.7	7.25	7.61	729	7.73		
	2	10	10	10	10	10	G	G	G	G	G																
80% Effluent	1	10	10	10	10	10	G	G	G	G	G	25.0	25.0	24.9	24.9	24.9	8.3	6.9	6.8	7.2	7.3	7.20	[illegible]	859	998		
	2	10	10	10	10	10	G	G	G	G	G																
100% Effluent	1	10	10	10	10	10	G	G	G	G	G	24.9	25.0	25.4	24.8	24.7	8.3	7.0	6.6	7.4	7.2	7.09	7.79	989	1083	257	86
	2	10	10	10	10	10	G	G	G	G	G																

Condition of Organisms: G = Good Condition; D = Dead; Im = Immotile; LE = Loss of Equilibrium; A = Atypical Behavior or Appearance; LA = Lab Accident (data not used in analyses). * = single bubble aeration started.

48-Hour *Ceriodaphnia dubia* Static Acute Toxicity Test Data

Industry/Toxicant: Steubenville WWTP Contact: ________
Address: Steubenville
Effluent Serial No: 001 Permit No.: 0PD00014*
Upstream Grab: collected 0800 AM/PM 4 / 4 / 12 (Date)
Effluent Comp.: collected from ____ AM/PM __/__/__ (Date)
to 0730 AM/PM 4 / 4 / 12 (Date)

Person(s) Conducting Test: AFA, PJC
Beginning: Date 4/5/12 Time 11:00
Ending: Date 4/7/12 Time 11:00
Test Organism: *Ceriodaphnia dubia*; Age: <24H 8/~
Dilution Water Used: Upstream

test vessel size = 30 mL; volume of test solution per cup = 25 mL; test temperature = 25 °C.

Concentration or % Effluent	Test Rep. No.	Number of Live Organisms			Condition of Organisms			Temperature (°C)			Dissolved Oxygen (mg/L)			pH (s.u.)		Conductivity (umho/cm)		Hard. (mg/L)	Alk. (mg/L)
		0 h	24 h	48 h	0 h	24 h	48 h	0 h	24 h	48 h	0 h	24 h	48 h	0 h	48 h	0 h	48 h	0 h	0 h
Lab Control	1	5	5	0	G	G	5D	25.1	25.0	24.1	8.5	–	8.5	7.63	7.72	350	364	120	103
	2	5	5	2	G	G	2G3D												
	3	5	5	0	G	G	5D												
	4	5	5	5	G	G	G												
Upstream Control	1	5	5	5	G	G	G	25.1	25.1	24.4	8.5	–	8.0	7.47	7.73	345	357	137	86
	2	5	5	5	G	G	G												
	3	5	5	5	G	G	G												
	4	5	5	5	G	G	G												
20% Effluent	1	5	5	5	G	G	G	25.0	25.1	24.4	8.3	–	7.9	7.37	7.74	476	479		
	2	5	5	5	G	G	G												
	3	5	5	5	G	G	G												
	4	5	5	5	G	G	G												
40% Effluent	1	5	5	5	G	G	G	25.0	24.8	24.4	8.2	–	8.0	7.29	7.75	607	636		
	2	5	5	5	G	G	G												
	3	5	5	5	G	G	G												
	4	5	5	5	G	G	G												
60% Effluent	1	5	5	5	G	G	G	25.0	24.9	24.4	8.2	–	7.9	7.25	7.63	729	736		
	2	5	5	5	G	G	G												
	3	5	5	5	G	G	G												
	4	5	5	5	G	G	G												
80% Effluent	1	5	5	5	G	G	G	25.0	25.0	24.4	8.3	–	7.9	7.20	7.50	859	865	~~757~~	~~86~~
	2	5	5	5	G	G	G												
	3	5	5	5	G	G	G												
	4	5	5	5	G	G	G												

Condition of Organisms: G = Good Condition; D = Dead; Im = Immotile; LE = Loss of Equilibrium; A = Atypical Behavior or Appearance; LA = Lab Accident (data not used in analyses).

48-Hour *Ceriodaphnia dubia* Static Acute Toxicity Test Data

Industry/Toxicant: Steubenville WWTP Contact: ______
Address: Steubenville, OH
Effluent Serial No: 001 Permit No.: OPD00014*
Upstream Grab: collected 0800 AM/PM (AM circled) 4/4/12 (Date)
Effluent Comp.: collected from ____ AM/PM ___/___/___ (Date)
to 0730 AM/PM (AM circled) 4/4/12 (Date)

Person(s) Conducting Test: AFR, PJL
Beginning: Date 4/5/12 Time 11:00
Ending: Date 4/7/12 Time 11:00
Test Organism: *Ceriodaphnia dubia*; Age: <24H #8/2
Dilution Water Used: upstream

test vessel size = 30 mL; volume of test solution per cup = 25 mL; test temperature = 25 °C.

Concentration or % Effluent	Test Rep. No.	Number of Live Organisms			Condition of Organisms			Temperature (°C)			Dissolved Oxygen (mg/L)			pH (s.u.)		Conductivity (umho/cm)		Hard. (mg/L)	Alk. (mg/L)
		0 h	24 h	48 h	0 h	24 h	48 h	0 h	24 h	48 h	0 h	24 h	48 h	0 h	48 h	0 h	48 h	0 h	0 h
100% Effluent	1	5	5	5	G	G	G	24.9	25.1	24.4	8.3	–	7.7	7.09	7.83	989	957	257	86
	2	5	5	5	G	G	G												
	3	5	5	5	G	G	G												
	4	5	5	5	G	G	G												
	1	5			G														
	2	5			G														
	3	5			G														
	4	5			G														
	1	5			G														
	2	5			G														
	3	5			G														
	4	5			G														
	1	5			G														
	2	5			G														
	3	5			G														
	4	5			G														
	1	5			G														
	2	5			G														
	3	5			G														
	4	5			G														

Condition of Organisms: G = Good Condition; D = Dead; Im = Immotile; LE = Loss of Equilibrium; A = Atypical Behavior or Appearance; LA = Lab Accident (data not used in analyses).

Appendix C

Parameter Data

PHYSICAL-CHEMICAL DATA OF SAMPLES WHEN RECEIVED BY LAB

Industry/Toxicant: Steubenville WWTP Type of Test: Acute
Address: Steubenville, OH Date of Test: 4/5-9/12
Effluent Serial No.: 001 Permit No.: 0PD00014*

Date Samples Collected: 4/4/12 Date Samples Read: 4/5/12 Analyst: BX

	100% Effluent	Upstream
Temperature (°C)	2.2	3.3
Dissolved Oxygen (mg/L)	11.2	12.5
pH (s.u.)	6.98	7.19
Conductivity (µmhos)	1008	346
Chlorine, total (mg/L)	<0.02	<0.02
Visual Description	light stain	light stain settled solids

Appendix D

SRT Data

48-H LC50 for *Pimephales promelas* Acute NaCl SRT

Date of Test	LC50	Cumulative LC50 Mean	SD	X+2SD	X-2SD	CV*	X+3SD	X-3SD	LC50 Method[1]
08/24/10	9.5	8.8	0.7	10.2	7.5	0.08	10.8	6.8	SK
09/01/10	8.8	8.8	0.7	10.1	7.5	0.08	10.8	6.8	SK
10/20/10	9.0	8.8	0.6	10.0	7.6	0.07	10.6	7.0	SK
11/01/10	9.6	8.8	0.6	10.1	7.6	0.07	10.7	7.0	SK
12/01/10	9.5	8.9	0.6	10.1	7.7	0.07	10.7	7.1	SK
01/04/11	9.5	9.0	0.5	10.1	7.9	0.06	10.6	7.4	SK
02/16/11	8.9	9.0	0.5	10.1	8.0	0.06	10.6	7.4	SK
03/02/11	10.1	9.1	0.5	10.2	8.0	0.06	10.7	7.5	SK
04/19/11	9.2	9.1	0.5	10.1	8.1	0.06	10.6	7.5	SK
05/17/11	8.5	9.0	0.5	10.1	8.0	0.06	10.6	7.5	SK
06/28/11	8.3	9.0	0.5	10.1	8.0	0.06	10.6	7.5	TSK
07/06/11	10.0	9.1	0.6	10.2	8.0	0.06	10.8	7.5	SK
08/24/11	8.4	9.1	0.5	10.2	8.0	0.06	10.8	7.5	SK
09/07/11	9.1	9.1	0.5	10.2	8.0	0.06	10.8	7.5	SK
10/25/11	8.2	9.1	0.6	10.3	7.9	0.06	10.8	7.3	SK
11/09/11	9.0	9.1	0.6	10.2	7.9	0.06	10.8	7.3	SK
12/09/11	8.3	9.0	0.6	10.2	7.8	0.07	10.8	7.2	SK
01/04/12	9.4	9.0	0.6	10.2	7.8	0.06	10.8	7.3	SK
02/15/12	9.3	9.1	0.6	10.2	7.9	0.06	10.8	7.3	SK
03/20/12	8.1	9.0	0.6	10.2	7.8	0.07	10.8	7.2	SK

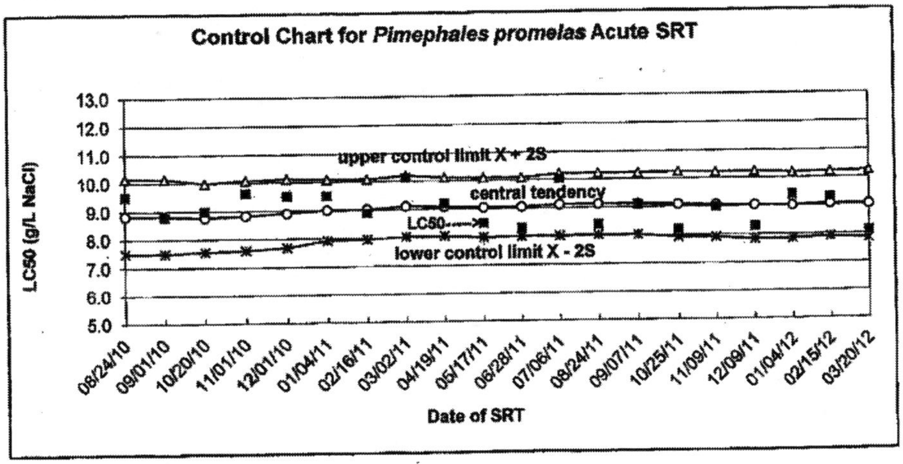

*- compare calculated CV to 75th percentile of CVs reported nationally in Table 3-3 in EPA/833/R-00/003 .
75th percentile of CVs for LC50 = 0.19
Calculated CV should not exceed 75th percentile of CVs.
$\overline{X}$ ± 3SD = Well outside expected range.

1 - P = Probit method, G = Graphical method, SK = Spearman-Karber method, TSK = Trimmed Spearman-Karber method.

48-H LC50 for *Ceriodaphnia dubia* Acute NaCl SRT

Date of Test	LC50	Cumulative LC50 Mean	SD	X+2SD	X-2SD	CV*	X+3SD	X-3SD	LC50 Method[1]
08/24/10	2.0	2.0	0.2	2.4	1.6	0.10	2.6	1.4	Probit
09/01/10	2.3	2.1	0.2	2.5	1.7	0.10	2.7	1.5	TSK
10/05/10	2.2	2.1	0.2	2.5	1.7	0.10	2.7	1.5	TSK
11/02/10	2.8	2.1	0.3	2.6	1.6	0.12	2.9	1.3	TSK
12/01/10	2.1	2.1	0.2	2.6	1.6	0.12	2.9	1.4	Probit
01/04/11	2.2	2.1	0.2	2.6	1.6	0.12	2.9	1.4	TSK
02/15/11	1.9	2.1	0.2	2.6	1.6	0.12	2.9	1.4	Probit
03/01/11	2.0	2.1	0.2	2.6	1.6	0.12	2.8	1.4	TSK
04/19/11	1.8	2.1	0.2	2.6	1.6	0.12	2.8	1.3	Probit
05/17/11	2.1	2.1	0.2	2.5	1.6	0.11	2.7	1.4	SK
06/28/11	2.0	2.1	0.2	2.5	1.6	0.11	2.7	1.4	Probit
07/06/11	2.6	2.1	0.2	2.6	1.6	0.12	2.8	1.4	SK
08/23/11	2.1	2.1	0.2	2.6	1.6	0.12	2.8	1.4	SK
09/07/11	2.8	2.1	0.3	2.7	1.6	0.13	3.0	1.3	Probit
10/25/11	1.8	2.1	0.3	2.7	1.5	0.14	3.0	1.3	Probit
11/08/11	2.0	2.1	0.3	2.7	1.6	0.14	3.0	1.3	TSK
12/21/11	2.0	2.1	0.3	2.7	1.6	0.14	3.0	1.3	SK
01/10/12	1.9	2.1	0.3	2.7	1.6	0.13	3.0	1.3	Probit
02/14/12	1.8	2.1	0.3	2.7	1.5	0.14	3.0	1.3	SK
03/20/12	2.0	2.1	0.3	2.7	1.5	0.14	3.0	1.2	TSK

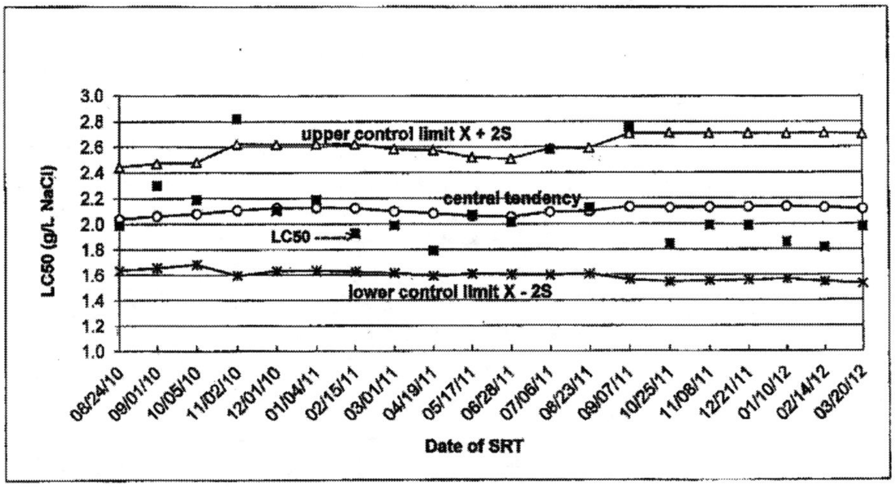

*- compare calculated CV to 75th percentile of CVs reported nationally in Table 3-3 in EPA/833/R-00/003 .
75th percentile of CVs for LC50 = 0.29
Calculated CV should not exceed 75th percentile of CVs.
$\overline{X}$ ± 3SD = Well outside expected range.

1 - P = Probit method, G = Graphical method, SK = Spearman-Karber method, TSK = Trimmed Spearman-Karber method.

Alloway

ANALYTICAL REPORT

Lab Project # L12-13260

Received: 04/18/2012
Reported: 04/24/2012
Date/Time Sampled: 04/18/2012 11:00
Sampled By: SG
Sampled Matrix: Wastewater
Containers: 1

Steubenville
Attn: Chuck Murphy
100 N. Water Street
Steubenville, OH 43952

Project Name: April

Sample ID: Fin. Head 041812 / Head
Lab Sample # L12-13260-02

Analyte	Results	Units	PQL	Method	Analyst	Extraction Date	Analysis Date
E. Coli	>2400	MPN/100 mL	1	SM-9223 B	KW		04/18/2012

Analysis Certified By: Lana L. Jackson

Alloway
Your Resource for Defensible Data

ANALYTICAL REPORT

Lab Project #	L12-13260
Received:	04/18/2012
Reported:	04/24/2012
Date/Time Sampled:	04/18/2012 11:00
Sampled By:	SG
Sampled Matrix:	Wastewater
Containers:	1

Steubenville
Attn: Chuck Murphy
100 N. Water Street
Steubenville, OH 43952

Project Name: April

Sample ID: Fin. Eff 04/18/12 / Final Out
Lab Sample # L12-13260-01

Analyte	Results	Units	PQL	Method	Analyst	Extraction Date	Analysis Date
E. Coli	82	MPN/100 mL	1	SM-9223 B	KW		04/18/2012

Analysis Certified By: Lana L Jackson

1101 N. Cole Street - Lima, Ohio 45805
419.223.1362 - Fax 419.227.3792
800.436.1243

508 Bissman Ct. - Mansfield, Ohio 44903
419.525.1644 - Fax 419.524.5575
800.635.3222

1776 Marion-Waldo Rd. - Marion, Ohio 43302
740.389.5991 - Fax 740.389.1481
800.873.2835

www.alloway.com

PERAGreen SOLUTIONS, LLC

ATTACHMENT C
Proxitane WW-12
Ohio River Water Sampling

Proxitane WW-12 - Ohio River Sampling

			PAA Residual (ppm)							Outfall	Outfall
			Upstream	Outfall	Outfall	Downstream	D.O.		Temp	Flow	PAA
Date	Test	Time	12-15'	0' Out	5.6' Out	10-12'	ppm	pH	C	MGD	(ppm)
4/11/2012	1	4:41PM		0.28			10.35	7	14.86	3.64	
	2	8:04PM		0.24			10.3	6.8	14.91	5.04	
	3	4:41PM			0.21		10.25	6.8	14.6	3.64	
	4	8:04PM			0.2		10.22	6.7	14.57	5.04	
4/12/2012	1	4:02PM	0.1				10.47	6.8	16.77	4.65	0.44
	1a	4:07PM	0.1				10.47	6.8	16.77	4.65	0.44
	2	4:10PM		0.42			9.93	6.8	17.11	4.65	0.44
	3	4:15PM			0.37		9.79	7	16.74	4.65	0.44
	4	4:20PM				0.03	10.79	6.8	16.64	4.65	0.44
	5	20:10PM	0.04				11.06	6.8	16.23	4.7	0.52
	6	20:15PM		0.42			10.29	6.9	16.69	4.7	0.52
	7	20:20PM			0.37		10.15	7	16.69	4.7	0.52
	8	20:25PM				0.06	10.75	6.8	16.12	4.7	0.52

* All samples collected 5-6' from river bank

Lessons Learned from St. Augustine WWTP – Comparison of VigorOx (Peracetic Acid) and Sodium Hypochlorite

Minh Tran, Commercial Development Manager, FMC Environmental Solution
1735 Market Street,
Philadelphia, PA 19103
water@fmc.com
(866) 860-4760

INTRODUCTION

Wastewater treatment plants are required to disinfect effluent prior to discharge in order to destroy any pathogenic organisms present and minimize public health concerns. Chlorine (Cl^-) has been the principle disinfection method in the wastewater industry despite the fact that disinfection with chlorine produces chlorinated disinfection by-products (DBPs), including trihalomethanes (TTHMs), that are toxic to aquatic life. In recent years, many municipalities have been required to install de-chlorination systems in order to address DBPs. However, upgrading an existing plant with a de-chlorination process typically results in significant capital cost and operational complexity.

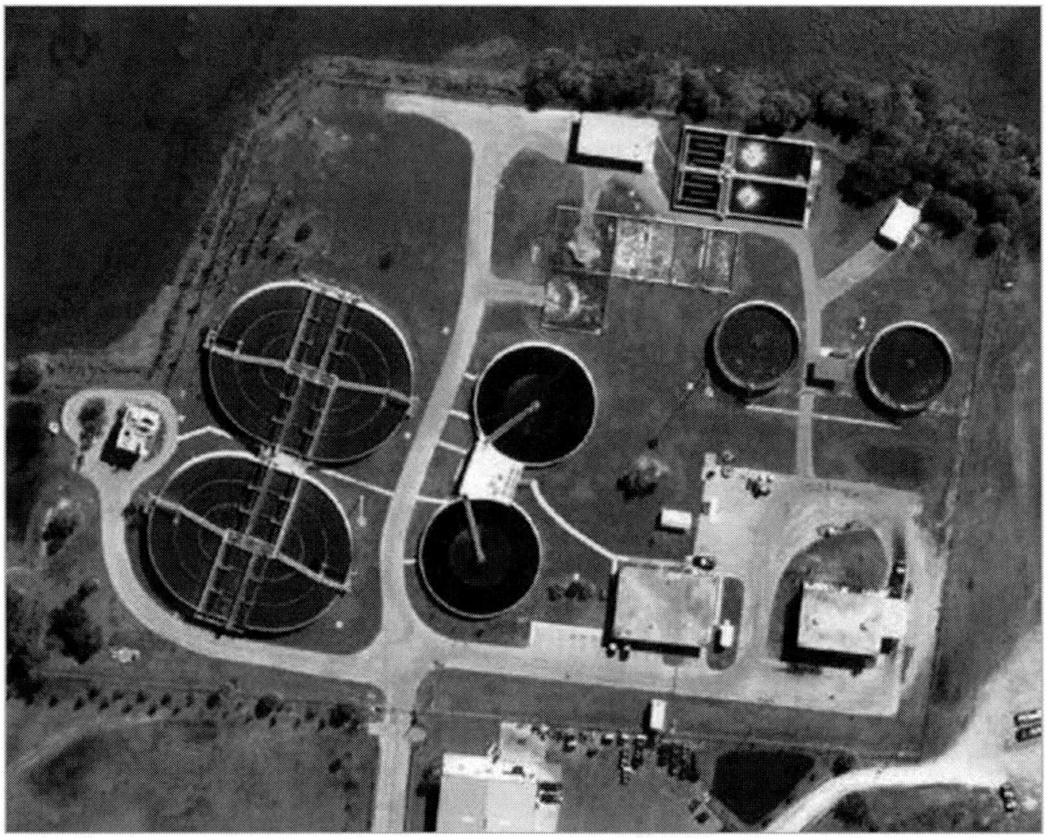

Figure 1: St. Augustine WWTP

The 2.7 MGD St. Augustine, Florida- wastewater treatment plant (WWTP) has a chlorination/de-chlorination disinfection system but decided to investigate an alternative disinfection process that did not generate DBPs, was cost effective, and simple to operate. The plant investigated other disinfection options such as ozone and UV disinfection. However, PAA disinfection was deemed the most attractive option based on the fact that it was the lowest capital approach since the plant's existing contact chamber could be used for the PAA technology, thereby providing large capital cost savings to the municipality compared to the other alternatives.

After successful bench scale testing with PAA, a full scale demonstration test was conducted to evaluate the performance of the PAA disinfection system compared to the existing chlorination/de-chlorination system based on several evaluation criteria,

- Disinfection performances
- Aquatic toxicity
- Disinfection by-products
- Chemical consumption/sustainability

METHODOLOGY

Disinfectant

VigorOx® WWTII, manufactured by FMC Corporation, was selected for the study. VigorOx® WWTII is a proprietary 15% peracetic acid formulation that is USEPA approved for wastewater disinfection, EPA Reg. No. 65402-8. The PAA solution breaks down to the non-harmful products water, oxygen and acetic acid (vinegar).

Disinfection Layout

The St. Augustine WWTP (Figure 1) has two identical disinfection contact tanks. Each tank receives fifty percent of the 2.7 MGD plant flow. After consulting with FMC, a PAA disinfection system consisting of a storage tank, dispensing system and a PAA residual monitor were deployed to treat the first contact tank. The second contact tank continued to operate using the existing chlorination/de-chlorination system. This enabled the plant to conduct a direct head-to-head comparison of the two technologies.

Both the disinfection systems were connected to the plant's SCADA system and identical flow pacing equipment was installed at each contact tank. The flow pacing equipment enabled chemical dosage and flow rates to be controlled in real-time. The plant's flow data, pathogen counts, chemical dosage rates and residual data were recorded during the trial.

RESULTS

Disinfection performances

A wastewater disinfectant's performance can be quantified based on two specific pathogens of concern, Enterococci and Fecal Coliform (FC). Effluent grab samples from both disinfection tanks were tested several times a week.

Enterococci

Figure 2 shows that an average VigorOx® WWTII concentration of 1.5 ppm [PAA] reduced the Enterococci count between 1 and 6 CFU/100mL. This is well below the plant discharge limit of 35 CFU / 100 mL. The average sodium hypochlorite dose of 7 ppm [OCL⁻] yielded similar Enterococci results to VigorOx® WWTII with one outlier at 29 CFU/100mL. Thus, the test demonstrated an opportunity for the plant to use a lower concentration of VigorOx® WWTII (versus sodium hypochlorite) and still meet Enterococci disinfection requirement.

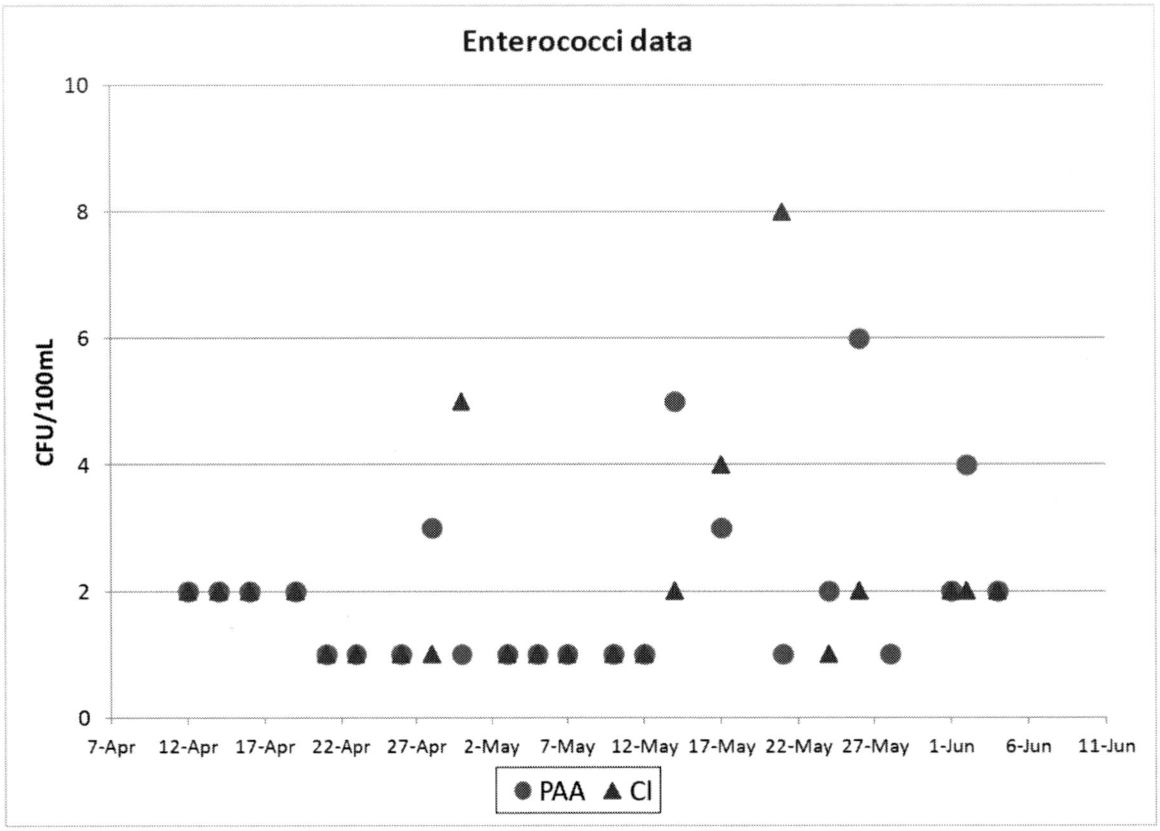

Figure 2: Enterococci graph

Fecal Coliform (FC)

The plant's discharge permit limit for FC is 200 CFU/100mL. Figure 3 shows the VigorOx® WWTII treatment performance side-by-side with the chlorination/de-chlorination treatment performance. The data showed that an average VigorOx® WWTII dosage of 1.5ppm [PAA] yielded similar disinfection performance to an average sodium hypochlorite dosage of 7ppm [OCL⁻] average, and that both disinfection processes were effective in keeping FC count below the discharge limit.

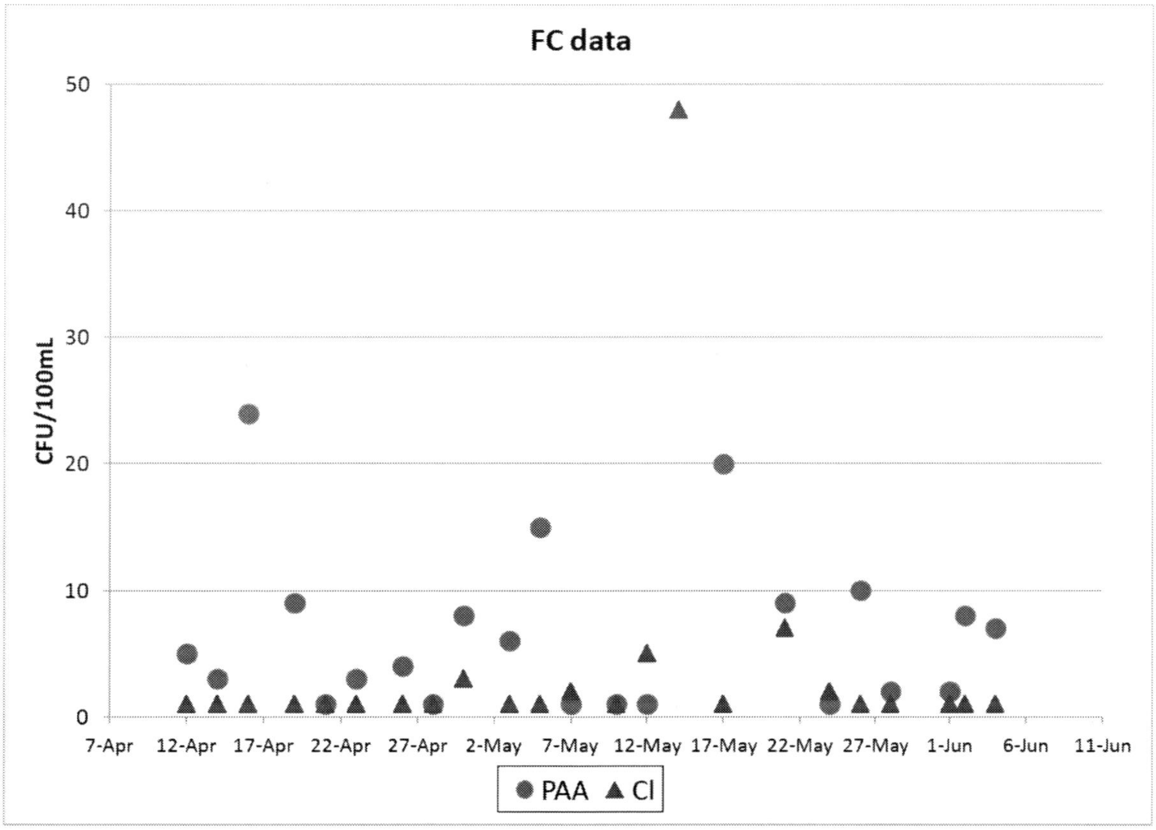

Figure 3: Fecal Coliform graph

Aquatic Toxicity Results

Wastewater can produce adverse effects on the biological system, damaging its infrastructure and causing death to aquatic life. The criteria for aquatic toxicity testing are mortality and reproduction rate. *Mysidopsis bahia* (mysid shrimp) and *Menidia beryllina* (tidewater silverside) were the species used for both acute and chronic toxicity testing.

During the trial, testing for acute and chronic toxicity at dilution levels ranging from 0% (control) to 100 % effluent. The first set utilized treated effluent from the chlorination/de-chlorination process (see Table 1). The second set utilized the VigorOx® WWTII treated effluent (see Table 2).

Table 1: Toxicity results from chlorination/de-chlorination

	M. bahia		*M. beryllina*	
%	**Survival**	**Growth**	**Survival**	**Growth**
Control	97.5	0.390	100	2.096
6.25	85	0.375	97.5	1.986
12.5	97.5	0.415	97.5	2.333
25	90	0.372	100	2.226
50	77.5	0.414	100	2.562
100	82.5	0.377	97.5	2.506

Table 2: Toxicity results from VigorOx® WWTII

	M. bahia		*M. beryllina*	
%	**Survival**	**Growth**	**Survival**	**Growth**
Control	92.5	0.353	100	2.374
6.25	90	0.383	100	2.487
12.5	95	0.380	100	2.491
25	95	0.396	100	2.426
50	90	0.377	100	2.563
100	97.5	0.358	97.5	2.337

The toxicity results show a higher survival rate for the *M. bahia* and the *M. beryllina* in the PAA treated effluent when compared to the chlorination/de-chlorination treated effluent. The growth rate results for both disinfection systems were comparable. The overall trend showed VigorOx® WWTII was slightly less toxic to aquatic life than chlorination/de-chlorination.

Disinfection By-Products (DBPs) Results

Disinfection by-products are chemicals formed as a result of the reaction between the disinfectant added to the water and naturally occurring organic material present in the water. The EPA has established discharge effluent limits for disinfection by-products (see Table 3 & 4).

Table 3: Disinfection by-products from a chlorination/de-chlorination treatment

Volatiles	**Effluent limit (annually)**	**Chlorination start**	**Chlorination end**	**After de-chlorination**
Bromodichloromethane [ug/L]	22	95.55	65.72	56.82
Bromoform[ug/L]	360	22.64	20.90	19.62
Chloroform[ug/L]	470.8	49.25	26.13	21.55
Dibromochloromethane[ug/L]	34	102.99	81.44	72.71
Total Trihalomethanes[ug/L]	-	270.43	194.19	170.70

Table 4: Disinfection by-products from PAA treatment

Volatiles	**Effluent limit (annually)**	**PAA start**	**PAA end**	**After de-chlorination**
Bromodichloromethane [ug/L]	22	0.6	0.6	0.6
Bromoform[ug/L]	360	0.6	0.6	0.6
Chloroform[ug/L]	470.8	0.64	0.64	0.64
Dibromochloromethane[ug/L]	34	0.75	0.75	0.75
Total Trihalomethanes[ug/L]	-	0.6	0.6	0.6

Disinfection by-product results from a chlorine treatment test are presented in Table 3. Untreated effluent grab samples were taken from the plant's side stream and tested for disinfection by-products once chlorine was added (chlorine start), after disinfection was completed (chlorine end) and after the de-chlorinating agent was added (after de-chlorination). Total Trihalomethanes (TTHMs) concentration was found to be 194.19ug/L after disinfection. After the de-chlorinating agent was added, TTHMs concentration decreased to 170.70ug/L. This suggested that the de-chlorinating agent had only a limited capability of reducing TTHMs concentrations.

The VigorOx® WWTII result show that there was virtually no TTHMs generated by this disinfection process. Currently, many WWTPs in North America are having difficulty meeting disinfection by –products permits. This trial test suggested that conversion to PAA disinfection could provide a low-capital solution to resolve any DBP issues.

Other results

The plants SCADA data revealed that the average chemical usage for chlorination/de-chlorination system was 235 gallon per day while the PAA system used only 23 gallons per day of chemical to meet the same disinfection requirements.

Of interested to note, is plant operators observed that the VigorOx® WWTII treatment was easier to adjust and maintain dosing in response to rapid charges in plant flow rate compared to the chlorination system. This may be attributed to the lower chemical feed-rate compared to that of the sodium hypochlorite.

Field studies have demonstrated that changes in wastewater characteristics can impact the dosage of VigorOx® WWTII required for effective pathogen control. In order to address this challenge, FMC developed a proprietary PAA analyzer unit that can accurately measure and record PAA residuals in real time and be integrated with a plant's automated control system and flow pacing equipment. This enables plant operators to adjust VigorOx® WWTII dosage rates based on real-time to help ensure consistent pathogen control while optimizing chemical consumption.

CONCLUSIONS

The objective of the St. Augustine trial was to compare a VigorOx® WWTII disinfection system to a chlorination/de-chlorination system at a full-scale over an extended period based on pathogen results, environmental impact, and cost.

At the St. Augustine WWTP, the VigorOx® WWTII system achieved equivalent disinfection performance for both *Enterococci* and *Fecal Coliform* compared to the chlorine treated system. The VigorOx® WWTII system was found to have environmental advantages compared to the chlorination/de-chlorination treated system since it did not generate disinfection byproducts and exhibited less toxicity to aquatic life. The VigorOx® WWTII disinfection system was found to be 10% less expensive compared to the chlorination/de-chlorination system due to the fact that VigorOx® WWTII was more efficient than chlorine with 1.5 ppm [PAA] concentration delivering equivalent microbial performance to 7.0 ppm of sodium hypochlorite. Finally, the VigorOx® WWTII system offered the St. Augustine plant some additional soft benefits including the opportunity to reduce its overall disinfection chemical consumption by 90% and respond more rapidly to changes in flow rates.

REFERENCES

[1] S. Rossi, M. Antonelli, V. Mezzanotte, and C. Nurizzo, "Peracetic Acid Disinfection: A Feasible Alternative to Wastewater Chlorination," *Water Environment Research*, vol. 79, 2007, pp. 341-350.

[2] A. Dellerba, D. Falsanisi, L. Liberti, M. Notarnicola, and D. Santoro, "Disinfecting behaviour of peracetic acid for municipal wastewater reuse," *Desalination*, vol. 168, 2004, pp. 435-442.

[3] M. Kitis, "Disinfection of wastewater with peracetic acid: a review," *Environment International*, vol. 30, 2004, pp. 47-55.

[4] D. Chen, X. Dong, and R. Gehr, "Alternative Disinfection Mechanisms for Wastewaters Using Combined PAA/UV Processes," *Water Environment Federation*, 2005, pp. 886-906.

[5] J. Koivunen and H. Heinonen-Tanski, "Inactivation of enteric microorganisms with chemical disinfectants, UV irradiation and combined chemical/UV treatments," *Water Research*, vol. 39, 2005, pp. 1519-1526.

[6] M. Wagner, D. Brumelis, R. Gehr, "Dinfection of Wastewater by Hydrogen Peroxide or Peracetic Acid: Development of Procedures for Measurement of Residual Disinfectant and Application to a Physicochemically Treated Municipal Effluent," *Water Environment Research*, vol. 74, 2002, pp. 33 – 50.

Kinetics and Mechanism of Coxsackievirus B5 Inactivation by Free Chlorine for Safe Global Water

Theresa Vonder Haar[1,3,*], Benito Mariñas[1,3], Joanna Shisler[2,3]

[1]Department of Civil and Environmental Engineering, University of Illinois at Urbana-Champaign
[2] Department of Microbiology, University of Illinois at Urbana-Champaign
[3] The Safe Global Water Institute at the University of Illinois at Urbana-Champaign
*4152 Newmark Civil Engineering Laboratory
205 N. Mathews Ave. ¦ Urbana, IL 61801
tvonder2@illinois.edu

ABSTRACT

In the field of drinking water treatment, viruses have emerged as pathogens of concern, both in the developed and developing world. Free chlorine is one of the most common disinfectants used globally and is typically quite effective for control of viruses in drinking water. However, Coxsackievirus (CV), a small, single-stranded RNA virus, has previously shown a unique resistance to free chlorine. Additionally, the inactivation by free chlorine had not been thoroughly characterized under a complete array of water quality conditions. In this study, the inactivation of Coxsackievirus B5 by free chlorine was fully characterized. Coxsackievirus, in the *Picornaviridae* family and *Enterovirus* genus, is highly virulent, with health effects including respiratory illnesses, diarrhea, fever, myocarditis, aseptic meningitis, and other adverse effects, especially in young children. As a result, it has been identified as a virus of concern in drinking water, thus, making the complete characterization of its inactivation kinetics by free chlorine an important venture.

Another important aspect to viral inactivation is the mechanism in which the virion becomes no longer infectious. Understanding the molecular target of disinfectants can lead to a better insight of viral inactivation, which is not well understood. This study is focused on using a variety of innovative molecular techniques that evaluate the mechanism of free chlorine disinfection, by specifically investigating the attachment of CVB5 to the receptor on the host cell before and after disinfection. For example, enzyme-linked immunosorbent assays (ELISAs) are used to determine the viral capsid integrity and the ability of the virus to bind to the receptor, and surface plasmon resonance (SPR) measures the affinity of virus binding. These results will not only advance the knowledge in the field of disinfection, but will also likely have important implications for the overall optimization of pathogen control in drinking water treatment worldwide.

KEYWORDS

free chlorine, coxsackievirus B5, viral inactivation, safe global water

INTRODUCTION

Control of pathogens continues to be one of the main goals of drinking water treatment globally. Viruses are of particular concern due to their persistence and prevalence in the environment, their resistance to many commonly used disinfectants, and the variety of adverse health effects they have on humans. Not only have viruses historically been of interest in regards to water treatment and public health, they are now re-emerging as pathogens of concern. Due to the formation of disinfection byproducts (DBPs), the use of free chlorine and ozone, two disinfectants commonly used for effective virus control, is decreasing. At the same time, alternative disinfectants that are less effective against viruses are being adopted. Combined chlorine, an alternate secondary disinfectant applied by utilities in countries such as the United States to reduce formation of regulated DBPs, does not perform effectively against viruses. Microfiltration and ultraviolet (UV) light, two technologies used to control *Cryptosporidium parvum* oocysts without forming DBPs, are also less effective in controlling viruses. Disinfection technologies currently recommended or implemented for safe drinking water in developing regions are not always completely effective in the challenging surface waters typical of these areas. For example, surface waters are commonly contaminated with inorganic and organic nitrogen from human and animal waste and upon addition of free chlorine, chloramines are formed. This study is focused on inactivation by free chlorine due to its extremely popular use world-wide and CVB5's unique resistance to it.

Objectives

Overall, the goal of this research is to contribute to the understanding of the mechanisms of viral inactivation, with a focus on Coxsackievirus B5 by free chlorine disinfection. In this study, we are investigating the ability of chlorine-treated, or non-infectious, CVB5 to attach to its host cell receptor versus infectious virions. This is being achieved by utilizing a variety of molecular techniques to elucidate the mechanisms of inactivation.

COXSACKIEVIRUS BACKGROUND

Prevalence & Virulence

Coxsackievirus was first discovered in the small town of Coxsackie, New York in the late 1940's (*Fields Virology*, 2007; Mena *et al.*, 2003). This virus has been found in a variety of waters globally including natural and recreational waters such as rivers and lakes, wastewaters, and drinking water supplies and is the most prevalent non-polio enterovirus (Mena *et al.*, 2003). For example, several outbreaks, causing Acute Hemorrhagic Conjunctivitis were reported in Puerto Rico, Tunisia, and Brazil. According to Mena *et al.* (2003), outbreaks in the United States causing such symptoms as abdominal pain and fever, as well as aseptic meningitis have been contracted from sources such as a high school football team's shared water bottles, a lake at a children's summer camp (with no source of sewage identified), and a well used for drinking water. It has also been detected in wastewater discharge and in raw sewage (Hsu, 2008; Mena *et al.*, 2003; Tavares, 2004; Triki *et al.*, 2007). Most viruses are fairly susceptible to free chlorine; however, the limited data available have shown that Coxsackievirus might be quite resistant to this disinfectant (Cromeans *et al.*, 2010; Engelbrecht *et al.*, Kahler *et al.*, 2010; 1980; Sobsey *et al.*, 1988).

Coxsackievirus has a median infection dose of 100 and a mortality ratio of 1% (MWH, 2005). Health effects can include and have been linked to myocarditis, aseptic meningitis, diabetes, miscarriages, hemorrhagic conjunctivitis, as well as vomiting, diarrhea, fever, and rash. The symptoms of infection vary depending on the serotype and overall health of the host. Coxsackievirus is transmitted via the fecal-oral route (*Fields Virology*, 2007; Ford, 1999; Mena *et al.*, 2003).

Structure

Coxsackievirus is in the genus *Enterovirus*, which is in the family *Picornaviridae* also containing viruses such as poliovirus and echovirus. Coxsackievirus has a ss RNA genome with positive-sense. It can be described as an icosahedral, unenveloped, small virus, about 30 nanometers in size. There are two types of Coxsackievirus: A (23 serotypes) and B (6 serotypes) (Hendry *et al.*, 1999; *Fields Virology*, 2007). One of the more prevalent forms of Coxsackievirus is B5 (Mena *et al.*, 2003). Serotype B5 has also been shown to be one of the more virulent and disinfectant resistant viruses (Engelbrecht *et al.*, 1980).

The capsid is composed of 60 copies of four different proteins, denoted as VP1, VP2, VP3, and VP4. VP1, VP2, and VP3 contain 270 amino acid residues and make up the exterior of the capsid, while the smaller VP4, with 70 amino acid residues, resides on the interior of the capsid (Hendry *et al.*, 1999). Many viruses, including CVB5, within the *Picornaviridae* family have a depression, called the canyon, within the fivefold axes of their icosahedral capsids, and this is said to be the site for binding to the receptor on the host cell (Acheson 2007; *Fields Virology*, 2007).

Coxsackie Adenovirus Receptor

The first step of the CVB5's life cycle is attachment of the virion to a receptor on the host cell. The primary receptor for CVB5 on host cells is thought to be the Coxsackie Adenovirus Receptor (CAR). CAR is a 46 kDa cellular membrane protein that is composed of two domains: D1 and D2 (He *et al.*, 2001). It has been proposed that CVB5 is able to bind to other receptors and even use them to enter and carry out their life cycles (Orthopoulos *et al.*, 2004), however, most studies indicate and have shown that CVB5 is the main receptor. This is also the receptor that the protruding fiber knob of Ad2 recognizes and binds to before infecting the cell (Acheson 2007; Ahn *et al.*, 2008; He *et al.*, 2001; *Fields Virology*, 2007). It is currently still under investigation where exactly CAR binds on the CVB5 capsid and which proteins play a role in binding, but it commonly proposed that binding occurs with D1 domain of CAR and theVP1 protein of CVB5 (He *et al.*, 2001; Hendry *et al.*, 1999).

MOLECULAR METHODS

The goal of this molecular work is to understand if the attachment step is inhibited causing loss of viability, or reduction of plaques, as a result of free chlorine disinfection. Therefore, the techniques are looking at the integrity of the viral capsid as well as the ability of Coxsackievirus to bind to CAR and the host cell before and after disinfection. Coupled with the chemical and conformational damage that a strong oxidant can cause to proteins, we believe that one mechanism of inactivation by free chlorine is damage to the capsid protein resulting in a lower ability and/or affinity to bind to CAR.

The Enzyme-Linked Immunosorbent Assay (ELISA) is a technique that can measure the ability of an antibody to bind to the virus, or the capsid integrity, via a colorimetric signal. In order to assess the ability of CVB5 to bind to the host cell receptor, a sandwich ELISA was used to determine the ability of the virus to attach to immobilized CAR. Doing both types of ELISAs alongside disinfection experiments, will allow us to compare both the capsid integrity and ability of binding to the host cell receptor against the virus being able to complete its life cycle. Because it has been reported that exposure to free chlorine causes oxidative damage to proteins and amino acids (Pattison and Davies, 2001), we expected to find a correlation in the decrease of viability with the decrease in integrity of the capsid proteins as well as the binding ability.

Another that we are utilizing have been shown to measure the interactions between biological molecules (Gambaryan and Matrosovich, 1992). One method that is designed to assess the affinity and reactivity of two biomolecules is Biospecific Interaction Analysis (BIAcore) (Suenaga *et al.,* 2012; vanCott *et al.,* 1992). We have immobilized the CAR protein to a biosensor plate and are measuring the affinity of CVB5 in a flow-through system. Another technique we are using is flow cytometry. In this procedure, the host cells are infected with virus (treated or untreated) for just enough time to attach, at 4C, and then incubated with the anti-CVB5 or anti-VP1 antibody. A fluorescent conjugated secondary antibody is used as a means of visualization of if the viruses have attached to the cell.

These techniques have the potential for demonstrating lowered capsid integrity and/or decreased affinity for binding between Coxsackievirus B5 and its host cell receptor, CAR, after inactivation. The methods and results that will be discussed will are contributing to the understanding of the mechanisms of viral inactivation, by focusing on CVB5 with free chlorine disinfection.

REFERENCES

N.H. Acheson. *Fundamentals of Molecular Virology,* Ch. 16: "Picornaviruses." p. 169-180. John Wiley & Sons, 2007.

J. Ahn, Y. Jee, I. Seo, S.Y. Yoon, D. Kim, Y.K. Kim, H. Lee. Primary Neurons Become Less Susceptible to Coxsackievirus B5 Following Maturation: The Correlation With the Decreased Level of CAR Expression on Cell Surface. *Journal of Medical Virology,* 80: 434–440; 2008.

T.L. Cromeans, A. Kahler, V.R. Hill. Inactivation of Adenoviruses, Enteroviruses, and Murine Norovirus in Water by Free Chlorine and Monochloramine. *Applied and Environmental Microbiology,* Vol. 76, No. 4, 1028-1033; 2010.

R.S. Engelbrecht, M. J. Weber, B. L. Salter, C. A. Schmidt. Comparative Inactivation of Viruses by Chlorine. *Applied and Environmental Microbiology*, 40, 249-256, 1980.

Fields Virology, 5th Edition. Editors-in-Chief: D.M. Knipe and P.M. Howley. Volume 1, Section 2, "Picornaviridae." p. 795-948. Lippincott Williams & Wilkins, a Wolters Kluwer Business, Philadelphia, PA, USA; 2007.

T. E. Ford. Microbiological Safety of Drinking Water: United States and Global Perspectives. *Environmental Health Prospectives*, 107, 191-202, 1999.

A.S. Gambaryan, M.N. Matrosovich. A Solid-Phase Enzyme-Linked Assay for Influenza Virus Receptor-Binding Activity. *Journal of Immunological Methods,* Vol. 146, Issue 2, 111-123; 1992.

C. P. Gerba, D. M. Gramos, N. Nwachuku. Comparative Inactivation of Enteroviruses and Adenovirus 2 by UV Light. *Applied and Environmental Microbiology,* 68, 5167-5169, 2002.

Y. He, P.R. Chipman, J. Howitt, C. Bator, M. Whitt, T.S. Baker, R.J. Kuhn, C.W. Anderson, P. Freimuth, M. G. Rossman. Interaction of Coxsackievirus B3 with the Full Length Coxsackievirusadenovirus Receptor. *Nature Publishing Group*, letters. <http://structbio.nature.com>; 2001.

E. Hendry, H. Hatanaka, E. Fry, M. Smyth, J. Tate, G. Stanway, J. Santti, M. Maaronen, T. Hyypia, D. Stuart. The Crystal Structure of Coxsackievirus A9: New Insights into the Uncoating Mechanisms of Enteroviruses. *Structure*, Vol. 7, No, 12, 1527-1538; 1999.

B-M Hsu, C-H Chen, M-T Wan. Prevalence of Enteroviruses in Hot Spring Recreation Areas of Taiwan. *Federation of European Microbiological Societies (FEMS) Immunol Med Microbiol,* 52, 253–259, 2008.

A. Kahler, T.L. Cromeans, J.M. Roberts, V.R. Hill. Effects of Source Water Quality on Chlorine Inactivation of Adenovirus, Coxsackievirus, Echovirus, and Murine Norovirus. *Applied and Environmental Microbiology,* Vol. 76, No. 15, 5159-5164; 2010.

K. D. Mena, C. P. Gerba, C. H. Haas, J. B. Rose. Risk Assessment of Waterborne Coxsackievirus. *American Water Works Association*, 95, 112-131, 2003.

G. Orthopoulous, K. Triantafilou, M. Triantafilou. Coxsackie B Viruses Use Multiple Receptors to Infect Human Cardiac Cells. *Journal of Medical Virology,* 74: 291-299; 2004.

D.I. Pattison, M.J. Davies. Absolute Rate Constants for the Reaction of Hypochlorous Acid with Protein Side Chains and Peptide Bonds. *Chem. Res. Toxicol, 14,* 1453-1464; 2001.

M. D. Sobsey, T. Fuji, P. A. Shields. Inactivation of Hepatitis A Virus and Model Viruses in Water by Free Chlorine and Monochloramine. *Water Science and Technology*, 20, 385-391, 1988.

E. Suenaga, H. Mizuno, K.K.R. Penmetcha. Monitoring Influenza Hemagglutinin and Glycan Interactions Using Surface Plasmon Resonance. *Biosensors and Bioelectronics,* Vol. 32, Issue 1, 195-201; 2012.

F.N. Tavares, E.V. Costa, S.S. Oliveira, C.C.A. Nicolai, M. Baran, E.E. da Silva. Acute Hemorrhagic Conjunctivitis and Coxsackievirus A24v, Rio de Janeiro, Brazil, 2004. *Emerging Infectious Diseases*, <www.cdc.gov/eid>, Vol. 12, No. 3, 2006.

H. Triki, D. Rezig, O. Bahri, N. Ben Ayed, A. Ben Yahia, A. Sadraoui, S. Ayed. Molecular Characterisation of a Coxsackievirus A24 that Caused an Outbreak of Acute Haemorrhagic Conjunctivitis, Tunisia 2003. *Clin Microbiol Infect*, 13: 176–182, 2007.

T.C. VanCott, L.D. Loomis, R.R. Redfield, D.L. Birx.. A Real-Time Biospecific Interaction Analysis of Antibody Reactivity to Peptides from the Envelope Glycoprotein, gp160, of HIV-1. *Journal of Immunological Methods,* Vol. 146, Issue 2, 163-176; 1992.

Water Treatment: Principles and Design, 2nd Edition- author MWH. ed by John Crittenden. New York, 2005.

Groundwater Coliform Occurrence and Disinfection Practices

Robert A. Ryder, P.E.
Kennedy/Jenks Consultants
303 Second Street, Suite 300 South
San Francisco, CA 94107
(415) 243-2150

ABSTRACT

Periodic occurrences of total coliform (TC) bacteria have been observed in public groundwater well supplies in California and other western states. These have been observed following earthquakes, after periods of non-use of wells upon restarting operation, following unusual wet or dry years of rising or falling water table and periodically at intermittent periods with no apparent physical change in condition.

Rarely is there more pollutant indicating fecal coliform, *E. coli*, or fecal streptococcus bacteria found with these incidences. Non-pathogenic bacteria of the Total Coliform group habitating soil and deep high organic aquifers, like Klebsiella, are typical. There are now means to quickly determine bacteria and viral types and to identify public health risk and disinfection needs.

Current health regulators often do not allow use of these wells unless they are individually disinfected and retested to indicate absence of total coliform. Incidences where rechlorination by AWWA Standards for Well Water Disinfection were not effective and more rigorous well rehabilitation methods with scrubbing, acids and sequestering chemicals were needed and the techniques used are described as are laboratory examination of the physical, chemical and biological properties of these deposits on well casings, screens and piping.

Also, continuous or periodic down-hole chlorination is not allowed due to apparent heightened concern of aquifer contamination, or even distribution system chlorination allowed by many current regulators. Down-well chlorination is a typical practice of aquifer storage and recovery projects to minimize clogging bioslimes. Options to provide effective chlorination, flow time and residual monitoring of pipeline or pressure vessels for contact time "CT"-chlorine residual concentrations conditions are described. These options are compared to the high costs and uncertainty of replacing the groundwater wells. There are now reliable residual chlorine analyzers that can constantly monitor for control chlorine dosage, "CT" and shut down wells and notify operating personnel if insufficient residual concentration.

There are often inconsistent well chemical regulations which do not allow for NSF-certified corrosion or scale inhibitors to be injected down hole in wells, again in concern for aquifer contamination. This is inconsistent with current insitu oxidation for iron and manganese treatment, well rehabilitation work, or aquifer storage and recovery practices.

There were developed past protocols for investigating groundwater under the direct influence of surface water (GUDI) that by Microscopic Particulate Analysis (MPA) monitor, protozoa, nematodes, organic matter, etc., and rate potential risk. If the MPA is a low numerical probability and turbidity already low, then chlorination alone is an acceptable practice to produce a safe potable water.

Reference to requirements of the Total Coliform Rule and the Ground Water Rule that require four logs of viral disinfection would provide criteria for well water chlorination.

The difficulties in obtaining public health regulator approval for an often more pragmatic approach to public health protection is discussed, as is the need for a more science-based approach to both public health and aquifer protection in times of increasing water shortages and high costs of water supply well replacement.

KEY WORDS

Total Coliform, Ground Water Rule, Chlorination, Contact Time, E. coli, Fecal Coliform, Virus, Klebsiella, Ground Water Wells, Biofilms, Well Disinfection, Well Cleaning and Rehabilitation, Mechanical Cleaning, Aquifer Recharge, Water Reclamation, Setback Distances, Soil Type Factors, Pathogen Transport, Ground Water Under Direct Influence of Surface Water, NSF Certification, Regulations, Health Risks, Economics

INTRODUCTION

Ground water supplies over a third of the U.S. drinking water supplies (EPA 2008). Although there are incidents of naturally occurring mineral contaminants, like arsenic or chrome, and manmade organic contaminants, like Nitrate, TCE, PCE, TPA, etc. Pathogenic microorganism contamination is considerably less likely than with surface water supplies.

Typically, ground water supplies, when there is adequate ground water availability, are far less expensive than surface water supplies, treatment and distribution. Advantages of ground water supplies include:

- Location Near Consumers
- Economical To Construct and Operate
- Minimal Treatment Or Surveillance Requirements
- Supply and Quality Less Effected By Weather Events, Drought Or Storms
- Less Subject To Aesthetic: Taste, Odor, Or Discoloration
- Less Environmental Impacts of Water Storage, River Flow, Fish and Wildlife Or Land Modification

There are also disadvantages, which include:

- Can Use More Energy For Pumping and Distribution

- Less Amenable To Water Quality Conditioning Or Improvements
- Increasingly Susceptible To Quality Deterioration By Land Use and Wastewater Disposal Practices
- Susceptible To Quality Deterioration By Agricultural Practices On Land Over Aquifer
- Typical Land Area and Access Limitations For Improved Treatment As Necessary
- Concern Of Water Supply Contamination By Aquifer Chemical Treatment Practices

There are significant costs to construct and develop well water supplies. Within the past few years in California, wells of 500 to 800 feet deep and of 1000-1500-gpm capacity cost from $750,000 to $1,000,000 to drill, provide well pump, connecting piping, emergency generator power supply, chlorination, and SCADA systems. Smaller wells of 50-150 gpm and 150 to 400 feet deep still can cost a third as much. Thus, for many smaller and medium size communities of less than 50,000 persons, these are considerable investments.

Periodically, well water facilities are typically replaced or rehabilitated more frequently than surface water facilities. Typical useful life of facilities for purposes of replacement (Div. Local Government Affairs 1979) provided to California water agencies are:

Ground Water Supplies

- Wells – 30 years
- Pumping Equipment – 25 years
- Emergency Power Equipment – 25 years
- Storage Tanks – 60 years
- Piping – 75 years
- Chlorination – 12 years

Surface Water Supplies

- Impounding Reservoirs – 100 years
- Canals – 100 years
- Intake – 50 years
- Supply Mains – 75 years
- Infiltration Galleries – 100 years

- Pumps – 50 years
- Water Treatment Plants
    - Structures – 40 years
    - Equipment – 30 years
    - Filters – 50 years

Overall, there is almost an expected double useful service life of surface water treatment facilities. However, when so often the initial capital and continuing operating costs are so favorable, ground water supplies are the choice of many municipal water agencies.

GROUND WATER DISINFECTION

The federal Ground Water Rule adopted in 2008 sets out the basic policy for ground water disinfection in the United States. It was a long and arduous process to achieve this rule as a part of the Safe Drinking Water Amendments of 1996, as there were many comments and opinions as to what could be agreed upon considering public health, economic, and technologically feasible considerations. Still this document provides considerable latitude to State and County regulators and regulatory agencies.

Its principal focus is the prevention of pathogenic microorganism contamination of the water supply and human illness. It cites that approximately a third of all ground water supplies have had or have detectable Total Coliform (TC). There are over 100 million people in the U.S. served by ground water supplies. In the ten-year period of 1990-2000, there were a total of 68 waterborne disease outbreaks affecting about 11,000 persons, or about 0.01 percent of the population. However, a much more conservative estimate of illness due to virus-contaminated ground water of 185,000 illnesses and three deaths per year were forecast, and by enactment of the provisions of this Rule, 42,000 illnesses and 0.7 deaths would be avoided, which is less than half. The discrepancy to the initial reported values are that so many illnesses are not reported.

The TC analysis is the initial analyses used to indicate possible fecal contamination of the water supply. Most of the organisms within the TC group are not pathogenic, but it has served for more than 100 years as an easily analyzed indicator organism of possible human pathogenic microorganisms. These are more precisely determined by analysis of fecal coliform, E. coli, or coliphage organisms if there are TC organisms initially detected. Even many organisms detected as fecal coliforms are not pathogens or cause illness. A much higher correlation is with E. coli, while coliphage relates more to potential enteric viruses. However, in recognition of the status of TC as only an indication of potential pathogenic organisms, the federal Total Coliform Rule allows as much as five percent of water distribution samples to be positive for this organism before additional measures to provide improved disinfection are required.

There is quite a different interpretation and requirement for ground water source detection and treatment requirement of the Ground Water Rule that prescribes four steps to be taken if there are positive tests:

1. A Sanitary Survey of potential contamination sources to be conducted every three years.
2. Source Water Monitoring of each individual well at a frequency based upon population served of TC, and if there is a positive within 24-hour period again for TC and further testing for E. coli or Enteric Virus. Then, if positive:
3. Treatment Technologies: Disinfection for 4-log virus reduction by chlorination and contact time "CT," membrane filtration, or other State-approved disinfection practice, could be U.V.
4. Compliance Monitoring

In practice, the Ground Water Rule is often more strictly applied by State and County regulators. Several recent incidents have arisen in California where regulators enforce considerably stricter requirements, and do not much consider either the corollary economic or practical requirements of the Ground Water Rule. These are:

- A city of 30,000 in the California Central Valley with one of six wells showing periodic TC
- A mobile home park serving 100 connections with one of two wells showing occasional coliform
- A water supply well of a college campus on the California Central Coast that showed TC following an earthquake after fifteen years' service.

These are described in the context of what occurred, the regulatory decisions, and actions taken to minimize public health risk.

CENTRAL VALLEY COMMUNITY WELL

Periodic Total Coliform (TC) were detected for several years in the early 1990's from this well in the spring after heavy rains beginning in 1990 after the well had been in operation for about five years. The well was 400 feet deep, had a capacity of 2000 gpm, was located within 100 feet of the first residential connection, and one of six in the community, with no other wells ever indicating any coliform. The well had a standard 50-foot sanitary seal, with a water table of about 80 feet below the surface.

Further testing of the well for fecal coliform and E. coli never found their occurrence. This well was tested weekly and then daily for several weeks after the initial positive test, which was an MPN usually of 5-10/100 ml. Typically, the well would be disinfected in accordance with AWWA C654 with a minimum of 50 mg/L of free chlorine and 12- to 24-hour exposure. Then the well would not have any TC in the water until the next season. The origin of the TC was never discerned as the soils were clayey for more than 100 feet deep to the first permeable aquifer. The well was in a sewered residential area. It was, however, surmised that the coliform were somehow in the organic matter of the aquifer, as there were reducing conditions. Total Organic Carbon concentrations of 2 to 4 mg/L and manganese concentrations of about

0.08 mg/L, which was being sequestered by addition of polyphosphate. At this time, none of the City's wells were being chlorinated, and there was no desire by the City at that time to chlorinate all of the well water supply.

After three or four episodes of springtime TC presence, the California Department of Public Health (CDPH) ordered the City to shut down the well. This well had cost over $800,000 to construct and was needed in the summer when there were large water demands in the City for landscape irrigation. It was proposed to the CDPH that the well water be chlorinated prior to distribution and then be dechlorinated to minimize breakdown of the polyphosphate manganese sequestering and possible residential paint staining and discoloration due to the elevated manganese concentrations.

This occurred prior to the Ground Water Rule, yet a credible chlorine dosage and contact time was proposed to the CDPH together with a size of a pressure contactor that could be constructed on a small available triangular site adjacent to a highway, a street, and a residence's fence.

Criteria for chlorine concentrations and contact time at various pH and temperature conditions for E. coli, various types of pathogenic enteric virus and protozoa had been printed in the *Handbook of Chlorination* (White 1986) that had plots from data presented by the *Journal of the American Water Works Association*(Baumann and Ludwig 1962).

A CT of 10 mg/L –minutes was proposed for a 1.5 mg/L free chlorine residual with 10 minutes of contact time to provide what was shown to be a 99.6-100 percent kill of Hepatitis virus at pH of 7.

A 12,000-gallon steel horizontal dis head pressure vessel, 10 feet in diameter by 20 feet long, divided into three baffle plate dividers with multiple small orifices to provide a near plug flow actual to theoretical contact efficiency of 80 percent.

Comparing that CT criterion to those criteria currently used for 4-log virus removal, the following CTs for free chlorine and chloramine residuals are relevant (MP-CWC HDR 1989):

	CT 4-log virus removal	
	At 10°C	At 15°C
Free Chlorine	6	4 mg/L – minutes
Chloramine	1491	994

This illustrates why it is so important to dose with chlorine sufficiently to provide a free chlorine residual.

The chlorine dosage was paced by flow and a free chlorine residual analyzer controller located at the exit of the contact tank. Polyphosphate was dosed before the chlorine into the well pump discharge pipe before the contactor to suppress any oxidation and precipitation of manganese. Then following the contactor and after the residual analyzer, sodium bisulfite solution is dosed at 1.1:1 ratio to free chlorine residual to provide what was then a zero chlorine residual concentration to deliver to the nearby residences. This system was approved by the CDPH and has been effective for more than twenty years although TC continue to periodically be present in the ground water supply.

Recently, however, this City installed a chlorine oxidation-anthracite green sand filtration system to remove manganese from this well and to provide a chlorine residual to all of the water wells of the City. There are also blends with supplemental treated surface water. Still, this well water has occasional TC detected.

There is now sufficient time in the filters to provide "CT" compliant with Ground Water Rule 4-log virus inactivation with filter discharge at 1 mg/L and no longer need for either the contactor or dechlorination chemical system.

MOBILE HOME PARK WELLS

A mobile home park serving 85 homes and a restaurant/bar is served by two wells of 60-gpm and 90-gpm by submersible pumps. The wells are 200 feet deep with static water levels of 13 to 18 feet deep and with typical 50 feet sanitary seals. They are located 200 to 500 feet upgradient of an aerated pond wastewater treatment facility, which receives approximately 12,000 gpd with disposal by combination of evaporation and percolation.

The soil at the park is very tight clay soil extending to a depth of ten to fifteen feet below the surface and with percolation rates of less than 1/8-inch per hour. The park is surrounded by irrigated agricultural land with crops of tomatoes, sweet corn, and alfalfa on three sides, and the San Joaquin River and its flood control levees to the west.

The ground water is moderately mineralized with Total Dissolved Solids of 600-650 mg/L, total hardness of about 350 mg/L $CaCO_3$, pH of about 7.5. The well water has high iron and manganese concentrations with iron as high as 3.5 mg/L in the East Well and manganese as high as 0.77 mg/L in the North Well. An iron and manganese filtration plant was constructed for the facility with hypochlorite addition to provide a 0.8 mg/L free chlorine residual after filtration into the distribution system.

TC samples of well water and of the distribution system are collected monthly, and these and other water supply distribution pressure chart data and chlorine residuals reported to the County Health Department.

Several times a year, the East Well will have a TC concentration of 4-20 MPN/100 ml, but upon repeat sampling a day later often none and only on one occasion an E. coli MPN of 4/100 ml.

The water treatment operator when this occurred, but without regulator approval, would dump about a pound of calcium hypochlorite crystals down into the well. When the County Health Department regulator heard of this practice, it was outlawed and ceased. Instead, it was insisted that the well be disinfected in accordance with AWWA standard practice, which could cost between $5,000 to $10,000. Typically, over 100 mg/L of hypochlorite was used for this periodic treatment, but then the well would be flushed before use.

The water treatment practice is to chlorinate the well water with dosages of chlorine of 2 to 5 mg/L. The minimum contact time through the treatment plant is 7 minutes, which provides a CT of over 5.6 mg/L –minute, and assure that the 4-log virus removal of the Ground Water Rule is being achieved.

It was found that an amperometric chlorine residual analyzer was too difficult for this facility to maintain in calibration. Instead a colorimetric analyzer is simpler and more reliable.

There is also an alarm signal to the park manager facilities to inform if chlorine residual becomes lower than 0.4 mg/L. In that case, a standby hypochlorite feed pump is placed into service.

Overall, it has been necessary to devote considerable expense to assure that there is sufficient disinfection to achieve the Ground Water Rule is being achieved, despite never precisely finding why there were occasional TC. However, it is quite likely that these TCs are naturally occurring bioslimes in the well water or attached to the well screens as described by Schnieders 2001. These organisms are facultative anaerobes, non-pathogenic, naturally occur particularly in waters that show reducing conditions by elevated concentration of iron and manganese.

These well pumps are intermittently operated and do not have initial start pump to waste features. Instead, upon pump starts and initiation of flow bioslime containing TC are quite likely dislodged. It has been estimated that there may be more than a million times the TC concentration in a bioslime than in the plantonic bacteria found in the water and detected by the laboratory (Schnieders 2003). However, there remains a dichotomy between the regulator prohibition of periodic down hole chlorination as compared to the much more expensive AWWA standard disinfection condition.

CAMPUS WELL

The primary water supply of a large college campus in California is a well that provides 1200 gpm of water from a 650-foot-deep, 16-inch-diameter casing well. There is also a backup system connection for a City surface and ground water supply, but the cost of that water is about four times more than the ground water supply, and it is chloraminated while the well water is chlorinated to a free chlorine residual of about 0.8 mg/L at the well, and above 0.3 mg/L at any far away building in the campus water distribution system. The mixture of chlorinated and chloraminated supplies often will cause breakpoint conditions, tastes, and odors.

This well had effectively served the campus for fifteen years and there had never been any positive TC found in the well water and at two dozen locations in the campus distribution system conducted weekly. An earthquake with a magnitude of 5.6 on the Richter scale occurred on 30 October 2007 with the epicenter about six miles away. Then on the following week of testing on 6 November, a positive TC was detected in the well water but none in the distribution water samples.

The CDPH was immediately contacted about the incident. The well was shut off and the alternative municipal supply turned on to supply the campus until the difficulty could be resolved.

Following the initial detection of coliform, a water well contractor was retained and a series of well water disinfections in accordance with AWWA standards were conducted. These included five disinfection trials of increasing dosage and detention time.

(1)	11/7/2007	50 mg/L chlorine, supply well; 12 hours flushed
	11/8/2007	Well tested TC positive
(2)	11/13/2007	50 mg/L chlorine, 50-gpm well, 24-hour retention
	11/15/2007	Well tested TC positive
(3)	11/26/2007	100 mg/L of chlorine, surge, 48-hour retention
	11/29/2007	Well tested TC positive
(4)	12/2/2007	100 mg/L of chlorine, surge, 72-hour retention, 24-hour pump to waste
	12/6/2007	Well tested TC positive
(5)	12/8/2007	200 mg/L of chlorine, 96-hour retention, 24-hour pump to waste
	12/14/2007	Well tested TC positive

All of the TC analyses were by 12-hour colilert test method.

It was then decided to remove the pump, video the well casing, and determine condition, which was done on 14 December 2007. There was some evident corrosion on the well screen, but extensive quantities of bioslime growth as well. Samples of the well water were obtained and sent to a laboratory specializing in well water difficulties. A summary of their findings is shown on Table 1, and although they did not find in either the casing sample taken a few seconds after initial startup or of the aquifer sample after an hour of pumping. There was considerable indication that much more rigorous well cleaning and disinfection than the AWWA standard would be required to eliminate TC presence in the well water.

Other options presented to the campus were chlorination to 4-log virus removal in conformance with the Ground Water Rule, construction of a new well, or well rehabilitation.

The estimated costs of these alternatives were:

(1) Ground Water Rule Compliance – 4-log virus reduction by improved chlorination, a contactor, residual analyzer, and automation of rating for supplemental supply if insufficient residual - $500,000

(2) Replacement well - $1,000,000

(3) Well cleanings and rehabilitations - $100,000 - $200,000.

The well cleaning and rehabilitation alternative was chosen if approved by the CDPH.

TABLE 1
Campus Well Water Quality Results

CHEMICAL			
Characteristic	**Units**	**Casing**	**Aquifer**
pH	-	7.5	7.5
TDS	mg/l	625	599
Total Hardness	mg/l $CaCO_3$	344	368
Total Alkalinity	mg/l $CaCO_3$	276	296
Langelier Index	LI	+0.09	+0.4
Iron	mg/l	ND	ND
Manganese	mg/l	ND	ND
Nitrate	mg/l	4.7	4.7
Phosphate	mg/l	0.3	0.3
TOC	mg/l	0.03	0.0
ORP	MV	74	101
BACTERIOLOGICAL			
Characteristic	**Units**	**Casing**	**Aquifer**
Plate Count	colonies/ml	25	3
Anaerobic Growth	%	10	10
Sulfate Reducing Bacteria	%	ND	ND
Fe and Mn Oxidizing Bacteria	%	ND	ND
ATP Initial(1)	cells/ml	304,000	23,000
ATP 24-hour	cells/ml	944,000	20,000
Total Coliform	Presence	ND	ND
E. coli Coliform	Presence	ND	ND
Pseudomonas mendocina	Presence	Present	Present
Achromobactic xylosxidons(2)	Presence	Present	Present
Gallionella	Presence	Trace	Trace

NOTES
(1) Adenosine tripolyphosphate – A measurement of biological activity
(2) A possible pathogenic bacteria

The well rehabilitation would follow methods of mechanical scraping of the well casing and screen, and cleaning of deposits, high concentration chlorination surging at low pH with scale removing additions by techniques described by Lennox in the AWWARF publication, "Application of Well Condition Assessment and Rehabilitation Techniques" (2006), Schnieder's "Chemical Cleaning Disinfection and Decontamination of Water Wells" (2003), and Cullimore's *Practical Manual of Groundwater Microbiology* (2008).

The CDPH regulators said they could approve of a water well rehabilitation if all of the chemicals used were NSF certified. Also, in their experience, TCs were not frequently found in California well waters. However, one other well water system on the Central Coast had reported TC following the 30 October earthquake. The proposed rehabilitation treatment included:

- Remove pump and column piping
- Inspect sanitary seal and upper well integrity
- Utilize a stiff nylon brush to mechanically scrub bioslime and other deposits from casing and screen
- Chemical treatment by phosphorus acids at a 7 percent concentration together with a polymer chemical disposal at 25 percent concentration. This mixture at 1.5 times the casing volume 10 is surged for 20 minutes, through each 20-foot section of the well, leave standing overnight at pH less than three, then pump to waste.
- Neutralize water in well to original pH 7.5 with soda ash and then chlorinate at 300 mg/L of chlorine at a pH between 6 and 6.5 to assure a predominance of hypochlorous acid. Surging this mixture at one minute per foot of casing depth for a day, then the next day surging at half that rate.

  Following the chlorination and surging, the permanent pump and column were reinstalled after spraying all parts with a 200 mg/L chlorine solution. Then the well would be reactivated and tested by a schedule requested by the CDPH to include after initial startup TC samples.

  > 5, 15, 30, and 60 minutes; 6, 24, and 48 hours, then if all samples indicted an absence of TC, it would be considered satisfactory and could be restored to service.

A contract was prepared and competitive bids for the well restoration were received from several prospective contractors with both cost and experience considerations for selection. A contract for $190,000 well rehabilitation plus $30,000 for chemicals was awarded. The work proceeded subsequently through the summer of 2008, was successful in that no TC were detected, and the well returned to service by September when classes began again at the campus.

The satisfactory service and negative TC analysis made each week has continued now for nearly five years. The question remains as to why TC occurs, the differences between enteric pathogens and non-pathogenic coliform bacteria, what are physical impacts, and when there is increasing emphasis on wastewater recycling by irrigation, indirect potable water reuse, aquifer recharge and recovery, and are regulators uninformed or lack experience in assessing conditions, sometimes over cautious, ignoring economic and practical considerations.

TC IN WELL WATER

TC issues are highly relevant to ground water supplies in a nation and world of increasing population, water shortage, and increased water reuse, and are discussed as follows:

- TC occurrence in ground water is common and frequently occurs for non-pathogenic bacteria.

- The increasing use of pass/fail TC testing rather than enumeration by MPN testing has led to an apparent but maybe not actual indication of increasing coliform concentrations in ground water (Schnieder 2001).

- Type of soils to a depth of twenty feet are important, i.e., more porous sand or gravel vs. clay tight soil

  Karst

  Fractured Rock

  Inadequate sanitary seals

  Leaky sewer lines

- Environmental Factors

  Temperature

  Rainfall

  Nutrient availability

  Earthquakes

  Flow surges

- Ground Water Under Direct Influence of Surface Water (Vasconcellos 1992)

  This was a criterion of microscopic particle analyses (MPA) for determination if ground water was under the direct influence of surface water that was not formally adopted by the EPA, yet provide a very good basis for contamination assessment of springs, infiltration galleries and water wells close to stream courses.

- Setback Distances for Potable Water Wells (Asano 1997)

  - Streams, Lakes 25 feet

  - Septic Tanks/Leach Fields 50 feet

    - Reclaimed Water - Irrigation – Public Access 100 feet
    - Reclaimed Water – Irrigation – Unrestricted Access 300 feet
    - Aquifer Recharge 500 feet
    - Indirect Potable Reuse 2000 feet
- Chlorine Effectiveness for TC Reduction
    - Free chlorine – 200-500 mg/L well disinfection
    - >500 mg/L less effective

CONCLUSIONS

The occurrence of TC in ground water is rather common. These are usually non-pathogenic soil bacteria types that occur in biofilms, which attach to soil particles,, well gravel packing, casings, and screens. Intermittent pumping, earthquakes or high rainfall events often dislodge some of the bioslime and then the presence of TC.

Often regulators will require disinfection of wells by AWWA Standard Techniques. Often this is insufficient to correct the TC difficulties. More rigorous mechanical scrubbing techniques, the use of acids, and dispersants are needed to restore a well to a condition that it does not persistently show TC.

The Ground Water Rule is formulated to provide a 4-log enteric virus reduction. Free chlorine is much more effective than chloramine to achieve a 4-log reduction. If a chloramine residual is needed in the distribution system for THM reduction or better biofilm control, then the ammonia should be added after initial chlorine contact.

Ground water supplies and treatment are usually much less costly than treatment and distribution of surface water supplies. However, with increasing emphasis on wastewater reuse and aquifer recharge, there is likely more rather than less opportunity for TC detection. The challenge will be to provide effective, affordable, and practical control of TC in well water.

REFERENCES

Asano, T (1998). *Wastewater Reclamation and Reuse*, Technomic, Lancaster, Pennsylvania.

Asano, T. (1995). *Artificial Recharge of Groundwater*, Butterworth, Boston, Massachusetts.

AWWA (2005). *Disinfection of Wells – Standard C654-06*, AWWA, Denver, Colorado.

Baumann, R.E. and Ludwig, D.D. (1962). "Free Available Chlorine Residuals for Small Non-Public Water Supplies," *JAWWA*, 59, 1379.

Cullimore, D.R. (2008). *Practical Manual of Groundwater Microbiology, 2nd Edition,* CRC, Boca Raton, Florida.

Derlon, N., et al (2010). "Stratification in the Cohesion of Biofilms Grown Under Various Environmental Conditions," *Water Research,* V. 42, pp. 2102-2110.

DeZuane (1997). *Handbook of Drinking Water Quality, 2nd Edition*, Wiley, New York.

EPA (1989). "Total Coliform Rule," *Federal Register*, June 29, 1989, Washington, D.C.

EPA (2006). "Ground Water Rule," *Federal Register*, November 8, 2006, Washington, D.C.

Hoff, J.C. (1985). "Inactivation of Microbial Agents by Chemical Disinfectants," EPA 60012 80/67, USEPA Water Engineering Research Laboratory, Cincinnati, Ohio.

LeChevallier, M.W. (1990). "Coliform Regrowth in Drinking Water – A Review," *JAWWA,* V. 82, 4 11, pp. 74-86.

LeChevallier, M.W., et al (1990). "Disinfecting Biofilms in a Model Distribution System," *JAWWA*, V. 12, 7, pp. 87-91.

Lennox, J.B. (2006). "Application of Well Condition Assessment and Rehabilitation Techniques," AWWARF, Denver, Colorado.

Malcolm Pirnie and CWCHDR (1989). "Guidance Manual for Compliance with the Filtration and Disinfection Requirements for Public Water Systems Using Surface Water Sources," USEPA, Washington, D.C.

McFeters, G.A., et al (1974). "Comparative Survival of Indicator Bacteria and Enteric Pathogens in Well Water," *Applied Microbiology*, V. 27, pp. 283-289.

Schnieders, J.H. (2001). "Coliforms and Disinfection of of Water Wells," *Water Well Journal*, October, pp. 12-16.

Schnieders, J.H. (2003). "Chemical Clearing Disinfection and Decontamination of Water Wells," Johnson Screens Inc., St. Paul, Minnesota.

Vasconcellos, J. and Harris, S. (1992). "Consensus Method for Determining Groundwater Under Direct Influence of Surface Water Using Microscopic Particulate Analysis (MPA)," EPA 91019-92-029, Manchester Environmental Laboratory, Port Orchard, Washington.

White, G.C. (1986). *The Handbook of Chlorination, 2nd Edition,* Van Nostrand Reinhold, New York.

Photochemical Activation of Free Chlorine to Reactive Oxygen Species for Enhanced Inactivation of Chlorine-Resistant Microorganisms

Jenna E. Forsyth[1], Quanxin Mao[1], John S. Meschke[2], and Michael C. Dodd[1,2,*]

[1]University of Washington, Department of Civil and Environmental Engineering
[2]University of Washington, Department of Environmental and Occupational Health Sciences
*Author to whom correspondence should be addressed, at: doddm@uw.edu

ABSTRACT

Recent findings by our group indicate that inactivation of chlorine-recalcitrant microorganisms may be substantially improved by harnessing sunlight-driven photolysis of free chlorine to yield various reactive oxygen species (ROS), including hydroxyl radical ($HO^{\bullet}$), atomic oxygen ($O(^3P)$), and ozone (O_3), during conventional chlorination. For example, our results indicate that a >2-fold increase in inactivation of chlorine-recalcitrant *Bacillus subtilis* spores can be achieved at pH 8 and 25° C in solutions dosed with 3-7 mg/L as Cl_2, following exposure to simulated sunlight for as little as 12 minutes. No enhancement of inactivation rates was observed in the presence of 50-mM *t*-BuOH (a $HO^{\bullet}$ and $O(^3P)$ scavenger), and no inactivation was observed under simulated sunlight alone, highlighting the key role of photochemically-generated ROS. Furthermore, the process was confirmed to function effectively under natural sunlight, as well as in the presence of DOC and alkalinity in natural water samples obtained from regional drinking water utilities.

KEYWORDS

Chlorine, photolysis, sunlight, reactive oxygen species, disinfection, *Bacillus subtilis*, spores

INTRODUCTION

Free available chlorine (FAC – typically comprising the sum of hypochlorous acid, HOCl and hypochlorite, OCl^-, in aqueous solution) remains the most widely used disinfectant in drinking water practice worldwide (e.g., its use was practiced at over 97% of drinking water facilities within the United States as of 2000 (2000)). It is cheap, easily utilized, and in general highly effective for achieving the inactivation of a wide variety of waterborne microbial pathogens. However, when used alone, it is known to have only moderate to marginal effectiveness as a disinfectant of such important pathogenic organisms as *Cryptosporidium parvum*, *Mycobacterium avium*, and *Giardia lamblia* at dosing levels practical for drinking water treatment. This has led to widespread adoption of more effective, but often substantially more capital-, equipment-, and energy-intensive alternative disinfectants such as UV light and ozone (O_3). However, inactivation of chlorine-recalcitrant microorganisms may be achievable at considerably lower expense by utilizing the UVA range ($300 < \lambda < 400$ nm) of sunlight to photolyze FAC residuals to reactive oxygen species such as hydroxyl radical ($HO^{\bullet}$) and atomic oxygen ($O(^3P)$) during conventional chlorination processes, according to the following pathways (where Φ = quantum yield, noting that Φ for eq 1a was obtained in the gas phase):

$$HOCl + h\nu \rightarrow HO^{\bullet} + Cl^{\bullet} \quad (\Phi_{310nm} = 1 \text{ (Molina et al., 1980)}) \quad (1a)$$
$$OCl^- + h\nu \rightarrow O^{\bullet -} + Cl^{\bullet} \quad (\Phi_{313nm} = 0.127,\ \Phi_{365nm} = 0.08 \text{ (Buxton and Subhani, 1972)}) \quad (1b)$$

$$O^{\bullet-} + H_2O \leftrightarrows HO^{\bullet} + OH^- \quad (2)$$
$$Cl^{\bullet} + H_2O \leftrightarrows HOCl^{\bullet-} + H^+ \quad (3)$$
$$HOCl^{\bullet-} \leftrightarrows HO^{\bullet} + Cl^- \quad (4)$$
$$OCl^- + h\nu \rightarrow Cl^- + O(^3P) \quad (\Phi_{313nm} = 0.075, \Phi_{365nm} = 0.28 \text{ (Buxton and Subhani, 1972)}) \quad (5)$$
$$O(^3P) + O_2 \rightarrow O_3 \quad (6)$$

As noted in eqs 1a and 1b, HOCl and OCl- are homolytically cleaved to form $HO^{\bullet}$ and $Cl^{\bullet}$ upon exposure to light with wavelengths between 300 and 400 nm. At conditions typical of natural freshwater resources (i.e., low Cl- concentrations, circumneutral pH), the $Cl^{\bullet}$ generated in each reaction will also equilibrate with H_2O to yield additional $HO^{\bullet}$ (eqs 3 and 4 (Yu, 2004)). By comparison, heterolytic cleavage of OCl^- under the same conditions yields ground-state atomic oxygen, $O(^3P)$, as per eq 5. The generated $O(^3P)$ can then react rapidly ($k = 4 \times 10^9\ M^{-1}s^{-1}$ (Klaening et al., 1984)) with O_2 under oxic conditions to form the highly-effective disinfectant O_3, as per eq 6.

Substantial research has highlighted the possible benefits of combining two or more disinfectants in sequence to yield enhanced pathogen inactivation beyond that achievable by an individual disinfectant alone. For example, pre-disinfection with O_3 has been shown to substantially enhance the susceptibility of microbial agents (including *C. parvum*) to inactivation by FAC (Cho et al., 2006; Rennecker et al., 2000). Additional studies indicate that the use of $HO^{\bullet}$ as a pre-oxidant could also enhance sensitivity of model microbial agents to subsequent inactivation by FAC (Cho et al., 2011; Cho and Yoon, 2006). In this context, it is quite notable that sunlight-driven FAC photolysis can result in the simultaneous exposure of microorganisms to FAC, $HO^{\bullet}$, *and* O_3, in turn providing a means of achieving the same benefits observed for sequential disinfectant application, but in a single step. The current investigation has been undertaken to evaluate the potential for applying this promising strategy to achieve enhanced inactivation of chlorine-resistant microorganisms during drinking water treatment, using *B. subtilis* spores as model microbial agents.

METHODOLOGY

Materials. FAC stocks were prepared by dilution of commercially-available NaOCl solutions of 7% or higher concentration (Fisher Scientific; Waltham, MA; USA), and standardized spectrophotometrically at 292 nm, using a molar extinction coefficient of 350 $M^{-1}cm^{-1}$ for OCl^- (Kumar et al., 1986). Organic chemical reagents were obtained from Sigma-Aldrich (St. Louis, MO; USA), and were of reagent-grade purity or higher. All reagent stocks and growth media utilized in disinfection experiments or for culturing or enumerating microbial agents were filter-sterilized or autoclaved as appropriate. Freeze-dried *B. subtilis* (ATCC 6633) was obtained from the American Type Culture Collection (Manassas, VA; USA). Natural water samples were obtained from two Seattle-area drinking water treatment plants, WTP 1 (pH 7.4, DOC = 0.6 mg C/L, and alkalinity = 16.3 mg/L as $CaCO_3$), and WTP 2 (pH 7.4, DOC = 1.2 mg C/L, and alkalinity = 3.5 mg/L as $CaCO_3$).

B. subtilis **Spore Stock Preparation and Viability Assay.** Freeze-dried *B. subtilis* cultures were revived in nutrient broth and incubated at 37° C under constant agitation for approximately 30 hours. Following centrifugation, rinsing in PBS, and resuspension in sterile water, vegetative cell suspensions were plated on AK#2 sporulation agar and incubated at 37° C for 5-7 days. Spores were then recovered from the agar and purified by repeated centrifugation and PBS wash steps, followed by heat treatment at 70° C for 15 minutes. Spore viability was quantified by spot-titer culturing on nutrient agar plates (Beck et al., 2009).

Disinfection Experiments. Ten-mM phosphate-buffered solutions contained within 60-mL crystallization dishes were dosed with 1×10^4 to 1×10^5 CFU/mL of *B. subtilis* spores, in addition to 1 μM of *p*-chlorobenzoic acid (pCBA) or cinnamic acid. pCBA was utilized as a FAC-recalcitrant, UVA-stabile probe for semi-quantitatively monitoring ROS exposure during irradiation of FAC containing solutions, as it is known to react rapidly with $HO^{\bullet}$ (Rosenfeldt and Linden, 2007), and can also be expected to react rapidly with $O(^3P)$ on the basis of reported reactivities of $O(^3P)$ toward other aromatic molecules (Bucher and Scaiano, 1994). Cinnamic acid, which forms benzaldehyde in stoichiometric yield upon reaction with O_3 (Leitzke et al., 2001), was utilized as a selective probe for O_3. The spore and ROS-containing solutions were then treated with FAC alone (dark controls), light alone (FAC controls), or combined light and FAC. A Newport-Oriel Model 66924 arc lamp source (Oriel Instruments; Stratford, CT; USA) equipped with a 450-W Xe arc lamp, focusing collimator, atmospheric attenuation filter (with a cutoff of 300 nm), and dichroic mirror was used to irradiate samples with simulated sunlight in the wavelength range of 300-400 nm, at 25 °C. Samples were collected from each solution at pre-defined times for measurement of (a) spore viability, (b) pCBA or benzaldehyde concentrations, and (d) residual FAC concentration.

RESULTS

Effects of FAC Photolysis on Spore Inactivation. A series of experiments was undertaken according to the protocol described above to examine effects of varying simulated sunlight irradiation times on enhancement of dark inactivation rates in systems dosed with the same initial FAC concentration. All samples were initially dosed with $[FAC]_0 = 3.5$ mg/L as Cl_2. Inactivation in a dark control series was then compared to inactivation in samples exposed to t = 3.75-, 15-, 30-minute, and continuous periods of irradiation with simulated sunlight generated by the Newport system described above. An additional experiment was undertaken with continuous irradiation of a matched sample dosed with [*t*-BuOH] = 50 mM. Figure 1 summarizes key findings from these experiments, with ordinate data plotted vs. FAC *Ct* values in mg/L*min (where *Ct* was determined as $\int_0^t [FAC]dt$, using measurements of residual [FAC]) to normalize for variability in [FAC] concentrations versus time.

ROS Probe Compound Measurements. pCBA present in FAC-containing solutions exposed to simulated and natural sunlight was observed to degrade according to pseudo-first-order kinetics. In addition, the formation of benzaldehyde during FAC photolysis was confirmed in solutions dosed with cinnamic acid (data not shown).

Brief Irradiation to a Target FAC Residual. Figure 2 summarizes findings from a separate series of experiments in which a solution initially dosed with $[FAC]_0 = 6.5$ mg/L as Cl_2 was irradiated with simulated sunlight for a time sufficient to lower [FAC] to the same level as in a dark control dosed with $[FAC]_0 = 3.25$ mg/L as Cl_2, following which inactivation rates in the irradiated solutions were monitored under dark conditions.

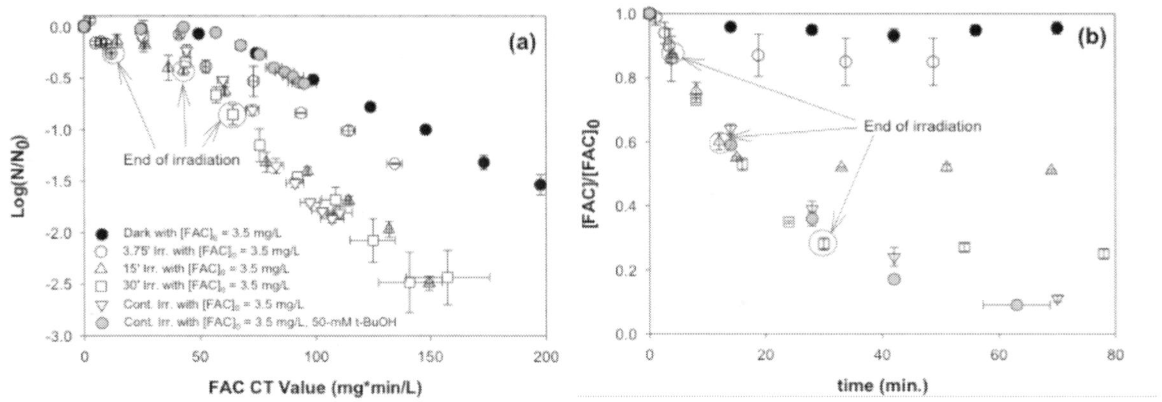

Figure 1. (a) Observed inactivation of *B. subtilis* 6633 spores in pH 8, 10-mM phosphate-buffered solutions dosed with $[FAC]_0 = 3.5$ mg/L as Cl_2 at 25° C, under the influence of varying times of irradiation by simulated sunlight (generated by the Newport system described above), where N is viable *B. subtilis* colony density in CFU/mL, and $N_0 = 5 \times 10^4$ CFU/mL. In samples irradiated for discrete times (as identified in the figure legend), irradiation was carried out for the pre-determined time, after which the lamp was turned off and spore inactivation followed under dark conditions. (b) Residual [FAC] in the same solutions.

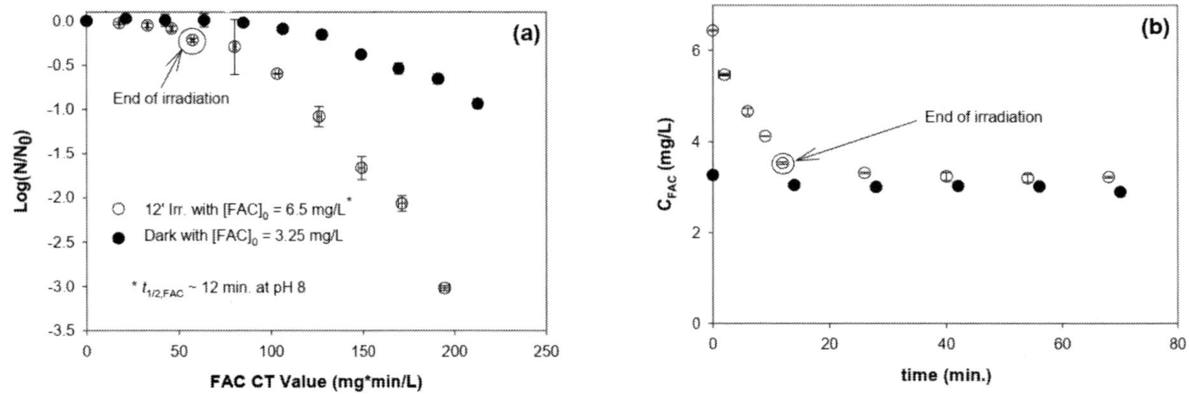

Figure 2. (a) Observed inactivation of *B. subtilis* 6633 spores at 25° C in pH 8, 10-mM phosphate-buffered solutions dosed with $[FAC]_0 = 6.5$ mg/L as Cl_2 and irradiated with simulated sunlight for 12 minutes, compared to a dark control dosed with $[FAC]_0 = 3.25$ mg/L as Cl_2. The former solution was irradiated for a time corresponding to $t_{1/2,FAC}$ under these conditions, in order to bring [FAC] to the same level as in the dark control, after which the lamp was turned off, and spore inactivation followed under dark conditions. N is viable *B. subtilis* colony density in CFU/mL, and $N_0 = 5 \times 10^4$ CFU/mL. (b) Residual [FAC] in the same solutions.

Effects of pH, Temperature, Light Source, Matrix, and $[FAC]_0$. The degree to which inactivation of spores is enhanced through sunlight-driven activation of FAC can be quantified by examining the percent reduction in Ct values required to achieve a given degree of inactivation during exposure to FAC under irradiated conditions, relative to exposure to FAC dark conditions. Table 1 summarizes percent reductions in Ct_{99} values (i.e., Ct values necessary to achieve 2-log inactivation of *B. subtilis* spores) observed during treatment of *B. subtilis* spores under a variety of conditions by the brief irradiation approach represented in Figure 2, including under natural sunlight and in two natural water samples. Enhancement effects were found to be independent of starting FAC concentration after normalization for Ct (data not shown).

Table 1. Summary of percent reduction in Ct_{99} values achievable by solar irradiation of FAC-containing samples under various conditions. Irradiated samples were initially dosed with twice the FAC concentration applied to dark controls, exposed to light for times sufficient to lower [FAC] to the same levels as in dark controls, and allowed to react further under dark conditions (as described above for the data shown in Figure 2). All values are given as ± 1 SE. WTP = Water Treatment Plant.

Light Source	Matrix	Temp. (°C)	pH	% Reduction in Ct_{99} under irradiated conditions
Simulated Sunlight	10-mM phosphate buffer	25	8	54.5 (±0.7)
			7	33.3 (±0.1)
			6	13.8 (±0.1)
Simulated sunlight	10-mM phosphate buffer	10	8	55.6 (±0.1)
			7	42.2 (±0.1)
			6	15.3 (±0.1)
Natural sunlight	10-mM phosphate buffer	33	8	53.5 (±0.1)
Simulated sunlight	WTP 1	10	7.4	38.7 (±0.1)
Simulated sunlight	WTP 2	10	7.4	28.6 (±0.1)

DISCUSSION

As apparent from the data in Figure 1a, a measurable increase in inactivation rate of chlorine-recalcitrant *B. subtilis* spores could be achieved at pH 8 and 25° C in solutions dosed with FAC, following exposure to simulated sunlight for as little as 3.75 minutes. The degree of enhancement also appeared to increase with increasing irradiation period up to a maximum under continuous irradiation, as evident from the inactivation rates for samples irradiated for $t \geq 15$ minutes. These data indicate that the generation of ROS during FAC photolysis enhances inactivation by both shortening lag phase, and by increasing the rate with which spores are inactivated during treatment. This suggests that pre-treatment with ROS in the presence of FAC results in degradation of some stoichiometric quantity of chlorine-recalcitrant biomolecules within the spore coat or cortex, which in turn exposes vital components of the spores to attack by FAC, in analogy to effects reported for pre-treatment of FAC-recalcitrant microorganisms by O_3

and ClO_2 (Cho et al., 2006; Rennecker et al., 2000). Furthermore, no enhancement of inactivation rates was observed in the presence of 50-mM *t*-BuOH ($HO^{\bullet}$ and $O(^3P)$ scavenger), and no inactivation was observed under simulated sunlight alone (data not shown). These observations provide strong evidence for the role of ROS in enhancing inactivation of *B. subtilis* spores during exposure of FAC-containing solutions to simulated sunlight.

Observations of pCBA loss and benzaldehyde formation confirm the continuous generation of ROS ($HO^{\bullet}$ and/or $O(^3P)$) and O_3 during solar irradiation of FAC-containing solutions under the conditions investigated here, and provide further support for the role of such species in enhancing inactivation of *B. subtilis* spores. In fact, estimates of O_3 yields generated in experiments conducted during continuous irradiation of a pH 8 solution with $[FAC]_0 = 8$ mg/L as Cl_2 indicate O_3 *Ct* values of 0.3-0.4 mg/L*min, which is on the order of practically-applied levels reported as sufficient to sensitize *C. parvum* to inactivation by FAC (Rennecker et al., 2000)

The scenario depicted in Figure 2 represents a potentially more practical means of implementing photochemical-enhancement of chlorination when maintenance of a target FAC residual post-irradiation is desired. Using this approach, a water supply can be dosed with an excess of FAC beyond that desired for maintenance of residual, irradiated for a time sufficient to photolyze the FAC excess (exposing microorganisms to ROS in the process), and allowed to react further in the dark at the desired target FAC residual in order to achieve enhanced inactivation compared to a non-irradiated system. The results summarized in Figure 2 for this approach also provide a direct comparison between photochemically-enhanced and dark inactivation rates for the same [FAC] value, illustrating that the rate of *B. subtilis* spore inactivation is markedly accelerated even after FAC-photolysis (and hence, exposure to ROS) ceases. This is consistent with the hypothesis that damage caused initially by ROS renders the spores more susceptible to attack by FAC.

The data summarized in Table 1 illustrate that the degree of enhancement of *B. subtilis* spore inactivation (as measured by % reduction in Ct_{99}) increases dramatically with increasing pH. This is likely due in part to the increased importance of heterolytic FAC photolysis (via eqs 5 and 6) over homolytic FAC photolysis (eqs 1-4). Under oxic conditions (such as utilized here), FAC heterolysis would be expected to lead to increased yields of O_3, which is a more selective and stable disinfectant than $HO^{\bullet}$, and thus likely to accumulate in solution to levels sufficient to yield greater enhancement than achievable by $HO^{\bullet}$ alone. Temperature, in contrast, appears to have relatively little effect on the degree of enhancement achieved during FAC photolysis at any of the pH values investigated. Substantial % reductions in Ct_{99} values were observed under natural sunlight and in the two natural water samples, WTP 1 and WTP 2. These latter results are of particular note, as they confirm the effectiveness of natural sunlight in driving this process, and illustrate that significantly enhanced inactivation can be achieved even in the presence of representative concentrations of DOC (which can lower ROS yields by scavenging and light screening) and alkalinity (which can also act as a ROS scavenger).

CONCLUSIONS

The work reported here confirms the utility of sunlight-driven photochemical activation of FAC for enhancing the inactivation of chlorine-resistant *B. subtilis* spores in aqueous solution. Future work will be directed toward evaluating the potential for enhanced inactivation of moderately to highly chlorine-resistant human pathogens (e.g., *C. parvum*, *M. avium*, Coxsackievirus B5). Such an approach could have substantial benefits over existing alternatives to chlorine-based disinfection processes. First, FAC is used in the vast majority water treatment facilities. Second, utilization of solar radiation as a light source could net significant cost and energy savings in comparison to processes utilizing O_3 or artificial UV light. Third, this approach could be implemented in a single treatment step, in contrast with existing approaches to combined disinfectant applications. Perhaps even more importantly than for the case of centralized municipal water treatment, photochemically-enhanced chlorination could also offer a reliable, low cost means of ensuring disinfection of chlorine-resistant pathogens during point-of-use applications, on account of its low costs, ease of implementation, and low materials requirements.

REFERENCES

Community Water System Survey. Volume II: Detailed Tables and Survey Methodology. United States Environmental Protection Agency: 2000.

Beck, N.K., Callahan, K., Nappier, S.P., Kim, H., Sobsey, M.D. and Meschke, J.S. (2009) Development of a spot-titer culture assay for quantifying bacteria and viral indicators. *J. Rapid Methods Autom. Microbiol.* 17(4), 455-464.

Bucher, G. and Scaiano, J.C. (1994) Laser flash photolysis of pyridine *N*-oxide: Kinetic studies of atomic oxygen $O(^3P)$ in solution. *J. Phys. Chem.* 98(48), 12471-12473.

Buxton, G.V. and Subhani, M.S. (1972) Radiation-chemistry and photochemistry of oxychlorine ions. 2. Photodecomposition of aqueous-solutions of hypochlorite ions. *J. Chem. Soc., Faraday Trans. 1* 68, 958-969.

Cho, M., Gandhi, V., Hwang, T.M., Lee, S. and Kim, J.H. (2011) Investigating synergism during sequential inactivation of MS2 phage and *Bacillus subtilis* spores with UV/H_2O_2 followed by free chlorine. *Water Res.* 45(3), 1063-1070.

Cho, M., Kim, J.H. and Yoon, J. (2006) Investigating synergism during sequential inactivation of *Bacillus subtilis* spores with several disinfectants. *Water Res.* 40(15), 2911-2920.

Cho, M. and Yoon, J. (2006) Enhanced bactericidal effect of O_3/H_2O_2 followed by Cl_2. *Ozone-Science & Engineering* 28(5), 335-340.

Klaening, U.K., Sehested, K. and Wolff, T. (1984) Ozone formation in laser flash photolysis of oxoacids and oxoanions of chlorine and bromine. *J. Chem. Soc., Faraday Trans. 1* 80, 2969-2979.

Kumar, K., Day, R.A. and Margerum, D.W. (1986) Atom-transfer redox kinetics: General-acid-assisted oxidation of iodide by chloramines and hypochlorite. *Inorg. Chem.* 25, 4344-4350.

Leitzke, A., Reisz, E., Flyunt, R. and von Sonntag, C. (2001) The reactions of ozone with cinnamic acids: formation and decay of 2-hydroperoxy-2-hydroxyacetic acid. *Journal of the Chemical Society, Perkin Transactions 2* (5), 793-797.

Molina, M.J., Ishiwata, T. and Molina, L.T. (1980) Production of $^{\bullet}OH$ from photolysis of HOCl at 307-309 nm. *J. Phys. Chem.* 84(8), 821-826.

Rennecker, J.L., Driedger, A.M., Rubin, S.A. and Marinas, B.J. (2000) Synergy in sequential inactivation of *Cryptosporidium parvum* with ozone/free chlorine and ozone/monochloramine. *Water Res.* 34(17), 4121-4130.

Rosenfeldt, E.J. and Linden, K.G. (2007) The $R_{\bullet OH,UV}$ concept to characterize and the model UV/H_2O_2 process in natural waters. *Environ. Sci. Technol.* 41(7), 2548-2553.

Yu, X.Y. (2004) Critical evaluation of rate constants and equilibrium constants of hydrogen peroxide photolysis in acidic aqueous solutions containing chloride ions. *J. Phys. Chem. Ref. Data* 33(3), 747-763.

Anticipating Challenges Associated with In-Building Disinfection for Control of Opportunistic Pathogens in Premise Plumbing

William J. Rhoads [a], Amy Pruden [a], Marc A. Edwards [a]

[a] Civil and Environmental Engineering, Virginia Tech, 418 Durham Hall, Blacksburg, VA, 24060

ABSTRACT

Opportunistic pathogens (e.g., *Legionella* and others) in premise (i.e., building) plumbing (OPPPs) are now the primary cause of waterborne disease in the United States and in many developed countries. A large body of experience has been gathered regarding possible engineering approaches that can influence the incidence of OPPP-related waterborne disease. These approaches include interventions implemented by water utilities and individual building owners. In order to reduce the risk of waterborne disease, building owners turn to regulations, standards, and guidelines for recommendations on management of their water systems. The new American Society of Heating, Refrigerating and Air-Conditioning Engineers (ASHRAE) standard for control of *Legionella*, ASHRAE 188, emphasizes use of thermal disinfection and dosing of disinfectants in buildings. This paper presents a review of in-building disinfection approaches, emphasizing their advantages and disadvantages related to their efficacy for controlling OPPP growth, potential interactions with premise plumbing material that may affect their efficacy, and ambiguities in implementation.

KEY WORDS

In-building disinfection, thermal disinfection, pipe scaling, waterborne pathogens, ASHRAE 188

INTRODUCTION

A key challenge of water treatment is to ensure that safe potable water produced by utilities will not damage water distribution system infrastructure. It has recently been recognized that the biological and chemical quality of water delivered to buildings often deteriorates within the building's premise plumbing system. Premise plumbing is defined as the portion of the distribution network on the building owner's side of the property line and not completely controlled by utilities. The National Research Council (NRC) has identified premise plumbing as a high priority area for research (NRC, 2006). While the direct focus of most research is to protect public health, any changes must carefully balance corrosion control, scaling, operation/maintenance, energy conservation, scalding, and other considerations along with public health goals (NRC, 2006; Brazeau and Edwards, 2012) given that privately owned plumbing systems represent an asset valued on the order of a trillion dollars (Edwards, 2004),.

Opportunistic premise plumbing pathogens (OPPPs) are now the leading cause of water borne disease in developed countries (CDC, 2011). Representative OPPPs include *Legionella* spp. (with emphasis on *L. pneumophila*), *Mycobacterium avium* complex (MAC), *Pseudomonas aeruginosa*, *Acanthamoeba* spp., and *Nagleria fowleri*. Growth of OPPPs generally occurs in the bulk water and biofilms within premise plumbing, although cooling towers and other systems can also be very important sources of these pathogens. Attempts to control OPPPs can include secondary disinfection applied by utilities, but OPPPs can also be controlled by treatment within buildings, as is evident in historical treatment to controls within hospitals (e.g., Stout et al., 2007; Srinivasan et al., 2003). Even if the utility employs aggressive levels of disinfectant residuals to control OPPPs, the strategy can be undermined either passively (e.g., normal chlorine decay) or actively (e.g., granular activated carbon treatment that removes chlorine) in buildings. The emerging shared responsibility for control of this public health concern complicates implementation of effective controls, and knowledge regarding their efficacy and secondary impacts on premise plumbing infrastructure is currently incomplete.

Premise plumbing systems can sometimes have low or no secondary chlorine residuals due to long water retention times and/or a high tendency of premise plumbing pipes to form corrosion scale. Corrosion scale and premise plumbing materials consume chlorine (Nguyen et al., 2011; 2012; Zhang et al., 2009). The rate of decay is dependent on various factors, including water chemistry, disinfectant method, plumbing system design, plumbing system operation, and consumer water use patterns. In some situations there is likely little the water utility can do to cost-effectively control OPPPs for the entire range of situations encountered within individual buildings.

In-building disinfection approaches that have demonstrated effectiveness include both remedial (i.e., one-time treatments) and continuous disinfection regimes. Remedial techniques include: 1) thermal and 2) chlorine shock. Continuous treatments include: 1) chlorine, 2) chloramine, 3) chlorine dioxide, 4) ozone, 5) copper-silver ionization, 6) ultra-violet light irradiation, 7) peroxide, 8) thermal control, and 9) proprietary controls. Field and case studies have demonstrated, mostly in hospital settings, that these methods can be effective for some OPPPs in some situations, but for reasons not well understood they are less effective in other situations. To begin to address knowledge gaps, a recent report synthesized the state of the knowledge of OPPPs and their control, summarizing several factors required for pathogen growth, routes of exposure, and rates of disease occurrence (Pruden et al., 2012). The inherent complexity and variability in premise plumbing systems often makes it difficult to predict, diagnose, and remediate outbreaks of waterborne disease originating in premise plumbing.

This review paper is focused on Legionnaires' disease (LD) because it is the most common waterborne disease in the United States (CDC, 2011). *L. pneumophila* is the main causal agent of LD and is illustrative of other OPPPs. Many strategies controlling LD outbreaks and mitigating *L. pneumophila* growth are expected to impact growth of other OPPPs as well.

GUIDELINES FOR CONTROL OF LEGIONNELLA

Several government and private entities have developed guidelines to follow in the event of a LD outbreak. The most prominent include recommendations of the World Health Organization (WHO) (Bartram, 2007), the Florida Department of Health (Florida DoH), Association of Water Technologies (AWT, 2003), the Allegheny County Health Department (ACHD, 1997), the Occupational Safety and Health Administration (OSHA, 1999) and the American Society for Heating, Refrigeration, and Air-Conditioning Engineers (ASHRAE, 2000; 2012). The focus of these guidelines, until recently, has been on remedial actions to be taken after an outbreak occurs. These recommendations often cover basic disinfection procedures, requirements for epidemiological studies, and how to maintain a safe plumbing system after disinfection. There has been little focus on guidelines as preventative measures and none recommend monitoring of target organisms as a long-term preventative or follow-up procedure.

A reluctance to recommend monitoring could be due, in part, to the variation in sampling procedures and sample analysis methods across institutions and research labs. A recently conducted expert workshop identified the difficulty of comparing results between methods and techniques as a research gap for detection and quantification of OPPPs (Pruden et al., 2012). Although there are standard sampling and analysis protocols for culture methods for *Legionella* (Barbaree, 1987), these are fraught with limitations and inconsistencies in results. While molecular-based methods are promising, consensus is still needed in their application. Therefore, while some of the sampling guidelines exclusively support culture methods as an important method of assessing *Legionella* risk, more quantitative measurements would be beneficial to further knowledge about dose-response relationships related to disease caused by OPPPs. In general, microbiological sampling for *Legionella* is ambiguous because presence according to culture-based detection of *Legionella* does not necessarily predict the occurrence of disease (Stout et al., 2007). While some guidelines refer to 30% of total samples positive for *L. pneumophila* as a trigger for remedial measures, the guidelines do not recommend where to take samples or how many should be taken. In addition, the accuracy and utility of the 30% action level is questionable. For example, in one study the majority of taps in a building (83%) were positive for *L. pneumophila* serotype 1, but were not associated with disease, while a low number of positive taps (30%) resulted in cases of LD (Stout et al., 2007). Therefore, there is a need to better understand sampling methods, relative to prevention of OPPP outbreaks.

ASHRAE Standard 188 is the most recent and first legally enforceable standard dealing with control of *Legionella*. The basic requirement of ASHRAE 188 is a written Hazard Analysis Critical Control Point (HACCP) plan that addresses all aspects of prevention of, response to, and long-term control of *Legionella* outbreaks. It uses the concept of the HACCP plan to require building owners and operators to have preventative and reactive measures in place before a outbreak occurs.

Overall, ASHRAE 188 is a three-tier standard (Figure 1). First, if there are no risk factors associated with a building, then a simple yearly survey is conducted to ensure none have developed.

Second, rather than collecting samples and monitoring *Legionella*, it identifies risk factors for outbreaks (Figure 1). If a building meets any risk factors, there must be a HACCP plan in place that meets all the HACCP plan requirements. These include: 1) Conducting a hazard analysis, 2) Determining the critical controls points, 3) Setting critical control limits for the critical control point, 4) Establishing a system to monitor control of the critical control points, 5) Establishing actions to be taken when monitoring results indicate critical control limits have been violated, 6) Establishing procedures to confirm the HACCP plan is working effectively, and 7) Establishing documentation for all procedures and records appropriate to the principles and application of the HACCP plan. In order to address water-specific requirements of the HACCP plan, the standard refers to ASHRAE Guideline 12 (2000) to prevent LD. That guideline, in turn, promotes the use of in-building disinfection systems for long-term prevention of LD.

The last tier of ASHRAE 188 is specific to buildings with cooling towers and/or evaporative condensers for the HVAC system as these systems have been linked with LD outbreaks (Cordes et al., 1980; Dondero et al., 1980a; 1980b; Nguyen et al., 2006). Details of this requirement of the standard are beyond the scope of this paper.

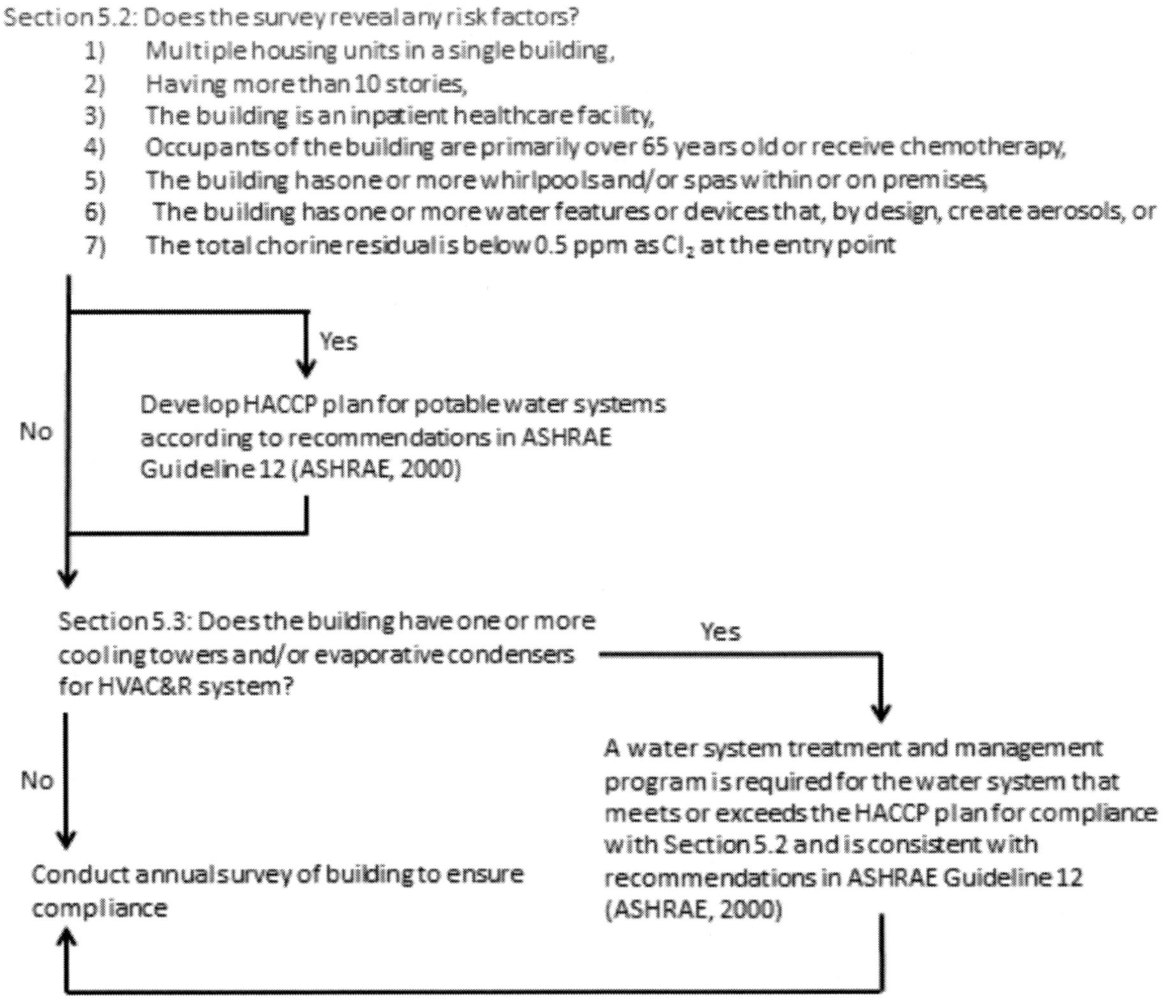

Figure 1. ASHRAE Standard 188 flow diagram of requirements.

Although ASHRAE 188 is a step toward assigning enforceable responsibility to preventing disease outbreaks, it is in some ways imperfect and unrealistic. For instance, one risk factor the standard identifies is having an influent chlorine residual below 0.5 mg/L as Cl_2. While it seems easy to measure the influent chlorine residual as a quick and simple check, in practice the residual will vary markedly with system water demand and season. In some systems there may be a residual present only after extensive flushing in the summer, but could be present at all times during the winter. It is unclear how this is addressed in the standard In addition, although the standard refers to a comprehensive guideline in preventing LD outbreaks (ASHRAE, 2000), neither the standard nor the guideline provide even a basic framework for determining which remedial action(s) or continuous disinfection systems are optimal for different water qualities, system design features, or intended uses. While it is important to maintain safe drinking water, some of the in-building techniques suggested in *Legionella* control guidelines and peer-reviewed literature are lacking basic research dealing with how different water chemistries or different premise plumbing materials affect disinfection efficacy. In general, complex premise plumbing networks may create unintended consequences for treatments that are not thoroughly understood.

This paper summarizes remedial (i.e., one-time thermal or chlorine shock) and long-term (i.e. continuous) disinfectant methods, with a focus on factors influencing site dependency, compatibility with plumbing systems, and knowledge gaps. While most guidelines provide the basic pros and cons of viable treatment options, at present there is no logical framework for holistic decision making. Only WHO includes sections with details on identifying and monitoring control measures based on the intended use of potable water in specific applications (Bartram, 2007). For example, a hospital intensive care unit has a much different risk associated with consumer exposure than does a cruise ship residential cabin. When choosing a remedial disinfection method, the intended use and consumers should play role in the decision-making process.

REVIEW OF REMEDIATION-BASED TECHNIQUES

The goal of remediation is to disinfect a system that is known or suspected to be contaminated with pathogenic bacteria, often in response to an outbreak of waterborne disease. Two categories of remediation include thermal and chlorine shock treatments.

Thermal Shock. For hot water systems, thermal shock is commonly the first choice because temperature inactivation is proven effective. Although guidelines vary slightly as to which temperature should be used and the duration of exposure (Table 1), thermal shock generally targets attaining water temperatures above 60°C for 20-30 minutes while flushing taps, or achieving a temperatures greater than 80°C at least momentarily at all points in the plumbing system. It can be logistically challenging to maintain high enough water temperatures throughout the entire plumbing network to be effective in large systems (Muraca et al., 1990). Plastic pipes used in hot water systems such as chlorinated polyvinyl chloride (CPVC) pipes are rated to withstand temperatures up to 82°C (180°F) for continuous use (e.g. Harvel, Gerog Fischer Piping Systems, Easton, PA); however, there has been no research on the effects of continuous or repeated heat exposure on physical integrity or leaching from other types of plastic pipes including PEX and HDPE. In additions, none of the guidelines that recommend thermal shock treatment for *Legionella* remediation warn building owners or operators against the potential for rapid scaling, or the precipitation of calcium carbonate [$CaCO_3$] catalyzed by elevated temperatures in hard waters, a factor discussed later in this paper.

Scalding potential must be taken into consideration. It has been estimated that 3,800 injuries and 34 deaths occur from tap water scald burns (CPSC, 2005). Young children are especially at risk for scald injuries (NSKC, 2004). Where applicable, mixing values that blend hot and cold water

Table 1. Summary of Thermal-based Guidelines for Control of Legionellosis

Entity	Shock Recommendation	Long-term Recommendation	Caution Against Scalding	Caution Against Pipe Damage in Long-Term Recommendations (e.g. scaling)	Recommend Repeated Thermal Shock
Florida DOH	Heat HWHs to 71-77°C for 24 hr, then flush each tap for 5-20 minutes	Set HWHs to 60°C; drain periodically to clean; hot-water recirculation pumps run continuously	Yes	No; allude to systems where thermal shock is not possible	Yes; when systems cannot be set at 60°C because of scalding (retrofit with mixing valves not possible)
OSHA	Heat HWHs to 70°C for 24 hrs; flush each tap for 20 minutes	Set HWHs to 60°C with minimum delivery temperature of 50°C; drain periodically to clean; hot-water recirculation pumps run continuously	Yes	No	No
WHO	No temperature or exposure time recommended for pasteurization; Periodic flushing with waters 50-60°C	Maintain all water temperatures > 50°C	Yes	No	No
ASHRAE	Periodically raising HWHs to 66°C followed by flushing	Store water at 60°C; Maintain minimum return temperature 51°C; Where not practical maintain all water temperature above 49°C	Yes	No	Yes; when long-term temperature maintenance not possible
AWT	Heat HWHs >60°C; preferably 66°C; flush for up to 30 minutes	Has no long-term recommendations, but states in review that at 50°C, there is 90% kill with 2 hour exposure	Yes	No	No

Florida DoH – Florida Department of Health; OSHA – Occupational Safety and Health Administrations; WHO – World Health Organization; ASHRAE – American Society for Heating, Refrigeration, and Air-Conditioning Engineers; AWT – Association of Water Technologies; HWH-hot water heaters can be used to prevent scalding during treatment. Otherwise, special means to inform occupants of the disinfection should be taken to avoid unnecessary scalding. Although this technique can be effective for controlling *Legionella* (Kim et al., 2002; Muraca et al., 1987; Lin et al., 1998b), recent work has demonstrated the potential for *Legionella* regrowth, presumably due to necrotrophic consumption of dead cell biomass resulting from thermal shock (Temmerman et al., 2006).

Chlorine Shock. Dosing a high concentration of free chlorine is another common remediation strategy. The recommended concentration for disinfection varies as well. The Florida DoH recommends that 20-50 mg/L free chlorine as Cl_2 be maintained in hot water tanks with a minimum concentration of 2 mg/L Cl_2 in the plumbing system for two hours. OSHA recommends 10 mg/L Cl_2 in hot water systems (OSHA, 1999). Others do not recommend a specific exposure time or target concentration, but require that there is no more than 30% reduction of the original concentration over the test period (AWT, 2003; ACHD, 1997).

There is ambiguity about protocols for "passing" shock chlorination recommendations listed above in terms of how and when to take residual measurements. For instance, new copper pipes in some waters will not pass the shock treatment requirements (BOCA, 1997; Edwards et al, 2011; International Code Council, 2000) due to rapid consumption of the chlorine by the pipe wall (Edwards et al., 2011; Nguyen et al., 2011). Repeated doses may not satisfy the demand and there are also concerns about initiating non-uniform pitting corrosion (Rushing & Edwards, 2004; Cong & Scully, 2010; Sarver et al., 2011). To combat this effect, pH may be maintained at lower levels (< pH 8) or a corrosion inhibitor may be introduced to reduce copper corrosion, adding further complexity to the method. High density polyethylene (HDPE) pipes subjected to repeated shock chlorination can damage the surface of the pipes, leading to reduced overall lifetime (Whelton and Dietrich, 2008; Whelton et al., 2011).

REVIEW OF CONTINUOUS DISINFECTION PRACTICES

After an outbreak has occurred and remedial treatments have been implemented, or if a plumbing system is considered conducive to pathogen growth, continuous in-building disinfection practices may be considered. These include 1) chlorine, 2) chloramine, 3) chlorine dioxide, 4) ozone, 5) copper-silver ionization, 6) ultra-violet light irradiation, 7) peroxide, 8) thermal control, and 9) proprietary controls. For some of these systems, factors mediating efficacy towards target pathogens and interaction with plumbing are unclear, especially in relation to application in different water qualities and across different premise plumbing materials.

There may also be regulatory and legal implications to installing additional disinfection practices for in-building applications. For example, the USEPA indicates that any building serving more than 25 individuals dosing a disinfectant is considered a public utility, triggering onerous drinking water standards and associated monitoring requirements. This makes installation of a secondary disinfection system considerably less attractive, or may encourage "unofficial" building-level disinfection systems.

High variability in case study designs, intervention and control measures taken, and application of successive disinfectant patterns makes cross comparison of studies and disinfection methods difficult. Most field studies have been conducted in hospitals or other special care facilities, and application to smaller private homes or small apartment buildings with less maintenance staff are not widely available. The following summarizes literature on each disinfection method, with a focus on implications for premise plumbing.

Chlorine and Chloramine. While there are over 190 years of experience using chlorine and chloramine as disinfectants (USEPA, 2011), it has only recently been recognized that these disinfectants can sometimes disappear relatively quickly via reactions within premise plumbing systems even when they are stable in the main distribution network (Zhang et al., 2009; Zhang and Edwards, 2009; Nguyen et al., 2011; 2012). Although both chlorine and chloramine are widely-used for utility distribution system applications (Seidel et al., 2005), dosing of these disinfectants within buildings may have maintenance and corrosion challenges that exceed the capabilities of many building staff. Unfortunately, there is not a large body of research on the application of chlorine and chloramine at the building-level, and it has only recently been recognized that higher levels of these disinfectants can sometimes cause severe pitting of copper pipe (Sarver et al., 2011; Rushing and Edwards, 2004; Marshall, 2004; Lytle and Schock, 2008).

Chlorine is a strong oxidant that can react with many common plumbing materials including lead, PEX, copper, and brass but apparently not PVC (Sarver et al., 2011; Nguyen et al., 2012; Whelton et al., 2011; Edwards and Dudi, 2004). It can be dosed at the building-level as a powder or liquid targeting concentrations of 1-2 mg/L free chlorine (Muraca et al., 1990). It is well known that certain water chemistries can accelerate or inhibit corrosivity of chlorine towards specific building plumbing materials. For instance, maintenance of a lower pH, presence of natural organic matter, dosing a corrosion inhibitor such as poly- or ortho-phosphate, or natural silica can be effective in limiting the rate of chlorine induced pitting of copper tubing (Rushing & Edwards, 2004; Sarver et al., 2011; Lytle and Schock, 2008). Chlorine can also benefit some copper pipes by converting soluble cupric hydroxide [$Cu(OH)_2$] to tenorite [CuO], reducing aqueous copper and chlorine concentrations markedly (Edwards et al, 1996; Edwards et al., 1999; Hidmi & Edwards, 1999; Lagos et al., 2001; Patterson et al., 1991; Schock et al., 1995; Nguyen et al., 2011). In some waters without disinfectant residuals the conversion of $Cu(OH)_2$ to tenorite or other stable species never occurs, resulting in very persistent blue water until at least a trace of chlorine residual is present (Edwards et al., 2000).

While chloramine has been used at some US utilities since the 1930s (USEPA, 2011), many utilities switched to chloramine as a secondary disinfect in response to the Disinfection Byproduct Rule passed in 2003 (Seidel et al., 2005). Target levels of chloramine (chlorine plus ammonia) are up to 4 mg/L as Cl_2 (USEPA, 2009). Chloramine only reacts weakly with natural organic matter to form disinfection byproducts such as trihalomethanes and for this reason it tends to be more persistent than chlorine in the main distribution system (Neden et al., 1992). However, in premise plumbing with longer stagnation times, greater surface area to volume ratios, more reactive materials, higher levels of microbes, and warmer waters, chloramines can decay rapidly (Nguyen et al., 2012). Nitrifying bacteria, in particular, can cause rapid decay of chloramine by via formation of nitrite in premise plumbing (Nguyen et al., 2012; Zhang et al., 2009; Zhang and Edwards, 2009).

Chlorine Dioxide. For chlorine dioxide there is a body of literature indicating that it is sometimes effective in hospital applications (Stout and Yu, 2003). However, even relative to the scarce field data on chloramine and chlorine as in-building disinfectants, there is even less research that examines the effects of water chemistry and pipe material on its efficacy, corrosivity or stability at distal taps. Each of these factors is important in understanding how chlorine dioxide may be utilized most effectively for OPPP control in a given premise plumbing system. For in-building applications, chlorine dioxide gas is mechanically or electrochemically generated from a sodium chlorite solution. Commercial kits (e.g. Halox Technologies Corporation, San Antonio, Texas) have been developed for small to medium applications and are relatively easy to install and run. Chlorine dioxide is injected based on flow rate to a final concentration not to exceed USEPAs 0.8 mg/L ClO_2 maximum level (USEPA, 2009). Special consideration must be taken with the formation of chlorite in certain hospitalized patients (Bartram, 2007). Although limited work was done in this area, animal and human studies have indicated that individuals susceptible to oxidative hymolysis (e.g., immunocompromised individuals whose red blood cells are G6PD deficient), can become anemic when using water treated with chlorine dioxide (Moore et al., 1978; Calabrese, 1978). Some systems using chlorine dioxide have failed to observe these adverse effects (Smith and Willhite, 1990; Ames and Stratton, 1987). However, when exposure to individuals at risk is likely, including anemic young children, and fetuses of pregnant women, the residual must be removed by activated carbon or other treatments (Srinivasan et al., 2003). Although it sometimes takes several months and continuous application to be effective, it is claimed to completely eliminate *Legionella* spp. even at residuals below USEPA maximum levels (Sidari et al., 2004). At these levels, there is no significant increase in corrosion rates (Srinvasan et al., 2003) and they can be effective for infrequently used branches of the distribution systems (Thomas et al., 2004). One field application study showed that over 17 months, distal taps colonized by *Legionella* were effectively decontaminated from 45% positive (no in-building control) taps to 4% (with chlorine dioxide) positive (Srinvasan et al., 2003). However, there is decreased persistence of the disinfectant residual with higher temperatures, increasing total organic carbon (TOC) concentrations (Zhang et al, 2006 as cited in Zhang et al., 2008), and iron and copper corrosion scale (Zhang et al., 2008).

Although chlorine dioxide is relatively non-reactive with to TOC, reactions with goethite, magnetite, and cuprite would theoretically consume more chlorine dioxide (Zhang et al., 2008). In addition to potentially rapid abiotic decay, chlorine dioxide is not effective in eliminating amoeba hosts in biofilm growth in cool (35°C) water systems (Thomas et al., 2004). Planktonic *Legionella* was removed in this study, but rapid regrowth occurred in the absence of continual treatment, hypothesized to be due to the host capabilities of amoeba for *Legionella* (Thomas et al., 2004). Some researchers have reported a resistant subpopulation of *L. pneumophila* and *E. coli* in controlled batch experiments (Berg et al., 1988), yet no follow up work has been done to confirm to confirm these observations.

Ozone. Ozone [O_3] is generated from oxygen [O_2] and dosed based on flow rate to a residual of 1-2 mg/L. It is a very potent biocide and has the capability to inactivate viruses (Katzenelson et al., 1974). It has to be produced on-site due to its extremely short half-life. As a result, there is an extra expense in the initial setup and retrofitting systems may encounter space limitations for equipment. The systems are technically challenging to maintain and can be very expensive

(Muraca et al., 1987). Although ozone requires a short contact time, no disinfectant residual beyond the point of treatment is present. The efficacy is not generally affected by temperature or pH changes in bench-scale studies (Domingue et al., 1988), but some research has shown that ozone reacts more rapidly and less discriminately at higher pHs (>7) (Hoigne and Bader, 1976; Peyton et al., 1982). Unlike chlorine dioxide, ozone is capable of reducing established biofilms at 0.5 mg/L (Thomas et al., 2004), but controlled studies on the corrosivity of ozone is lacking. It has been shown that there can be 99% reduction of *L. pneumophila* with only a 5 minute reaction period in laboratory, batch-reactor studies (Domingue et al., 1988), but long term efficacy is unknown (Muraca et al., 1987).

Copper-Silver Ionization. Copper-silver biocides lyse pathogenic bacteria and protozoa and denature their proteins (Friedman and Dugan, 1968; Bitton and Feihofer, 1978; Slawson et al,. 1990; Lin et al, 1998a). The recommended dose of the ions is 0.2-0.4 mg/L copper and 0.02-0.08 mg/L silver, depending on water quality (Cachafeiro et al., 2007). In some systems, these levels of copper can be maintained naturally by corrosion and dissolution from existing copper pipes. The WHO recommends levels up to 2 mg/L copper (WHO, 2006a) and 0.1 mg/L silver (WHO, 2006b). This concentration of copper exceeds USEPA's action level in cold water, but because hot water is not intended for human consumption this may not have the same health implications. However, the concentrations that would be required to be effective in certain water qualities may conflict with goals of low copper in sewage influent and effluent (Cachafeiro et al., 2007; Boulay and Edwards, 2000). In addition, cooper and silver levels above 0.02-0.04 mg/L may cause discoloration of water or plumbing devices in some waters (Stout et al., 2003). Basic, lab-scale research on its effectiveness is also lacking.

Liu et al. (1998) conducted a case study with a building with an ionization system versus a control building with no ionization system. Initially, the test and control buildings had 50% and 20% taps positive for *Legionella*, respectively, before the installation. After four weeks of treatment, 0% of the taps in the test building were colonized and remained at 0% until week 22, six weeks after copper-silver injection had been stopped. The control building fluctuated between 20%-100% positive taps throughout the study, with a mean of 83%. Alternatively, Blanc et al. (2005) observed that before installation of a copper-silver ionization system, 90% of taps in a test building were colonized with *Legionella* while 66% taps in a control building were colonized. After installation and deployment of the ionization system, 93% and 56% of the test and control buildings, respectively, were positive, indicating that the approach was not effective in this water. Copper silver ionization is most effective when concentrations can continually be monitored and adjusted (Bartram, 2007), requiring special equipment and expertise. Although there is no one optimum concentration of copper and silver ions, there is a synergistic effect when they are used at the upper range of the allowable concentrations (Lin et al., 2002).

There is also some evidence that copper-silver ionization is not able to penetrate established biofilms (Thomas et al., 2004) or can be consumed by the plumbing (Loret et al., 2005). Knowledge, such as how water quality parameters affect the efficacy of the treatment, would be helpful when evaluating treatment options. It is likely that pH, alkalinity, phosphate and other water constituents affect efficacy for controlling OPPPs, as has been observed for other premise plumbing microorganisms, such as nitrifying bacteria (Zhang et al., 2009). Copper ions can also cause severe deposition corrosion of galvanized or steel plumbing systems (Kentworthy, 1943;

Cruse, 1971), and silver ions might also attack copper pipes and other metals by the same mechanism (Clark et al., 2011).

Ultraviolet Light. Ultraviolet (UV) light irradiation at 254 nm damages DNA and disrupts its replication, preventing growth of pathogens. For in-building use, there are many commercial systems that range from 3.7 L/min to 1900 L/min that require only basic plumbing skills and an electrical outlet, with a wide price range for the initial setup. There is minimal maintenance involved, including replacing the UV bulbs approximately annually and cleaning the quartz sleeve that the water flows through quarterly. The efficacy of UV treatment is dependent on low turbidity water and temperature fluctuations are known to alter its efficiency of the treatment (Muraca et al., 1990). There is the opportunity for breakthrough of microbes shielded by small particles or for amoeba present in the protective cyst stage of their life-cycle. Like ozonation, there is no disinfectant residual thus there is no inactivation of microbes that may proliferate downstream of the treatment and no biofilm disinfection. On the basis that necrotrophic growth occurs after short-term thermal disinfection, generation of dead cells might be expected to be problematic with UV treatment. If downstream growth is of concern, UV may be most effective in combination with other treatment options (Bartram, 2007).

Thermal Disinfection. Perhaps the simplest continuous disinfection method to employ is thermal control. In general, it is thought that bacterial growth is limited in cold water systems below 20°C (Bartram, 2007, AHSRAE, 2000, Florida DOH) and in hot water there is nearly immediate eradication of bacteria above 60°C (Muraca et al., 1987). The USEPA currently recommends a hot water heater set point of 48°C for domestic systems to reduce the risk of scalding consumers and for the assumed energy savings of cooler water temperature (USEPA, 2004). While 48°C is less likely to scald consumers or foul plumbing systems with scale, it is also less likely to reduce risk of exposure to heat-resistant pathogens (NRC, 2006; Brazeau and Edwards, 2012). In general, there is a major need for practical research aimed at aiding basic decision- and policy-making logic in hot water infrastructure. While certain localities have adopted a tax-credit system for the installation and specific operation of certain hot water systems for in-home and –building applications (Brazeau and Edwards, 2012), there may be long-term, unintended negative consequences associated with these across-the-board recommendations. The new ASHRAE 188 standard for control of *Legionella* growth recommends maintaining water tanks above 60°C and 51°C throughout the entire water network.

Overall Costs. Few economic evaluations or total life cycle cost estimates have been conducted on the various methods of disinfection. Murcaca et al. (1987), included a comparative cost table (Table 2) in a broad review of the disinfection methods. This analysis, however, is limited in scope to application in a hospital setting with a fixed number of beds and water heaters. The costs of each method have a wide range and are very general in consideration. A more precise and up-to-date price comparison that accounts for variation in plumbing materials, implementation, and type of building would be useful. More cross-comparisons of in-building disinfection types are needed with specific focus on the implications and interactions the treatment technique has on the premise plumbing system.

Table 2. Representative cost estimates from one case study (data from Muraca et al, 1987) and status of knowledge regarding corrosion control and scaling.

Disinfection Method	Start Up Costs ($)	Annual Maintenance Costs ($)	Scaling Implications	Corrosion Implications
Ozonation	35,000-60,000	6,000-9,000	No	Higher DO
Hyperchlorination	30,000-45,000	5,000-8,000	No	Cl_2 is corrosive
Ultraviolet Light	18,000-35,000	1,000-2,000	No	No
Instantaneous Thermal	15,000-30,000	2,000-4,000	Yes	Plastics degradation concern
Metal Ionization	20,000-35,000	1,500-4,000	Might alter type of scale	Deposition corrosion of Ag, Cu
Continuous Thermal	3,000-5,000	500-800	Yes	Could protect pipe or create non-uniform corrosion

*Normalized for a 500-bed hospital with 13 floors and a recirculating hot water distribution system with two how water tanks

CRITICAL EVALUATION OF THERMAL DISINFECTION METHODS

While the USEPA's recommendation of 48°C in water heaters may not be aggressive enough to reduce the risk for certain populations, such as immunocompromised consumers, ASHRAE's recommendation to require maintenance of 51°C throughout the system may also be unrealistic. For instance, in most premise plumbing systems, flow will remain stagnant long enough for the water in the pipes to cool to ambient room temperature, often very quickly. In order to meet this requirement, the building owners would need to have continuously recirculating systems in place. Some municipalities are mandating these systems for their purported energy/water efficiency of offering tax credits for their installation (MCWD, 2005). Recirculating systems, however, can consume more energy than traditional systems. Thus, recirculation to meet AHSRAE recommendations may sometimes conflict with energy conservation goals (Brazeau and Edwards, 2013). Even with recirculation in most systems, water to distal taps (e.g. from the recirculating loop to a shower head – Figure 2) may cool to room temperature. Thus, maintaining water temperatures above 51°C throughout systems may not be practical or possible.

Thermal treatment may also prove ineffective in eliminating some pathogens in certain scenarios. Optimal growth for *Legionella* is between 32-42°C (Yee & Wasowsky, 1982). Evidence of growth has been observed at 51.6°C (Kusnetsov et al., 1996) and viable *Legionella* have been isolated from water with temperatures up to 66°C (Dennis et al., 1984). In addition, pathogenic *Legionella* have been shown to survive successive thermal (70°C for 30 minutes) and chemical (hydrogen peroxide) treatments, with the overall biofilm mass intact and the microbial community only temporarily affected (Farhat et al., 2012). Certain bacterial pathogens also have the distinct capability of surviving and replicating within an amoeba host organism, e.g., such as *Legionella* inside *Hartmonella vermiformis*. This relationship can aid in the survival of the pathogenic bacteria by shielding it from treatment (Greub and Raoult, 2004).

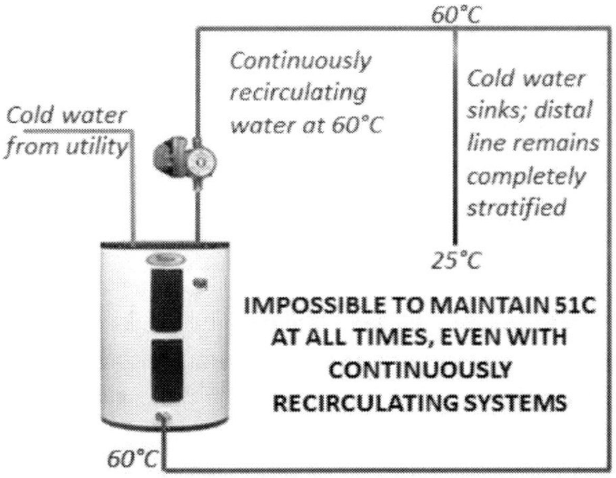

Figure 2. Practical scenario demonstrating limitation to maintaining 51°C in continuously recirculating hot water systems.

Increasing the water temperature in premise plumbing also increases the propensity of certain water chemistries to rapidly precipitate scale such as calcium carbonate or magnesium silicate. This could require increased energy demand in heating and recirculating the water and cause permanent damage to plumbing systems (Paul et al., 2010). Researchers at the Water Quality Research Council (WQRC) and New Mexico State University found that there was a 30% increase in energy demand associated with hard water after just 14 days in a gas water heater (Isaacs and Stockton, 1981). The Langelier Saturation Index (LSI), while widely discredited for use in determining the relative corrosivity of water and estimating the leaching potential of lead, zinc, and copper from pipes and other premise plumbing materials (AWWARF, 1996; Edwards et al, 1996; Schock, 1999), is very useful in predicting the potential for certain waters to precipitate scale (Langelier, 1936; Elfil and Hannachi, 2006). The LSI is a calculation based on the calcium ion concentration, pH, and alkalinity of the water in question. It compares the measured pH of the water to a theoretically calculated saturation pH (Equations 1 and 2).

(Eq. 1) $$LSI = pH - pH_s$$

(Eq. 2) $$pH_s = pK'_2 - pK'_s - \log[Ca^{2+}] - \log[Alkalinity]$$

Where pH is the measured pH of the solution, pH_s is the saturation pH, pK'_2 is the equilibrium constant at a given ionic strength for the second carbonate equilibrium dissociation constant, and pK'_s is the solubility product for calcium carbonate. While there are alternatives to the LSI (Ryznar, 1944; Stiff and Davis, 1952; Hissel and Salensos, 2002) and even variations on it (Elfil & Hannachi, 2006), each has its own limitation. Two main limitations of calcium carbonate saturation indices for pressurized potable water include (Elfil & Hannachi, 2006): 1) only considering calcite thermodynamics and not less soluble forms of calcium carbonate such as aragonite or monohydrate calcium carbonate; 2) the lack of thermodynamic data to incorporate other solids into the models. Despite these downfalls, the LSI is still commonly used, with the following interpretation:

LSI > 0; $CaCO_3$ has a tendency to precipitate

LSI < 0; $CaCO_3$ does not have the tendency to precipitate.

It is important for building owners and operators to take the propensity for the water to scale into account before deciding to employ thermal disinfection to control pathogen growth (Figure 3). In some circumstances, it may possible to develop scaling inhibition or softening strategies to avoid scaling. If the water is subject to scaling, and inhibitors are not viable, the other disinfection methods should be considered (Figure 4).

It is important to note that the LSI predicts nothing about the kinetics of the reactions taking place. Constituents in the water, the degree of saturation, and other factors will determine if the rate is sufficient to cause damage. Even when kinetics dictate rapid precipitation of calcite, there are several constituents in water that can slow down the precipitation to negligible rates. For instance, Lin and Singer (2005) demonstrated in lab-scale studies that orthophosphate dramatically inhibited the precipitation of calcium carbonate by the adsorption of phosphates on the calcite surfaces. Further, the inhibitory effects were altered by carbonate:calcium ratios and system pH – as each variable decreased, the effectiveness of orthophosphate inhibition increased. In addition, the capacity of polyphosphates proved two orders of magnitude more effective than orthophosphate in inhibiting precipitation of calcite (Lin and Singer, 2005; 2006). However, more quantitative research on how water quality parameters and phosphate dosing affect the efficacy of OPPP controls is desired.

Other issues that could arise from increased water temperatures include increased corrosion rates of metallic pipes and scalding of consumers. Increased water temperature is known to increase leaching propensity of lead from plumbing devices (Sarver and Edwards, 2011) and increase pinhole corrosion rates in copper tubing (Rushing, 2002). Several researchers have observed markedly higher corrosion rates at higher temperatures for copper pipes under moderate to continuous flow conditions simulating drinking water systems (Obrecht and Quill 1960a-f). The degradation of the physical attributes of the plumbing system could necessitate using a lower flow rate in premise plumbing hot water systems due to erosion corrosion.

Measures to avoid scalding are listed in all *Legionella*-related control guidelines (Table 1). Instead of reducing the temperature of the entire hot water heater system to avoid scald burns, thermostatic mixers at each tap to blend both hot and cold water to cool the hot water tap to a non-scalding "hot" water temperature immediately before use could be installed. Some Australian standards mandate the use of mixing valves in hot water systems with 60˚C water heater temperatures (Spinks et al., 2003). These valves would be useful in waters that are capable of being heated to 60˚C, but not all buildings can easily be retrofitted with these mixing devices. In addition, no research has been conducted on their longevity or effect on bacterial growth taps. However, several studies have implicated metered and hands-free faucets as suspect for supporting OPPP growth due to several factors that include the water temperatures delivered to the taps and materials used in the faucet design (Halabi et al., 2001; Yapicioglu et al., 2011). Similar concerns may emerge with thermal mixing valves as more research is conducted.

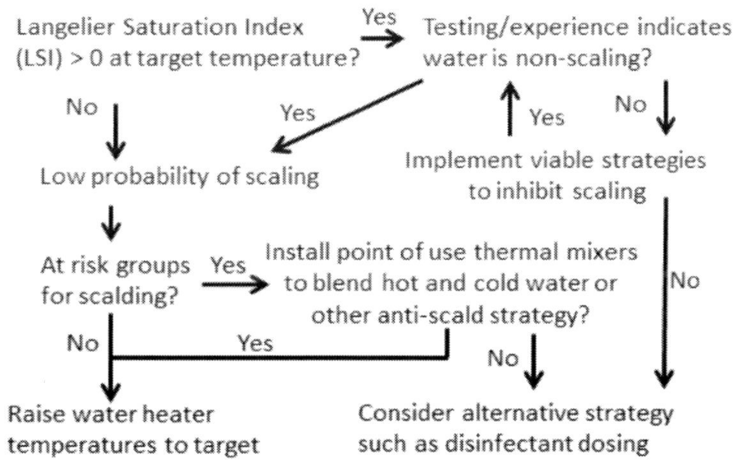

Figure 3. Logical framework/decision tree to determine if thermal treatment (shock or continuous) is likely to damage pipes at 60°C.

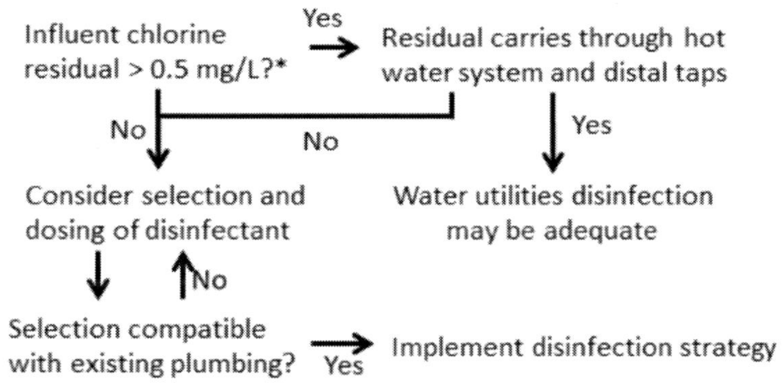

Figure 4. Logical framework/decision tree for implementing a treatment technique other than thermal disinfection.

CONCLUSIONS

There is a need to assign roles and responsibilities to the multiple stakeholders in premise plumbing-associated waterborne disease. While ASHRAE 188 is a critical step in this direction, requiring development of a plan for preventing, responding to, and following up on LD outbreaks, the standard brings attention to numerous research needs. Many systems will be unable to raise water temperatures to 60° C due to rapid scaling and other limitations. In the years ahead, practical experience and guidance will need to be developed. The Langelier Saturation Index (LSI) provides a useful starting point for considering whether scaling will be problematic, but more practical tests will be needed given the capability of inhibitors to reduce rates of scaling.

REFERENCES

Allegheny County Health Department (ACHD). (1997). Approaches to Prevention and Control of *Legionella* Infection in Allegheny County Health Care Facilities. Accessed at: http:// www.legionella.org/achd_guideline.pdf

American Society for Heating Refrigeration, and Air-Conditioning Engineers (ASHRAE). (2000). ASHRAE 12-2000 Guideline: Minimizing the Risk of Legionellosis Associated with Building Water Systems.

American Society for Heating Refrigeration, and Air-Conditioning Engineers (ASHRAE). (2012). ASHRAE Standard 188: Prevention of Legionellosis Associated with Building Water Systems.

Ames, R. G., & Stratton, J. W. (1987). Effect of chlorine dioxide water disinfection on hematologic and serum parameters of renal dialysis patients. *Archives of Environmental Health: An International Journal*, 42(5); 280-285.

Association of Water Technologies (AWT). (2003). *Legionella* 2003: An Update and Statement by the Association of Water Technologies. Access at: http://www.legionella.com/images/awtlegionella2003.pdf

Barbaree, J. M., Gorman, G. W., Martin, W. T., Fields, B. S., & Morrill, W. E. (1987). Protocol for sampling environmental sites for legionellae. *Applied and Environmental Microbiology, 53*(7), 1454-1458.

Bartram, J. (2007). *Legionella* and the prevention of Legionellosis. World Health Organization (WHO).

Berg, J. D., Hoff, J. C., Roberts, P. V., & Matin, A. (1988). Resistance of Bacterial Subpopulations to Disinfection by Chlorine Dioxide. *Journal American Water Works Association*, 80(9), 115-119.

Bitton, G., & Freihofer, V. (1977). Influence of extracellular polysaccharides on the toxicity of copper and cadmium toward *Klebsiella aerogenes*. *Microbial Ecology*, 4(2), 119-125.

Boulay, N., Edwards, M. (2000). Copper in the Urban Water Cycle. Critical Reviews in Environmental Science and Technology. V. 30, No. 3, 297-326.

Brazeau, R., Edwards, M. (2013). Water and Energy Savings from On-Demand and Hot Water Recirculating Systems, To be published in the Journal of Green Building.

Brazeau, RH., Edwards, MA. (2012). A Review of the Sustainability of Residential Hot Water Heater Infrastructure: Public Health, Environmental Impacts, and Consumer Drivers. *J of Green Buildings* 6(4): 77-95.

Brown, C. M., Nuorti, P. J., Breiman, R. F., Hathcock, A. L., Fields, B. S., Lipman, H. B., Cetron, M. (1999). A community outbreak of Legionnaires' disease linked to hospital cooling towers: an epidemiological method to calculate dose of exposure. International Journal of epidemiology, 28(2), 353-359.

Building Officials and Code Administrators (BOCA), (1997). BOCA National Plumbing Code, nineth ed. BOCA, Country Club Hills, IL.

Cachafeiro, S. P., Naveira, I. M., & García, I. G. (2007). Is copper–silver ionisation safe and effective in controlling *Legionella*?. Journal of Hospital Infection, 67(3), 209-216.

Calabrese, E. J. (1978). The Health Effects of Chlorine Dioxide as a Disinfectant in Potable Water: A Literature Survey. *Journal of Environmental Health*, 41(1), 26-31.

Centers for Disease Control and Prevention (CDC). (2005). Legionnaires Disease Associated with Potable Water in a Hotel --- Ocean City, Maryland, October 2003--February 2004. *Morbidity and Mortality Weekly Report (MMWR)*. 54(07); 165-168.

Centers for Disease Control and Prevention (CDC). (2011). Surveillance for Waterborne Disease Outbreaks Associated with Drinking Water --- United States, 2007--2008. *Morbidity and Mortality Weekly Report (MMWR)*. 60(ss12); 38-68.

Clark, B.N., A.L. Hernandez, and M. Edwards, (2011). Deposition Corrosion of Water Distribution System Materials, in *Water Quality Technology Conference*, Phoenix, AZ.

Cong, H., & Scully, J. R. (2010). Effect of chlorine concentration on natural pitting of copper as a function of water chemistry. *Journal of The Electrochemical Society,* 157(5), C200-C211.

Consumer Product Safety Commission (CPSC). 2005. Tap water scalds. Document #5098. Accessed at http://www.cpsc.gov/cpscpub/pubs/5098.html.

Cordes, L.G., Fraser, D.W., Skaliy, P. Perlino, C.A., Elsea, W.R. Mallison, G.F., Hayes, P.S. (1980). Legionnaires' Disease Outbreak at an Atlanta, Georgia, Country Club: Evidence for Spread from an Evaporative Condenser. *American Journal of Epidemiology*, 111(4), 425-431.

Cruse, H., (1971). Dissolved-Copper Effect on Iron Pipe. *Journal American Water Works Association*, 1971. 63(2): p. 79-81.

Dennis, P. J., Green, D., & Jones, B. P. C. (1984). A note on the temperature tolerance of *Legionella*. *Journal of Applied Microbiology*, 56(2), 349-350.

Domingue, E. L., Tyndall, R. L., Mayberry, W. R., & Pancorbo, O. C. (1988). Effects of three oxidizing biocides on *Legionella pneumophila* serogroup 1. *Applied and Environmental Microbiology*, 54(3), 741-747.

Dondero Jr, T. J., Rendtorff, R. C., Mallison, G. F., Weeks, R. M., Levy, J. S., Wong, E. W., & Schaffner, W. (1980a). An outbreak of Legionnaires' disease associated with a contaminated air-conditioning cooling tower. New England Journal of Medicine, 302(7), 365-370.

Dondero Jr, T. J., Rendtorff, R. C., Mallison, G. F., Weeks, R. M., Levy, J. S., Wong, E. W., & Schaffner, W. (1980b). An outbreak of Legionnaires' disease associated with a contaminated air-conditioning cooling tower. *New England Journal of Medicine*, 302(7), 365-370.

Edwards, M. (2004). Corrosion Control in Water Distribution Systems. One of the Grand Engineering Challenges for the 21st Century. Edited by Simon Parsons, Richard Stuetz, Bruce Jefferson and Marc Edwards. *Water Science and Technology* p. 1-8 V. 49, N. 2.

Edwards, M., & Dudi, A. (2004). Role of chlorine and chloramine in corrosion of lead-bearing plumbing materials. *Journal American Water Works Association*, 96(8); 69-81.

Edwards, M., Jacobs, S., & Dodrill, D. (1999). Desktop guidance for mitigating Pb and Cu corrosion by-products. *Journal of the American Water Works Association*, 91(5); 66-77.

Edwards, M., Jacobs, S., & Taylor, R. J. (2000). The blue water phenomenon. *Journal American Water Works Association*, 92(7); 72-82.

Edwards, M., Schock, M. R., & Meyer, T. E. (1996). Alkalinity, pH, and copper corrosion by-product release. *Journal American Water Works Association*, 88(3), 81-94.

Edwards, M.A., Parks, J., Griffin, A., Raetz, M., Martin, A., Scardina, P., Elfland, C., (2011). Lead and Copper Corrosion Control in New Construction. Water Research Foundation, Denver, CO.

Elfil, H., Hannachi, A. (2006). Reconsidering water scaling tendency assessment. *AIChE Journal*, 52(10), 3583-3591.

Farhat, M., Moletta-Denat, M., Frère, J., Onillon, S., Trouilhé, M. C., & Robine, E. (2012). Effects of Disinfection on *Legionella* spp., Eukarya, and Biofilms in a Hot Water System. *Applied and Environmental Microb*iology, 78(19), 6850-6858.

Florida Department of Health (Florida DOH).Guidelines for the Surveillance, Investigation, and Control of Legionnaires' Disease in Florida. Accessed at: http://www.doh.state.fl.us/environment/medicine/foodsurveillance/investigation_docs/Legionella_Guidelines.pdf.

Friedman, B.A., Dugan, P.R., (1968). Concentration and accumulation of metallic ions by the bacterium Zoogloea. *Dev Ind Microbiol*. 1968;9:381–8.

Greub, G., & Raoult, D. (2004). Microorganisms resistant to free-living amoebae. *Clinical Microbiology Reviews*, 17(2), 413-433.

Halabi, M., Wiesholzer-Pittl, M., Schoberl, J., & Mittermayer, H. (2001). Non-touch fittings in hospitals: a possible source of Pseudomonas aeruginosa and Legionella spp. *Journal of Hospital Infection*, 49(2), 117-121.

Halox Technologies Corpoartion, Accessed at: http://www.haloxtech.com/default.asp

Harvel Webpage. (2013). Product Specifications. CPVC Industrial Pipe: Schedule 40 & 80. Accessed at: http://www.harvel.com/sites/www.harvel.com/files/documents/Specifications-CPVC_Pipe.pdf

Hidmi, L., & Edwards, M. (1999). Role of temperature and pH in Cu (OH) 2 solubility. Environmental science & technology, 33(15), 2607-2610.

Hissel J, Salengos P. (2002). New graphical method for the presentation of calcium-carbonate equilibria. Application to the water treatment. *J Water Sci.* 15;435-458.

International Code Council, 2000. International Plumbing Code. International Code Council, Falls Church, VA.

Isaacs, W.P., Stockton, G.R. (1981). Softened Water Energy Savings Study Controlled Experiemntal Testing Program on Household Water Heaters, New Mexico State University, Las Cruces, NM in association with Water Quality Research Council (WQRC).

Katzenelson, E., Kletter, B., & Shuval, H. I. (1974). Inactivation Kinetics of Viruses and Bacteria in Water by Use of Ozone. *Journal American Water Works Association*, 66(12), 725-729.

Kenworthy, L., (1943). The Problem of Copper and Galvanized Iron in the Same Water System. The *Journal of the Institute of Metals*, 69, 67-90.

Kim, B. R., Anderson, J. E., Mueller, S. A., Gaines, W. A., & Kendall, A. M. (2002). Literature review—efficacy of various disinfectants against *Legionella* in water systems. *Water Research*, 36(18), 4433-4444.

Kusnetsov, J. M., Ottoila, E., & Martikainen, P. J. (1996). Growth, respiration and survival of *Legionella pneumophila* at high temperatures. *Journal of Applied Microbiology*, 81(4), 341-347.

Lagos, G. E., Cuadrado, C. A., & Letelier, M. V. (2001). Aging of copper pipes by drinking water. *Journal American Water Works Association*, 93(11)94-103.

Langelier, W.F. (1936). The analytical control of anticorrosion water treatment. *Journal American Water Works Association*, 28; 1500-1521.

Lin, Y. P., Singer, P. C. (2005). Inhibition of calcite crystal growth by polyphosphates. Water research, 39(19), 4835-4843. 2539.

Lin, Y. P., Singer, P. C. (2006). Inhibition of calcite precipitation by orthophosphate: Speciation and thermodynamic considerations. *Geochimica et Cosmochimica Acta*, 70(10), 2530-

Lin, Y. S. E., Vidic, R. D., Stout, J. E., McCartney, C. A., & Yu, V. L. (1998a). Inactivation of *Mycobacterium avium* by copper and silver ions. *Water Research*, 32(7), 1997-2000.

Lin, Y. S. E., Vidic, R. D., Stout, J. E., Yu, V. L. (2002). Negative effect of high pH on biocidal efficacy of copper and silver ions in controlling *Legionella pneumophila*. *Applied and Environmental Microbiology*, 68(6), 2711-2715.

Lin, Y-S.E., Stout, J.E., Yu, V.L., Vidic, R.D. (1998b). Disinfection of water distribution systems for *Legionella*. *Semin Respir Infect.* 13(2):147–59.

Loret, J. F., Robert, S., Thomas, V., Cooper, A. J., McCoy, W. F., & Lévi, Y. (2005). Comparison of disinfectants for biofilm, protozoa and *Legionella* control. *Journal of Water and Health: A Journal of the International Water Association*, 3(4), 423-33.

Lytle, D. A., & Schock, M. R. (2008). Pitting corrosion of copper in waters with high pH and low alkalinity. *Journal American Water Works Association*, 100(3);115-129.

Marina Coast Water Distric (MCWD). (2005). Water conservation rules – new requirements: Engineering procedures guidelines, and design requirements. Updated October, 2005.Accessed at: http://www.mcwd.org/docs/conservation/HotWaterRecirc Systems_100105.pdf

Marshall, B. J. (2004). Initiation, propagation, and mitigation of aluminum and chlorine induced pitting corrosion. Master's Thesis, Virginia Polytechnic and State Univ., Blacksburg, VA.

Moore, G. S., Calabrese, E. J., DiNardi, S. R., & Tuthill, R. W. (1978). Potential health effects of chlorine dioxide as a disinfectant in potable water supplies. *Medical Hypotheses*, 4(5); 481-496.

Muraca, P. W., Yu, V. L., & Goetz, A. (1990). Disinfection of water distribution systems for *Legionella*: a review of application procedures and methodologies. Infection Control and Hospital Epidemiology, 79-88.

Muraca, P., Stout, J. E., & Yu, V. L. (1987). Comparative assessment of chlorine, heat, ozone, and UV light for killing *Legionella pneumophila* within a model plumbing system. *Applied and Environmental Microbiology*, 53(2), 447-453.

National Research Council (NRC). (2006). Alternatives to premise plumbing, in: Drinking Water Distribution Systems: Assessing and Reducing Risk, 316-340.

National SAFE KIDS Campaign (NSKC). 2004. Burn injury fact sheet. Washington, DC: National SAFE KIDS Campaign.

Neden, D. G., Jones, R. J., Smith, J. R., Kirmeyer, G. J., & Foust, G. W. (1992). Comparing chlorination and chloramination for controlling bacterial regrowth. *Journal American Water Works Association*, 84(7), 80-88.

Nguyen, C. K., Powers, K. A., Raetz, M. A., Parks, J. L., & Edwards, M. A. (2011). Rapid free chlorine decay in the presence of $Cu(OH)_2$: Chemistry and practical implications. *Water Research*, 45(16), 5302-5312.

Nguyen, C., Elfland, C., & Edwards, M. (2012). Impact of advanced water conservation features and new copper pipe on rapid chloramine decay and microbial regrowth. *Water Research*. 46(3); 611-621

Nguyen, T. M. N., Ilef, D., Jarraud, S., Rouil, L., Campese, C., Che, D., ... & Desenclos, J. C. (2006). A community-wide outbreak of legionnaires disease linked to industrial cooling towers—how far can contaminated aerosols spread? *Journal of Infectious Diseases*, 193(1), 102-111.

Obrecht M. F. and Quill L. L. (1960a) How temperature, treatment, and velocity of potable water affect corrosion of copper and its alloys in heat exchanger and piping systems. *Heat. Piping Air Cond.* January 1960,165-169.

Obrecht M. F. and Quill L. L. (1960b) How temperature, treatment, and velocity of potable water affect corrosion of copper and its alloys; cupro-nickel, admiralty tubes resist corrosion better. *Heat. Piping Air Cond.* September 1960, 125-133.

Obrecht M. F. and Quill L. L. (1960c) How temperature, treatment, and velocity of potable water affect corrosion of copper and its alloys; different softened waters have broad corrosive effects on copper tubing. *Heat. Piping Air Cond.* July 1960, 115-122.

Obrecht M. F. and Quill L. L. (1960d) How temperature, treatment, and velocity of potable water after corrosion of copper and its alloys; monitoring system reveals effects of different operating conditions. *Heat. Piping Air Cond.* April 1960, 131-137.

Obrecht M. F. and Quill L. L. (1960e) How temperature, treatment, and velocity of potable water affect corrosion of copper and its alloys; tests show effects of water quality at various temperatures, velocities. *Heat. Piping Air Cond.* May 1960, 105-113.

Obrecht M. F. and Quill L. L. (1960f) How temperature, treatment, and velocity of potable water affect corrosion of copper and its alloys; what is corrosion? *Heat. Piping Air Cond.* March 1960, 109-116.

Occupational Safety and Health Administration (OSHA). (1999). OSHA Technical manual TED 1-0.15 A. US Dept. of Labor.

Patterson, J. W., Boice, R. E., & Marani, D. (1991). Alkaline precipitation and aging of copper from dilute cupric nitrate solution. *Environmental Science & Technology*, 25(10); 1780-1787.

Paul, D.D., Gadkari, V.V., Evers, D.P., Goshe, M.E., Thornton, D.A. (2010). Final Report: Study on Benefits of Removal of Water Hardness (Calcium and Magnesium Ions) from a Water Supply. Battelle Memorial Institute, Columbus, OH. Accessed at: http://www.flawater treatment.com/pdfs/Battelle_Final_Report.pdf.

Pruden, A., Edwards, M., Falkinham III, J. (2012). Research Needs for Opportunistic Pathogens in Premise Plumbing: Experimental Methodology, Microbial Ecology and Epidemiology. Water Research Foundation Project 4379, Denver, CO.

Rushing, J. C. (2002). Effects of chlorine and aluminum on copper pitting, temperature gradients on copper corrosion, and silica on iron release (Doctoral dissertation, MS thesis, Virginia Polytechnic Institute and State Univ., Blacksburg, Va).

Rushing, J. C., & Edwards, M. (2004). Effect of aluminium solids and chlorine on cold water pitting of copper. *Corrosion Science*, 46(12), 3069-3088.

Ryznar, J. W. (1944). An Index for Determining Amount of Calcium Carbonate Scale Formed by a Water. *Journal American Water Works Association*, 36(4), 472.

Sarver, E., Dodson, K., Scardina, P., Lattyak-Slabaugh, R., Edwards, M., & Nguyen, C. (2011). Copper Pitting in Chlorinated, High pH Potable Waters. *J. Am. Water Works Assoc*, 103(3); 86-98.

Schock, M. R., Lytle, D. A., & Clement, J. A. (1995). Effect of pH, DIC, orthophosphate and sulfate on drinking water cuprosolvency (No. PB--95-269270/XAB; EPA--600/R-95/085). National Risk Management Research Lab., Cincinnati, OH (United States).

Schock, M.R. (1990). Internal Corrosion and Deposition Control. Water Quality and Treatment, 5th Ed. New York: McGraw-Hill.

Seidel, C. J., McGuire, M. J., Summers, R. S., & Via, S. (2005). Have utilities switched to chloramines?. *Journal American Water Works Association*. 97(10); 87-97.

Sidari, F. P. I., Stout, J. E., Van Briesen, J. M., Bowman, A. M., Grubb, D., Neuner, A., ... & Yu, V. L. (2004). Keeping *Legionella* Out of Water Systems. *Journal American Water Works Association,* 96(1), 111-119.

Slawson, R. M., Lee, H., & Trevors, J. T. (1990). Bacterial interactions with silver. *Biometals*, 3(3), 151-154.

Smith, R. P., & Willhite, C. C. (1990). Chlorine dioxide and hemodialysis. Regulatory *Toxicology and Pharmacology*, 11(1); 42-62.

Spinks, A. T., Coombes, P., Dunstan, R. H., & Kuczera, G. (2003). Water quality treatment processes in domestic rainwater harvesting systems. In 28th International Hydrology and Water Resources Symposium: About Water; Symposium Proceedings (p. 2). Institution of Engineers, Australia.

Srinivasan, A., Bova, G., Ross, T., Mackie, K., Paquette, N., Merz, W., & Perl, T. M. (2003). A 17-month evaluation of a chlorine dioxide water treatment system to control *Legionella* species in a hospital water supply. *Infection Control and Hospital Epidemiology*, 24(8), 575-579.

Stiff Jr, H., Davis, L. (1952). A method for predicting the tendency of oil field waters to deposit calcium carbonate. *Journal of Petroleum Technology*, 4(9), 213-216.

Stout, J. E., & Yu, V. L. (2003). Experiences of the first 16 hospitals using copper-silver ionization for *Legionella* control: implications for the evaluation of other disinfection modalities. *Infection Control and Hospital Epidemiology*, 24(8); 563-568.

Stout, J. E., Muder, R. R., Mietzner, S., Wagener, M. M., Perri, M. B., DeRoos, K., Yu, V. L.. et al. (2007). Role of environmental surveillance in determining the risk of hospital-acquired legionellosis: a national surveillance study with clinical correlations. *Infection Control and Hospital Epidemiology: The Official Journal of the Society of Hospital Epidemiologists of America, 28*(7), 818-824.

Temmerman, R., Vervaeren, H., Noseda, B., Boon, N., & Verstraete, W. (2006). Necrotrophic growth of Legionella pneumophila. *Applied and Environmental Microbiology*, 72(6), 4323-4328.

Thomas, V., Bouchez, T., Nicolas, V., Robert, S., Loret, J. F., & Levi, Y. (2004). Amoebae in domestic water systems: resistance to disinfection treatments and implication in *Legionella* persistence. Journal of applied microbiology, 97(5), 950-963.

United States Environmental Protection Agency (USEPA). (2009). National Primary Drinking Water Regulations. EPA 816-F-09-004. Cincinnati, OH.

United States Environmental Protection Agency (USEPA). (2011). Information about Chloramine in Drinking Water. Accessed at: http://www.epa.gov/ogwdw/disinfection /chloramine/pdfs/chloramine2.pdf

Whelton, A. J., Dietrich, A. M., & Gallagher, D. L. (2011). Impact of chlorinated water exposure on contaminant transport and surface and bulk properties of high-density polyethylene and cross-linked polyethylene potable water pipes. *Journal of Environmental Engineering*, 137(7); 559-568.

Whelton, A.J., and A.M. Dietrich. 2008. Critical Considerations for the Accelerated Ageing of High Density Polyethylene Potable Water Materials. *Polymer Degradation and Stability*, 94(7): 1163-1175.

World Health Organization (WHO). (2006a). Copper. In: Guidelines for Drinking Water Quality: Recommendations. 3rd ed.Geneva: World Health Organization. p. 335-337.

World Health Organization (WHO). (2006b). Silver. In: Guidelines for Drinking Water Quality: Recommendations. 3rd ed.Geneva: World Health Organization. p. 434-435.

Yapicioglu, H., Gokmen, T. G., Yildizdas, D., Koksal, F., Ozlu, F., Kale-Cekinmez, E., . . . Candevir, A. (2011). Pseudomonas aeruginosa infections due to electronic faucets in a neonatal intensive care unit. *Journal of Paediatrics and Child Health*. 48(5); 430-437.

Yee, R. B., & Wadowsky, R. M. (1982). Multiplication of *Legionella pneumophila* in unsterilized tap water. *Applied and Environmental Microbiology*, 43(6), 1330-1334.

Zhang, Y., & Edwards, M. (2009). Accelerated chloramine decay and microbial growth by nitrification in premise plumbing. *Journal AWWA*, 101(11), 51-62.

Zhang, Y., Love, N., Edwards, M. (2009). Nitrification in drinking water systems. Critical Reviews in *Environmental Science and Technology*. 39 (3), 153-208.

Zhang, Z., Stout, J. E., Yu, V. L., & Vidic, R. (2008). Effect of pipe corrosion scales on chlorine dioxide consumption in drinking water distribution systems. *Water Research*, 42(1), 129-136.

Effects of UV-Based Treatment on Water and Air Chemistry in Chlorinated, Indoor Pools

Mehrnaz Zare Afifi and Ernest R. Blatchley III
School of Civil Engineering, Purdue University, West Lafayette, Indiana 47907-2051

ABSTRACT

Combined UV/chlorine treatment is commonly used in treatment of swimming pool water because UV irradiation is known to be effective against (some) chlorine-resistant microbial pathogens (*e.g.*, *C. parvum*), and it has the potential to promote photodecay of some DBPs (*e.g.*, inorganic chloramines). However, it has been reported in the literature that UV-based treatment has the potential to promote formation of some DBPs (*e.g.*, CNCl and $CNCHCl_2$). Promotion of these reactions also results in increases of free chlorine consumption rates. To further examine this issue and to explore the dynamics of UV-induced effects on water chemistry in chlorinated indoor swimming pools, a three-year field experiment has been initiated at a high school, indoor swimming pool under operating conditions of: conventional chlorination (1st year), chlorination augmented by MP UV irradiation (2nd year), and chlorination augmented by LP UV irradiation (3rd year). The first year of sampling was recently completed and the MP UV system has been installed. Water samples were collected from the pool and analyzed for free and combined chlorine, urea and volatile DBPs. Eleven volatile DBPs were identified: monochloramine (NH_2Cl), dichloramine ($NHCl_2$), trichloramine (NCl_3), chloroform ($CHCl_3$), bromoform ($CHBr_3$), dichlorobromomethane ($CHBrCl_2$), dibromochloromethane ($CHBr_2Cl$), cyanogen chloride (CNCl), cyanogen bromide (CNBr), dichloroacetonitrile ($CNCHCl_2$), and dichloromethylamine (CH_3NCl_2). Of these eleven compounds, ten were identified as regularly occurring, with dichlorobromomethane ($CHBrCl_2$) only appearing sporadically. Also since many DBPs are volatile, air quality within indoor swimming pool facilities has become an important issue with regard to human health. A large fraction of the current research regarding swimming pool air quality has focused on trichloramine (NCl_3) exposure. Trichloramine in the gas phase is analyzed by air sparging-DPD/KI analytical method. The results showed a correlation between bather loading and gas phase trichloramine concentration in the 1st year of experiment under chlorination operating condition. The results of this study are expected to yield fundamental improvements in our understanding of how UV systems can be used to improve water chemistry in swimming pool facilities. At the end of these experiments the following questions will be addressed: ultimately, should we be using UV for water treatment in chlorinated, indoor swimming pool facilities? If so, MP UV or LP UV? The results of these experiments are also expected to benefit UV/chlorine applications in other water treatment settings, such as treatment of drinking water, wastewater, or in water reuse facilities.

KEYWORDS
Chlorine; UV irradiation; Volatile disinfection by-products (DBPs); Swimming.

INTRODUCTION

Swimming in pools is a common recreational activity worldwide and is believed to provide health and fitness benefits. Chlorination is the predominant disinfection method applied in swimming pools due to its cost-effectiveness, ease of application and ability to inactivate a wide variety of pathogenic microorganisms. Chlorination promotes oxidation of some dissolved organic compounds. However, it is also known to produce disinfection by-products (DBPs), mostly via reactions between free chlorine (defined as the sum of the concentrations HOCl, OCl^-, and Cl_2) and natural organic matter in the source water, as well as inputs from bathers through sweat, urine, hair, skin particles, and cosmetics (Kristensen *et al.*, 2009). Human sweat and urine appear to be two important sources of nitrogenous compounds that are introduced to swimming pools by humans which can cause the formation of volatile DBPs (Weaver *et al.*, 2009; Li and Blatchley, 2007). WHO (2006) indicated a typical range of 25 - 77.5 mL of urine per bather released into swimming pool waters. Weng and Blatchley (2011) measured concentrations of DBPs and urea in a natatorium before and during a large swimming competition and their results demonstrated strong diurnal patterns in the concentration of urea and several DBPs, with a concentration of urea and NCl_3 increasing by a factor of more than four during a period of intensive pool usage.

Human exposure to volatile DBPs has been linked to adverse health effects and it can occur via inhalation, ingestion and dermal absorption. For some DBPs (*e.g.,* $CHCl_3$), the dominant pathway for uptake is dermal absorption due to contact of large surface area of skin with water (Lindstrom *et al.*, 1997; Weisel *et al.,* 1994). For other compounds (*e.g.,* NCl_3), inhalation can be important route of exposure. These pathways for human exposure are potentially problematic due to the toxic nature of some DBPs. Acute and chronic human health problems associated with exposure to DBPs in swimming pools include: promotion of asthma (Bernard *et al.*, 2003; 2007); increased incidence of rhinitis and hay fever (Bougault *et al.*, 2010; Bernard *et al.*, 2009); as well as skin (contact dermatitis) and eye irritation (Fantuzzi *et al.*, 2010; Safranek *et al.*, 2007). DBP exposure in swimming pools has also been associated with an increased risk of bladder cancer (Villanueva *et al.,* 2007).

In recent years ultraviolet (UV) irradiation has been applied as a secondary disinfection process in chlorinated swimming pools. UV-based systems have been reported to yield improvements in swimming pool water and air chemistry. UV is a broad-spectrum antimicrobial agent, and has been demonstrated to be effective for control of a wide range of microbial pathogens, including *C. parvum* and *G. Lamblia* (Craik *et al.*,2001; Hijnen *et al.*, 2006), and it has a potential to promote photodecay of some DBPs. However, to date much of the available information regarding the effects of UV on water chemistry in recreational water facilities is anecdotal and incomplete. There is conflicting information in literature regarding the effects of UV-based treatment on DBP precursors in chlorinated water that are common in pools and other settings. UV irradiation at germicidally-active wavelengths has been shown to promote direct photolysis of inorganic chloramines DBPs and to increase the rate of free chlorine consumption (Li and Blatchley, 2009). DeLaat *et al.* (2010) also demonstrated that molecular oxygen plays an important role in UV-induced decay of NH_2Cl. On the other hand, Liu *et al.* (2006) and Weng *et al.* (2012) demonstrated that the formation of THMs, HAAs, and CNCl compounds have been promoted as a result of secondary UV exposure. Moreover, Cassan *et al.* (2006) reported increases in trihalomethane (THM) formation as a result of installation of a UV system at a pool, as compared with before UV installation in the same pool; while Beyer *et al.* (2004) reported a decrease in THM concentrations in a similar study. Kristensen *et al.* (2009) did a long term study including matched control periods with and without different types of UV treatment sources in a public pool, and their results showed no effect on THM concentrations in a swimming pool.

Results presented by Hansen *et al.* (2013) indicated that MP UV lamps are able to degrade chloroform to some extent. According to the studies of Li and Blatchley (2009) and Weng *et al.* (2012), UV exposure results in photodecay of dichloromethylamine (CH_3NCl_2), and chloramines (*e.g.* NCl_3); however, UV irradiation promotes the formation of dichloroacetonitrile ($CNCHCl_2$) and cyanogen chloride (CNCl) and also promotes free chlorine consumption.

It is hypothesized that volatile DBPs (*e.g.*, NCl_3, THMs, HAAs, CNCl and $CNCHCl_2$) will behave as intermediates in pools when UV/chlorine treatment is applied. The formation of these DBPs appears to be promoted by the inclusion of UV in the chlorination process. However, for each of these compounds, at least one mechanism of loss from solution is known to exist. Loss mechanisms include hydrolysis (sometimes promoted by free chlorine), photolysis, and liquid→ gas transfer. Therefore, it is expected that the concentrations of these compounds will increase immediately after inclusion of UV-based treatment, followed by a period in which the concentrations of these compounds decrease.

However, since swimming pools are complex systems with many parameters that will influence DBP dynamics (*e.g.,* pH, temperature, precursors concentrations, bather load pattern, UV dose per treatment cycle, hydraulic residence time in pool and residence time in the system, etc.), it is difficult to assess the overall effect of the UV treatment. In addition, there are uncertainties about which source of UV irradiation (*i.e.*, low-pressure [LP] or medium-pressure [MP] Hg lamps) would be more effective as a secondary swimming pool water treatment process.

Since many DBPs are volatile, air quality within indoor swimming pool facilities has become an important issue with regard to human health. A large fraction of the current research regarding swimming pool air quality has focused on trichloramine (NCl_3) exposure. Trichloramine (NCl_3) has a high Henry's law constant, so it has relatively high potential to escape from liquid to gas phase. NCl_3 is known to undergo photodecay when subjected to germicidal UV radiation (Li and Blatchley, 2009; De Laat *et al., 2010*). However, recent work has also demonstrated that combined application of UV/chlorination can promote NCl_3 formation from some organic-N compounds that are likely to be present in pools. Therefore, NCl_3 is likely to exist as an intermediate in swimming pools where UV and chlorine are used together. A study using MP UV treatment in the recirculation system of an indoor pool found up to 32% reduction of NCl_3 in air (Cassan *et al.*, 2011).

Therefore, the aims of this research are to define the dynamics of volatile DBPs in a chlorinated, indoor pool, to examine the effects of UV-based treatment on water and air chemistry in indoor pool settings, and also to compare the nature of differences between LP and MP UV for alteration of water and air chemistry in chlorinated, indoor swimming pools.

Experiments are to be conducted to characterize water and air quality at a chlorinated, indoor pool facility under the following operating conditions:

- *Conventional operation, based on chlorination* (1st yr), as a control for comparison with subsequent measurements (2011-2012);
- *Conventional chlorination augmented with medium-pressure (MP) UV irradiation* (2nd yr), A MP UV system has been installed at the high school indoor pool facility as a

secondary water treatment. MP lamps are polychromatic with output that spans virtually the entire UVA, UVB, and UVC spectra. Measurements taken during this period will allow for quantitative evaluation of the effects of MP UV on water and air chemistry in a chlorinated pool (2012-2013);

- *Conventional chlorination augmented with low-pressure (LP) UV irradiation* (3rd yr) A LP UV system will be installed at the high school indoor pool as a secondary water treatment. LP lamps provide essentially monochromatic output (λ =253.7 nm). Measurements taken during this period will allow for quantitative evaluation of the effects of LP UV on water and air chemistry in a chlorinated pool. The LP UV system will be selected and operated in a manner that is comparable to the MP UV system. (2013-2014).

For each operating condition, samples of water and air are collected over a twelve-month period, and subjected to a battery of analytical procedures to allow characterization and quantification of the chemical constituents present in these samples. The twelve-month time frame for each operating condition was selected to allow examination of seasonal changes (*i.e.,* HVAC system operation), diurnal changes, and different patterns of pool use (*i.e.,* bather load).

METHODS AND MATERIALS

Air and water chemistry are monitored at the high school, indoor swimming pool facility daily during the academic year and once per week during summer, under each of the operating conditions described above. In the 2nd and 3rd year of this experiment, UV reactors (MP and LP UV) which will be provided by a UV system manufacturer - Engineered Treatment System, LLC (ETS, Beaver Dam, WI), will be installed at the high school indoor pool as a secondary water treatment process and the changes in air and water chemistry will be monitored during this period. The first year of sampling under chlorination operating condition has recently been completed and the MP UV system as a secondary treatment was installed at the pool on 11/27/12 (Figure 1a).

As indicated in Figure 1b, this pool facility has an L-shaped pool that contains 770 m^3 of water. One leg of the "L" is a six-lane, 22.86 m (25-yard) pool, while the other leg of the "L" is a six-lane, 25 m pool. The air space of this facility has a volume of approximately 3200 m^3, and the HVAC system at the facility is operated at roughly nine air changes per hour. All of the air brought into this facility by the HVAC system was outside air along with heat.

This pool is used by competitive swimmers on high school teams and an age-group swimming club. It is also used for swimming lessons and for physical education classes for students at the school.

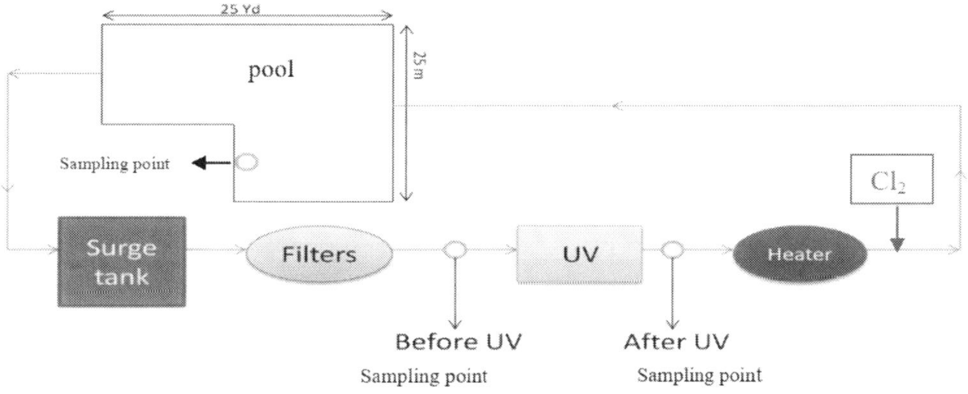

(a)

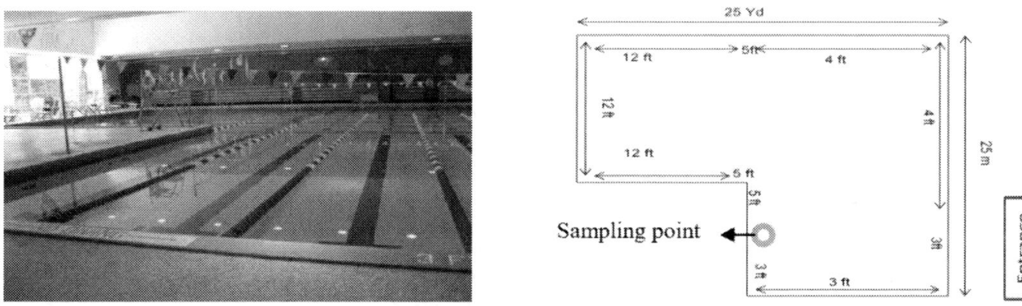

(b)

Figure 1. (a) Schematic of UV/chlorination as a disinfection process treatment at high school (b) High School indoor swimming pool facility. The open orange circles indicate locations of water sample collection.

Swimming pool water samples are collected in 125 mL Nalgene polyethylene bottles with polypropylene screw caps. Samples are collected from approximately 30 cm below the pool water surface without head-space. Since installation of the MP UV system, water samples have also been collected from taps located immediately before and after the UV reactor (Figure 1a). Water samples are transported to the Environmental Engineering Laboratories at Purdue University and then subjected to analytical procedures (Table 1).

Sample collection was conducted five days per week from 9/14/11 until 5/21/12; however, sampling was switched to once per week during the period from 5/23/12 until 9/30/12 because the pool was not used in this period. Five-day-per week sampling was reinitiated 10/02/12 and continues to present.

Students from the Advanced Placement (AP) Chemistry classes taught at the high school are involved in measurements of residual chlorine (free and combined) by using DPD/KI portable photometer, as well as measurements of gas-phase NCl_3 concentration by using an air sparging-DPD/KI method (described below).

Various analytical procedures are being applied in this experiment for quantification of volatile DBPs and some precursors that are known to be common to pool water and air (see Table 1).

Table 1. Analytical methods to be used for characterization water and air chemistry at a high school indoor swimming pool.

Methods	**Target analytes**
Membrane Introduction Mass Spectrometry (MIMS)	11 volatile DBPs
Colorimetric DPD/KI with portable photometer	Free and combined chlorine compounds (measured mostly by AP high school students)
Colorimetric DPD with UV/vis spectrometry	Free chlorine
Antipyrine digestion/ colorimetric method	Urea
Air sparging- DPD/KI method	Gas-phase NCl_3 concentration (measured mostly by AP high school students)
Portable pH meter	pH and temperature
Alkaphot (Palintest) tablets with portable photometer	Total alkalinity
Ion Chromatography (IC)	Anions (NO_3^-, NO_2^-, Cl^-)

Volatile DBP analysis by Membrane Introduction Mass Spectrometry (MIMS)

MIMS is applied for differentiation and quantification of volatile DBP compounds in aqueous samples. The MIMS system that is used in this research is based on an Agilent 6850 bench top GC/MS, containing an Agilent 5975C quadrupole mass selective detector (MSD) with electron (70 eV) ionization. The membrane interface was constructed around small-diameter silicone tubing, as described in Shang and Blatchley (1999), with liquid flow rate of 0.7 mL/min, and gas (Helium) flow rate at 0.5 mL/min (Figure 2). Volatile DBPs are identified in mass spectra scan mode (m/z: 49-400 amu), while selected ion monitoring (SIM) mode is applied for quantification of target DBPs precursors.

Eleven volatile DBPs: monochloramine (NH_2Cl), dichloramine ($NHCl_2$), trichloramine (NCl_3), chloroform ($CHCl_3$), bromoform ($CHBr_3$), dibromochloromethane ($CHBr_2Cl$), dichlorobromomethane ($CHBrCl_2$), cyanogen chloride (CNCl), cyanogen bromide (CNBr), dichloroacetonitrile ($CNCHCl_2$), and dichloromethylamine (CH_3NCl_2)) are analyzed and identified by MIMS within 2 hours of sample collection. Weaver *et al.* (2009) illustrated that these eleven compounds are common to chlorinated indoor swimming pool water samples, with ten regularly occurring and dichlorobromomethane ($CHBrCl_2$) only appearing sporadically.

Standards of volatile compounds were made following the methods in Li and Blatchley (2007). All chemicals and standards solutions were of analytical grade purchased from Sigma-Aldrich. Each volatile DBP was quantified using a unique m/z peak. Ions at m/z 53, 61, 74, 88, 89, 98, and 105 were used to quantify monochloramine (NH_2Cl), cyanogen chloride (CNCl), dichloroacetonitrile ($CNCHCl_2$), trichloramine (NCl_3), dichloramine (NHCl2), dichloromethylamine (CH_3NCl_2), and cyanogen bromide (CNBr), respectively. Chloroform ($CHCl_3$), dichlorobromomethane ($CHBrCl_2$), dibromochloromethane ($CHBr_2Cl$), and bromoform ($CHBr_3$) were quantified from ions at m/z 83, 118, 129, and 173. Interpretation, in terms of conversion of abundance signals to corresponding concentration, and accounting for variability in instrument sensitivity followed the method in Weaver *et al.* (2009). Monitoring of these 11 volatile DBPs compounds is applied for all liquid phase experiments from swimming pool water samples.

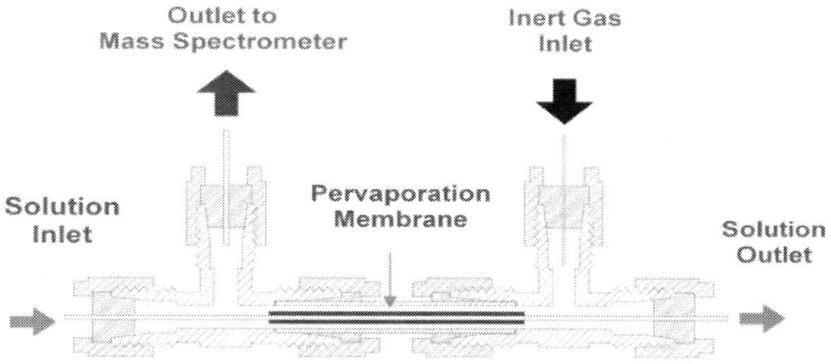

Figure 2. Schematic of membrane interface for MIMS systems used in this project adapted from Shang and Blatchley (1999).

Weaver *et al.* (2009) defined the limits of detection for each of 11 DPBs analytes based on the basis of a 2:1 signal: noise ratio (see Table 2).

Table 2. MIMS detection limits (Weaver *et al.*, 2009).

Compound	**NH_2Cl**	**CNCl**	**$CNCHCl_2$**	**$CHCl_3$**	**$CHBrCl_2$**	**NCl_3**
m/z ratio	53	61	74	83	83	88
Detection Limit (µg/L)	82	0.8	2	3.7	4.3	3.6
Compound	**$NHCl_2$**	**CH_3NCl_2**	**CNBr**	**$CHBr_2Cl$**	**$CHBr_3$**	
m/z ratio	89	98	105	129	173	
Detection Limit (µg/L)	38	1.6	0.8	0.25	0.71	

Gas-phase NCl_3 concentration measurements in indoor chlorinated swimming pool facilities

A modified version of the Predieri and Giacobazzi sparging method (Aggazzotti *et al.*, 2007) was applied for quantifying trichloramine (NCl_3) in gas-phase samples. This method is designed to trap gas-phase inorganic combined chlorine in a DPD/KI solution. In this method, a known volume of air is pumped through two sequential gas-washing bottles, equipped with glass stems possessing course fritted ends. Figure 3, provides a schematic illustration of the apparatus used in this method.

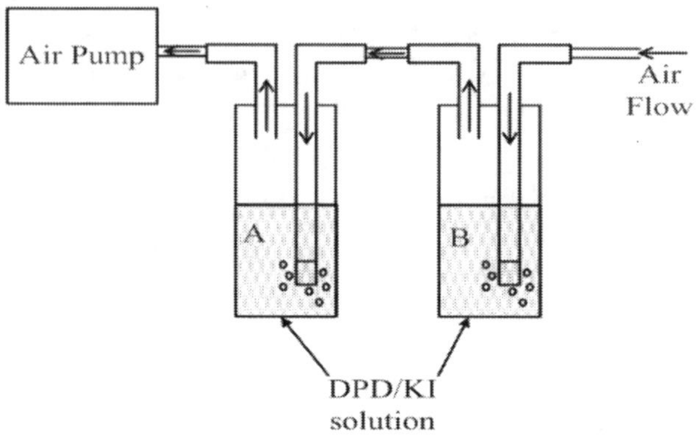

Figure 3. Schematic of sparging method for gas-phase NCl_3 measurement. The arrows indicate the air flow pattern (Weng *et al.*, 2011).

Each gas-washing bottle contains 30 mL of an aqueous solution of *N*, *N*-diethyl-*p*-phenylenediamine/potassium iodide (DPD/KI), which is developed using two Palintest DPD No.1 test kit tablets and 0.5 g KI (Aldrich) into Barnstead nano-pure deionized water. Transfer of gas-phase constituents into the aqueous phase is promoted by the fritted glass ends of the gas-washing bottles. For each measurement, air is pumped by VSS-5 sampling pump (A.P. Buck, Inc., Orlando, Florida) through the gas-washing bottles at a volumetric flow rate of 1.0 L/min for a period of 30 minutes (V_{air}). Gas-phase NCl_3 and other oxidizing compounds are trapped by the DPD/KI solution and quantified photometrically by a Palintest portable photometer. Quantification of gas-phase NCl_3 is accomplished using a mass-balance approach, as described by Weng *et al.*, (2011).

RESULTS AND DISCUSSION

Pool usage pattern: The high school pool was used for swimming lessons and physical education (PE) classes by students at the high school from 17th of October 2011 until 10th of December 2011 (the total number of swimmers for swimming lessons and PE classes in this period were approximately 161 in each day) and again from 26th of March 2012 until 23rd of May 2012 (the total number of swimmers were between 110 to 142). Also, high school swimming teams used the pool from 23rd of October 2011 until 23rd of February 2012 (in total approximately 54 swimmers per day in this period). The pool was not used or rarely used from 23rd of February 2012 until 26th of March 2012 and the pool was unused from the 23rd of May 2012 until 23rd of September 2012. The pool was re-opened on the 24th of September 2012 (Table 3).

Table 3. Pool usage pattern and bather loading.

Date	Oct 17th - Dec 10th 2011	March 26th-May 23rd 2012	Sep 24th 2012- present
Total bather loading per day	161 (PE class) +54 (swimming team)	between 110-140 (just PE class)	112 (PE class) +30 (swimming team)

The data presented in Table 4 include the arithmetic mean concentrations for each DBP compound, along with their high and low concentration under chlorination operating condition (1st year). Implications of these measurements will be discussed below.

Time-course summaries of liquid-phase concentration measurements are presented in Figures 4 through 19. These Figures should be interpreted as follows: the dashed grey vertical line in all of these Figures indicates alternate Mondays (two-week periods). As appropriate, the solid horizontal lines represent concentration limits or guidelines for each compound. The dashed pink horizontal line in Figures indicates MIMS detection limits for each compound. The dark green solid vertical line indicates the date when the MP UV was installed at the pool as a secondary disinfection process.

As indicated in Figure 1a, after the MP UV reactor was installed at the high school pool, water samples were collected from three locations: swimming pool, before, and after the UV reactor.

Table 4. Arithmetic mean volatile DBP concentrations plus maximum and minimum concentration value at high school chlorinated, indoor swimming pool for year one (chlorination operating condition).

Compounds	Mean (mg/L) (n=163)	High Conc(mg/L)	Low Conc (mg/L)
Free Cl_2 - DPD-KI method	2.6	8.0	0.05
Combined chlorine - DPD-KI method	0.46	1.80	0.05
NH_2Cl (as Cl_2) - MIMS	0.089	0.620	0.015
$NHCl_2$ (as Cl_2) - MIMS	0.06	0.25	0.00
NCl_3(as Cl_2) - MIMS	0.42	2.2	0.003
$CHCl_3$ - MIMS	0.08	0.21	0.02
$CHBr_3$ - MIMS	0.001	0.032	0
$CHBr_2Cl$ - MIMS	0.002	0.01	0
$CHBrCl_2$ - MIMS	0.000019	0.003	0
TTHM - MIMS	0.08	0.28	0.03
CNCl - MIMS	0.024	0.2	0.001
CNBr - MIMS	0.005	0.024	0
$CNCHCl_2$ - MIMS	0.009	0.03	0.001
CH_3NCl_2 - MIMS	0.22	1.9	0.01

Free and combined chlorine

Based on the National Swimming Pool Foundation (NSPF) (2010) guideline, 1 mg/L (as Cl_2) represents the accepted minimum free chlorine concentration and 5 mg/L (as Cl_2) represents the accepted maximum in swimming pools. As reported by the NSPF (2010) the ideal range for free chlorine is within 2 - 4 mg/L (as Cl_2). Also, the combined chlorine concentration as measured using DPD/KI is suggested to be below 0.2 mg/L (as Cl_2) and 0.5 mg/L (as Cl_2) in pools and spas, respectively (NSPF, 2010).

As shown in Figure 4, the free chorine concentration exceeded the NSPF recommended maximum value (5 mg/L (as Cl_2)) eight times in year 1; free chlorine fell below the recommended minimum value (1 mg/L (as Cl_2)) 15 times in this same period. From when MP UV source was installed (11/27/12) the free chlorine concentration exceeded the NSPF maximum value once and fell below the recommended minimum value four times up to the present. According to the Li and Blatchley, (2009) and Weng *et al.* (2012) studies, UV irradiation increases the rate of free chlorine consumption. The operators of this pool have indicated that the chlorine use pattern has changed, such that free chlorine is consumed more rapidly with the UV system installed; however, no data regarding chlorine use are available for this facility.

Based on the NSPF (2010) guideline, the combined chlorine concentration in swimming pools should be below 0.2 mg/L as Cl_2; however, from Figure 4, it is evident that the combined chlorine concentration regularly exceeded this recommended limit, which is quite common in swimming pools (Weaver *et al.*, 2009). Kristensen *et al.* (2009) indicated that despite the continuous removal of combined chlorine by UV-treatment, combined chlorine formation is observed during pool usage. Kristensen *et al.* (2009) demonstrated that combined chlorine concentration increased in the early parts of a day when UV exposure was included in treatment, but the long-term trend in the UV system was a decrease in combined chlorine concentration, relative to a same facility that did not use UV.

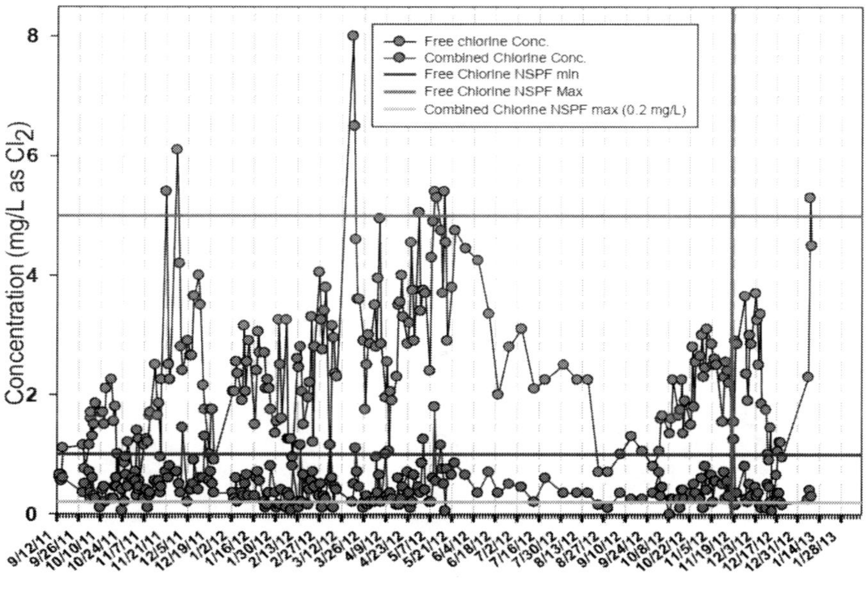

Figure 4. Free and combined chlorine concentration as measured by the DPD/KI spectrophotometric method.

Monochloramine (NH_2Cl)-MIMS

From Table 4 and Figure 5, the lowest and highest concentrations of monochloramine at this pool were 0.015 (below detection limit of 0.08 mg/L) and 0.62 mg/L, respectively, with mean concentration of 0.089 mg/L as Cl_2 during the first year of the experiment.

Monochloramine reached its highest concentration of 0.62 mg/L as Cl_2 on a date that coincided with a free chlorine excursion in the pool (at the end of December 2011). The free chlorine excursion of March 2012 also yielded on increase in monochloramine concentration, but it was smaller than the increase observed in December 2011.

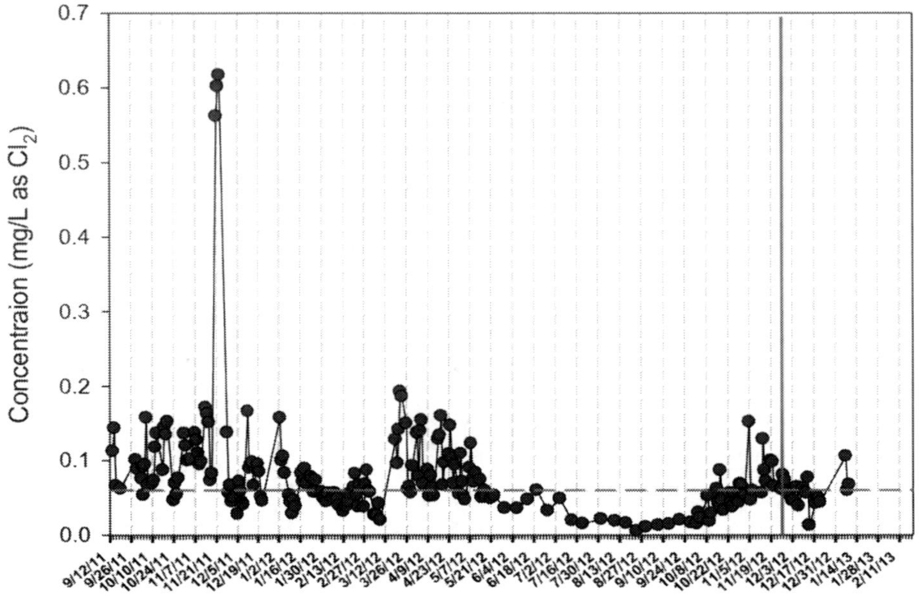

Figure 5. Monochloramine (NH_2Cl) range at high school measured by MIMS.

Also evident in Figure 5 is a weekly pattern of monochloramine concentration. In most weeks the concentration of monochloramine was high on Mondays and it started to decrease during the following days until the next Friday. It is hypothesized that this weekly pattern was due to pool usage. During the weekend, particularly on Sundays, nobody was using the swimming pool; this provides a mechanism for accumulation of volatile DBPs during the weekend because of no usage and activity in the pool. Kristensen *et al.* (2010) performed an on-line monitoring experiment with THMs in swimming pools with a similar pattern of behavior for these compounds. As will be demonstrated in the following sections, this weekly pattern was observed with some of the other volatile DBPs.

There was no usage or rare usage of the pool from the end of February 2012 until mid-March 2012, and also no usage of the pool from the end of May 2012 until end of September 2012. During these periods of low use, the concentration of NH_2Cl tended to be relatively low and stable. However, no obvious trends in liquid-phase NH_2Cl concentration were observed following installation of the MP UV system at the pool.

Figure 6, shows the monochloramine concentration at three sampling locations for the period immediately following installation of the UV system. The concentrations of NH_2Cl in water immediately before and after the UV reactor followed a consistent pattern, with the "after UV" sample often containing NH_2Cl at a slightly lower concentration than the "before UV" sample; this behavior is consistent with the slow photodecomposition of NH_2Cl that has been reported by Li and Blatchley (2009). However, the corresponding samples collected from the pool consistently contained substantially higher NH_2Cl concentration than these other two samples. This implies that NH_2Cl is being formed largely in the pool.

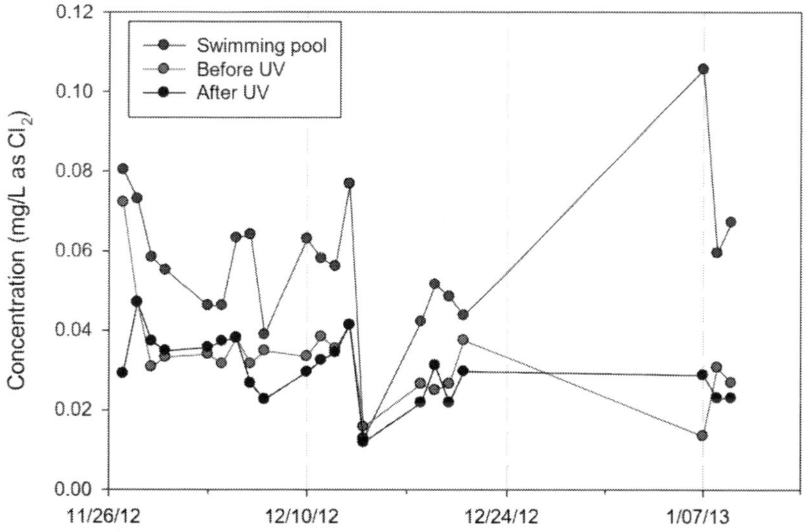

Figure 6. Monochloramine concentration at three different locations: swimming pool, before and after UV source.

Dichloramine ($NHCl_2$)-MIMS

The lowest and highest concentration of dichloramine were undetectable and 0.25 mg/L, respectively, with the mean concentration of 0.065 mg/L as Cl_2, as shown in Figure 7 and Table 4, during the first year of the experiment.

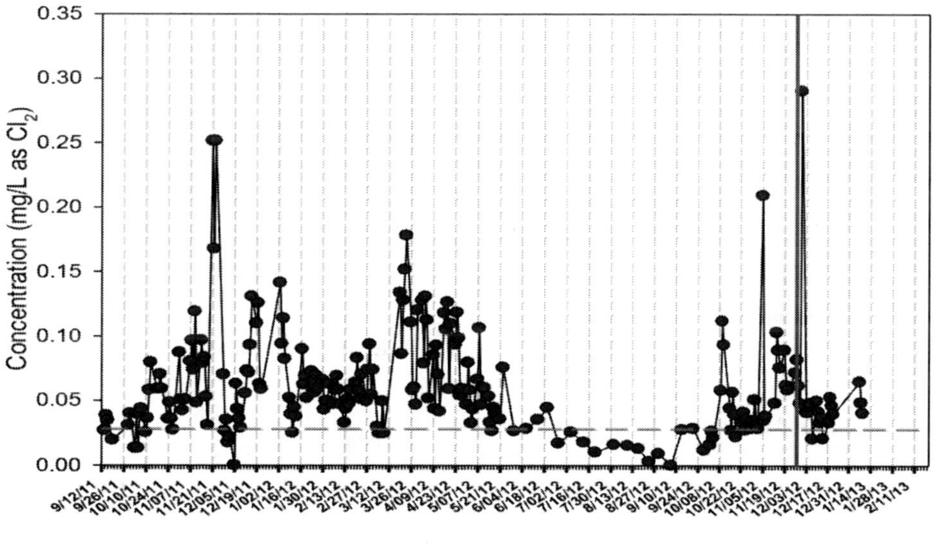

Figure 7. Dichloramine ($NHCl_2$) range at high school measured by MIMS.

As with NH_2Cl, the two highest concentrations of $NHCl_2$ coincided with the free chlorine excursions of December 2011 and March 2012. Dichloramine is generally less stable than

NH_2Cl (Jafvert and Valentine, 1992), which may have contributed to the relatively erratic behavior of this compound. The weekly pattern was less consistent than that of monochloramine. During the periods of low pool usage, there was less fluctuation in $NHCl_2$ concentration than in periods if heavy use. However, no obvious changes in the trends of $NHCl_2$ concentration have been observed to result from installation of MP UV at the pool.

According to the results presented in Figure 8, the dichloramine concentration immediately downstream of the MP UV system has been consistently higher than the other two locations. This implies that $NHCl_2$ formation may be promoted by MP UV. However, the $NHCl_2$ concentration in the pool is consistently low and has not indicated any trend of increasing or decreasing following initiation of MP UV treatment.

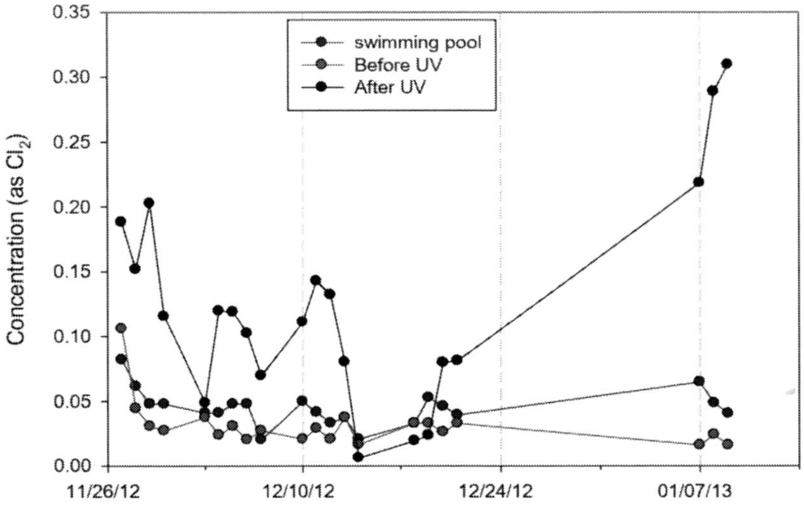

Figure 8. Dichloramine concentration at three different locations: swimming pool, before and after UV source.

Trichloramine (NCl_3)-MIMS

Measurements of inorganic combined chlorine, particularly trichloramine, are often used as an indicator of swimming pool water and air quality. As cited by the IARC (2004), Kirk and Othmer (1993) identified that trichloramine aqueous concentrations of 0.02 mg/L (as Cl_2) can cause strong and unpleasant odors. At this pool, the highest and lowest concentrations of liquid-phase trichloramine were 2.2 and 0.003 mg/L (as Cl_2) (near to DL of 0.0036 mg/L) with a mean concentration of 0.42 mg/L as Cl_2, as shown in Figure 9 and Table 4 (1st year).As shown in Figure 9, the peak NCl_3 concentrations coincided with the free chlorine excursions. A weekly pattern of relatively high concentration on Monday, followed by steady decline during the week, was often observed for trichloramine.

By comparing Figures 5, 7 and 9, it appears that the NCl_3 concentration is consistently higher than the other inorganic chloramines in this swimming pool, which it is consistent with results of Blatchley and Cheng (2010) and Li and Blatchley (2007). These previous studies indicated that

NCl_3 is formed directly from reaction between free chlorine and organic-N compounds (*e.g.*, urea, creatinine, amino acids). On the other hand, NCl_3 will also undergo hydrolysis to yield $NHCl_2$ and NH_2Cl, and it is quite volatile. Therefore, NCl_3 exists as an intermediate in swimming pools. The high Cl:N molar ratio that characterizes this and (most) other chlorinated pools also favors preferential NCl_3 formation, relative to the other inorganic chloramines (Blatchley and Cheng, 2010).

No obvious changes in the trends of liquid-phase NCl_3 concentration have been observed to date to have resulted from inclusion of the MP UV system.

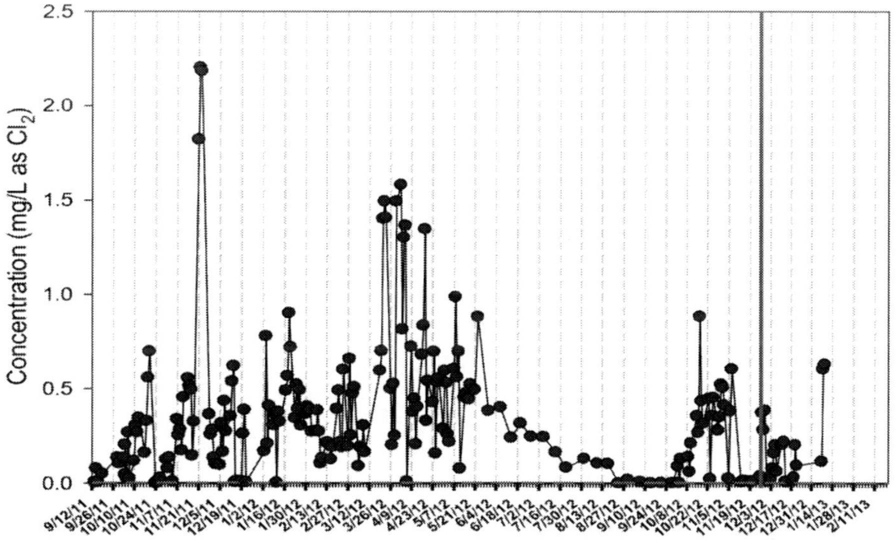

Figure 9. Trichloramine (NCl_3) range at high school measured by MIMS.

It has been estimated that trichloramine is 1000 times more volatile than monochloramine (Holzwarth *et al.*, 1984), and NCl_3 has been associated with occupational asthma, acute lung injury, and it has been indicated to cause irritation of the eyes and nose (Hery *et al.*,1995; Massin *et al.*, 1998; Bernard *et al.*, 2003). Therefore, if the concentration of trichloramine in the liquid phase increased, one would expect that the concentration of trichloramine in the gas phase would also increase. However, gas-phase NCl_3 concentration also depends on the efficiency of gas-phase transfer and the behavior of the HVAC system. In general, it is expected that high liquid-phase NCl_3 concentration offers the potential for high gas-phase NCl_3 concentration. However, the relationship between these two parameters is complicated by other factors.

According to Figure 10, trichloramine concentration immediately after MP UV irradiation is generally lower than immediately after UV, and generally lower than in the pool as well. This behavior is consistent with UV-induced photodecay of this compound, which has been reported previously (Li and Blatchley, 2009). For the conditions of operation used with this UV system, which were selected to be typical of UV installations in pools, the extent of NCl_3 photodecay appears to be relatively small for each pass of the water through the system. However, it is important to recognize that UV systems in pools are generally included as part of the recirculating system for pool water. As such, water in a pool will be exposed to UV radiation repeatedly, and the accumulated exposure to UV has the potential to decrease the concentrations of some photolabile compounds, such as NCl_3. The effect of UV on NCl_3 concentration in a pool will also be influenced by liquid-phase (non-photochemical) reactions that take place in the pool itself. At present, the balance of the combined effects of photodecomposition and non-photochemical reactions relative to NCl_3 remains incompletely defined.

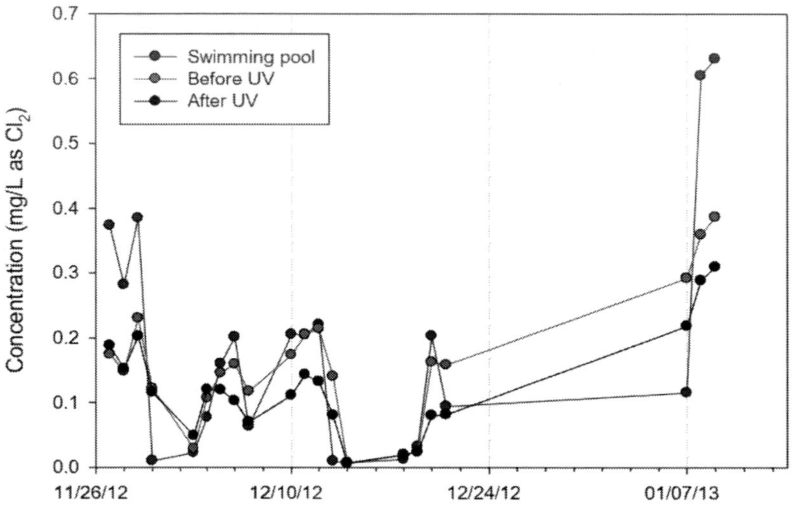

Figure 10. Trichloramine concentration at three different locations: swimming pool, before and after UV source.

Dichloroacetonitrile ($CNCHCl_2$)-MIMS

According to Table 4 and Figure 11 the highest and lowest dichloroacetonitrile (DCAN) concentration at this pool were 0.03 and 0.001 (below the detection limit of 0.002 mg/L), respectively, with a mean value of 0.0086 mg/L for the first year of the experiment. As illustrated in Figures 4 and 11, the free chlorine excursions appear to promote DCAN formation. A pattern of DCAN decline from Monday through Friday was often observed.

The formation and decomposition rates of DCAN are slow in swimming pools. It is hypothesized that DCAN concentration depends largely on pool usage. As shown in Figure 11, the concentration of DCAN started to increase gradually from October 2011 until December 2011, when competitive swimmers and students from physical education classes were using the pool (heavy usage led to net production during this period). Then from February 2012 through March 2012, a period of minimal usage of pool, the concentration of DCAN began to decrease (light usage led to net consumption during this period). This pattern was repeated in the period from the 3rd week of March 2012 when students from physical classes started to use pool until the 3rd week of May 2012. After the third week of May 2012, DCAN concentration decreased gradually because of no usage of pool and again DCAN concentration began to gradually increase from end of September when pool was opened. The pattern of DCAN concentration in the pool since the inclusion of the MP UV system does not appear to have changed markedly to date.

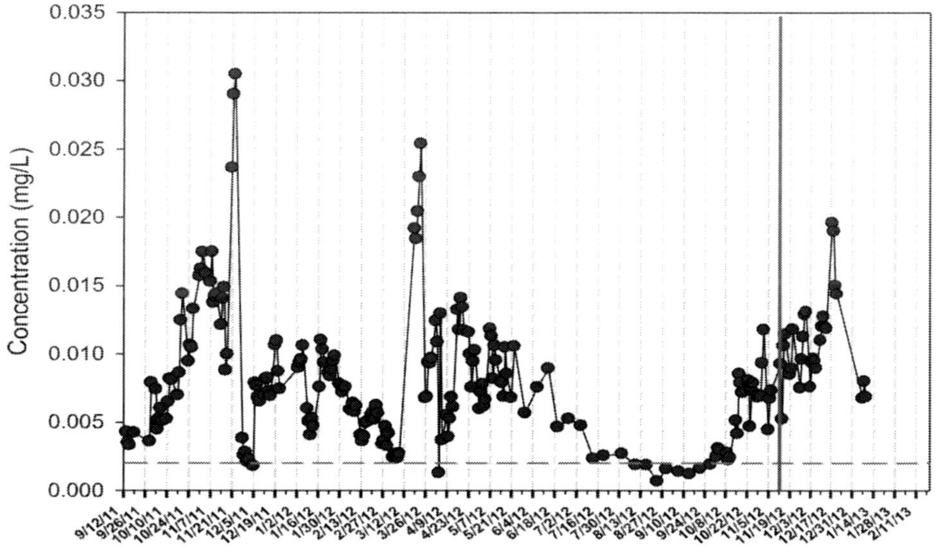

Figure 11. Dichloroacetonitrile ($CNCHCl_2$) range at high school measured by MIMS.

Figure 12 shows DCAN concentration at three different locations in the period immediately following inclusion of MP UV. The DCAN concentration at all three locations follows a similar pattern, which is consistent with the slow rates of DCAN formation and decay. The concentrations of DCAN before and after UV are almost identical.

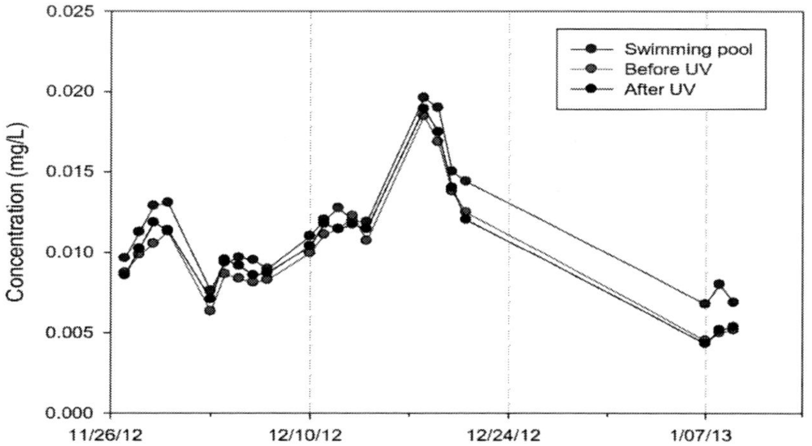

Figure 12. DCAN concentration at three different locations: swimming pool, before and after UV source.

Chloroform ($CHCl_3$)-MIMS

According to Table 4 and Figure 13, the highest and lowest chloroform concentration at this pool were 0.21 and 0.02 mg/L, respectively, with a mean value of 0.081 mg/L during the first year.

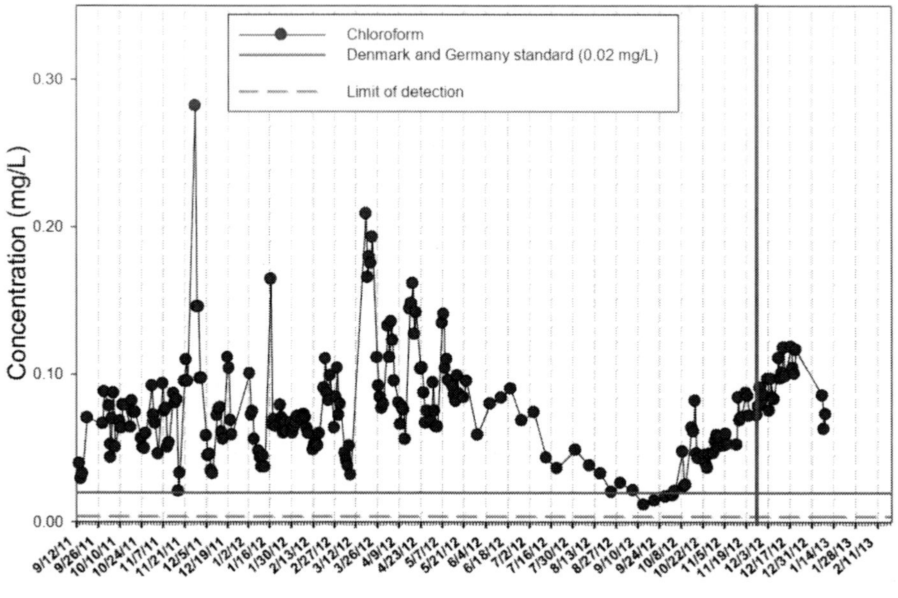

Figure 13. Chloroform ($CHCl_3$) range at high school measured by MIMS.

The horizontal solid red line in Figure 13 indicates the Denmark and Germany standard for the maximum concentration of chloroform in swimming pools (0.02 mg/L). As shown in Figure 13, the chloroform concentration at this pool exceeded this standard for almost all samples. The USA has not been established any regulation or guideline for chloroform in swimming pools.

The peaks of $CHCl_3$ concentration in Figure 13 generally coincided with the free chlorine excursions. A time-dependent pattern was evident for chloroform, similar to the observations of Kristensen *et al.* (2010), wherein the THM concentration started to increase in the evening when the pool closed down and kept increasing until the pool opened again the following morning, when it began to decline during the period of swimming activity. Kristensen *et al.* (2010) suggested that because THMs have a relatively high Henry's Law constant, activity in the water will promote THM transfer from the water to the air. Also they suggested that by increasing the bather load and their activity, THMs will transfer faster from the swimming pool water to the air.

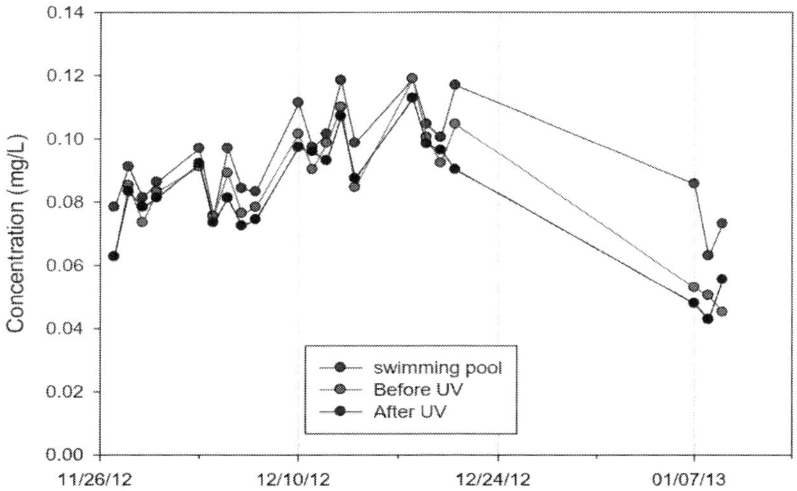

Figure 14. Chloroform concentration at three different locations: swimming pool, before and after UV source.

Figure 14 shows chloroform concentration at three different locations. The chloroform concentration for the sample collected from the swimming pool was consistently slightly higher than other two locations, but no obvious effect of MP UV has been observed. Overall, this suggests that the majority of chloroform formation takes place in the pool, and the dynamics of this compound do not appear to have been influenced by UV exposure.

Cyanogen Chloride (CNCl)-MIMS

Cyanogen chloride is known to be produced and oxidized rapidly from reactions involving amino acids and free chlorine (Na and Olson, 2004); hence CNCl is an intermediate in swimming pools. In the absence of free chlorine at near neutral pH, the half-life of CNCl due to OH-assisted hydrolysis is greater than 19 hr (Na and Olson, 2004). The decomposition of CNCl is known to be catalyzed by presence of free chlorine (Na and Olson, 2004; Edwards *et al.,*1986; Baily *et al.,*1973). Na and Olson (2004 and 2006) proposed mechanisms to describe the formation of CNCl from glycine and the OCl^--catalyzed decomposition of CNCl.

From Table 4 and Figure 15, the highest and lowest cyanogen chloride concentrations at this pool were 0.2 and 0.001 mg/L, respectively, with a mean value of 0.024 mg/L during a 1st year period. The toxicology of cyanogen chloride has not yet been defined for swimming pool exposure; however, the American Conference of Governmental Industrial Hygienists (ACGIH) has identified an aqueous Threshold Limit Value for CNCl of 0.3 mg/L (Weaver *et al.*, 2009). The data represented in Figure 15 indicate that the concentration of CNCl in this pool is consistently below this limit. The weekly pattern of concentration decrease from Monday through Friday was observed often. On dates with free chorine excursions, the CNCl concentration was low because as mentioned earlier, CNCl will undergo OCl^- - catalyzed oxidation (Na and Olson, 2004).

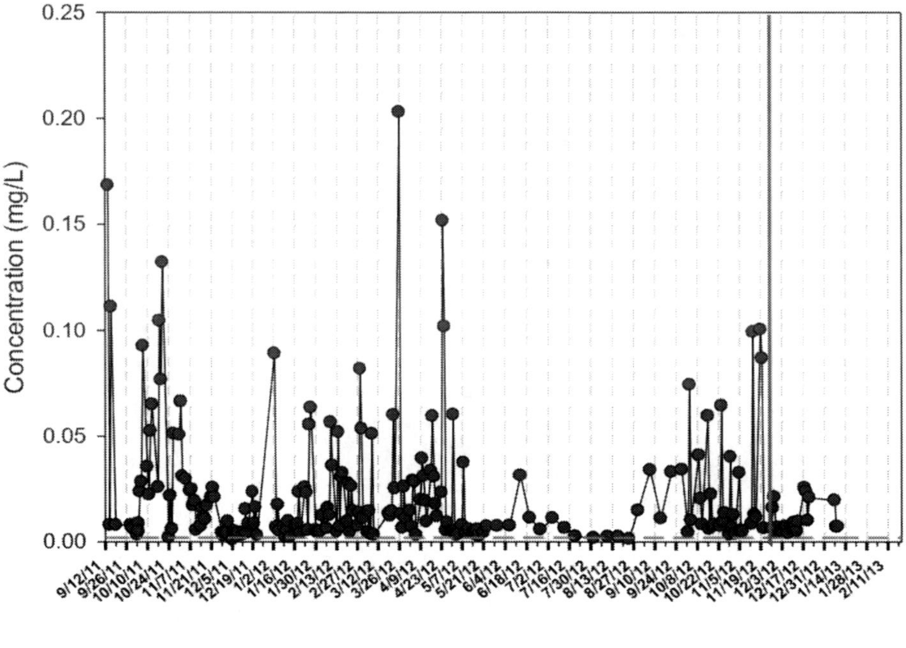

Figure 15. Cyanogen chloride (CNCl) range at high school pool measured by MIMS.

As indicated in Figure 16, the concentration of CNCl was usually higher in the "after UV" sample than in the "before UV" sample, which is consistent with the observation of Weng *et al.* (2012) that indicated that UV promotes CNCl formation. However, the concentration in the pool was consistently higher than either sample, thereby suggesting that (net) formation in the pool was faster than reactions that were attributable to UV exposure.

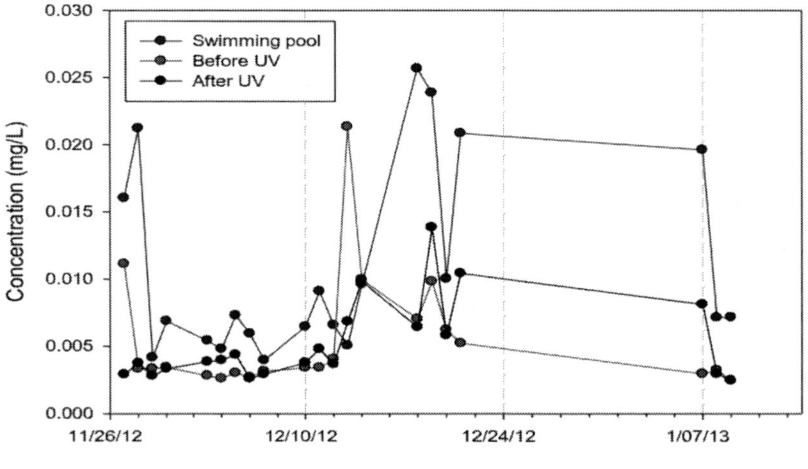

Figure 16. Cyanogen chloride concentration at three different locations: swimming pool, before and after UV source.

Dichloromethylamine (CH_3NCl_2)

From Table 4 and Figure 17, the highest and lowest dichloromethylamine concentrations at this pool were 1.9 mg/L and undetectable, respectively, with a mean value of 0.223 mg/L during the first year of sample collection.

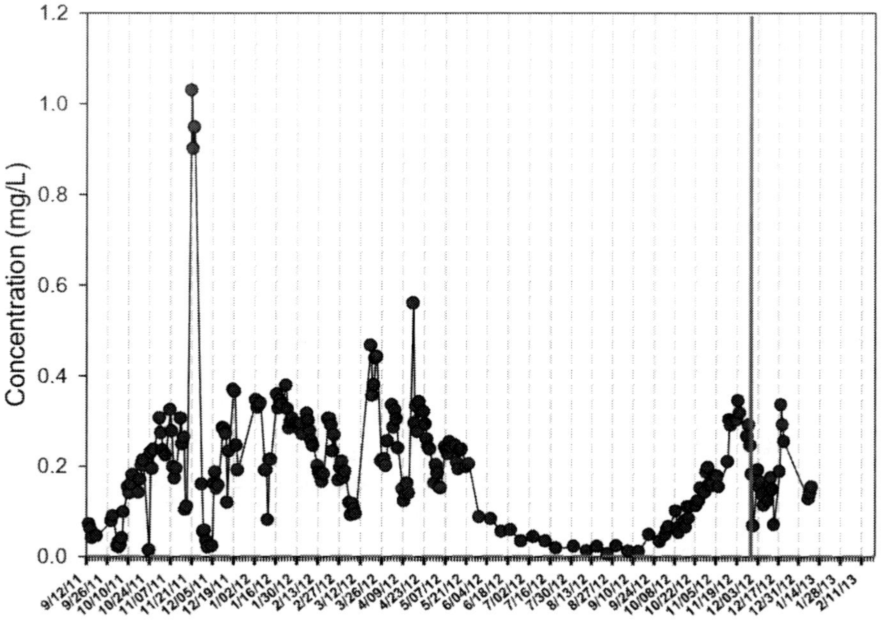

Figure 17. Dichloromethylamine (CH_3NCl_2) range at high school pool measured by MIMS.

As shown in Figure 17, CH_3NCl_2 concentrations increased during the free chlorine excursions. The weekly pattern of concentration decrease from Monday through Friday was observed often. The long-term pattern of CH_3NCl_2 was qualitatively similar to the DCAN concentration pattern described in previous section. No obvious changes in the trends of CH_3NCl_2 were observed to result from inclusion of the MP UV system at the pool.

As illustrated in Figure 18, the concentration of dichloromethylamine has followed a similar pattern at all three locations since the inclusion of the MP UV system. The "after UV" samples typically contain a CH_3NCl_2 concentration that is slightly lower than the "before UV" sample, but no clear trend of CH_3NCl_2 concentration has emerged from the data to date.

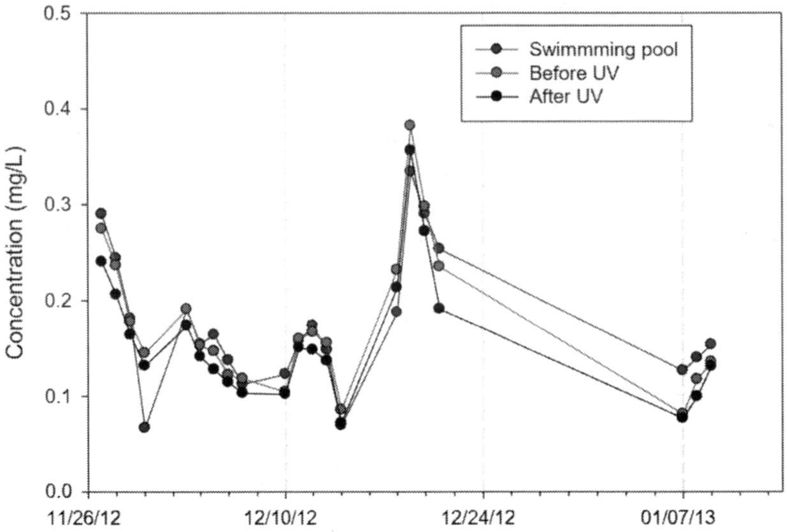

Figure 18. Dichloromethylamine concentration at three different locations: swimming pool, before and after UV source.

Gas-phase trichloramine concentration measurements in indoor chlorinated swimming pool facilities

Figure 19 illustrates the relationships between bather load, gas-phase trichloramine concentration. For interpretation of Figure 19, the vertical blue bars indicate the bather load (right axis) at the high school swimming pool for each date at the time of gas-phase trichloramine measurements. For dates that do not show a bather loading bar, no swimmers were in the pool at the time of measuring. However, sometimes the pool was used heavily before the NCl_3 gas-phase measurement and no swimmers were present at the time of measuring, hence the gas-phase concentration of NCl_3 was high at the pool during the period of measurements. The highest and lowest gas-phase trichloramine concentration at this pool were 0.62 mg/m^3 and undetectable, respectively, with the mean value of 0.15 mg/m^3 during the first year of the study.

As shown in Figure 19, from 12/05/12 (a week after MP UV was installed at the pool) the trichloramine concentration in the gas-phase began to give the negative results consistently, based on the analytical method that is used in this experiment for measuring NCl_3 concentration in the gas-phase (the negative result never occurred in the 1st year of experiment when only chlorination was used as a water treatment process in the pool). The method is based on the difference in colorimetric responses between the blank and sample gas-washing bottles. This signal is assumed to be attributable to the gas-phase NCl_3 concentration in this analytical method, and for pools where UV is not used, this method has been demonstrated to provide accurate measurements (Weng *et al.*, 2011). It is assumed that gas-phase NCl_3 (which is presumably dominant in the sample gas-washing bottle) and other oxidizing compounds (which are assumed to be trapped in the blank gas-washing bottle) are trapped by the DPD/KI solution and quantified photometrically. Test results that yield apparent "negative" concentration are attributable to a condition in which the blank bottle has a stronger colorimetric signal (the

Würster dye gave pinker color) than the sample bottle. It is hypothesized that the MP UV somehow changed the water chemistry at the pool, so that the volatile DBPs that are transferred from liquid to the gas phase behave differently in this analytical method than NCl_3. An alternative explanation is that by applying the MP UV as a secondary treatment, the NCl_3 concentration in the gas-phase was reduced to the point that it could not be reliably quantified by this analytical method; however, this is not entirely consistent with the measurements of liquid-phase NCl_3 reported above. The pool patrons have indicated (anecdotally) that the strength of the "chlorine odor" has diminished since inclusion of the MP UV system.

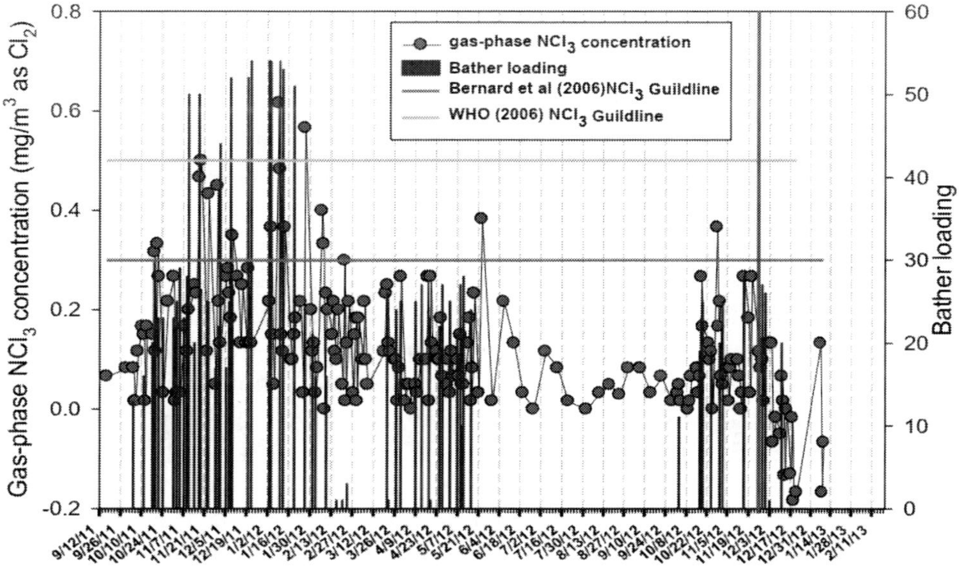

Figure 19. Gas-phase trichloramine concentration range (left y-axis) measured by air sparging DPD/KI method and bather loading (right y-axis).

The horizontal pink and cyan lines in Figure 19 indicate the suggested gas-phase limits of Bernard *et al.* (2006): 0.3 mg/m^3, and WHO (2006): 0.5 mg/m^3, respectively. These guidelines were established to limit irritation of the eyes and respiratory system. As shown in Figure 19, the concentration of NCl_3 in the gas-phase appeared to be influenced by increases in bather load and/or swimmer activity, which is consistent with the finding of Weng *et al.* (2011). The highest NCl_3 concentration in the gas phase happened when the number of swimmers was around 54, when the competitive swimmers practiced at the pool.

CONCLUSIONS

During the first twelve months of monitoring at the high school pool, free chlorine concentration exceeded the maximum recommended concentration of the NSPF (2010) of 5 mg/L (as Cl_2), eight times. These free chlorine excursions had a direct and immediate effect on the concentrations of a number of volatile DBPs. The peak all volatile DBP concentrations (except CNCl) coincided with the free chlorine excursions. The CNCl concentration was low on dates with free chorine excursions, probably because CNCl will undergo OCl^- - catalyzed oxidation.

A weekly pattern of volatile DBP concentrations was observed, in which the concentrations of DBPs were typically high on Mondays and decreased gradually through the remainder of the work week, when the pool was experienced heavy use; pool use on weekends was less than during the week. It is hypothesized that this weekly pattern was due to pool usage.

Among the inorganic chloramines, the NCl_3 concentration was consistently higher than the other inorganic chloramines (also higher than other volatile DBPs) in this swimming pool. This is believed to be because in swimming pools, NCl_3 will be formed directly from reactions between free chlorine and organic-N compounds (*e.g.*, urea). On the other hand, NCl_3 will also decay by hydrolysis to yield $NHCl_2$ and NH_2Cl, and it is quite volatile. Therefore NCl_3 exists as an intermediate in swimming pools.

The behavior of DCAN was quite dynamic during this sampling period. DCAN showed relatively slow increases during periods of heavy pool use, presumably because the pool received inputs of the organic-N precursors that are responsible for formation of DCAN, and other DBPs. In contrast, when the pool experienced low or zero bather loads, the DCAN concentration showed a slow, steady decay.

Regarding air quality, trichloramine concentration in the gas-phase was measured in this pool facility. Results of the first year of analysis under the chlorination operating condition, demonstrated that NCl_3 transfer to the gas phase was influenced by bather loading and swimmer activity, as well as aqueous NCl_3 concentration. Gas-phase trichloramine concentration in the pool appears to have changed markedly since the inclusion of the MP UV system. The reason for this change is not well-defined, based on data available at this time.

According to Weng *et al.* (2012), UV irradiation results in enhancement of the formation of several DBPs, including DCAN, CNCl, and CH_3NCl_2, all of which are expected to behave as intermediates in pools. However, no obvious trends in the concentrations of DCAN, CNCl, and CH_3NCl_2 have been observed (to date) to have resulted from inclusion of the MP UV system. UV irradiation has been shown to increase the rate of free chlorine consumption in pools.

ACKNOWLEDGEMENTS

Support for this work was provided by grants from the National Swimming Pool Foundation (NSPF), the American Chemistry Council, and Engineered Treatment System, LLC (ETS). The cooperation and support of pool operators and facility managers involved in this study is also greatly appreciated. AP Chemistry students at the high school who are involved in this research include; in 1st year: Kay Krueger, Gauri Shastri, Rebecca Gwin, Parker Rhoads, Xiaoqin Zhu, Rett Berry, and Krithika Subramaniam, in 2nd year: Anuj Patel, Megan Chiu, Stuart Kurtz, Darian Williams, Molly Towns, Brody Forbes, Melinda Crane, Phoebe Moh, and Peter VonWerder (Teacher).

REFERENCES

Aggazzotti, G.; Giacobazzi, P.; Predieri, G. (2007) "Nitrogen trichloride detection in indoor swimming pool air. In P. E. R. B. III (Ed.)," University of Modena and Reggio Emilia.

Bailey,P.L; Bishop, E. (1973) "Hydrolysis of Cyanogen Chloride," *J. Chem. Soc. Dalton Trans.*, **9**, 912-916.

Bernard, A.; Nickmilder, M.; Voisin, C.; Sardella, A. (2009) "Impact of Chlorinated Swimming Pool Attendance on the Respiratory Health of Adolescents," *Pediatrics*, **124**, 1110-1118.

Bernard, A.; Carbonnelle, S.; Dumont, X.; Nickmilder, M. (2007) "Infant Swimming Practice, Pulmonary Epithelium Integrity, and the Risk of Allergic and Respiratory Diseases Later in Childhood," *Pediatrics*, **119** (6), 1095-1103.

Bernard, A.; Carbonnelle, S.; Michel, O.; Higuet, S.; de Burbure, C.; Buchet, J.-P.; Hermans, C.; Dumont, X.; Doyle, I. (2003) "Lung hyperpermeability and asthma prevalence in schoolchildren: Unexpected associations with the attendance at indoor chlorinated swimming pools," *Occup. Environ. Med.*, **60**, 385-394.

Beyer, A; Worner, H; Van Lierop R. (2004) "The use of UV for destruction of combined chlorine," Technical note, Wallace & Tiernan. Retrieved Jan 1, 2013, from http://www.pwtag.org/researchdocs/Used%20Ref%20docs/25%20the_use_of_uv_for_destruction_of_combined_chlorine%20Wallace%20and%20Tiernan.pdf

Blatchley III, E. R. and Cheng, M., (2010) "Reaction Mechanism for Chlorination of Urea," *Environ. Sci. & Tech.*, **44**, 22, 8529-8534.

Bougault, V.; Turmel, J.; Boulet, L.P. (2010) "Effects of Intense Swimming Training on Rhinitis in High-Level Competitive Swimmers," *Clinical and Experimental Allergy*, **40**, 1238-1246.

Cassan, D.; Mercier ,B.; Castex, F.; Rambaud, A.(2006) " Effects of medium-pressure UV lamps radiation on water quality in a chlorinated indoor swimming pool," *Chemosphere*, **62**, 1507–13.

Cassan, D.; Mercier ,B.; Castex, F.; Rambaud, A. (2011) "Nitrogen trichloride levels in air in chlorinated indoor swimming pools treated bymedium-pressure UV radiation," Proceeding of IOA, IUVA World Congress & Exhibition, Paris, France.

Craik, S.A.; Weldon, D.; Finch, G.R.; Bolton, J.A.; Belosevic, M. (2001) "Inactivation of *Cryptosporidium parvum* oocysts using medium- and low-pressure ultraviolet radiation," *Wat. Res.*, **35**, 6, 1387-1398.

DeLaat, J; Boudiaf, N; Dossier-Berne, F. (2010) "Effects of dissolved oxygen on the photodecomposition of monochloramine and dichloramine in aqueous solution by UV irradiation at 253.7 nm," *Wat. Res.*, **44**, 10, 3261-3269.

Edwards, J.O; Erstfeld, T.E.; Ibne-Rasa, K.M.; Levey, G; Moyer, M. (1986) "Reaction rates for nucleophiles with cyanogen chloride: Comparison with two other diagonal carbon compounds," *Int. J. Chem. Kinet.*, **18**, 2, 165-180.

EPA (1992). Monochloramine (CASRN 10599-90-3) Retrieved July 17, 2012, from http://www.epa.gov/iris/subst/0644.htm

Fantuzzi, G.; Righi, E.; Predieri, G.; Giacobazzi, P.; Mastroianni, K.; Aggazzotti, G. (2010) "Prevalence of Ocular, Respiratory and Cutaneous Symptoms in Indoor Swimming Pool Workers and Exposure to Disinfection By-Products (DBPs)," *Inter. J. of Envi. Res. and Pub. Health*, **7**, 1379-1391.

Harp, D. (2002) *Current Technology of Chlorine Analysis for Water and Wastewater*, Technical Information Series- Booklet No. 17.

Hansen, K. M.S.; Zortea, R.; Piketty, A.; Vega, S.R.; Henrik, R.A. (2013) "Photolytic removal of DBPs by medium pressure UV in swimming pool water," *Science of the Total Environment*, **443**, 850-856.

Hery, M.; Hecht, G.; Gerber, J. M.; Gender, J. C.; Hubert, G.; Rebuffaud, J. (1995). "Exposure to chloramines in the atmosphere of indoor swimming pools," *Ann. Occup. Hyg.*, **39**(4), 427-439.
Hijnen, W.A.M.; Beerendonk, E.F.; Medema, G.J. (2006), "Inactivation credit of UV radiation for viruses, bacteria and protozoan (oo)cysts in water: a review," *Wat. Res.*, **40** (1), 3-22.

Holzwarth, G.; Balmer, R. G.; Soni, L. (1984) "The fate of chlorine and chloramines in coolingtowers– Henrys law constants for flashoff," *Wat. Res.*, **18**, 11, 1421-1427.

IARC (2004) "IARC Monographs on the evaluation of carcinogenic risks to humans," from http://monographs.iarc.fr/ENG/Monographs/vol84/mono84-8.pdf

Jafvert, C.T.; Valentine, R.L. (1992) "Reaction Scheme for the Chlorination of Ammoniacal Water," *Envi. Sci. & Tech.* **26**, 3, 577-586

Kristensen, G.H.; Kausen, M. M.; Hansen, V. A.; Lauritsen, F.R. (2009) "Online monitoring of the dynamics of trihalomethane concentrations in a warm public swimming pool using an unsupervised membrane inlet mass spectrometry system with off-site real-time surveillance," *Rapid Commun. Mass Spectrom.*, **24**, 30-34.

Kristensen, G.H.; Klausen,M.M.; Andersen, H.R.; Erdinger, L.; Lauritsen, F.R.; Arvin, E. (2009) "Full scale test of UV-based water treatment technologies at Gladsaxe Sportcentre — with and without advanced oxidation mechanisms," *Proceeding of the Third International Swimming Pool and Spa Conference*, March 17–20; London, United Kingdom. Retrieved Jan. 3, 2013, from http://www.pwtag.org/researchdocs/Used%20Ref%20docs/6%20Paper%204.1%20UV%20&%20THMs%20Denmark.pdf

Kransner, S. W.; McGuire, M. J.; Jacangelo, J. G.; Patania, N. L.; Reagan, K. W.; Aieta, E. M. J. (1989) "The occurrence of difinfection by-products in US drinking water," *Americ. Water. Work Assoc.*, **81**, 8, 41-53.

Li, J.; Blatchley, E. R. III, (2009) "UV Photodegradation of Inorganic Chloramines," *Envi. Sci. Tech.,* **43**, 1, 60-65.

Li, J.; Blatchley III, E. R. (2007) "Volatile disinfection by-product formation resulting from chlorination of organic-nitrogen precursors in swimming pools," *Envi. Sci. Tech.*, **41**, 6732-6739.

Lindstrom, A. B.; Pleil, J. D.; Berkoff, D. C. (1997) "Alveolar breath sampling and analysis to assess trihalomethane exposures during competitive swimming training," *Envi. Health Persp.*, **105**, 6, 636-642.

Liu, W.; Cheung, L.;Yang, X.; Shang, C. (2006). "THM, HAA and CNCl formation from UV irradiation and chlor(am)ination of selected organic waters," *Wat. Res.*, **40**, 10, 2033-2043.

Massin, N.; Bohadana, A. B.; Wild, P.; Héry, M.; Toamain, J. P.; Hubert, G., (1998) "Respiratory symptoms and bronchial responsiveness in lifeguards exposed to nitrogen trichloride in indoor swimming pools," *Occup. and Envi. Med*, **55**, 4, 258-263.

Na, C.; Olson, T. M. (2006) "Mechanism and Kinetics of Cyanogen Chloride formation from the chlorination of glycine," *Envi. Sci. Tech.*, **40**, 1469-1477.

Na, C.; Olson, T. M. (2004) "Stability of Cyanogen Chloride in the Presence of Free Chlorine and Monochloramine," *Envi. Sci. Tech.*, **38**, 22, 6037-6043.

National Swimming Pool Foundation (NSPF) (2010), NSPF Certified Pool-Spa Operator Handbook National Swimming Pool Foundation, Colorado Springs, CO.

National Institute of Occupational Safety and Health (NIOSH) (2003), *The Emergency Response Safety and Health Database: CYANOGEN CHLORIDE,* NIOSH, CDC Atlanta GA, from http://www.cdc.gov/NIOSH/ershdb/EmergencyResponseCard_29750039.html

Periera, M.A.; Daniel, F.B.; Lin, E.L.C. (1985) "Relationship between metabolism of haloacetonitriles and chloroform and their carcinogenic activity," *Wat. Chlorin.,* **5**, Chemistry, Environmental Impact and Health Effects, Lewis, Chelsea, MI, 229-236.

Safranek, T.; Semerena, S.; Huffman, T.; Theis, T.M.; Magri, J.; Török, T.; Beach, M.J.; Buss, B. (2007) "Ocular and Respiratory Illness Associated with an Indoor Swimming Pool - Nebraska, 2006," *MMWR*, **56**, 36, 929-932.

Shang, C.; Gong, W.-L.; Blatchley, E. R. (2000) "Breakpoint chemistry and volatile byproduct formation resulting from chlorination of model organic-N compounds," *Envi. Sci. Tech.,* **34**, 9, 1721-1728.

Shang, C.; Blatchley, E. R. III, (1999) "Differentiation and Quantification of Free Chlorine and Inorganic Chloramines in Aqueous Solution by MIMS," *Envi. Sci. & Tech.*, **33**, 13, 2218-2223

Villanueva, C.M; Cantor, K.P.; Grimalt, J.O;Malatas,N; Silverman, D.; Tardon, A.; Garcia-Closas, R.; Serra, C.; Carrato, A.; Catano-vinyals, G.; Marcos, R.; Rothman, N.; Real, F.X.; Dosemeci, M.; Kogevina, M. (2007) " Bladder cancer and exposure to water disinfection by-products through ingestion, bathing, showering, and swimming in pools," *American Journal of Epidemiology,* **156**, 2, 148-156.

Weng S.C.; Li, J.; Blatchley III, E.R. (2012) "Effects of UV_{254} irradiation on residual chlorine and DBPs in chlorination of model organic-N precursors in swimming pools," *Wat. Res.*, **46**, 2674-2682.

Weng S.C.; Weaver,W.A; Zare Afifi M.; Blatchley, T.N.; Cramer, J. S.; Chen, J.; Blatchley III, E. R. (2011) "Dynamics of gas-phase trichloramine in chlorinated, indoor swimming pool facilities," *Indoor Air*, **21**, 391-399.

Weng, S.C.; Blatchley III, E.R. (2011) "Disinfection by-product dynamics in a chlorinated, indoor swimming pool under conditions of heavy use: national swimming competition," *Wat. Res.,* **45**,16, 5241 5248.

Weaver, W. A.; Li, J.; Wen, Y.; Johnston, J.; Blatchley, M. R.; Blatchley, E. R. III, (2009) "Volatile disinfection by-product analysis from chlorinated indoor swimming pools," *Wat. Res.*, **43**, 13, 3308-3318.

Weaver, W.A. (2008) "Volatile Disinfection Byproduct Analysis from Chlorinated Indoor Swimming Pools," MS Thesis, School of Civil Engineering, Purdue University, West Lafayette, IN.

Weisel. C.P; Shepard, T.A. (1994) "Chloroform exposure and the body burden associated with swimming in chlorinated pools," Wang RGM, ed. *Wat. Contam. and Health*. New York: Marcel Dekker.

World Health Organization (WHO), (2006) *Guidelines for Safe Recreational Water Use,* **2***: Swimming Pools and Similar Environments*, WHO, Geneva. Retrieved November 24, 2010, from http://www.who.int/water_sanitation_health/bathing/srwe2chap4.pdf.

Influence of Biofilm on DBP Formation in Simulated Water Distribution System

Zhikang Wang[1], Kimberly M. Coburn[2], Cassondra Jennings[3], and Youngwoo Seo*[1,2]
1 Department of Chemical and Environmental Engineering, University of Toledo, Mail Stop 307, 3048 Nitschke Hall, Toledo, OH 43606, USA
2 Department of Civil Engineering, University of Toledo, Mail Stop 307, 3006 Nitschke Hall, Toledo, OH 43606, USA
3 Department of Bioengineering Engineering, University of Toledo, Mail Stop 303, 5051 Nitschke Hall, Toledo, OH 43606, USA

ABSTRACT

In order to decrease the formation of disinfection by-products (DBPs) in water distribution systems, significant efforts have been made for the reduction of DBP precursors entering to the distribution system. However, considering the regulatory presence of residual chlorine and ubiquitous presence of microbial biofilm in distribution systems, traditionally studied untreated NOM may not be the sole contributor to DBP formation. This study investigated the influence of biofilm on DBP formation in simulated water distribution systems under chlorine disinfection (Cl_2). Lab scale annular reactors fed by treated natural organic matter (NOM) were operated for six month to establish preformed biofilm. After establishing biofilm formation, three different Cl_2 residual concentrations (0.5, 2 and 4 mg/L) were applied to the reactors and DBP formations were monitored. Obtained results suggest that biofilm and biofilm associated NOM can react with residual Cl_2 and play a significant role on both the formation and decay of DBPs.

KEYWORDS: Biofilm, Bioreactor, Disinfection By-Products (DBPs), Chlorine Disinfection, and Natural Organic Matter (NOM)

INTRODUCTION

Since Disinfection By-Products (DBPs) were discovered in the 1970s, the presence of DBPs in drinking water has become a prominent concern for public health [1]. While significant efforts have made to understand DBP formation at the water treatment plant, DBP formation in the water distribution system is not well understood yet. It has been reported that concentrations of DBPs present in the distribution system tap water were even higher than those of treated water at the water treatment plant [2]. Untreated natural organic matter (NOM) entering the distribution system has been commonly acknowledged as major DBP precursors in water distribution systems [3, 4]. However, studies also reported increased DBP formations exceeding expected DBP formation from the reaction between NOM and chlorine alone, which may be attributed to biofilm or biofilm associated NOM reacting with residual disinfectants [5]. Biofilm could provide a dynamic repository for various organic materials present in the bulk fluid and subsequently utilize them as a primary nutrient source [6, 7]. Heterogeneous structure of biofilm may also facilitate the sorption of available untreated NOM in distribution system [8]. Furthermore, extracellular polymeric substances (EPS) of biofilm contain a significant concentration of organic carbon and nitrogen rich organic compounds [proteins/amino acids, polysaccharides, lipids, and nucleic acids], which possess similar chemical composition and reactivity to traditional DBP precursors (NOM).

Recent studies reported the impact of algal species on DBP formation in water treatment processes and results indicated that algal organic matter could serve as a significant DBP precursor along with NOM, suggesting that biomolecules have a considerable role in DBP formation [9, 10]. Thus, structural characteristics of biofilm EPS and their reactivity with disinfectants may greatly contribute to DBP formation. However, the role of biofilm on DBP formation is not well understood due to the complexity of water distribution systems.

The primary objective of this study was to elucidate the role of biofilm on DBP formation in a simulated drinking water distribution system. Chlorine was used as a model disinfectant and both C-DBP and N-DBP formations were monitored from the reactor effluents.

MATERIALS AND METHODS

Annular reactor operation

Mixed species bacteria obtained from a drinking water treatment plant were inoculated and grown in continuously flowing annular reactors (CDC reactor, Boltzmann, MT) for six months to form preformed biofilm. The feed solution to the annular reactors was made from tap water from the City of Toledo that was filtered by a granular activated carbon (GAC) filter to remove both chlorine residual and any existing DBPs. Then, humic acid (Sigma Aldrich, USA) was added to the filtered water and jar tests with aluminum sulfate were conducted to simulate water treatment processes mimicking coagulation, flocculation, and sedimentation processes. After jar tests, ozonation was applied to supernatant solution. The total organic carbon (TOC) concentration of treated humic acid solution was maintained to approximately 2 mg/L, which is a commonly found TOC concentration in water distribution systems. The feed solution was delivered to reactors with the flow rate of 0.5 mL min^{-1} using peristaltic pumps (Masterflux, Cole-Parmer, IL) for an hydraulic retention time of 12 hours [11]. Pretreated humic acid solution was used as the sole carbon source for biofilm formation in this study. Cl_2 stock solution was prepared by adding bleach to DI water. Concentrations of Cl_2 solution were determined by DPD method via a spectrophotometer (HACH, USA, DR 2800). Disinfectant stock solutions were stored in a gas tight glass syringe and delivered to reactor by a syringe pump (Fisher Scientific, USA). Three different chlorine residual concentrations (0.5, 2, and 4 mg L^{-1}) were applied to annular reactors during three experimental phases.

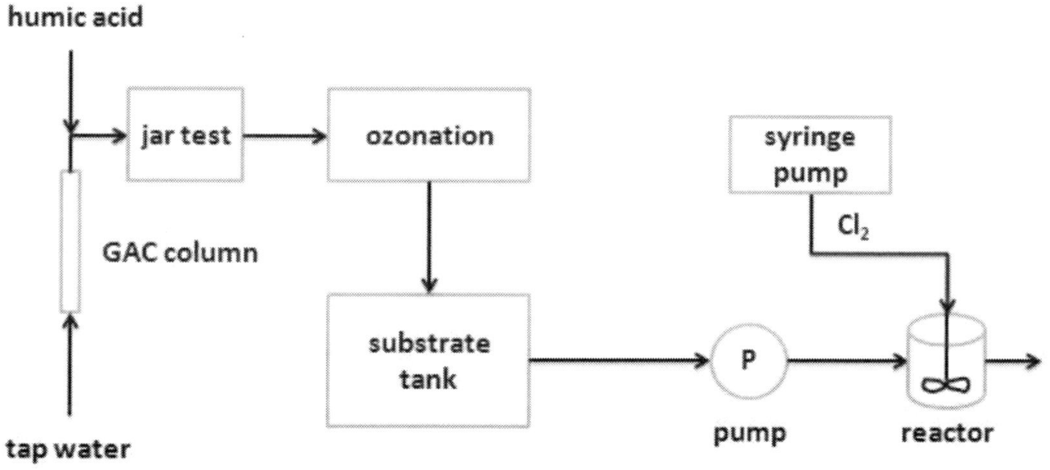

Fig. 1: The experimental setup.

DBP Analysis

For DBP formation tests, samples from reactor effluents were collected in 40 mL amber glass vials and prepared for DBP analysis. Both regulated and emerging DBPs consisting THM_4 [chloroform (CF), bromodichloromethane (BDCM), dibromochloromethane (DBCM) and bromoform (BF)], HAA_9 [monochloroacetic acid (MCAA), monobromoacetic acid (MBAA), dichloroacetic acid (DCAA), bromochloroacetic acid (BCAA), trichloroacetic acid (TCAA), bromodichloroacetic acid (BDCAA), dibromoacetic acid (DBAA), chlorodibromoacetic acid (CDBAA), and tribromoacetic acid (TBAA)], HAN_4 [trichloroactetonitrile (TCAN), dichloroacetonitrile (DCAN), bromochloroacetonitrile (BCAN), and dibromoacetonitrile (DBAN)], HK_2 [1,1-dichlro-2-propanone (1,1-DCP) and 1,1,1-trichloropropanone (1,1,1-TCP)], trichloronitromethane (TCNM) and chloral hydrate (CH) were analyzed in this study. Detailed analysis method could be found from our previous study [5]. Spike recovery was calculated to verify the validity of entire analysis in this study. Both average spike recoveries and minimum detection limits (MDL) for all tested DBPs were determined.

RESULTS AND DISCUSSION

During the first phase of the experiment, where low concentration of Cl_2 residual (0.5 mg L^{-1}) was maintained in reactors, minimal THMs and HAAs formations were observed. In addition, the formation of emerging DBPs was almost negligible, which may be attributed to both volatilization and biodegradation of DBPs by biofilm. To confirm this hypothesis, batch experiments were also conducted under the same conditions in reactors (0.5 mg L^{-1} of Cl_2 residual and 12 h reaction time) and results were comparable to the

DBP formation results from reactors confirming the volatilization of THMs, HANs, HKs, and TCNM [12, 13] and the biodegradation of HAAs [14, 15].

During the second phase, with higher Cl_2 residual, significantly increased DBP formation was observed. The levels of THMs and HAAs increased dramatically and then reached a plateau after 20 days. The formations were dramatically increased to 63 μg L^{-1} and 80 μg L^{-1} of THMs and HAAs, respectively. However, after 20 days, a decrease of DBP formation was observed at the end of the experimental phase. The formations of emerging DBPs (HANs and CH) also showed the similar trend with those of THMs and HAAs having the highest concentrations of approximately 20 μg L^{-1}. For the second phase, the trend of DBP levels (rapid increase followed by decrease) suggested that biofilm and biofilm associated NOM may initially facilitate DBP formation. However, under higher Cl_2 residual, biofilm formation decreased and biofilm was further removed by constantly applied hydraulic shear.

For the third phase, the reactor was exposed to 4 mg L^{-1} of Cl_2 residual. However, higher amounts of DBPs were not detected compared to the previous phases. In contrast to the second phase, DBP levels kept constant during this phase. Constant DBP formation during the end of second phase and the entire third phase (4 mg L^{-1} Cl_2 residual) further confirmed that treated humic acid was the main precursor in this phase due to the eradication of biofilm. Especially, compared to DBP formation results from the second phase experiments, lower formation of HANs were observed during the third phase. Since organic nitrogen is more enriched in biofilm and EPS rather than treated humic substances [5, 16-18], the increases of N-DBPs indicated that biofilm contributed to N-DBP formation in the water distribution system.

CONCLUSION

Results from this research demonstrate that NOM associated biofilm could play an important role for DBP formation and decay in water distribution systems under the presence of residual chlorine. It appears that the higher organic nitrogen content in biofilm especially influences toxic N-DBPs formation.

ACKNOWLEDGEMENT

The authors greatly appreciate the funding support from NSF (CBET0932288 and CBET1236433)

REFERENCES

[1] J.J. Rook, Chlorination reactions of fulvic acids in natural waters, Environmental Science & Technology, 11 (1977) 478-482.
[2] H. Baribeau, L. Boulos, H. Haileselassie, G. Crozes, P.C. Singer, C. Nichols, S.A. Schlesinger, R.W. Gullick, S.L. Williams, R.L. Williams, L. Fountleroy, S.A. Andrews,

E. Moffat, Formation and Decay of Disinfection By-Products in the Distribution System, American Water Works Association, Washington D.C., 2006.
[3] H. Gallard, U. von Gunten, Chlorination of natural organic matter: kinetics of chlorination and of THM formation, Water Research, 36 (2002) 65-74.
[4] A.D. Nikolaou, T.D. Lekkas, The Role of Natural Organic Matter during Formation of Chlorination By-products: A Review, Acta hydrochimica et hydrobiologica, 29 (2001) 63-77.
[5] Z. Wang, J. Kim, Y. Seo, Influence of Bacterial Extracellular Polymeric Substances on the Formation of Carbonaceous and Nitrogenous Disinfection Byproducts, Environmental Science & Technology, 46 (2012) 11361-11369.
[6] A.K. Camper, Involvement of humic substances in regrowth, International Journal of Food Microbiology, 92 (2004) 355-364.
[7] A. Rodrigues, A. Brito, P. Janknecht, J. Silva, A. Machado, R. Nogueira, Characterization of biofilm formation on a humic material, Journal of Industrial Microbiology & Biotechnology, 35 (2008) 1269-1276.
[8] C.S. Laspidou, B.E. Rittmann, A unified theory for extracellular polymeric substances, soluble microbial products, and active and inert biomass, Water Research, 36 (2002) 2711-2720.
[9] J. Fang, J. Ma, X. Yang, C. Shang, Formation of carbonaceous and nitrogenous disinfection by-products from the chlorination of *Microcystis aeruginosa*, Water Research, 44 (2010) 1934-1940.
[10] J. Fang, X. Yang, J. Ma, C. Shang, Q. Zhao, Characterization of algal organic matter and formation of DBPs from chlor(am)ination, Water Research, 44 (2010) 5897-5906.
[11] H. Baribeau, S.W. Krasner, R. Chinn, P.C. Singer, Impact of biomass on the stability of HAAs and THMs in a simulated distribution system, Journal-American Water Works Association, 97 (2005) 69-81.
[12] W.A. Weaver, J. Li, Y. Wen, J. Johnston, M.R. Blatchley, E.R. Blatchley Iii, Volatile disinfection by-product analysis from chlorinated indoor swimming pools, Water Research, 43 (2009) 3308-3318.
[13] S.W. Krasner, J.M. Wright, The effect of boiling water on disinfection by-product exposure, Water Research, 39 (2005) 855-864.
[14] H.-h. Tung, Y.F. Xie, Association between haloacetic acid degradation and heterotrophic bacteria in water distribution systems, Water Research, 43 (2009) 971-978.
[15] P. Zhang, T.M. LaPara, E.H. Goslan, Y. Xie, S.A. Parsons, R.M. Hozalski, Biodegradation of Haloacetic Acids by Bacterial Isolates and Enrichment Cultures from Drinking Water Systems, Environmental Science & Technology, 43 (2009) 3169-3175.
[16] A. Jahn, T. Griebe, P.H. Nielsen, Composition of *Pseudomonas putida* biofilms: Accumulation of protein in the biofilm matrix, Biofouling: The Journal of Bioadhesion and Biofilm Research, 14 (1999) 49 - 57.
[17] J. Wingender, T.R. Neu, H. Flemming, What are bacterial extracellular polymeric substances? , in: J. Wingender, Neu, T.R., Flemming, H. (Ed.) Microbial Extracellular Polymeric Substances: characterization, structure and function, Springer-Verlag, New York, NY, 1999, pp. 1-17.
[18] B.G. Oliver, Dihaloacetonitriles in drinking water: algae and fulvic acid as precursors, Environmental Science & Technology, 17 (1983) 80-83.

Dynamic Behavior of Volatile DBPs and their Precursors in Chlorinated, Indoor Swimming Pools

ShihChi Weng[1], Mehrnaz Zare Afifi[1], Robert W. Staton[1], Zhe Sun[1], Rett Berry[2], Xiaoqin Zhu[2], Ernest R. Blatchley III[1]

[1]School of Civil Engineering, Purdue University, West Lafayette, IN

[2]West Lafayette High School, West Lafayette, IN

ABSTRACT

The nature of swimmer activities has been linked to changes in air and water quality at swimming pool facilities. Previous studies have demonstrated correlations between bather loading, gas-phase NCl_3 concentration, liquid-phase NCl_3, CH_3NCl_2, $CHCl_3$, and urea concentrations under conditions of heavy use. Although these studies have addressed the overall influences of swimmers on air and water quality, the dynamic behavior of volatile disinfection byproducts (DBPs) and their precursors remain largely undefined. To address this issue, experiments were conducted at a chlorinated, indoor swimming pool at a public high school to examine dynamic changes in gas- and liquid-phase composition during regular practice sessions for the school's swimming team.

During the experiments, gas-phase NCl_3 concentration increased immediately after swimming activity commenced. In the liquid phase, corresponding decreases of NCl_3 and other volatile DBPs (*e.g.,* $CHCl_3$ and $CNCHCl_2$) were observed in near-surface water. The liquid-phase NCl_3 concentration for the water samples collected from the near-surface location displayed dynamic behavior that was strongly linked to swimmer activity. The concentration at this location steadily decreased as the practice session continued, resulting in almost total depletion of liquid-phase NCl_3 from the near-surface area by the end of a 3-hour practice session. However, the NCl_3 concentration in the deeper water locations (middle and bottom) remained steady during the experiment. In contrast, urea concentration showed a steady increase in the near-surface water during the practice session; however, urea concentration at the deeper sampling locations remained largely unchanged.

These results suggest that the swimmers promoted transfer of NCl_3 (and presumably other volatile DBPs) from near-surface water to the gas-phase, and introduced substantial quantities of urea (and presumably other precursors associated with human body fluids) into near-surface water. These results also indicate clear evidence of stratification of pool water under a heavy use condition. The results also have implications with respect to water and air quality in swimming pool settings, particularly under conditions of heavy use. In addition, these findings suggest that water and air quality in pools may be strongly affected by water and air circulation patterns, as well as the hygiene habits of users.

Key Words: Disinfection Byproducts, Swimmer activity, NCl_3, Urea

INTRODUCTION

Swimming is popular among all ages, in part because it is believed to promote overall health and fitness. Chlorine is added to pools to promote oxidation of reduced organic compounds, and to provide protection against waterborne microbial pathogens. In chlorinated, indoor swimming pool facilities, air and water qualitiesboth have the potential to adversely affect human health because of DBP toxicity and volatility. Swimmer activity has been demonstrated to promote liquid→gas phase transfer of volatile DBPs (Hsu *et al.*, 2009; Weng *et al.*, 2011), and to lead to introduction of DBP precursors (De Laat *et al.,* 2011). For example, NCl_3 is a volatile DBP that is produced via chlorination of a number of organic-N precursors (Li and Blatchley, 2007; Blatchley and Cheng, 2010). The only source of gas-phase NCl_3 in indoor swimming pool facilities is liquid→gas mass transfer from a pool. Gas-phase NCl_3 concentration is likely to be influenced by liquid-phase NCl_3 concentration, temperature, air circulation (as governed by HVAC system design and operation), and transport behavior at the liquid:gas interface (Bessonneau *et al.*, 2011; Schmalz *et al.*, 2011). Previous retrospective studies have shown positive correlation between irritation symptoms among swimmers and patrons and high gas-phase NCl_3 concentration at indoor pool facilities (Bowen *et al.*, 2007; Kaydos-Daniels *et al.*, 2008). Impaired respiratory behavior has been observed among elite swimmers, which suggests that adverse health implications may result from frequent visits to a swimming pool environment (Clearie, *et al.*, 2010; Bougault and Boulet, 2012). Moreover, anecdotal evidence suggests that water and air qualities in chlorinated, indoor pools both deteriorate under conditions of heavy use.

Regarding water quality, the main sources of DBP precursors in swimming pools are substances introduced by bathers (Keuten *et al.,* 2012; Hansen, *et al.*, 2012). In particular, nitrous organic compounds (N-organics) from human sweat and urine are largely responsible for formation of chloramines and other nitrogenous DBPs. For example, urea is the dominant source of organic nitrogen (organic-N) in both sweat and urine. In previous research, it has been estimated that an average of roughly 70 mL of urine is introduced by each swimmer per swimming event (Erdinger *et al.*, 1997). De Laat *et al.* (2011) reported the urea concentration in swimming pool ranged from 0.12 mg/L to 3.6 mg/L. Kim *et al.* (2002) examined the effects of human tissues (hair and skin), human fluids (salvia and urine), and personal care products (lotion) on DBP formation in bench scale experiments. The results of these experiments demonstrated that human materials were important for DBP formation in pools. Li and Blatchley (2007) examined reactions of four constituents of urine (*L*-histidine, *L*-arginine, creatinine, and urea) with free chlorine. NCl_3 observed as a common by-product of chlorination of each of these compounds. CH_3NCl_2 and $CHCNCl_2$ were identified as products of chlorination of creatinine and *L*-histidine, respectively. Blatchley and Cheng (2010) elucidated the mechanism of chlorine reaction with urea; urea reacts slowly with free chlorine in pools, but is an efficient precursor to NCl_3 formation. Zwiener *et al.* (2006) demonstrated a positive relationship between THM concentration and bather loading, which suggested that bathers are an important source of DBP precursors.

From a public health perspective, information regarding the dynamics of volatile DBPs, both in liquid and gas phases, is important for swimming pool patrons. However, the physics and chemistry responsible for liquid↔gas transfer of volatile compounds in pools are largely undefined. Therefore, this study was designed to elucidate the relationship between air or water quality and swimmer activity during an intense use period.

METHODS AND EXPERIMENTS

Experiments were conducted at a chlorinated, indoor swimming pool at a public high school on January 28, 2012 from 7 AM to 12 PM to examine dynamic changes in gas- and liquid-phase composition during regular practice sessions for the school's swimming teams. The record of sampling time and swimmer activity in the pool is shown in Table 1. Samples were collected from an L-shaped pool with 770 m^3 water, as shown in Figure 1. One leg of the "L" is a six-lane, 22.86 meter (25-yard) pool, while the other leg of the "L" is a six-lane, 25-meter pool. Two sampling stations were set up for gas- and liquid-phase sampling on the pool deck. Section 1 was located near an unused portion of the pool (calm zone), while station 2 was near the part of the pool that was used by swimmers (practice zone). Involvement of high school students in the experiments allowed gas-phase NCl_3 and liquid-phase DBPs to be monitored as a function of time at both sampling positions.

Gas-phase NCl_3 measurements were conducted once per hour at each position using the impinger method described in Weng *et al.* (2011). Liquid-phase samples were collected on an hourly basis at both positions from three different depths (near-surface, middle of the pool depth, and pool bottom) and transported to laboratory immediately (roughly 5 min travel time) for analysis. Free chlorine and total chloramine concentrations in surface water samples were measured on-site by the DPD/KI colorimetric method (APHA-AWWA-WEF, 1998) with a portable photometer (Model 7500; Palintest, KY). Analysis of liquid phase samples included volatile DBPs (*e.g.*, inorganic chloramines, $CHCl_3$, $CNCHCl_2$, and CNCl) by Membrane Introduction Mass Spectrometry (MIMS, Weaver *et al.*, 2009) and urea by a digestion/colorimetric method (Prescott and Jones, 1969).

Table 1. Sample collection schedule.

Time	Note
7:00 AM	Arrive
7:30 AM	Air sampling initiated (once per hour: 7:30, 8:30, 9:30, 10:30, 11:30 AM)
8:00 AM	Girl's team practice starts, 14 swimmers, lap swim, First water sample at 8 AM (before practice start)
8:45 AM	Second water sample at 8:45 AM (in the middle of the air sampling) Liquid phase sampling schedule: 8:00, 8:45, 9:45,10:45, 11:45 AM
9:00 AM	Girl's team practicing ended
9:20 AM	Boy's team start practicing, 17 swimmers, lap swim
11:00 AM	Boy's team practice ends, 7 swimmers, less aggressive
11:10 AM	Boy's team departs, 1 swimmer in practice, lap swim
12:00 AM	Experiment end

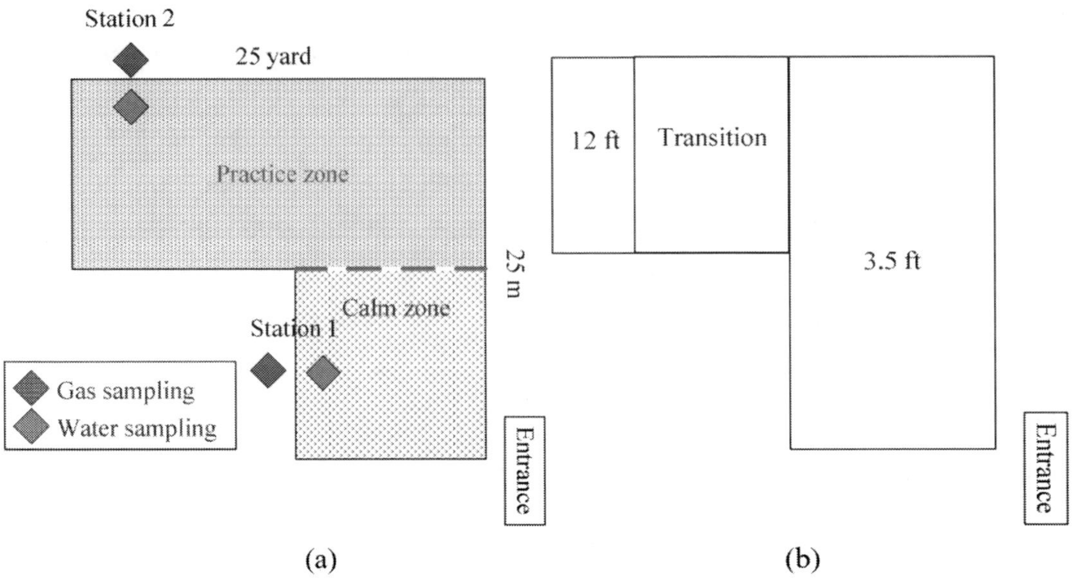

Figure 1. Schematic illustration of (a) sampling locations and (b) the corresponding pool depth of the target pool.

RESULTS AND DISCUSSIONS

Behavior of NCl_3 in Gas and Liquid Phases

Gas-phase NCl_3 concentration increased at station 2 immediately after swimming activity commenced, as shown in Figure 2. The gas-phase NCl_3 concentration appeared to be related to bather loading and swimmer activity, in that gas-phase NCl_3 concentration increased steadily during swimmer practicing period (from 8 - 11 AM) and decreased after practice ended (after 11 AM). This behavior is consistent with previous observations (Weng *et al.*, 2011). On the other hand, the gas-phase NCl_3 concentration at station 1 showed less influence by swimmer activity.

A corresponding decrease of liquid-phase NCl_3 concentration was observed at in near surface samples collected at station 2 (Figure 3). Vertical stratification of pool water was also observed at this location. For both locations, liquid-phase NCl_3 concentration in deeper water (bottom) was relatively stable. In contrast, the liquid-phase NCl_3 concentration for the water samples collected from the near-surface location displayed dynamic behavior that was strongly linked to swimmer activity, especially for location 2. From Figure 3(b), the concentration at this location steadily decreased as the practice session continued, resulting in almost total depletion of liquid-phase NCl_3 from the near-surface area by the end of a 3-hour practice session. The NCl_3 that was transferred from water presumably entered the air space and contributed to the increase of gas-phase

NCl_3 concentration (Figure 2). However, the NCl_3 concentration in the deeper water locations (middle and bottom) remained relatively steady during the experiment. On the other hand, the liquid-phase NCl_3 concentration at Station 1, which was an area of essentially no swimmer activity (station 1, Figure 3(a)) showed less variability during the practice session.

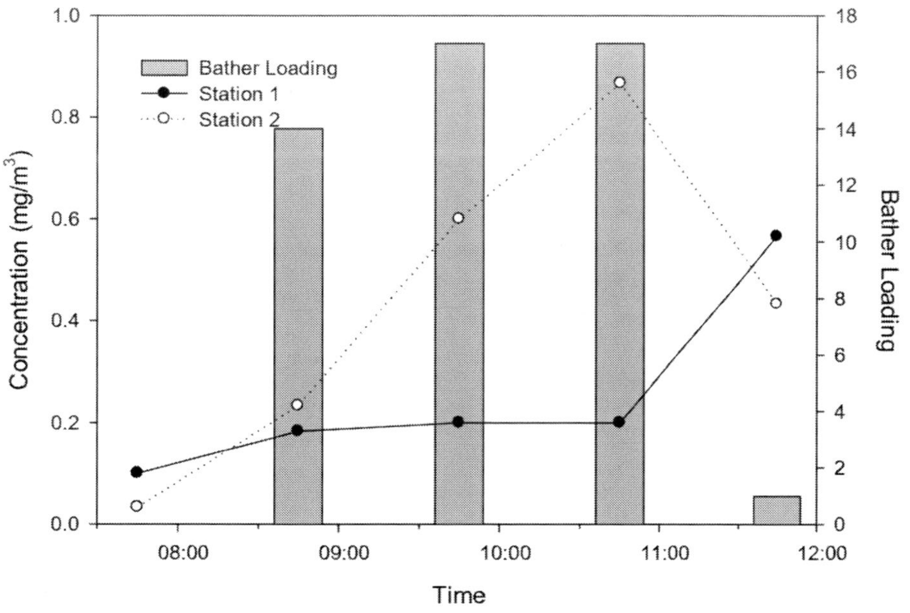

Figure 2. Dynamics of gas-phase NCl_3 concentration and bather loading

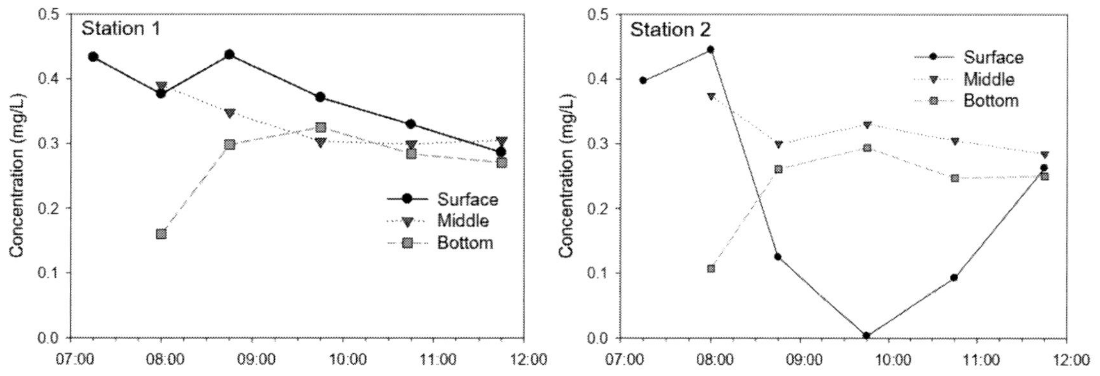

(a) (b)

Figure 3. Dynamics of liquid-phase NCl_3 concentration at (a) station 1 and (b) station 2.

Behavior of Volatile DBPs

The dynamic behavior of $CHCl_3$ and $CNCHCl_2$ in the liquid phase is illustrated shown in Figure 4 and Figure 5. From Figure 4, a decrease of $CHCl_3$ concentration was observed in near-surface water. The loss of $CHCl_3$ from near surface water during period of swimming activity is consistent with previous research (Kristensen *et al.*, 2010), and is believed to be attributed to liquid→gas transfer, as described Weng *et al.* (2011). On the other hand, $CNCHCl_2$ concentration was relatively steady (compared with $CHCl_3$). $CNCHCl_2$ is less volatile than $CHCl_3$. $CNCHCl_2$ formation and decay reactions also tend to be fairly slow. At location 1, there was essentially no change in $CNCHCl_2$ concentration during this experiment, and less vertical stratification was evident than with $CHCl_3$ and NCl_3.

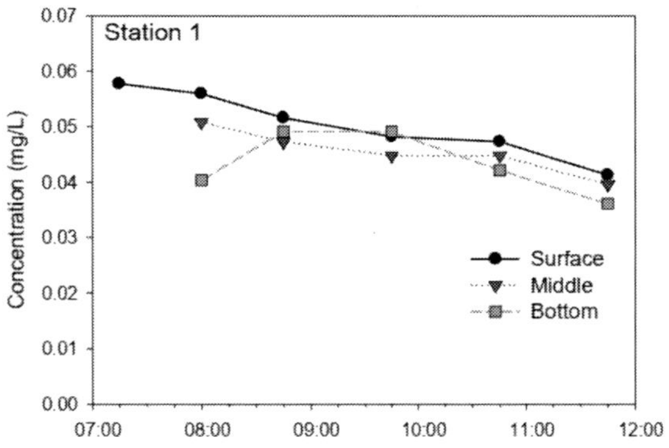

(a)

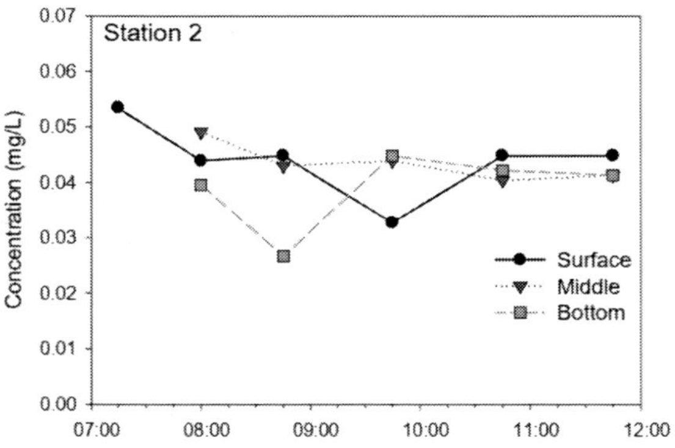

(b)

Figure 4. The dynamics of liquid-phase $CHCl_3$ concentration at (a) station 1 and (b) station 2

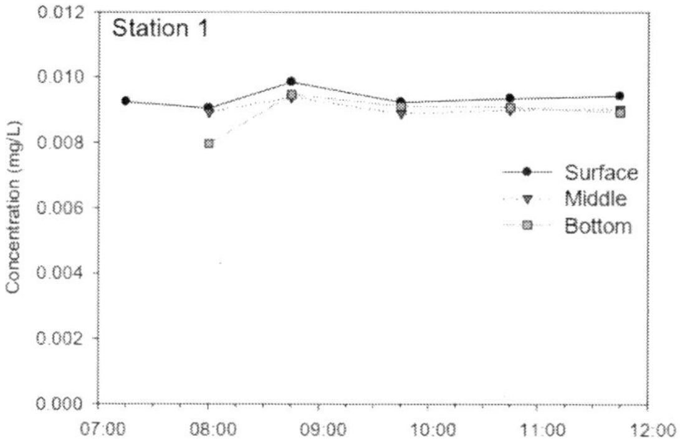

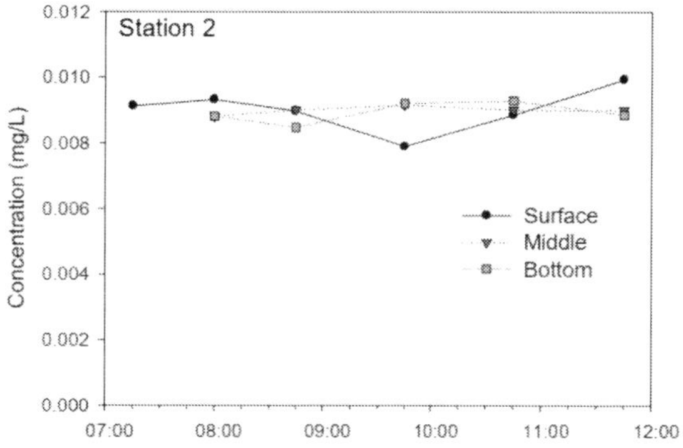

Figure 5. Dynamics of liquid-phase $CNCHCl_2$ concentration at (a) station 1 and (b) station 2

CNCl functions as a disinfection intermediate in chlorinated pools, with generation from chlorination of N-precursors (*e.g.*, glycine) and decay by oxidation of free chlorine (Lee *et al.*, 2006; Na and Olson, 2004). Liquid-phase concentration of CNCl is believed to be influenced by a combination of physical process (volatilization from liquid phase to gas phase) and chemical process (formation/decay by chlorination). The liquid-phase concentration of CNCl demonstrated dynamic behavior during this experiment, and the changes were larger at location 2 than at location 1, as shown in Figure 6. For the near-surface water at location 2, an increase of CNCl concentration was evident as soon as practice began. During the period of heavy use, CNCl concentration in near-surface water dropped (at 9:45 AM) probably because of promotion of liquid-gas transfer by swimmer activity. Unlike the other DBPs described above, CNCl concentration increased to 5 times the initial concentration immediately after the practice session ended. This result suggested rapid net formation of CNCl following a period of heavy use, probably due to reaction between free chlorine and organic-N compounds introduced by swimmers. For example, glycine is known to be a major component of human urine and sweat, and is also an efficient precursor of CNCl formation in terms of both production rate and quantity (Lentner, 1981; Na and Olson, 2004).

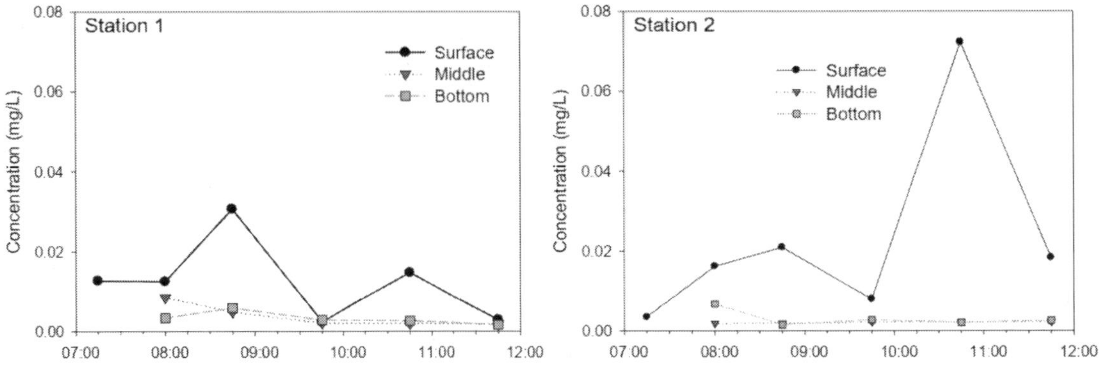

(a) (b)

Figure 6. Dynamics of liquid-phase CNCl concentration at (a) station 1 and (b) station 2.

Behavior of Urea

Urea is an important precursor to NCl_3 formation in chlorinated swimming pools. However, the reaction rate between free chlorine and urea is relative slow at near-neutral pH (Samples, 1959; Blatchley and Cheng, 2010). Therefore, urea concentration may be used as an index of reduced-N compounds introduced via human body fluids in pools.

In this experiment, the change of urea concentration was most evident among near-surface samples collected from location 2 than other locations. At location 2, urea concentration showed a steady increase in the near-surface water during the practice session; however, urea concentration in the deeper water remained largely unchanged. The results indicated that human body fluids introduced to the pool by swimmers remained in the near-surface water during the experiment. The concentration of urea increased steadily with pool use and decreased after activities ceased, probably because of the circulation pattern (mixing or overflow into the peripheral gutter).

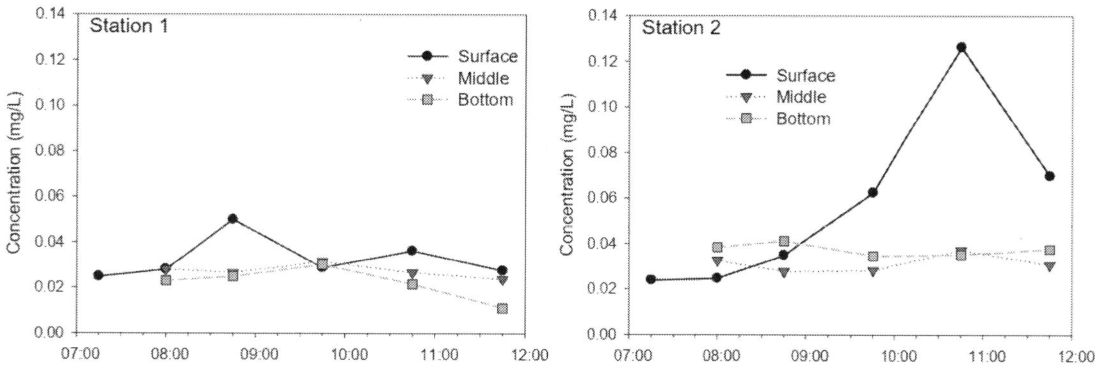

(a) (b)

Figure 7. Dynamics of liquid-phase urea concentration at (a) station 1 and (b) station 2

In a short-term monitoring experiment, the activity of swimmers promoted transfer of NCl_3 (and presumably other volatile DBPs) from near-surface water to the gas-phase, and introduced substantial quantities of urea (and presumably other precursors associated with human body fluids) into near-surface water. These results also indicated vertical stratification of pool water under a heavy use condition and localized impacts on water and air qualities at the area where the activities took place.

CONCLUSIONS

Collectively, these results suggest that the swimmer activity influenced air quality as well as water quality, especially in near-surface layer or areas where swimming activity took place. These findings suggest that water and air quality in pools may be strongly affected by water and air circulation patterns, as well as the hygiene habits of users (Keuten *et al.,* 2012). In addition, the results of this study indicated that development of a swimming pool model of air and water quality needs to account for horizontal and vertical variations in chemistry and mixing behavior.

ACKNOWLEDGEMENT

Support for this work was provided by grants from the National Swimming Pool Foundation, Engineered Treatment Systems, LLC and American Chemistry Council-Chlorine Chemistry Division.

REFERENCES

APHA-AWWA-WEF (1998) *Standard Methods for the Examination of Water and Wastewater,* 20thed., Washington. DC.

Bessonneau, V.; Derbez M.; Clément, M.; Thomas, O. (2011) Determinants of chlorination by-products in indoor swimming pools, *International Journal of Hygiene and Environmental Health* 215(1), 76-85.

Blatchley III, E.R. and Cheng, M. (2010) Reaction Mechanism for Chlorination of Urea, *Environmental Science & Technology* 44(22), 8529-8534.

Bond, T.; Huang J.; Templeton, M.R.; Graham, N. (2011) Occurrence and control of nitrogenous disinfection by-products in drinking water – A review *Water Research* 45(15), 4341-4354.

Bougault, V. and Boulet, L.P. (2012) Airway dysfunction in swimmers, *British Journal of Sports Medicine* **46**(6), 402-406.

Clearie, K.L.; Vaidyanathan, S.; Williamson, P.A.; Goudie, A.; Short, P.; Schembri, S.; Lipworth, B.J. (2010) Effects of chlorine and exercise on the unified airway in adolescent elite Scottish swimmers, *Allergy* 65(2), 269-273.

De Laat, J.; Feng, W.; Freyfer, D.A.; Dossier-Berne, F. (2011) Concentration levels of urea in swimming pool water and reactivity of chlorine with urea, *Water Research* 45(3), 1139-1146.

Erdinger, L.; Kirsch, F.; Sonntag, H.G. (1997) Potassium as an indicator of anthropogenic contamination of swimming pool water, *Zentralbl Hyg Umweltmed.* 200(5-6), 297–308.

Hansen, K.M.S.; Willach, S.; Mosbæk, H.; Andersen, H.R. (2012) Particles in swimming pool filters – Does pH determine the DBP formation? *Chemosphere* **87**(3), 241-247.

Hsu, H.T.; Chen, M.J.; Lin, C.H.; Chou, W.S.; Chen, J.H. (2009) Chloroform in indoor swimming-pool air: Monitoring and modeling coupled with the effects of environmental conditions and occupant activities, *Water Research* **43**(15), 3693-3704.

Kaydos-Daniels, S.C.; Beach, M.J.; Shwe, T.; Magri, J.; Bixler, D. (2008) Health effects associated with indoor swimming pools: A suspected toxic chloramine exposure, *Public Health* 122(2), 195-200.

Keuten, M.G.A.; Schets, F.M.; Schijven, J.F.; Verberk, J.Q.J.C.; van Dijk, J.C. (2012). Definition and quantification of initial anthropogenic pollutant release in swimming pools, *Water Research* 46(11), 3682-3692.

Kim, H.; Shim, J.; Lee, S. (2002) Formation of disinfection by-products in chlorinated swimming pool water, *Chemosphere*, 46(1), 123-130.

Kristensen, G.H.; Klausen, M.M.; Hansen, V.A.; Lauritsen, F.R. (2010) On-line monitoring of the dynamics of trihalomethane concentrations in a warm public swimming pool using an unsupervised membrane inlet mass spectrometry system with off-site real-time surveillance. *Rapid Communications in Mass Spectrometry* 24(1), 30-34.

Lee, C.; Choi, W.; Kim Y.G.; Yoon, J. (2005) UV Photolytic Mechanism of N-Nitrosodimethylamine in Water: Dual Pathways to Methylamine versus Dimethylamine, *Environmental Science & Technology*, 39(7), 2101-2106.

Lentner, C., Ed. *Geigy Scientific Tables, Volume 1: Units of measurement, Body Fluids, composition of the Body Nutrition*, 8th edition; Ciba-Geigy, Basle, Switzerland, 1981.

Li, J. and Blatchley III, E.R. (2007)Volatile Disinfection Byproduct Formation Resulting from Chlorination of Organic–Nitrogen Precursors in Swimming Pools, *Environmental Science & Technology* 41, 6732-6739.

Na, C. and Olson, T.M. (2004) Stability of Cyanogen Chloride in the Presence of Free Chlorine and Monochloramine. *Environmental Science & Technology* 38(22), 6037-6043.

Prescott, L.M. and Jones, M.E. (1969) Modified methods for the determination of carbamyl aspartate, *Analytical Biochemistry* 32(3), 408-419.

Samples, W.R. (1959) *A study on the chlorination of urea.* Ph.D. Dissertation; Harvard University, Cambridge, MA.

Schmalz, C.; Frimmel, F.H.; Zwiener, C. (2011) Trichloramine in swimming pools – Formation and mass transfer, *Water Research* 45(8): 2681-2690.

Weaver, W.A.; Li, J.; Wen, Y.; Johnston, J.; Blatchley, M.R.; Blatchley III, E.R. (2009) Volatile disinfection by-product analysis from chlorinated indoor swimming pools. *Water Research* 43(13), 3308-3318.

Weng, S.C. and Blatchley III, E.R., (2011) "Disinfection by-product dynamics in a chlorinated, indoor swimming pool under conditions of heavy use: National swimming competition." *Water Research* 45(16), 5241-5248.

Weng, S.C.; Weaver, W.A.; Afifi, M.Z.; Blatchley, T.N.; Cramer, S.J.; Chen, J.; Blatchley III, E.R. (2011) "Dynamics of Gas-phase Trichloramine (NCl_3) in Chlorinated, Indoor Swimming Pool Facilities," *Indoor air*, 21(5), 391-399.

Zwiener, C.; Richardson, S.D.; De Marini, D.M.; Grummt, T.; Glauner, T.; Frimmel, F.H. (2007) Drowning in Disinfection Byproducts? Assessing Swimming Pool Water, *Environmental Science & Technology* 41(2), 363-372.

Bigger, Better, Brighter: UV Inactivation with New Technology Integral to Melbourne's Advanced Tertiary Treatment Upgrade

Bryan Townsend, Black & Veatch
Robert Hulsey, Black & Veatch
Gary Hunter, Black & Veatch
John Mieog, Melbourne Water Corporation

Black & Veatch
10715 David Taylor Drive, Suite 240
Charlotte, NC 28262
(704) 510-8473

ABSTRACT

Melbourne's Eastern Treatment Plant was upgraded through the implementation of an advanced tertiary treatment plant. The existing activated sludge secondary treatment plant was augmented by additional treatment comprising pre-ozone, biological media filtration, post-ozone, UV and chlorine. The UV facility, designed to provide a 4-log inactivation of *Cryptosporidium*, is equipped with the TrojanUVFlex™ disinfection system, whose innovative design and validation strategy were key elements in the successful implementation of UV disinfection at the Eastern Treatment Plant. The advanced tertiary upgrade significantly improved the discharge quality of the facility, reducing its impact on the receiving marine environment and creating a high quality recycled water resource suitable for a broad range of non-potable applications. The Eastern Treatment Plant's advanced tertiary treatment plant is amongst the largest and most sophisticated of its kind in the world.

KEYWORDS

Ultraviolet, UV, disinfection; reuse; tertiary treatment; validation

INTRODUCTION

The Eastern Treatment Plant is located in Bangholme, Victoria, 40 km (24.9 miles) south east of Melbourne city centre, as is presented in Figure 1. It is the largest activated sludge wastewater treatment plant in Australia, and treats approximately 40 percent of Melbourne's sewage, servicing around 1.5 million people. The plant was constructed in 1975 and its sewerage catchment comprises predominantly domestic waste with some trade waste inputs.

All flows entering the treatment facility receive secondary treatment via activated sludge. The average current daily flow is around 350 MLD (90 MGD) and peak wet weather flows can reach 1,750 MLD (460 MGD). A buffering storage volume of approximately 3,000 ML (792 Mgal) exists downstream of the activated sludge process, with downstream treatment limited to a maximum of 700 MLD (180 mgd) due to the outfall pump station capacity. An aerial photograph of the Eastern Treatment Plant, circa 2008, is presented in Figure 2.

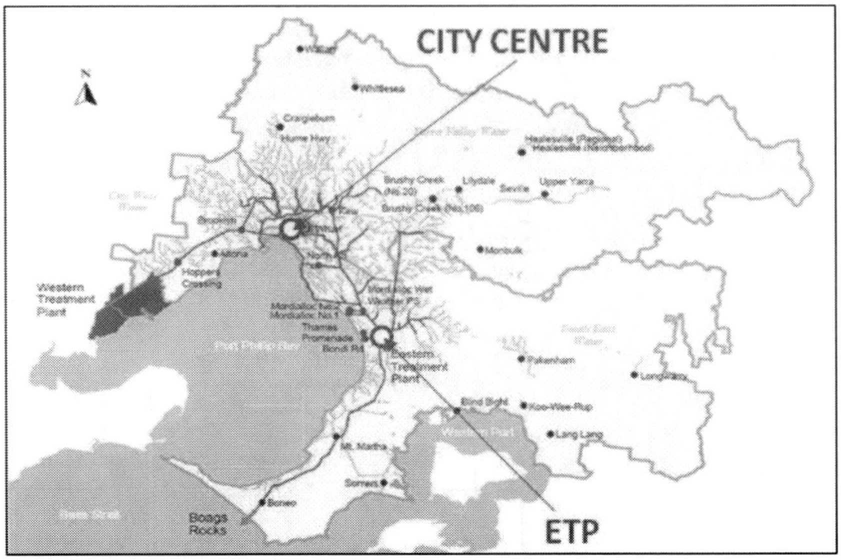

Figure 1 Location of the Eastern Treatment Plant

Figure 2 Eastern Treatment Plant – Circa 2008

Prior to the addition of the advanced tertiary treatment plant, the secondary effluent was disinfected through the use of monochlormaine disinfection in order to achieve an EPA annual median discharge limit of 200 org/100 ml and a 90th percentile of 1,000 org/100ml. Approximately 20 percent of the chlorinated secondary effluent was used as a 'Class C' recycled water resource, limited to non-potable applications having exposure controls. These applications consisted of site service water, the watering of sports fields and golf courses with access controls, and irrigation of food crops not eaten raw. Approximately 4 percent of the 'Class C' recycled water has been further treated by a third party to 'Class A' recycled water which can be used without access controls, including residential applications such as garden watering and toilet flushing. The majority of the treated secondary effluent was discharged to the marine environment via a near-shore ocean outfall. As presented in Figure 3, the discharge location is in close proximity to several popular local beaches used for public recreation, including swimming, surfing and fishing.

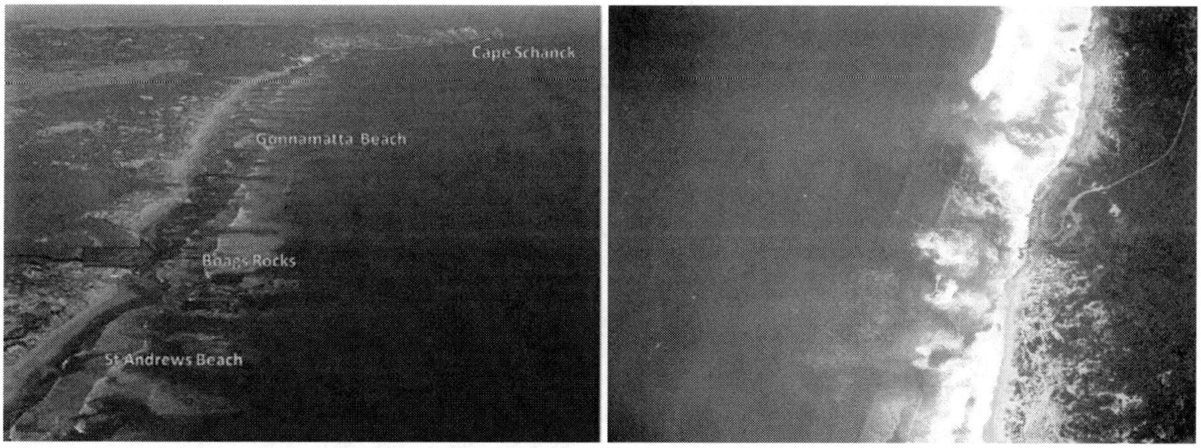

Figure 3 South East Outfall Discharge Point and Effluent Plume at Boags Rocks

ADVANCED TERTIARY TREATMENT PLANT

In an effort to significantly improve the discharge water quality, the Eastern Treatment Plant was upgraded to include advanced tertiary treatment. The upgrade was undertaken by the Eastern Tertiary Alliance (ETA), consisting of construction partners Baulderstone and UGL Infrastructure; design partners Black & Veatch and KBR; and Melbourne Water. The goals of the advanced tertiary treatment plant upgrade were twofold: (1) reduce the impacts of the effluent on the receiving marine environment; and (2) provide a 'Class A' high quality fit-for-purpose recycled water resource suitable for a broader range of non-potable applications than the historical 'Class C' recycled water, including residential (toilet flushing, garden watering); agricultural (including food crops eaten raw); and open space watering without public access restrictions.

The process train of the advanced tertiary treatment plant is presented in Figure 4. Secondary effluent enters the plant and is pre-ozonated for color and odor treatment. Following pre-ozonation, water flows through biological media filters for ammonia, oil and grease, foam, litter and solids reduction. The filtered effluent is then subjected to three separate disinfection processes, including ozone; UV and chlorine prior to treated water recycling or discharge.

An approach based on quantitative microbial risk assessment was used to determine the pathogen reduction requirements to produce fit-for-purpose recycled water in accordance with current Australian recycled water regulation and guidance. Ozone disinfection was designed to provide 4-log virus and bacteria inactivation in addition to a minimum 0.6-log inactivation of protozoa. The UV system was designed to achieve 4-log inactivation of *Cryptosporidium*, and is the primary barrier to address protozoa risks to the recycled water product and associated protection of public health.

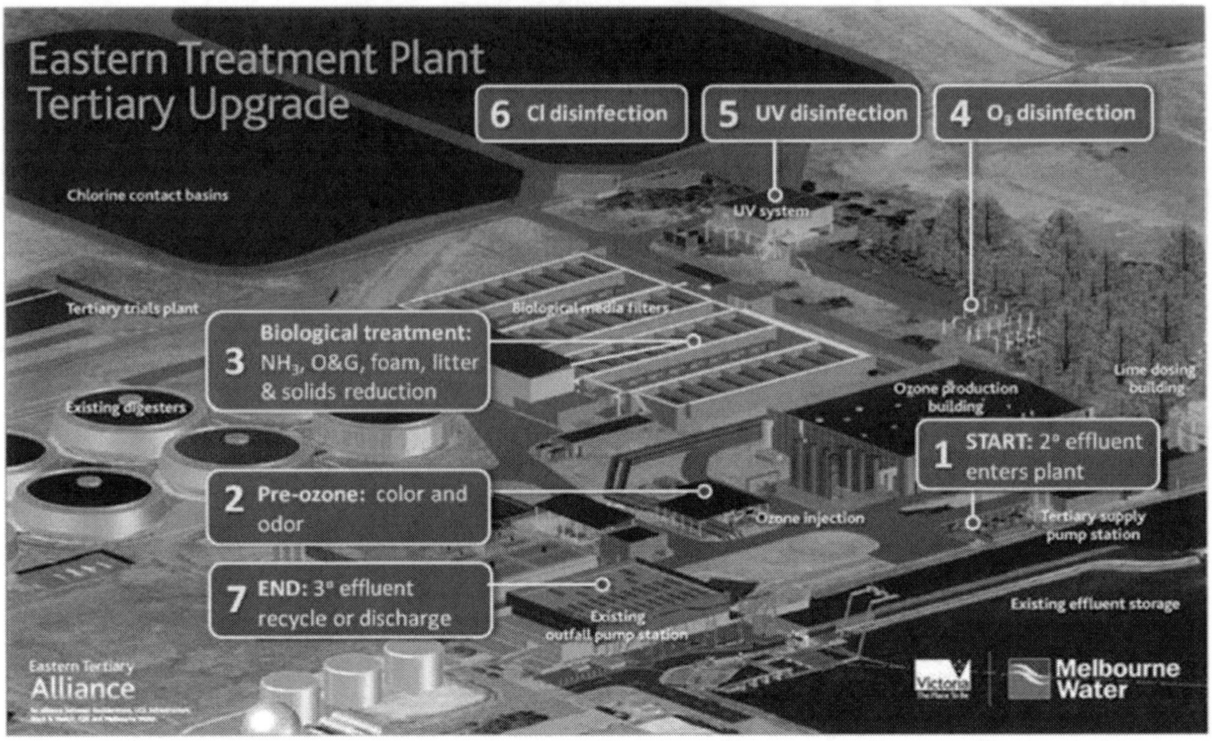

Figure 4 Eastern Treatment Plant Advanced Tertiary Treatment Process Train

Chlorine disinfection can operate as either free chlorine or monochloramine. In free chlorine mode, the plant is designed to provide up to 4-log virus and bacteria inactivation, while in monochloramine mode it will provide a minimum of 0.9-log inactivation of virus and bacteria. While the chlorine disinfection will provide a low residual, since most of the discharge will end at sea, the plant is typically operated so that the total chlorine residual at the outfall discharge is less than 1 mg/L. Sodium bisulphite dosing is available to trim any excess chlorine residual in the treated water discharged from the plant.

The Eastern Treatment Plant secondary effluent has a very low UV transmittance (UVT) and is highly colored due to industrial inputs. Tertiary treatment prior to UV treatment plays a crucial role in improving the UVT to a point more typical for UV disinfection applications and provides the ability to control the UVT and maintain it within the UV system's validated limits, thus maximizing recycled water availability. The secondary effluent enters the advanced tertiary treatment plant with a UVT of approximately 40 to 50 percent. Pre-ozonation and biological media filtration increases the UVT to around 60 percent, with post-ozonation increasing the UVT even further, between 70 and 80 percent.

UV DISINFECTION SYSTEM DESIGN AND SELECTION

The UV system was designed to provide a 4-log inactivation of *Cryptosporidium* in accordance with the procedures of the U.S. EPA's UV Disinfection Guidance Manual (UVDGM: EPA, 2006). Although the UVDGM was developed for validation, design and operation of UV systems for the disinfection of potable water, the recommended procedures are not application specific and have been successfully applied to the validation of reactors for the disinfection of secondary and tertiary treated wastewater. However, the application of the UVDGM's UV dose requirements for *Cryptosporidium*, which are derived from the U.S. EPA's Long Term 2 Enhanced Surface Water Treatment Rule, to the disinfection of a wastewater effluent, is somewhat unique. Due to the existence of particles in wastewater effluents, microbial dose-requirements are typically higher as compared to the same microbes in a drinking water source, in order to overcome the effects of particle shielding. However, since the Eastern Treatment Plant secondary effluent is subjected to advanced tertiary treatment prior to UV disinfection, the water quality is greatly improved, resulting in filtered water with typical turbidities of 0.5 to 1 NTU and a maximum of 5 NTU. Particle size distribution analysis demonstrated that filtration was also effective at removing particles larger than 5 microns, the same approximate size as *Cryptosporidium* oocysts. As a result, particle shielding is not a concern for the tertiary treated influent to the UV system and the adoption of the UVDGM dose requirements was considered to be appropriate for this application.

In order to optimize the UV system design, the availability and redundancy requirements for the UV system were critically reviewed. In particular, it was recognized that during wet weather events which contribute to peak flow conditions, a degree of dilution occurs which increases the UVT. This relationship was evaluated carefully for the development of the design UVT and required validated range at average and peak flow conditions. The tender for the UV system was open to both in-channel and closed vessel UV reactor configurations employing either medium pressure (MP) or low pressure high output (LPHO) lamp technologies. UV system design requirements are provided in Table 1.

TABLE 1 UV SYSTEM DESIGN REQUIREMENTS	
UV Reactor Requirements	
Reactor Type	Open Channel or Closed Vessel
Lamp Type	MP or LPHO
Disinfection Requirements	
Target LogI, pathogen	4-LogI *Cryptosporidium*
Required dose (mJ/cm^2)	22.0
Water Quality / Hydraulics	
Flow rate (L/s)	
Design / Average	8,200 / 4,400 (187 / 100 mgd)
Range	2,000 – 8,200 (46 – 187 mgd)
UVT (%)	
Design / Average	67.5 / 72.5
Range	62.5 – 80.0
Maximum Allowable Head Loss (mm)	750 (30 inches)

UV system bids were received from three manufacturers for a total of five proposed UV system designs. Proposed designs included 1 LPHO closed vessel reactor (7 trains, 1 reactor per train 672 lamps total), 1 medium pressure closed vessel reactor (5 trains, 2 reactors per train, 180 lamps total) and 3 LPHO open channel UV systems (5 channels, 2 banks per channel, approximately 1,000 to 2,000 lamps total). Evaluation of the provided bids was based on a multi-criteria assessment including a net present worth cost analysis. The associated capital costs for the proposed UV systems ranged from \$2.8 to \$7.8 million AUD (\$2.4 to \$6.8 million USD) for the UV equipment and \$7.5 to \$10.4 million AUD (\$6.5 to \$9.1 million USD) for the UV facility (normalized to a common scope with respect to inclusions and exclusions encompassing equipment supply and provision of services). The 20 year net present value operating and maintenance costs of the proposed designs ranged from \$7.0 to 20.5 million AUD (\$6.1 – 17.9 million USD). All values are reflective of the projected costs and conversion rates at time of the UV system bid evaluations, conducted in May, 2010. Design details for the selected UV system are presented in Table 2.

TABLE 2 UV SYSTEM DESIGN DETAILS

UV System Design	
Manufacturer	Trojan Technologies
Reactor Model	TrojanUVFlex™
Reactor Type	Closed vessel
Flange Diameter (mm)	1,200 (48 inch)
No. of Reactors	7
Flow Rate per Reactor (L/s)	1171 (27 mgd)
Reactor Components	
UV Lamp	TrojanUV Solo™ Lamp
Type	LPHO
Input Power (W)	1000
No. of Lamp Modules per Reactor	6
No. of Lamps per Module / Reactor / Total	16 / 96 / 672
Ballast Type	Electronic
No. of Lamps Run per Ballast	2
No. of Ballasts per Reactor / Total	48 / 336
Power Turndown (%)	100 – 30
UV Sensor Type	Dry, germicidal
No. of Sensors per Module / Reactor / Total	1 / 6 / 42
Cleaning System	Acticlean™ On-line Chemical-Mechanical Cleaning System
Reactor Validation	
Validation Status	Not pre-validated at time of bid

The selected UV system for The Eastern Treatment Plant's advanced tertiary treatment plant was the TrojanUVFlex™, manufactured by Trojan Technologies. The TrojanUVFlex™ is a closed vessel LPHO UV system with a 48 inch flange and of a total of 96 lamps per reactor. The lamps are equally divided between 6 discrete lamp banks (16 lamps per bank) with a single UV sensor monitoring each bank. Each ballast operates 2 lamps, allowing for variable power between 30 and 100 percent in 1 percent increments.

Further to being competitive in terms of present worth cost, the innovative design of the TrojanUVFlex™ was considered to offer significant project benefits. The reactor is equipped with TrojanUV Solo™ lamps, which operate at an input power of 1 kW, significantly higher than the lamps in other available LPHO UV reactors. This results in a reactor that has operating efficiency of a LPHO UV system, but with reduced footprint requirements as compared to a traditional LPHO reactor. In addition, the lamp and sensor sleeves are continuously cleaned by the Acticlean™ on-line chemical-mechanical cleaning system. On-line cleaning is a necessity for wastewater UV systems and a feature that is not available with other existing large scale LPHO closed vessel reactors.

UV REACTOR VALIDATION

As was the case for all of the proposed designs, the TrojanUVFlex™ had not been pre-validated at the time of the bid, with the design and sizing of the UV facility based on computational fluid dynamic (CFD) modeling by the UV manufacturer. As a result, the selection of any of the proposed UV system designs required reactor validation following the selection of the UV equipment. The need for any on-site validation testing was considered a serious risk to meeting the project schedule. Although the open channel UV systems could be validated at an off-site facility, there existed a potential need to conduct "spot check" bioassays on the equipment once installed at the plant in order to determine the impact of site-specific hydraulics on the validated performance models. These "spot check" bioassays would not be required for the closed vessel UV designs since proper approach hydraulics could be assured by having the necessary length of straight pipe upstream of the UV reactor to comply with UVDGM recommendations. As a result, closed vessel UV reactor designs, such as the TrojanUVFlex™, presented less of a risk to the project schedule, allowing for UV system validation testing to be conducted independently of full-scale construction and commissioning.

Selection of UV equipment prior to reactor validation was carefully considered during the selection process due to the potential for design issues and a reduced treatment capacity should the validated performance models not support the predicted performance used to size the proposed UV system design. The risk associated with the use of non-validated models for the design of the UV system was transferred to the manufacturers via the commercial contract, through the implementation of performance and power consumption guarantees. While there is added risk in the design of UV equipment based on non-validated models, the need for validation of the TrojanUVFlex™ following the selection of the UV system was not solely viewed in a negative light. This order of events presented a unique opportunity for the Eastern Tertiary Alliance to be intimately involved in the validation process and allowed for the protocols and validated range for the UV reactor to be specifically tailored to meet the Eastern Treatment Plant project requirements.

The TrojanUVFlex™ was validated at the Johnstown, NY UV Validation Facility in October of 2011. The Eastern Tertiary Alliance's role in the validation of the TrojanUVFlex™ reactor included: (1) Review and approval of the validation test plans and protocols, including confirmation that testing would fulfill the requirements of the specification and Eastern Treatment Plant design objectives while complying with the UVDGM; (2) Observation of the validation testing, including confirmation that the test train configuration and testing methods were in line with the approved protocols and verification of proper calibration of test equipment; (3) Review and approval of the validation testing report, including an in depth review of the data analysis procedures, model development and compliance with UVDGM recommended procedures; and (4) Confirmation of proper sizing of the TrojanUVFlex™ using the validated models. Photographs of the TrojanUVFlex™ validation are presented in Figure 5.

Figure 5 Validation of the TrojanUVFlex™ at the Johnstown, NY UV Validation Facility

Witnessing of the validation activities also provided the Eastern Tertiary Alliance with the ability see the UV equipment prior to delivery at the Eastern Treatment Plant, providing a preview of UV system operation and an opportunity to investigate the quality and workmanship of the UV system and associated components. Following review of the validation report by the Eastern Tertiary Alliance, analyses conducted with the approved validated equations confirmed that the TrojanUVFlex™ was properly sized and has adequate capacity to meet the design requirements for the UV facility.

PROJECT STATUS

The site works for the advanced tertiary treatment upgrade at the Eastern Treatment Plant commenced in early 2010. Commissioning of the advanced tertiary treatment plant is complete with the contract proof of performance testing successfully concluded at the end of December 2012. The project is currently in the final stages of completion as at January 2013.

The treated water quality benefits to the receiving marine environment have been realized throughout the commissioning period and 100 percent of the Eastern Treatment Plant flow has been treated by the advanced tertiary treatment plant since December 2012. This timing achieved the project objectives and coincided with the onset of the hot Australian summer and peak bathing season.

Figure 6 Advanced Tertiary Treatment Plant UV Disinfection Facility

Approval has been received from the Department of Health and EPA Victoria, granting the Eastern Treatment Plant the ability to produce a 'Class A' recycled water. Melbourne Water is currently in the process of obtaining independent certification of its Hazard Analysis and Critical Control Points (HACCP) system approval which represents the final step in the recycled water certification process. Although not currently certified as 'Class A' recycled water until the HACCP approval is obtained, existing 'Class C' recycled water customers are already enjoying the benefits of the advanced tertiary treatment through the resulting improved treated water quality. The 'Class A' recycled water from the Eastern Treatment Plant will be rolled out to customers over the course of 2013.

REFERENCES

EPA (2006) *Ultraviolet Disinfection Guidance Manual for the Final Long Term 2 Enhanced Surface Water Treatment Rule; EPA 815-R-06-007.* U.S. Environmental Protection Agency, Office of Water, Washington DC.